D1035246

FROM NEURON TO BRAIN

FROM NEURON TO BRAIN

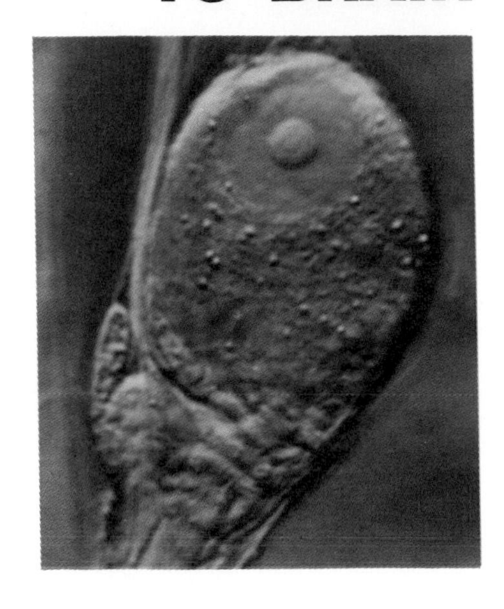

A Cellular Approach to the Function of the Nervous System

STEPHEN W. KUFFLER
HARVARD MEDICAL SCHOOL

JOHN G. NICHOLLS
STANFORD UNIVERSITY
SCHOOL OF MEDICINE

SINAUER ASSOCIATES, INC.
PUBLISHERS
SUNDERLAND, MASSACHUSETTS

We wish to thank our colleagues who kindly provided original illustrations from published and unpublished work. We also thank the editors of the Journal of Physiology and the Journal of Neurophysiology, from which many of the illustrations were taken.

Cover design by Laszlo Meszoly.

FROM NEURON TO BRAIN:
A Cellular Approach to the Function
of the Nervous System

First printing

© 1976 by Stephen W. Kuffler and John G. Nicholls
All rights reserved.
No part of this book may be reproduced
in any form or by any means
without permission in writing
from the publisher.
For information, address
Sinauer Associates, Inc.,
Sunderland, Massachusetts 01375.

Manufactured in U.S.A.

Library of Congress
Catalog Card Number: 75-32228

ISBN: 0-87893-442-1 (Cloth)
ISBN: 0-87893-441-3 (Paper)

To our colleagues
in the Department of Neurobiology
at Harvard Medical School

PREFACE

Our aim is to describe how nerve cells go about their business of transmitting signals, how these signals are put together, and how out of this integration higher functions emerge. This book is directed to the reader who is curious about the workings of the nervous system but does not necessarily have a specialized background in biological sciences. We illustrate the main points by selected examples, preferably from work in which we have first-hand experience. This approach introduces an obvious personal bias and certain omissions.

We do not attempt a comprehensive treatment of the nervous system, complete with references and background material. Rather, we prefer a personal and therefore restricted point of view, presenting some of the advances of the past few decades by following the thread of development as it has unraveled in the hands of a relatively small number of workers. For example, in Part One (Neural Organization for Perception) we emphasize the approach used by Hubel and Wiesel, which we were fortunate to witness step by step in laboratories next to our own. Similarly, Part Two (Mechanisms for Neuronal Signaling) leans heavily on the work of Hodgkin, Huxley, Katz, Miledi, and their colleagues, and omits comprehensive treatment of many other aspects. A survey of the table of contents reveals that many essential and fascinating fields have been left out: subjects like the cerebellum, the auditory system, eye movements, motor systems, and the corpus callosum, to name a few. Our only excuse is that it seems preferable to provide a coherent picture by selecting a few related topics to illustrate the usefulness of a cellular approach.

We describe the more complex functions first, because the visual systems of the cat and the monkey lend themselves well to an initial presentation of the neuronal events that are clearly correlated with such higher functions as perception. This approach puts in perspective the subsequent discussion in Parts Two and Three of the cellular machinery that is used to bring about the brain's more complex activity. Throughout, we describe experiments on single cells or analyses of simple assemblies of neurons in a wide range of species. In several instances the analysis has now reached the molecular level, an advance that enables one to discuss some of the functional properties of nerve and muscle membranes in terms of specific molecules.

Fortunately, in the brains of all animals that have been studied there is apparent a uniformity of principles for neurological signaling. Therefore, with luck, examples from a lobster or a leech will have relevance for our own nervous systems. As physiologists we must pursue that luck, because we are convinced that behind each problem that appears extraordinarily complex and insoluble there lies a simplifying principle that will lead to an unraveling of the events. For example, the human brain consists of over 10,000 million cells and many more connections that in their detail appear to defy comprehension. Such complexity is at times mistaken for randomness; yet this is not so, and we can show that the brain is constructed according to a highly ordered design, made up of relatively simple components. To perform all its functions it uses only a few signals and a stereotyped repeating pattern of activity. Therefore, a relatively small sampling of nerve cells can sometimes reveal much of the plan of the organization of connections, as in the visual system.

In Part Three and especially in Part Six, we discuss "open-ended business," areas that are developing and whose direction is therefore uncertain. As one might expect, the topics cannot at present be fitted into a neat scheme. We hope, however, that they convey some of the flavor that makes research a series of adventures.

From Neuron to Brain expresses our approach as well as our aims. We work mostly on the machinery that enables neurons to function. Students who become interested in the nervous system almost always tell us that their curiosity stems from a desire to understand perception, consciousness, behavior, or other higher functions of the brain. Knowing of our preoccupation with the workings of isolated nerve cells or simple cell systems, they are frequently surprised that we ourselves started with similar motivations, and they are even more surprised that we have retained those interests. In fact, we believe we are working toward that goal (and in that respect probably do not differ from most of our colleagues and predecessors). Our book aims to substantiate this claim and, we hope, to show that we are pointed in the right direction.

S. W. K.
J. G. N.
Woods Hole
August 1975

ACKNOWLEDGMENTS

We are fortunate to live in a stimulating environment, closely associated with colleagues who are also our friends and whose work interests us greatly. Our writing reflects their influence and ideas which we have assimilated over the years. For example, our first chapters on the neural organization for perception could not have been written without the influence of David Hubel and Torsten Wiesel, whose pioneering work forms the backbone of our story. To Edwin Furshpan and David Potter we owe much of what is rigorous in our treatment of neural mechanisms; to Edward Kravitz and Zach Hall we are indebted for any insight we have into neurochemistry; and Jack McMahan's influence on our discussions of structure has been great. A strong additional influence that has shaped our thinking has come from our joint teaching enterprises. Our lives over the past 15 years have been spent in an atmosphere of ferment, full of lively discussion and critical questioning not only by our senior colleagues but also by our younger coworkers and students. We hope to reflect much of their constructive influence.

The main body of this book was written during two summers at the Salk Institute in La Jolla and several summer periods at the Marine Biological Laboratory in Woods Hole. We are grateful to our colleagues Drs. Eric Frank, Andrew Szent-Györgyi, and Bruce Wallace, who critically read the whole work, and to Drs. Denis Baylor, Zach Hall, Jan Jansen, Ed Kravitz, Jack McMahan, John Moore, Rami Rahamimoff, Brian Salzberg, Carla Shatz, Ann Stuart, David Van Essen and Doju Yoshikami, who discussed portions of this book with us. Dr. Judith Mannix, Mrs. Bonnie Bambara, Mrs. Diantha Faherty, and Ms. Suzanne Kuffler gave us assistance at various stages, and Mrs. Marion Kozodoy provided essential and unfailing help throughout that made it possible to complete the manuscript.

Our collaboration with Lászlo Mészöly, who has done most of the artwork, was a pleasure throughout, and so was our association with Joseph Vesely, who handled all production matters, and Andy Sinauer, our editor.

CONTENTS

PART ONE

NEURAL ORGANIZATION FOR PERCEPTION

CHAPTER ONE

ANALYSIS OF SIGNALS IN THE CENTRAL NERVOUS SYSTEM

The brain uses stereotyped electrical signals to process all the information it receives and analyzes. The signals are virtually identical in all nerve cells. They are symbols that do not resemble in any way the external world they represent, and it is therefore an essential task to decode these signals. There is good evidence that the origins of nerve fibers and their destinations within the brain determine the content of the information they transmit. Thus fibers in the optic nerve carry visual information exclusively, while similar signals in another type of sensory nerve—for example, one arising in the skin—convey a quite different meaning. Individual neurons can encode complex information and concepts into simple electrical signals; the meaning behind these signals is derived from the specific interconnections of neurons.

From neuronal signals to perception

The brain is an unresting assembly of cells that continually receives information, elaborates and perceives it, and makes decisions. At the same time, the central nervous system can also take the initiative and act upon various sense organs to regulate their performance.

To carry out its task of determining the many aspects of behavior and controlling directly or indirectly the rest of the body, the nervous system possesses an immense number of lines of communication provided by the nerve cells (NEURONS). These are the fundamental units or building blocks of the brain, and it is our task to find out the meaning behind their signaling. A start has already been made in studying behavior in terms of the meaning of signals in nerve cells. It is not enough, however, simply to register impulses, the sign that a neuron is conveying information. When, for example, a picture is presented to the eye, an understanding is required of the specific, unique relation of the signals to higher visual functions, such as the perception of color, form, or depth. For this, neuronal signals must be traced from their origin in sense organs, as they progress through a succession of relays that includes the cerebral cortex.

That neurobiologists who study single nerve cells should be in a position to discuss higher functions of the brain, such as perception, is a very recent, somewhat unexpected development. It has long been realized, of course, that knowledge of the cellular properties of neurons is essential for any detailed study of the brain. Nevertheless, there were no clear indications of how an understanding of membrane properties, signaling, or connections could help to explain psychological phenomena such as depth perception or pattern recognition. It seemed quite possible that the workings of the cerebral cortex would still remain a mystery even if a great deal was known about signaling in individual nerve cells. We hope to show that such a pessimistic view has lost much of its force.

A comparison of the brain with other organs of the body throws the problem into sharper focus. The brain as an "organ" is much more diversified than, for example, the kidney or the liver. If the performance of relatively few liver cells is known in detail, there is a good chance of defining the role of the whole organ. In the brain, different cells perform different specific tasks. Frequently, such specificity is linked with a distinct chemistry of individual cells; a good example is provided by neurons that excite or start signals as contrasted to neurons that inhibit or suppress signaling. In addition to their different chemistry, inhibitory and excitatory neurons obey different plans of connections. Only rarely can aggregates of neurons be treated as though they were homogeneous. Above all, the cells in the brain are connected with one another according to a complicated but specific design that is of far greater complexity than the connections between cells in other organs.

Fortunately, there are many simplifying features in the nervous system. First, it has only two basic types of signals, one for short and the other for long distances. Second, these signals are virtually identical in all nerve cells of the body, whether they carry messages to or from centers, or are the result of painful stimuli or touch, or simply interconnect various portions of the brain. Better still, signals are so similar in different animals that even a sophisticated investigator is unable to tell with certainty whether a photographic record of a nerve impulse is derived from the nerve fiber of whale, mouse, monkey, worm, tarantula, or professor. In this sense, nerve impulses can be considered stereotyped units. They are the universal coins for the exchange of communication in all nervous systems that have been investigated.

The neurophysiologist has learned to deal reasonably well with the initial stages of sensory signaling that eventually lead to perception and also with certain "control" or "executive" tasks of the nervous system, particularly in relation to movement of skeletal muscle; but problems of different scope and magnitude arise in connection with questions concerning the neural basis of perception. The complications introduced by "higher functions" can no longer be avoided. For example, the functions of the cerebral cortex cannot be considered without reference to consciousness, sensation, perception, and recognition. Are the available physiological tools appropriate for finding a meaningful answer?

First, is it useful to study complex problems by recording the activity of individual cells or small groups of cells? To account for the neural events involved in the perception of touch, we can start by recording signals from a neuron that terminates in the skin and be generally satisfied about the meaning of its signals. These consist of brief electrical pulses, about 0.1 V in amplitude, that last for about 0.001 sec (1 msec). They move along the nerve at a speed of up to 270 miles/hr (120 meters/sec). Although the impulses in a cell may appear identical with those in other nerve cells, the significance and meaning are quite specific for that cell; they indicate that a particular part of the skin has been pressed. A most important generalization, first made by Adrian,[1] is that the frequency of firing in a nerve cell is a measure of the intensity of the stimulus. In the above example, the stronger the pressure applied to the skin, the higher the frequency and the better maintained the firing of the cell. So far, then, there seems little difficulty in interpretation, and we can go a step further and discuss a simple reflex involving two or three sequential steps (Chapter 4). The whole cycle of events in the pathway from the sensory stimulus to the motor response can be traced because the role the individual nerve cells are performing is known.

What type of information does an individual neuron convey?

E. D. Adrian

However, much less is known about the meaning of signals generated by a neuron deep in the brain that receives its input from many cells and in turn supplies many others. Before the analysis can be started, a great deal of information is needed. Does the neuron under study handle information derived from the skin, the eye, the ear, or all three? If it is influenced by the eye, does it regulate the size of the pupil, does it move the eye, or is it involved in perception of form?

Analyzing the situation within the brain is somewhat similar to examining a small portion of a circuit in a large, complex computer. This analogy is overworked but nevertheless still appropriate. The properties of some of the basic circuit elements are known, and the electrical signals can be recorded at various stages; but unless the design of the instrument is known, there is no clue to the role the circuit serves. The meaning of the measurements may therefore be very limited, akin to recognizing letters in a foreign language without understanding the words. And yet, the remarkable lesson of the last two decades is that, in spite of such difficulties, considerable progress can be made in understanding higher functions of the brain by correlating the activity of individual nerve cells with complex behavioral or perceptual activities.

Pattern of neuronal connections determines meaning of signals

At first, it may seem surprising that the nervous system uses only two types of electrical messages. The signals themselves cannot be endowed with special properties because they are stereotyped and much the same in all nerves. The mechanisms by which signals are generated are also similar, though with interesting variants. Thus, the

[1]Adrian, E. D. 1946. *The Physical Background of Perception.* Clarendon Press, Oxford.

brain deals with symbols of external events, which do not resemble the real objects any more than the letters *d o g*, taken together, resemble a spotted Dalmatian. Rather, a particular set of signals must have a precise and special relation to an event.

Theoretically, there is no reason why a great deal of information could not be conveyed by any agreed-upon symbol, including a code made up of different frequencies. In the nervous system, however, the frequency or pattern of discharges will not do on its own as a code, for the following reason: even though impulses and frequencies are the same in different cells responding to light, touch, or sound, the content of information is quite different. The quality or meaning of a signal depends on the origins and destinations of the nerve fibers, that is, on their connections. Various types of sensory modalities (light, sound, touch) are linked to different parts of the brain; even within each modality and in each area of the cortex, specific stimuli, such as lines or rectangles in the visual system, act selectively on specific populations of neurons. This organization is brought about by strictly determined connections. Frequency coding is used by the nervous system to convey information about the intensity of a stimulus rather than its quality.

Many lines of evidence, some gross but nevertheless convincing, bring home the essential point of the importance of connections. A blow to the eye or a current passed across the eyeball produces the sensation of a light flash. Another familiar example is the phantom limb phenomenon. Amputees frequently report sensations in a missing member, vividly referring to a specific region, such as the toe or knee. The sensations usually arise in the nerve stump when severed sensory axons are irritated, owing to scar formation or the development of a swelling produced by a disorderly growth of nerve fibers at the cut end of the stump.

Judicious electrical stimulation of sensory axons along their course in the body can also produce sensations that vary in modality according to the origin of the sensory fibers. At present it is not possible to reproduce with electrodes, or by other artificial means, the natural discharge pattern in a nerve composed of many fibers. This is what would be required to evoke complex sensations artificially while bypassing sense organs. In the same way, a complex message cannot be sent through a bundle of wires making up a telegraph cable by simply passing currents through it somewhere along its course.

An understanding of the importance of the pattern of connections is enhanced by examples that show how the information about external events is inherent in connections. A radio, a computer, and a TV set use commonplace components and stereotyped signals, yet perform a variety of tasks. The specialization resides in the design of the wiring. It is the diversity of connections, not the types of signals, that increases the complexity of the tasks that can be undertaken. In much the same way the nerve cells in the brain are made up of "standard" commonplace chemical materials. What endows them with their diverse capabilities and gives meaning to their signals is the manner in which they are linked to each other.

A further requirement that goes with complex, satisfactory computer performance is an adequate number of components; this condition is also met by the nervous system. The numbers of cells in the cortex are so great (probably more than 20,000/cu mm) that they do not as yet present a limitation to speculation. The brain, then, is an instrument, made of 10^{10} to 10^{12} components of rather uniform materials, which uses a few stereotyped signals. What seems so puzzling is how the proper assembly of the parts endows the instrument with the extraordinary properties that reside in the brain.

It is worth pointing out that the conclusions presented here were expressed in 1868 by the German physicist-biologist Helmholtz. Starting from first principles, long before the facts, as we know them, were available, he reasoned:[2]

> The nerve fibres have often been compared with telegraphic wires traversing a country, and the comparison is well fitted to illustrate the striking and important peculiarity of their mode of action. In the network of telegraphs we find everywhere the same copper or iron wires carrying the same kind of movement, a stream of electricity, but producing the most different results in the various stations according to the auxiliary apparatus with which they are connected. At one station the effect is the ringing of a bell, at another a signal is moved, at a third a recording instrument is set to work. . . . In short, every one of the hundred different actions which electricity is capable of producing may be called forth by a telegraphic wire laid to whatever spot we please, and it is always the same process in the wire itself which leads to these diverse consequences. . . . All the difference which is seen in the excitation of different nerves depends only upon the difference of the organs to which the nerve is united and to which it transmits the state of excitation.

Visual perception: analysis in terms of organization

The preceding sections point out some of the difficulties that enter into considerations of conscious perception. These difficulties are much reduced in the visual system, particularly when responses in cortical cells are analyzed. This is illustrated in Chapter 2, which deals mainly with recent experiments made by recording from single cells in the visual pathways. Although much information is available about the auditory and other sensory systems in the body, the mammalian visual system has several advantages, owing particularly to the relative technical simplicity of many of the experiments and their direct relevance to perception. It offers clues for an understanding of the code used by neurons to transmit not just simple information about light and darkness but also sophisticated concepts. For example, reasonable hypotheses can now be formulated, in terms of neural signals and organization, relating to the following questions: What neural mechanisms can explain the recognition of shapes, such as light edges or corners of certain

---

[2]Helmholtz, H. 1889. *Popular Scientific Lectures.* Longmans, London.

dimensions, positioned at one angle rather than another in the visual field? How can triangles or squares be recognized independently of their position on the retina or of their brightness and size? How can we perceive with both eyes one image rather than two, even though it is known that each eye really sees a somewhat different part of the world?

With regard to the problems just mentioned, the visual system is different from sensory systems dealing with input from the skin. For example, mechanical stimuli, like a gentle touch, applied to the skin of two corresponding areas on both sides of the body do not usually give rise to a single fused sensation in the midline. Similarly, the visual system is different from a photographic plate in that it takes account of contrast or differences rather than the absolute level of brightness; visual perception ignores information about absolute levels and can detect subtle differences even if the background illumination changes over a range of many orders of magnitude. An apparent paradox pointed out by David Hubel is that black newsprint seen in sunshine on the beach reflects more light than the white part of a page seen with an ordinary electric lamp; yet the print appears black and the paper white in both situations. The visual system also provides one of the most favorable systems for studying fundamental questions relating to the development and maturation of the nervous system. Are the neural circuits used for perception already present at birth or are they formed as a result of visual experience? What sort of stimuli must be present in the environment to prevent sensory systems from becoming atrophied and useless? These questions are considered in Chapter 19.

The advances that make it possible to discuss these problems in physiological terms have resulted from the discovery that in the visual system and in other sensory systems there exists a highly specific set of cell connections. The arrangement of these connections accounts for the almost infinite wealth of information that reaches us as we look at the world around us. This is not to deny that we still remain profoundly ignorant about higher functions such as perception. But it seems reasonable to expect that the processes that lead to perception can be analyzed by the use of the same principles that govern the other functions of the nervous system.

BACKGROUND INFORMATION

For an easier understanding of the material in Chapters 2 and 3, we first present a few basic facts about the structure of neurons, their interconnections, and the methods of recording from them. Some key terms and definitions appear at the end of this chapter. (A fuller description of signaling is given in Part Two, and other terms used are defined in the glossary.)

Shapes of neurons

The shape of a neuron, as well as information about its position, origin, and destination in the neural network, supplies valuable clues to its function. For example, the arborization provides a notion of how

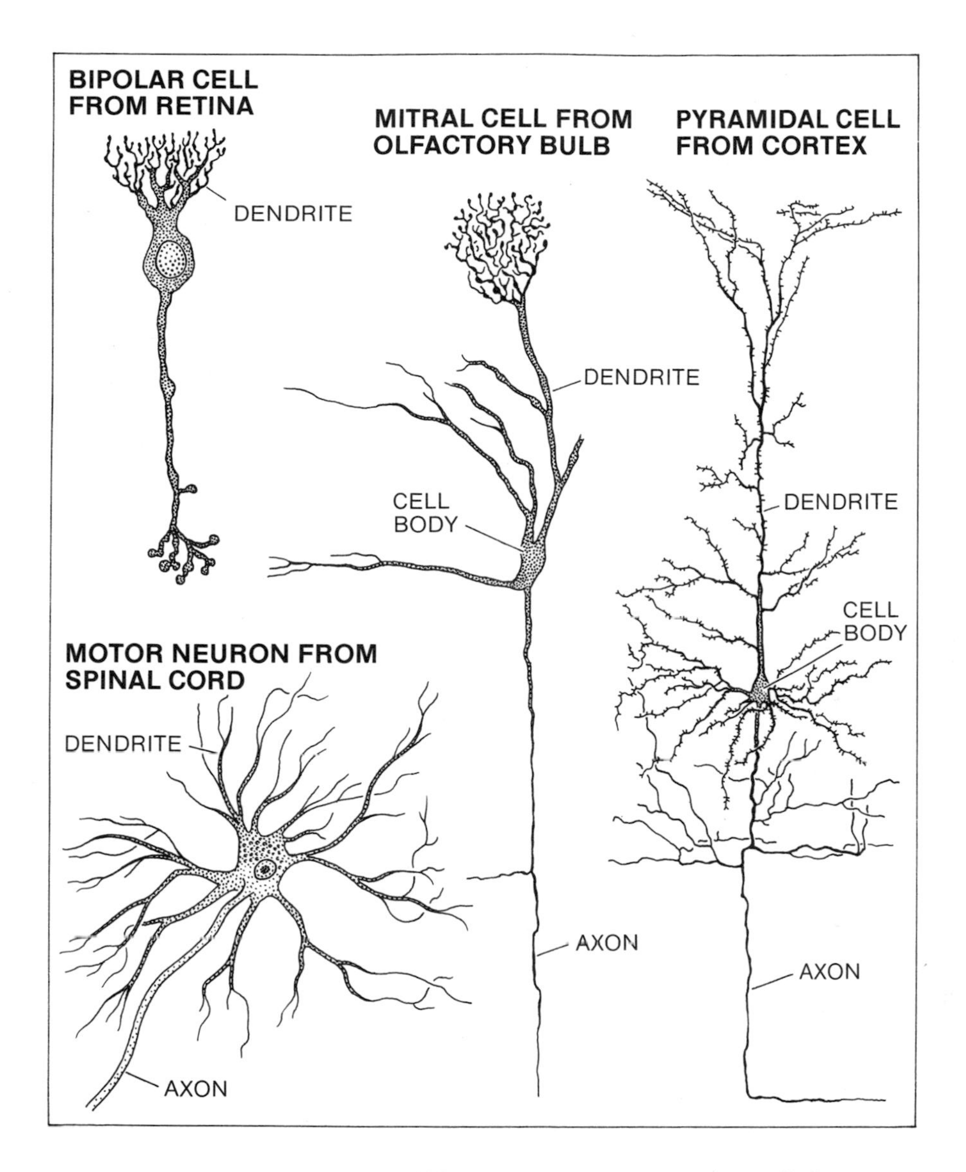

1

SHAPES AND SIZES OF NEURONS. The cells have processes, the dendrites, upon which other neurons form synapses. Each cell in turn makes connections with other neurons. The motor neuron, from a drawing by Deiters in 1869, was dissected from a mammalian spinal cord. The other cells, stained by the Golgi method, were drawn by Ramón y Cajal. The bipolar cell is from the retina of a dog, the pyramidal cell from the cortex of a mouse, and the mitral cell from the olfactory bulb (a relay station in the pathway concerned with smell) of a cat.

many connections a cell can accommodate and to how many sites it sends its own processes.

In practice it is quite difficult to find out about the configuration of neurons, because they are so densely packed. Early anatomists had to tease nervous tissue apart to see individual neurons. Figure 1 shows

Ramón y Cajal, about 1914.

a spinal motoneuron dissected by Deiters more than 100 years ago. Staining methods that impregnate all neurons are virtually useless for investigating cell shapes and connections because a structure like the cortex appears as a dark blur of intertwined cells and processes. Many of the pictures in Figures 1 and 2 were made with the Golgi staining method, which has become an essential tool because by some unknown mechanism it stains just a few neurons out of the whole population. Furthermore, the technique tends to stain individual cells in their entirety. The Golgi method, therefore, provides a random sampling. In recent years greater selectivity has been obtained through the method of injecting cells with a dye such as procion yellow, or by filling them with a metal such as cobalt or with the enzyme horseradish peroxidase. These methods enable the investigator to obtain the entire outline and geometry of cells from which he has recorded and whose physiological performance he knows.

Many of the illustrations in this chapter are based on the work of Ramón y Cajal, done before the turn of the century. Ramón y Cajal was one of the greatest students of the nervous system, selecting samples from a wide range of the animal kingdom with an almost unfailing instinct for the essential. The illustrations show several distinct cell types, some relatively simple, such as the bipolar cell, others with a highly complex arborization.

Cytology has demonstrated that what appears at first sight a staggering array of shapes and processes can be divided into meaningful groupings; thus cells can be recognized and classified in much the same way as trees. Although differences within a group can be considerable, one can always tell a birch tree from a Christmas tree, just as one can distinguish a spinal motor neuron from a pyramidal cell.

Design of connections as exemplified by the cerebellum

The cerebellum stands as one of the best examples of the orderliness of the design of the neuronal connections in the nervous system; the retina is a close competitor. While the cerebellum has only five distinct kinds of cells, the numbers of its neurons are truly staggering. The cell population for the entire human brain is frequently given as 10,000 million. This number is not likely to include the cells of the cerebellum, in which one cell type alone, the granule cell (Figure 2), is supposed to number around 10^{10} to 10^{11}.[3]

Cerebellar cytology came to life with the work of Ramón y Cajal[4] almost a century ago and received a new lease on life in the past two decades with advances that arose from developments in electron microscopy and new studies by physiologists in the laboratories of Eccles, Szentágothai, Palay, Ito, Llinás, and others.[5–7]

[3]Braitenberg, V., and Atwood, R. P. 1958. *J. Comp. Neurol.* *109*:1–33.

[4]Ramón y Cajal, S. 1955. *Histologie du système nerveux.* II. C.S.I.C., Madrid.

[5]Eccles, J. C., Ito, M., and Szentágothai, J. 1967. *The Cerebellum as a Neuronal Machine.* Springer Verlag, Berlin.

[6]Palay, S. L., and Chan-Palay, V. 1974. *Cerebellar Cortex.* Springer Verlag, Berlin.

[7]Llinás, R. R. 1975. *Sci. Am.* *232*:56–71.

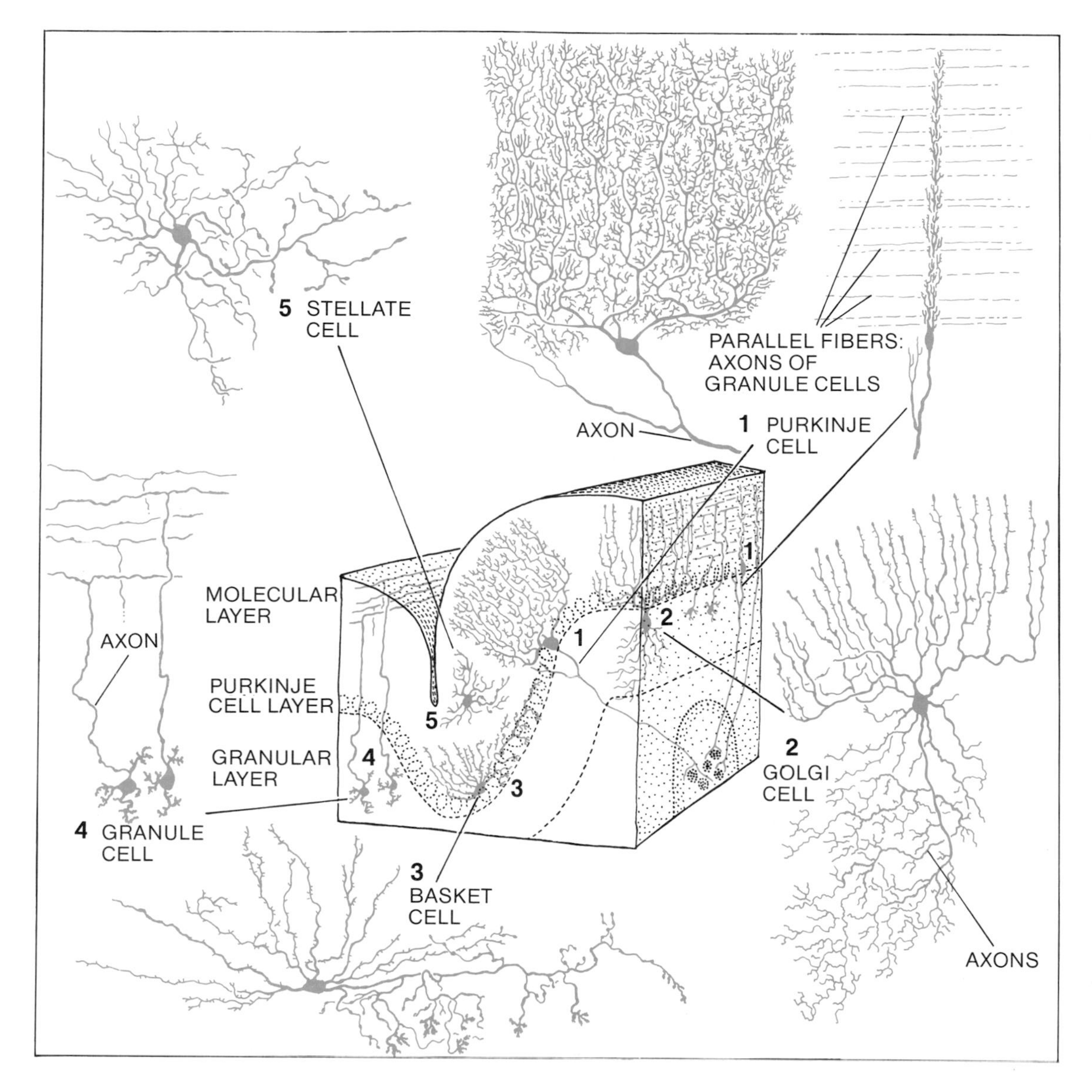

CEREBELLAR NEURONS. The human cerebellum has over 10^{10} cells, but only five neuronal types, whose cell bodies are confined to distinct layers; they are shown in drawings made from Golgi-stained preparations. The Purkinje cell (1), whose axons constitute the only output from the cerebellum, has its processes aligned in one plane. The axons of granule cells (4), traverse and make connections with the Purkinje cell process in the molecular layer. The Golgi cell (2), the basket cell (3), and the stellate cell (5) have characteristic positions, shapes, branching patterns, and synaptic connections. See Figure 1, Chapter 2, for the position of the cerebellum in the brain (1,2 after Ramón y Cajal, 1955; 3–5 after Palay and Chan-Palay, 1974)

2

The input to the cerebellum comes from the various sensory structures in muscles, skin, and joints; from the visual and auditory cortex, and so on. All the information is handled by repeating groupings of a few cell types—granule cells, basket cells, stellate cells, and Golgi cells—all of which directly or indirectly act upon the Purkinje neurons. These provide the only output from the cerebellar cortex.

The schematic presentation in Figure 2 gives an idea of the orderly arrangement of neurons. Within the cortex of the cerebellum the cell bodies of the various neuron types lie in distinct layers, sending their processes to other specific regions and to particular portions of target cells. The Purkinje cell dendrites (processes receiving synaptic inputs) are aligned in one plane, like a many-pronged candelabrum, reaching close to the cerebellar surface. The granule cells send their axons from their own layer, near the white matter straight toward the surface, where they bifurcate and run parallel to the cerebellar surface. They traverse the arborization of the Purkinje neurons at right angles and make synapses with its processes. It has been estimated that one Purkinje cell in the monkey receives synapses from about 80,000 granule cells, all of them restricted to small spiny protrusions. Other synaptic contacts on Purkinje cells are made by basket and stellate cells, mainly in the region of the cell body. The Golgi cells do not make contact directly, but do so indirectly by acting on granule cells. A single neuron may accommodate as many as 200,000 synapses.

The cerebellum is concerned with the regulation of movement. Although its overall task and its functional, well-ordered wiring diagram are known, information is lacking about what types of analyses the cerebellar cortex performs and how it does the analyses.

Recording techniques

The electrical activity of a neuron can be monitored by an electrode placed outside of the cell membrane (extracellular recording) or by an

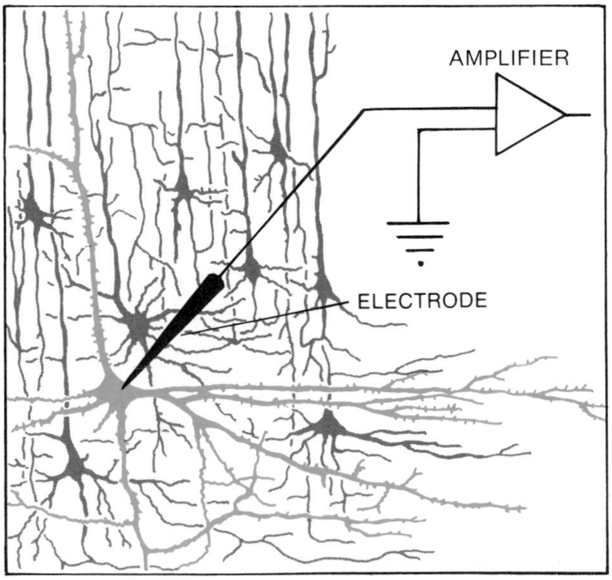

AMPLIFIER

ELECTRODE

3

EXTRACELLULAR RECORDING with a fine wire electrode. The electrode tip has been drawn close to a nerve cell in the cortex.

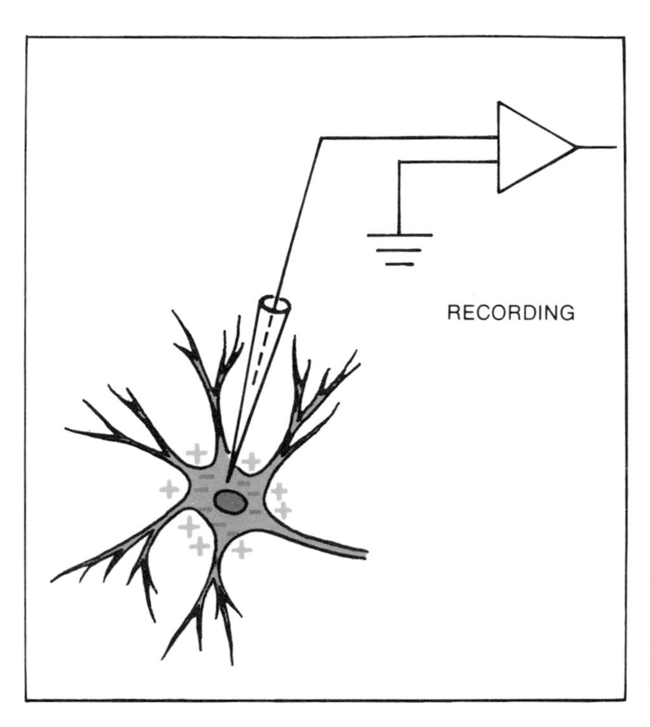

4
INTRACELLULAR RECORDING. The tip of a microelectrode has been inserted into a nerve cell. In a neuron that is at rest there is a potential difference of about 70 mV; the inside is negative with respect to the outside.

electrode that actually penetrates the cell (intracellular recording). Nearly all the recordings of electrical activity presented in the next two chapters were made with extracellular electrodes. The electrode itself can be either a fine wire insulated to its tip, or a capillary tube filled with salt solution. Figure 3 depicts diagramatically the arrangement for extracellular recording. This technique supplies information about whether a cell is firing or is quiescent, and whether the rate of firing is increasing or decreasing. With care it is possible to identify the signals from a single neuron.

Intracellular recording is used to obtain information about the processes of excitation, inhibition, and the mechanisms that initiate nerve impulses. The tip of a microelectrode is inserted into the cell as shown in Figure 4. The electrode is a fine glass capillary with a tip 0.1μm in diameter or smaller, containing a salt solution such as 3 M potassium chloride or 4 M potassium acetate. It measures the potential difference between the inside and the outside of the cell. The same electrode can also be used for passing electrical currents into or out of the cell.

SUGGESTED READING

Peters, A., Palay, S. L., and Webster, H. de F. 1976. *The Fine Structure of the Nervous System.* Saunders, Philadelphia.

Truex, R. C., and Carpenter, M. B. 1969. *Human Neuroanatomy,* 6th ed. Williams & Wilkins, Baltimore.

Williams, P. L., and Warwick, R. 1975. *Functional Neuroanatomy of Man.* Saunders, Philadelphia.

See overleaf for Review of Key Terms.

REVIEW OF KEY TERMS

IMPULSES ON A FAST TIME SCALE
mV
0.6
0.0
0 2 4 MSEC

The IMPULSE in a nerve fiber is fixed in size. It is a brief, stereotyped electrical event lasting for about 1 msec. It moves rapidly along the nerve from one end to the other. ACTION POTENTIAL is another term for impulse.

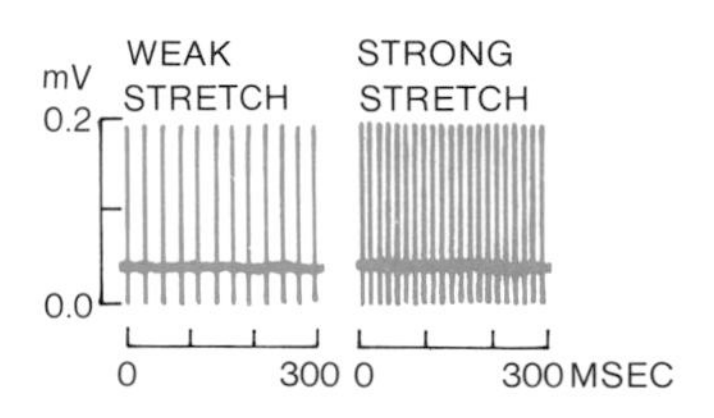

Stronger stimuli produce HIGHER FREQUENCIES of impulse firing. For example, a sensory nerve responding to stretch of a muscle fires at a rate proportional to the stretch.

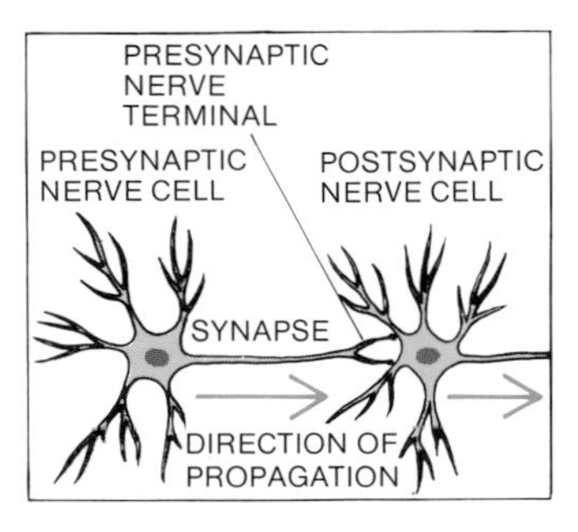

The junctions between nerve cells are called SYNAPSES. These are the sites at which cells transfer signals.

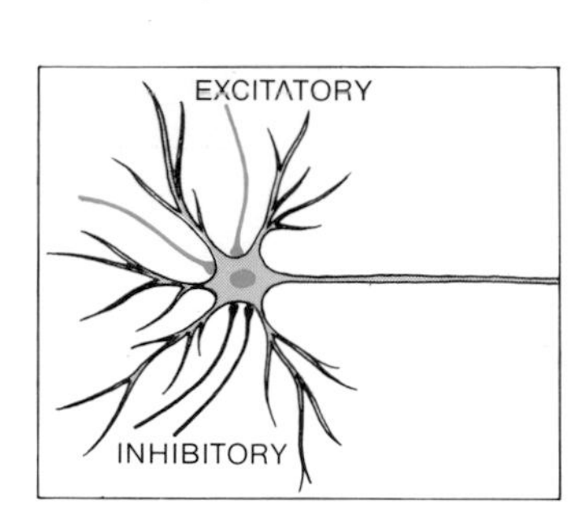

Nerve cells influence each other by (1) EXCITATION (shown in blue), that is, they produce impulses in another cell; and (2) INHIBITION, that is, they prevent impulses from arising in another cell.

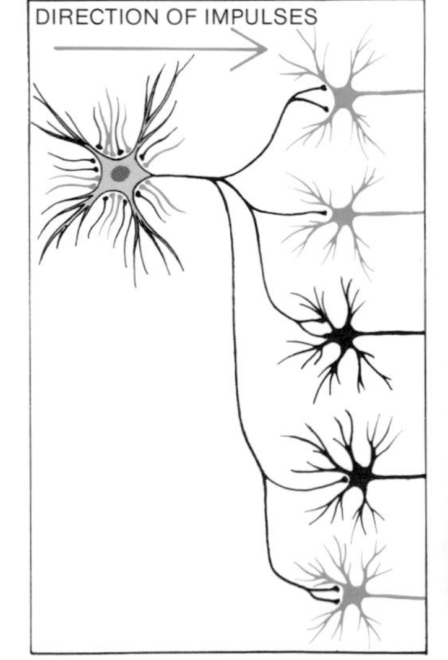

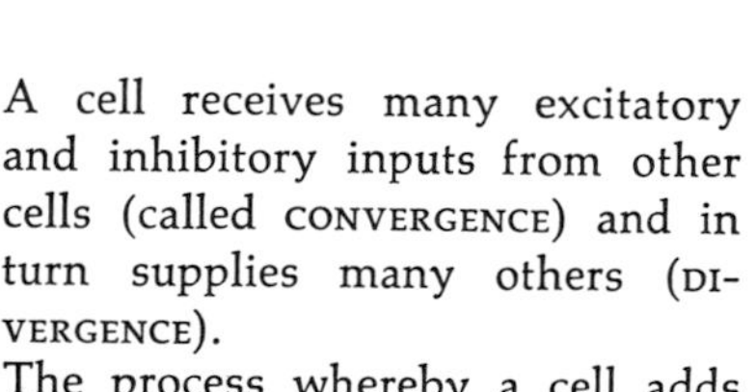

A cell receives many excitatory and inhibitory inputs from other cells (called CONVERGENCE) and in turn supplies many others (DIVERGENCE).
The process whereby a cell adds together all the incoming signals that excite and inhibit it is known as INTEGRATION.

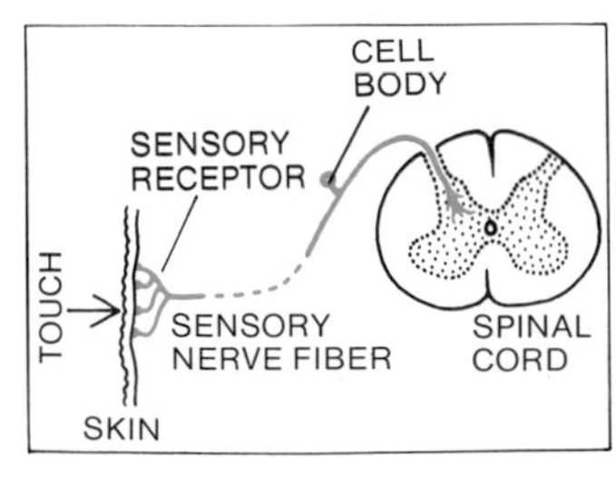

The specialized ending of a sensory nerve that responds to an external stimulus (touch, light, heat) is called a sensory RECEPTOR.

REVIEW OF KEY TERMS

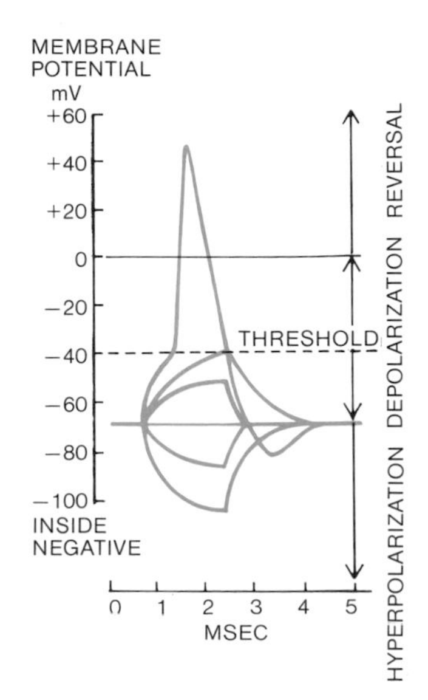

DEPOLARIZATION is a reduction of the membrane potential toward zero mV, the inside becoming more positive; HYPERPOLARIZATION is an increase in the potential, the inside becoming more negative.

Depolarization to a critical potential level, the THRESHOLD, causes the initiation of an impulse. At its peak the inside of the cell becomes positive with respect to the outside.

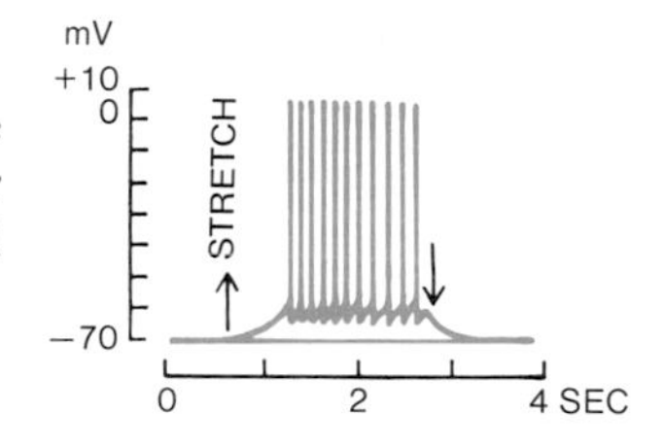

In a typical sensory nerve the effective or ADEQUATE STIMULUS (for example, stretch) depolarizes and sets up impulses.

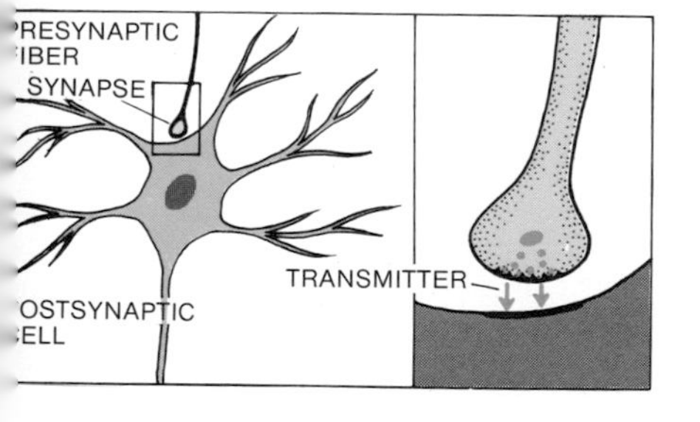

At the vast majority of synapses the presynaptic terminal liberates a chemical TRANSMITTER SUBSTANCE in response to a depolarization.

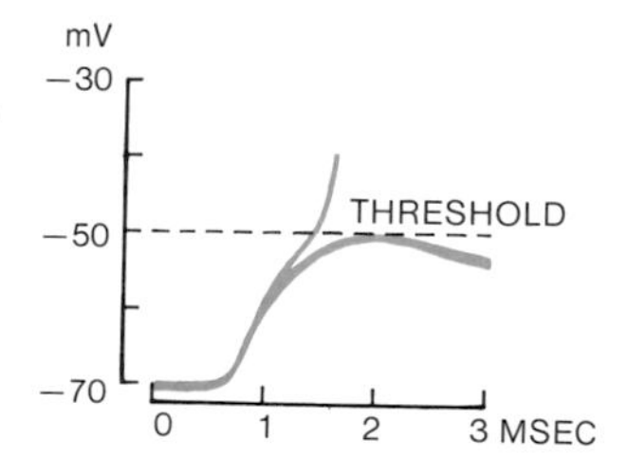

At an EXCITATORY SYNAPSE the chemical liberated by the presynaptic ending depolarizes the postsynaptic cell, driving its membrane potential toward threshold.

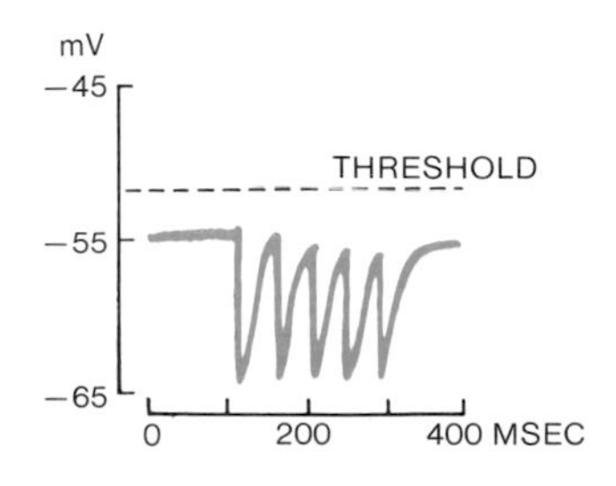

At an INHIBITORY SYNAPSE the transmitter tends to keep the membrane potential of the postsynaptic cell below threshold.

CHAPTER TWO

THE VISUAL WORLD: CELLULAR ORGANIZATION AND ITS ANALYSIS

Much progress has been made in determining how the activity of an individual neuron in the visual system is related to specific features of visual perception. Visual information is processed in successive stages: the neural responses to light start at the receptors, progress through the sequential layers of cells in the retina, ascend through a relay station (the lateral geniculate nucleus), and continue through a series of stages in the cerebral cortex. The transformation or integration that occurs at each level is best analyzed in terms of the receptive fields of neurons. This term refers to the restricted area on the retinal surface that influences, upon illumination, the signaling of an individual neuron in the visual system. Receptive fields are the building blocks for the synthesis and perception of the complex visual world. They demonstrate a general rule in vision: that the system is designed to perceive differences in intensity (that is, contrast) rather than absolute intensities of light.

To reveal the functional organization of connections, one must find for each neuron the characteristic light stimulus to which it will give the best response. This has been done most completely for the various cell types in the retina. For example, a ganglion cell, which represents the output of a small fraction of the retina, responds best to a small spot of light in the center of its circular receptive field or to a line or edge that passes through it. It responds relatively poorly to diffuse illumination.

The requirement for contrast is still more emphasized in cortical neurons, which practically ignore uniform illumination. Their activation requires highly specific shapes or forms—in particular, lines or edges with a certain orientation and position on the retina. Some categories of neurons are specialized to respond to angles or corners or to movements in one direction but not in another. According to the type of information they carry, neurons in the cortex have been provisionally classified as simple, complex, hypercomplex, etc.

A scheme that assumes a hierarchically ordered series of ascending connections explains many features of how neurons respond selectively to specific stimuli, for example, to a bar of light, a corner, or a square. At each stage the cells with relatively simple properties combine to form fields of progressively greater complexity and visual content. So far, transformation of visual information has been worked out only for the first seven stages between peripheral receptors and nerve cells within the visual cortex.

RETINA AND LATERAL GENICULATE NUCLEUS

To convey the relevance of the cellular approach to higher functions, this chapter deals principally with the performance of nerve cells at successive stages or relays of the visual systems of the cat and the monkey. However, because the initial aspects of processing in the retina leading to perception have not been so well studied in mammals, this aspect is illustrated by examples from work in a variety of species.

A comprehensive critical treatment is not possible within the scope of this book; the past few years have provided an overwhelming body of work on the structure and function of the visual system alone. In addition, psychophysics, color vision, dark adaptation, retinal pigments, transduction, and the organization of the retina could each form the basis of a self-contained monograph (see references at end of chapter). The same applies to comparative aspects of the visual system in invertebrates (such as the horseshoe crab, *Limulus*) and in lower vertebrates (fish, frog, turtle), as well as in mammals (rabbits and squirrels).

Nevertheless, a brief description of studies on cat and monkey, chiefly along the lines of Hubel and Wiesel's work, provides a clear, continuous thread extending from signaling to perception. These studies also help to relate the structure of the nervous system to questions of genetics and development (discussed in Part Six).

A crucial factor in the physiological analysis is the use of stimuli that mimic those occurring under natural conditions. For example, edges, contours, and simple patterns presented to the eye reveal features of the organization that could never be detected by shining bright flashes without form.

Another key to the success of Hubel and Wiesel's approach lies in asking not simply what stimulus evokes a response in a particular neuron, but what is the most effective stimulus. Since the impulses in any one cell have the same amplitude and time course, the "optimal" stimulus is defined as the one that produces the highest frequency discharge. Pursuit of this question through the various stages of the visual system has elicited many surprising and remarkable results.

Finally, the orderly, layered arrangement of neurons in the retina and within the brain itself suggests on its own that information processing is carried out in hierarchically arranged levels, proceeding from one functionally related group of cells to the next. Thus, it is reasonable to suppose that the behavior of one group of neurons can be explained

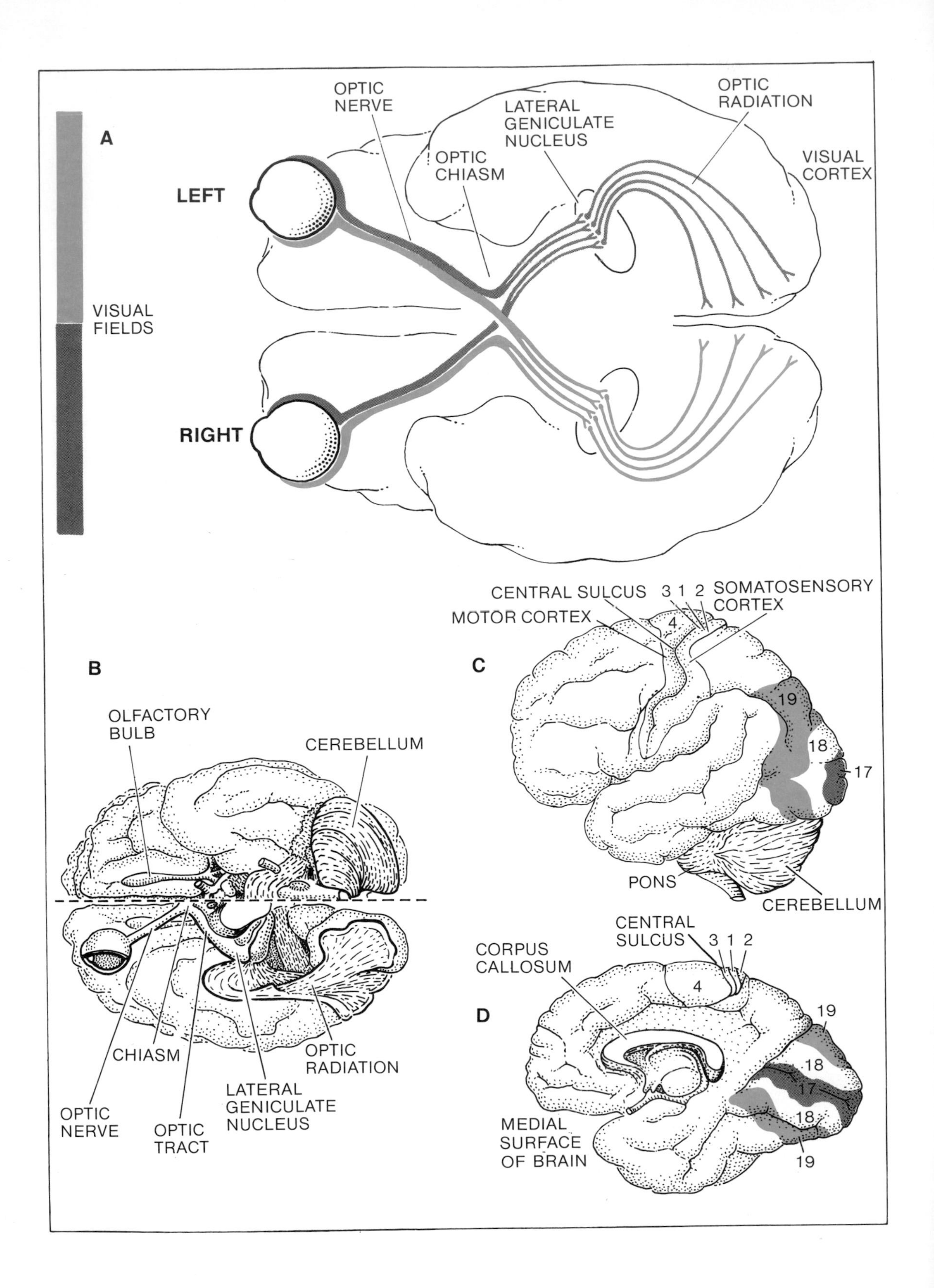
A
OPTIC NERVE
LATERAL GENICULATE NUCLEUS
OPTIC RADIATION
OPTIC CHIASM
VISUAL CORTEX
LEFT
RIGHT
VISUAL FIELDS
B
OLFACTORY BULB
CEREBELLUM
CHIASM
OPTIC RADIATION
LATERAL GENICULATE NUCLEUS
OPTIC NERVE
OPTIC TRACT
C
CENTRAL SULCUS
3 1 2
SOMATOSENSORY CORTEX
MOTOR CORTEX
4
19
18
17
PONS
CEREBELLUM
D
CENTRAL SULCUS
3 1 2
CORPUS CALLOSUM
4
19
18
17
18
19
MEDIAL SURFACE OF BRAIN

by understanding the effects produced by converging fibers which supply it.

This section describes first the principal anatomical features of the visual pathway and then the signals recorded at the successive stages.

Anatomical pathways in the visual system

The eye acts as a self-contained outpost of the brain. It collects information, analyzes it, and hands it on for further processing by the brain through a well-defined tract, the optic nerve.

The pathways from the eye to the cerebral cortex are illustrated in Figure 1*A*, and Figure 1*B* to *D* depicts some of the major landmarks of the human brain that are useful in the context of the following discussion. The optic nerve fibers arise from ganglion cells in the retina and end on cells in a relay station (the lateral geniculate nucleus) whose axons in turn project through the optic radiation to the cerebral cortex. From here on the progression becomes ever more complex, with no end station in sight.

Figure 1*A* also shows how the output from each retina divides in two at the optic chiasm to supply the geniculate nucleus and cortex on both sides of the animal. As a result, the right side of each retina projects to the right cerebral hemisphere. Figure 1 also shows that the right side of each retina receives the image of the visual world on the left side of the animal. Each cerebral hemisphere, therefore, "sees" the visual field of the opposite side of the world.

The other pathways that branch off to the midbrain are not described here. In higher vertebrates they are primarily concerned with regulating eye movements and pupillary responses and are not directly relevant for the types of pattern recognition considered here.

By merely examining the cellular anatomy of the various structures in the visual pathway, one can exclude the possibility that information is handed on unchanged from level to level. The neurons CONVERGE and DIVERGE extensively at any stage; that is, each cell makes and receives connections with a number of other cells. For example, the human eye contains over 100 million primary receptors, the rods and cones, but a smaller number of cells in the second layer; these make connections on ganglion cells which send only about 1 million optic nerve fibers into the brain. In the monkey and cat the same principle holds: a stepdown in neuronal numbers from receptors to ganglion cells. Therefore, within

1

VISUAL PATHWAYS. A. Outline of the visual pathways seen from below (base of the brain) in primates. The right side of each retina (shown in color) projects to the right lateral geniculate nucleus and the right visual cortex receives information exclusively from the left half of the visual field. B. Visual pathways in a partially dissected human brain seen from below. C, D. Lateral and medial views of the cortical surface. Area 17 is also known as the striate cortex or visual area I, while areas 18 and 19 are visual areas II and III. In addition, area 4 (the motor cortex) and areas 3, 1, and 2 of the sensory cortex are labeled.

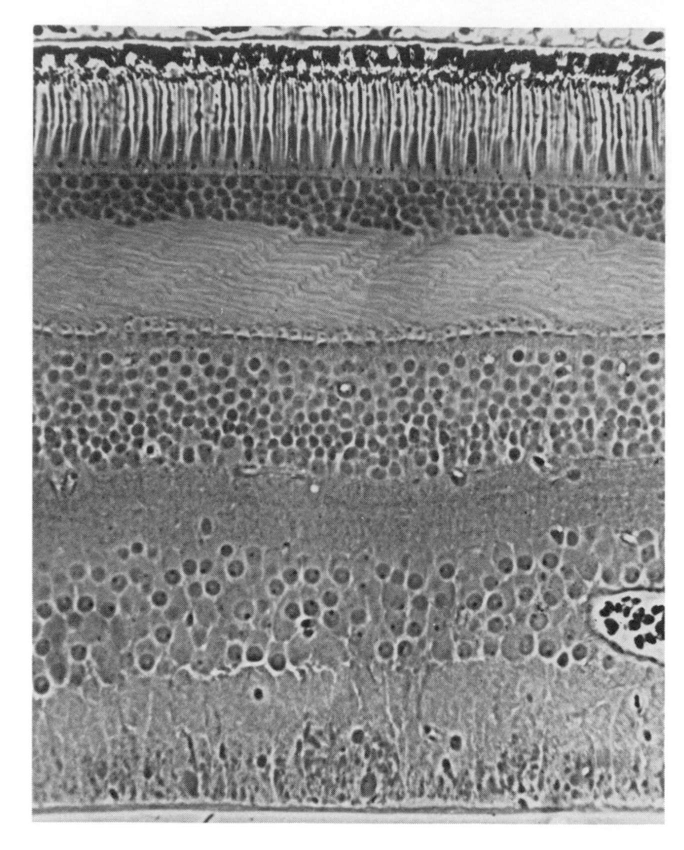

2 **SECTION THROUGH A HUMAN RETINA, showing the five principal cell types arranged in layers. Light enters the retina at the ganglion cell layer and reaches the photoreceptors (rods and cones), where it is absorbed, starting excitation in the outer segments. (From Boycott and Dowling, 1969)**

the eye as a whole there occurs a funneling of information. As a result, an individual neuron that receives impulses from several incoming nerve fibers cannot reflect separately the signals of any one of them. Instead, converging impulses of different origin are combined at each stage into an entirely new message that takes account of all the inputs. This process is called INTEGRATION (Chapter 1).

What makes the retina so specially inviting for physiological research is the neat layering and orderly repetition of the relatively few cell types—there are only five. The arrangement and the typical positions of various cells are illustrated in Figure 2, which shows a cross section of a human retina. On the deep surface, farthest from the incoming light, lie the RODS and CONES, which are concerned with night and daytime vision, respectively. They are connected to the BIPOLAR CELLS, which in turn connect to the GANGLION CELLS and so to the optic nerve fibers. Apart from this through-line, there are other cells that make predominantly lateral or side-to-side connections. These are the

HORIZONTAL and AMACRINE CELLS. The role played by these cells and their interconnections is discussed later. We shall see that the aggregation of comparable cells into layers is also a feature of the lateral geniculate nucleus and the visual cortex, although the arrangement is not so homogeneous and the boundaries not so sharp.

Retinal ganglion cells and the concept of receptive fields

A number of developments set the stage for the single-neuron analysis of the mammalian visual system. Among these was the elegant and lucid work of Adrian.[1] Hartline's[2] pioneering work on the horseshoe crab, *Limulus,* foreshadowed intriguing and exciting developments. If so much could be gleaned from a relatively simple invertebrate eye, would the vertebrate visual apparatus not yield corresponding insights, if similar methods could be used? Hartline himself advanced his studies into the retina of the frog, and Granit and his colleagues[3] went further by recording from single neurons of the mammalian eye.

These were the main points of departure for the next stage of exploration of the mammalian retina in cellular terms. The principal new approach was not so much a matter of technique; rather, it consisted of formulating the following question: What is the best way to stimulate individual ganglion cells whose axons carry information to the higher centers through the optic nerve? This question led logically to the use of discrete circumscribed spots and patterns of light for stimulation of selected areas of the retina.

A methodological feature of the experiments that became essential was the use of the practically intact eye whose normal refracting channels served as pathways for stimulation. Not only was it technically easier to record the signaling of ganglion cells, but the initial analysis was also simplified by beginning with the end result of activity in the eye. This presentation, therefore, discusses first the impulse patterns emerging from the eye and then the preceding steps that have occurred within the retina.

Illumination of selected areas of the retina introduces the important concept of the receptive field, which provided the key to understanding the significance of the signals. Applied to the visual system, the RECEPTIVE FIELD of a neuron can be defined as the AREA ON THE RETINA FROM WHICH THE DISCHARGES OF THAT NEURON CAN BE INFLUENCED.[4] For example, a record of the activity of one particular fiber in the optic nerve of a cat (Figure 4) shows that that fiber increases or decreases its rate of firing only when a defined area of retina is illuminated. This is its receptive field. By definition, illumination outside the field produces no effect at all. The area itself can be subdivided into distinct regions, some of which act to produce firing and others to suppress impulses in the cell.

---

[1]Adrian, E. D. 1946. *The Physical Background of Perception.* Clarendon Press, Oxford.

[2]Hartline, H. K. 1940. *J. Opt. Soc. Am. 30*:239–247.

[3]Granit, R. 1947. *Sensory Mechanisms of the Retina.* Oxford University Press, London.

[4]Hartline, H. K. 1940. *Am. J. Physiol. 130*:690–699.

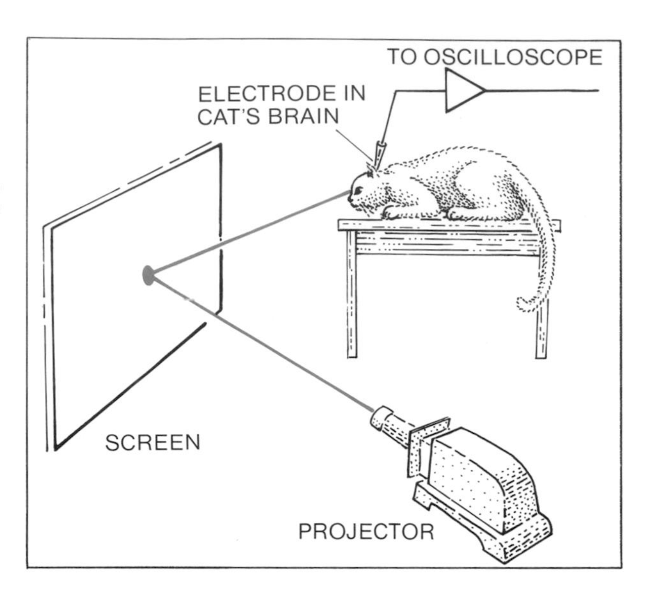

3

STIMULATION OF RETINA with patterns of light. The eyes of an anesthetized, light-adapted cat (or monkey) focus on a screen onto which various patterns of light are projected. An electrode records the responses from a single cell in the visual pathway. Light (or shadow) falling onto a restricted area of the screen may accelerate (excite) or slow (inhibit) the signals given by a neuron. By determining the areas on the screen from which a neuron's firing is influenced, one can delineate the receptive field of the cell. The positions of cells in the brain and the tracks of electrode penetrations can be reconstructed histologically after the experiment.

Receptive fields of ganglion cells and optic nerve fibers

The best way of illuminating particular portions of the retina is to anesthetize the animal lightly and have it face a screen at a distance for which its eye is properly refracted. If one then shines spots or patterns of light onto the screen, these will be well focused on the retinal surface (Figure 3). The experiments described below were made in the light-adapted state on immobilized eyes, but neuronal discharges have also been recorded from unrestrained, waking cats with chronically implanted microelectrodes.

When recording from a particular nerve cell, the first task is to find the location of its receptive field. Characteristically, most cells in the eye, and throughout the visual system, show continued discharges at rest even in the absence of illumination. Appropriate stimuli do not necessarily initiate activity, but may modulate the background firing; the responses can therefore consist of either an increase or a decrease of ongoing discharges.

At first it may seem puzzling and unexpected that uniform illumination of the eye by flashes of light is not the best way to influence discharges of ganglion cells. Uniform illumination, in fact, produces a bewildering array of responses, usually a transient burst of signals when the light is flashed on or turned off. A small spot of light, 0.2 mm in diameter, shone onto a receptive field is generally far more effective. Furthermore, the same spot of light can have opposite effects, depending on the exact position of the stimulus within the receptive field. For example, in one area the spot of light excites a ganglion cell for the duration of illumination. Such an "on" response can be converted into an inhibitory "off" response by simply shifting the spot by 1 mm or less across the retinal surface. The same spot of light, therefore, suppresses the firing of the same ganglion cell. When small spots are used to map large numbers of receptive fields, a constant and simplifying feature of the neural organization of the cat's retina emerges. There are only two basic receptive field types: the "ON" CENTER and the "OFF" CENTER. The receptive fields are all roughly concentric, with the ganglion cell in the geometrical central region of any field.[5] Figure 4 shows these features and some of the variants of responses that can be obtained. In an "on" center receptive field, light produces the most vigorous response if it completely fills the center, while for most effective inhibition it must cover the entire ring-shaped surround (annular illumination in Figure 4). Inhibition is always followed, when the light is turned off, by an "off" discharge. Another uniform feature is that the spotlike center and its surround are always antagonistic; therefore, if they are illuminated simultaneously, they tend to cancel each other's contribution (diffuse illumination in Figure 4). There then occurs merely a relatively weak "on" component and a similar "off" discharge when the light is turned off. The "off" center field has a converse organization, with inhibition arising in the circular center.

All these properties of receptive fields explain the initially puzzling finding that illumination covering an entire field has a much weaker action in arousing a ganglion cell than does a well-placed small spot or a line or an edge passing through the center.

A remarkable and simplifying consideration therefore emerges about the performance of the retina with its 100 million neurons. Their STEREOTYPED ARRANGEMENT with repeating units GIVES ONLY A FEW BASIC TYPES OF SIGNALS. Several variations in the types of concentric receptive fields are noted below.

Sizes and characteristics of receptive fields

Neighboring ganglion cells collect information from very similar, but not quite identical, areas of the retina. Even a small (0.1 mm) spot of light on the retina covers many receptive fields that have diverse responses, some ganglion cells being inhibited, others excited. This characteristic organization, with neighboring groups of receptors projecting onto neighboring ganglion cells in the retina, is retained at all levels in visual pathways. The systematic analysis of receptive fields demonstrates the general principle that NEURONS PROCESSING RELATED INFORMATION ARE CLUSTERED TOGETHER. In sensory systems this means that the central neurons dealing with a particular area of the surface can communicate with each other over short distances. This appears to be an economical arrangement, as it saves long lines of communication and simplifies the making of connections.

Size of receptive fields

The size of the receptive field of a ganglion cell depends on its location in the retina. The receptive fields situated in the central areas of the retina have much smaller centers than those at the periphery. In the cat, field centers range from 0.125 to 2.0 mm (about 0.5° to 8°) in diameter, being smallest in the area centralis (a region that corresponds to the fovea in the human eye), where the acuity, or resolving power, of vision is highest.[6]

As an example of the generalization of the functional importance

[5]Kuffler, S. W. 1953. *J. Neurophysiol.* *16*:37–68.
[6]Wiesel, T. N. 1960. *J. Physiol.* *153*:583–594.

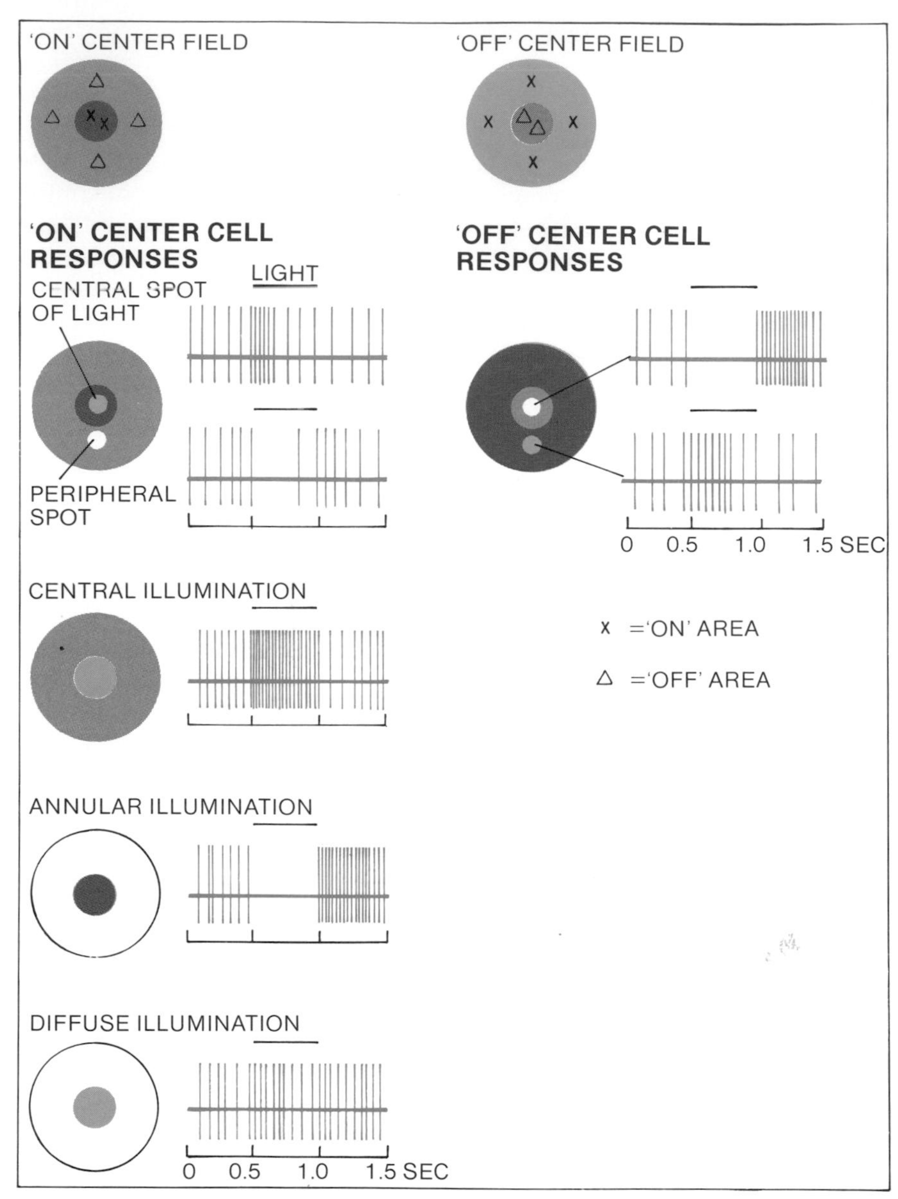

4 **RECEPTIVE FIELDS OF GANGLION CELLS in the retinas of cats and monkeys are grouped into two main classes: "on" center and "off" center fields. "On" center cells respond best to a spot of light shone onto the central part of their receptive fields. Illumination (indicated by bar above records) of the surrounding area with a spot or a ring of light reduces or suppresses the discharges and causes responses when the light is turned off. Illumination of the entire receptive field elicits relatively weak discharges because center and surround oppose each other's effects. "Off" center cells slow down or stop signaling when the central area of their field is illuminated and accelerate when the light is turned off. Light shone onto the surround of an "off" center receptive field area excites. (After Kuffler, 1953)**

of receptive fields of various sizes, consider sensory areas in the skin, activated by touch or pressure. There is a strikingly similar gradation of receptive field size in relation to fine resolution or discrimination. A higher order sensory neuron in the brain responding to fine touch applied to the skin of the fingertip has a very small receptive field compared with a neuron whose field is on the skin of the upper arm. To discern the form of an object, we use our fingertips and foveas, not the less discriminating regions on the receptor surfaces with poorer resolution.

In recent years the classification of concentric receptive fields has been extended and refined, and there is every expectation that additional features of receptive field performance are yet to be uncovered. Thus, two types of "on" center and "off" center fields in the cat have been distinguished, one giving a sustained discharge (X cells), the other a transient one (Y cells).[7,8] The axons of the X ganglion cells conduct more slowly than those of the Y cells. Presumably, the sustained X cells provide information about the position of relatively steady patterns, while the transient Y cells are used for objects moving across the visual field. In addition to the typical "on" and "off" center cells are others with more complex receptive field properties; for example, several laboratories have reported ganglion cells with centers that can be aroused by light *or* dark spots. These neurons, now called W cells,[9] have the slowest conduction velocity so far observed in any axon arising from a ganglion cell.

The initial choice of the cat for receptive field analyses was a lucky one; in the rabbit, for example, the situation would have been more complicated. As shown by Barlow,[10] the ganglion cells in the rabbit have more elaborate receptive fields and can respond specifically to such complex features as edges or to movement in one direction rather than another. Equally complex are lower vertebrates, such as frogs, that were investigated by Barlow and later by Lettvin, Maturana, Michael, and their colleagues.[11,12] As one might expect, cells whose receptive fields are specifically color-coded have been noted in various animals, including the monkey, the ground squirrel, and some fishes. These animals, in contradistinction to the cat, possess excellent color vision and an intricate neural mechanism for processing color.

How are retinal neurons connected to form receptive fields of ganglion cells?

The receptive field studies on ganglion cells have led logically into the exploration of higher centers. They also have stimulated an examination of the neural machinery that synthesizes the ganglionic receptive fields within the retina. The fields come about by synaptic interaction in the maze of retinal connections. One would like to know, therefore,

---

[7]Enroth-Cugell, C., and Robson, J. G. 1966. *J. Physiol.* *187*:517–552.
[8]Cleland, B. G., Dubin, M. W., and Levick, W. R. 1971. *J. Physiol.* *217*:473–496.
[9]Stone, J, and Hoffmann, K. P. 1972. *Brain Res.* *43*:610–616.
[10]Barlow, H. B., Hill, R. M., and Levick, W. R. 1964. *J. Physiol.* *173*:377–407.
[11]Maturana, H. R., Lettvin, J. Y., McCulloch, W. S., and Pitts, W. H. 1960. *J. Gen. Physiol.* *43*:129–175.
[12]Michael, C. R. 1973. *N. Eng. J. Med.* *288*:724–725.

both the wiring diagram and the mechanisms of synaptic action within the network.

Owing to methodological difficulties, a number of years passed before serious inroads were made on the problems. Among the essential improvements of methods was the development of finer capillary electrodes that would penetrate the various cells. However, identification of neurons from which intracellular recordings had been made necessitated marking them with dyes (see later, Figure 6). The introduction of improved methods through the pioneering work of Svaetichin,[13] Tomita,[14] and their colleagues gave the impetus for a quiet, but real revolution in retinal physiology. This development promises to gain further momentum by the use of cellular neurochemistry, a methodology that in the past 15 years has contributed greatly to the steady progress in understanding chemical communication between nerve cells. One has the impression, however, that much remains to be done before the diverse pieces fall into place for a coherent understanding of the entire retinal system.

To find out how the receptive fields of ganglion cells are produced, it is plainly necessary to examine, besides the ganglion cells, the performance of the other cell types within the retina: these are the photoreceptor, bipolar, horizontal, and amacrine cells—the neurons that occupy the bulk of the retina and form the input to the ganglion cells. But even in the absence of detailed information, on the basis of the physiological results discussed above one could predict some general features of intraretinal organization. To form concentric receptive fields, a set of specially arranged inhibitory and excitatory lines of communication must run between the photoreceptors and the ganglion cells. The crucial problem, therefore, is to find out what is happening in the two intraretinal synaptic stations, in the OUTER and INNER PLEXIFORM LAYERS.

A general guide for this discussion is supplied by Dowling and Boycott's[15] diagrammatic representation of a primate retina (Figure 5). The scheme is based on structural and physiological observations. The pathway from receptors through bipolar cells to ganglion cells is far more complex than a simple through-line. There exists, in addition, a variety of interconnections among cells in the various layers. For example, the horizontal cells shown in Figure 5 receive synapses from many receptors and in turn feed back onto them as well as onto bipolar cells. Similarly, amacrine cells, which receive their inputs from bipolar cells, send synapses back to them as well as to the ganglion cells. From such structural and physiological considerations, one can conclude that horizontal and amacrine cells modify or influence the transfer of information through the retina. The receptive field of a ganglion cell, therefore, is a composite constructed by the receptive fields of the cells along the lines leading to it.

---

[13]Svaetichin, G. 1953. *Acta Physiol. Scand.* *29*:565–599.
[14]Tomita, T. 1965. *Cold Spring Harbor Symp. Quant. Biol.* *30*:559–566.
[15]Dowling, J. E., and Boycott, B. B. 1966. *Proc. R. Soc. Lond.* B *166*:80–111.

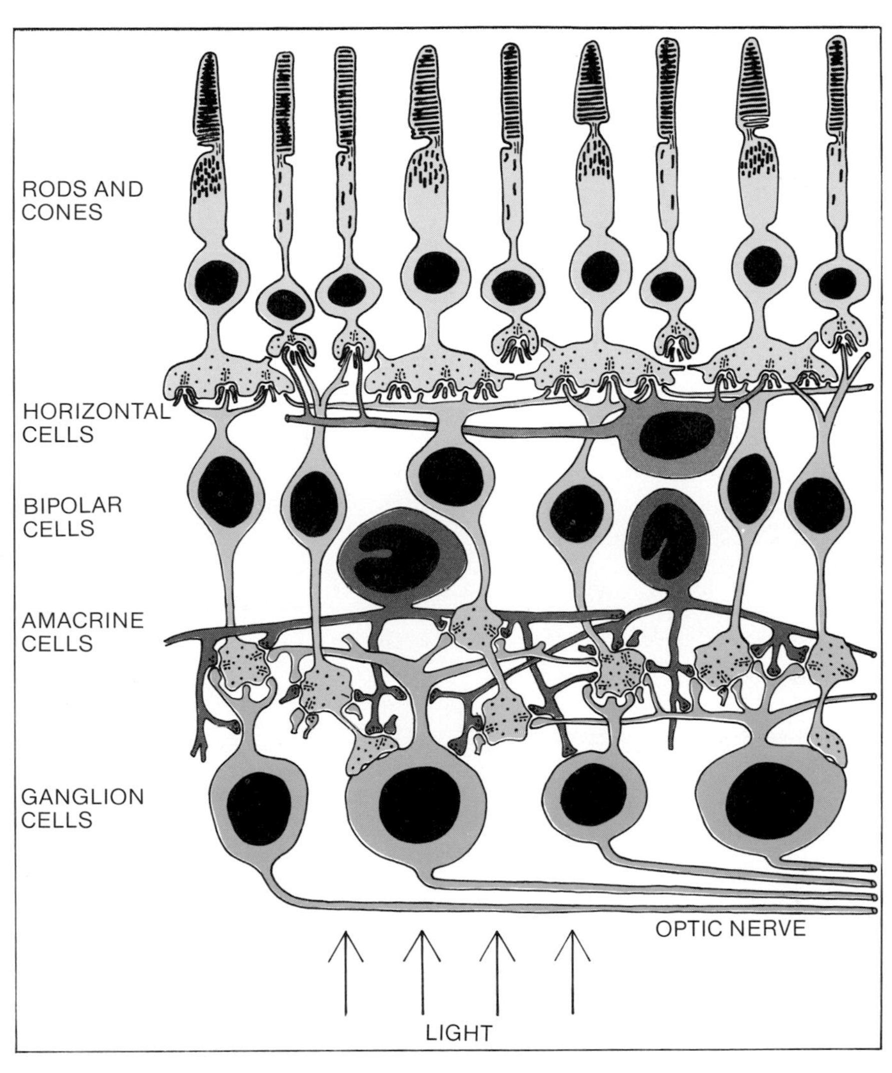

ORGANIZATION OF PRIMATE RETINA. (After Dowling and Boycott, 1966) 5

For technical reasons, lower vertebrates are more suitable than the cat for studying the properties and receptive fields of the various types of cells within the retina. In these animals the cell types appear similar and are easier to impale with microelectrodes. The brief account that follows does not deal with the mammalian retina, but is taken largely from the work of Tomita, Kaneko, Dowling, Werblin, Naka, Baylor, Fuortes, and their colleagues, who used a variety of fish, amphibian, and turtle retinas.[14,16–19]

[16]Kaneko, A. 1970. *J. Physiol.* *207*:623–633.

[17]Dowling, J. E., and Werblin, F. S. 1971. *Vision Res.* *3*:1–15.

[18]Naka, K. I., and Witkovsky, P. 1972. *J. Physiol.* *223*:449–460.

[19]Baylor, D. A., Fuortes, M. G. F., and O'Bryan, P. M. 1971. *J. Physiol.* *214*: 265–294.

The techniques of intracellular recording and dye injection are described in detail later. In brief the procedure is to record from a cell with an intracellular microelectrode that measures the potential across the surface membrane. By appropriate illumination one can determine the membrane properties, the responses to light, and the receptive field organization of that cell, and then inject it with a small amount of dye. Subsequently the tissue is examined histologically and the cell that had been impaled by the microelectrode is identified. Examples of various injected cell types are shown in Figure 6. Fortunately, the difficult and tedious technique of marking cells is necessary only in the beginning to establish a correlation of structure with function, because the different types of cells give highly distinctive responses which are sufficient for reliable identification.

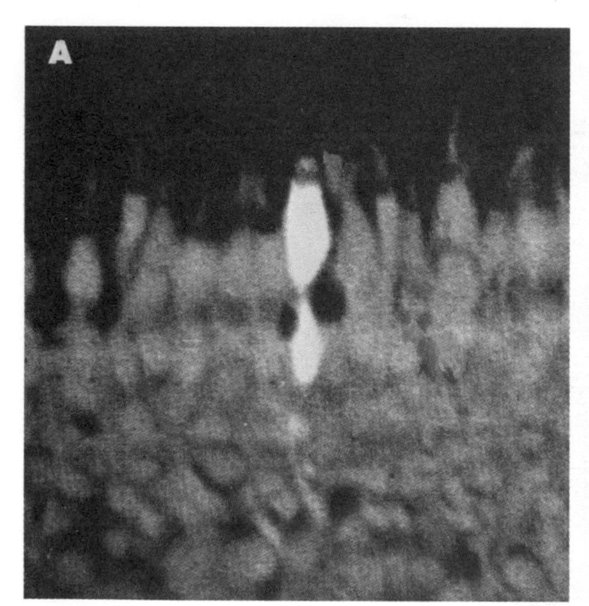

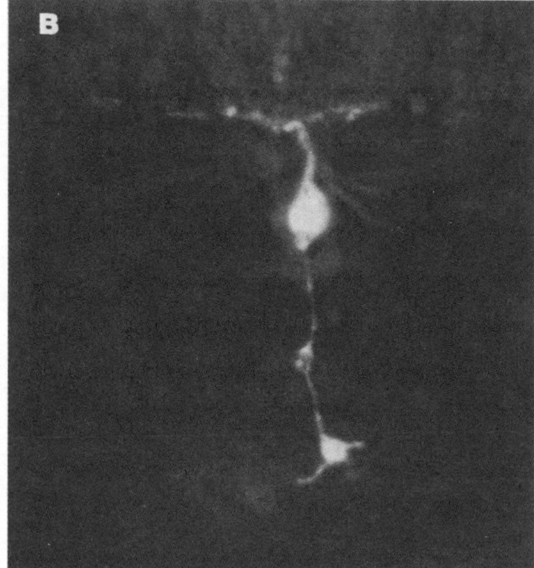

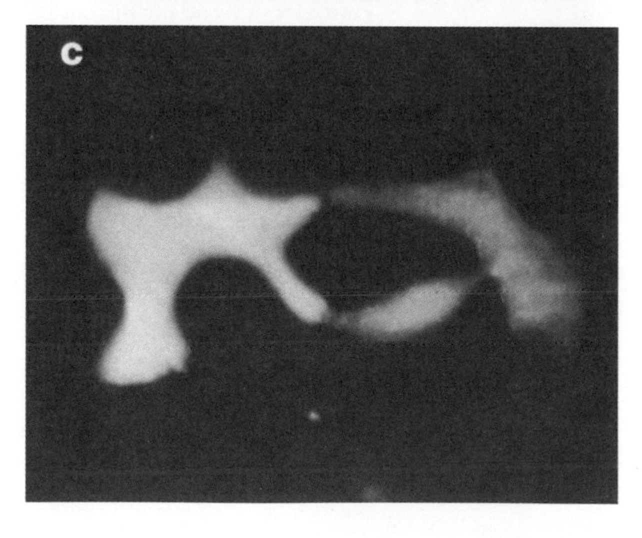

6

MARKING AND IDENTIFICATION OF NEURONS. A. Cone in retina of turtle injected with the fluorescent dye procion yellow. B. Bipolar cell of goldfish injected with procion yellow. C. Two horizontal cells in dogfish retina injected with different dyes. (A from Baylor, Fuortes, and O'Bryan, 1971; B from Kaneko, 1970; C from Kaneko, 1971)

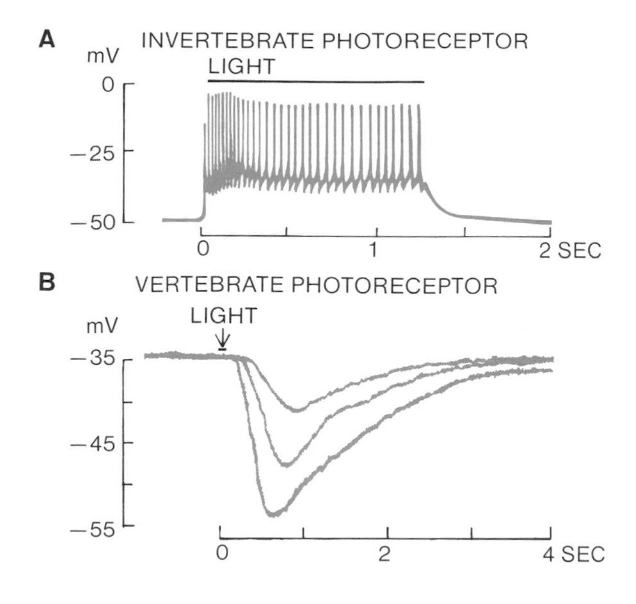

7
RESPONSES OF PHOTORECEPTORS. A. Photoreceptors of an invertebrate (horseshoe crab) respond to light (indicated by bar above record) by a depolarization that gives rise to impulses. This is the usual type of response elicited from vertebrate sensory receptors activated by various stimuli such as touch, pressure or stretch. B. Photoreceptors of a vertebrate (turtle) respond by a hyperpolarization that is graded according to the intensity of the light flash. (A after Fuortes and Poggio, 1963; B after Baylor, Fuortes, and O'Bryan, 1971)

From the outset, the intracellular records revealed that cells in the vertebrate retina gave different signals from what might have been predicted on the basis of knowledge of impulses in invertebrates and in the central nervous system of vertebrates. Only ganglion cells and amacrine cells gave impulses of the type usually observed in typical neurons (Chapter 4). The other cell types responded to illumination or darkness with relatively slow, graded potentials that in most cells are found only in synaptic regions. Particularly surprising and somewhat confusing, when first observed, was the electrical response given by receptors to illumination: HYPERPOLARIZATION (an increase in internal negativity). The sign of this potential was the opposite of what is seen in other sensory receptors of the body.

What is the interpretation of this result and how does it influence signaling? Chapter 15 describes in detail how sensory cells outside the eye respond to stimulation and how one cell hands on information to the next by secreting a chemical transmitter. For the moment it is convenient to illustrate critical points about the difference between signaling in photoreceptors in vertebrate and invertebrate eyes. Readers who are unfamiliar with this kind of material may prefer to read Chapter 4 before continuing.

Figure 7*A* shows the response of a photoreceptor cell in the eye of the horseshoe crab (*Limulus*).[20] The sequence of signals is similar to that in all other sensory cells that have so far been studied. They all become DEPOLARIZED by the appropriate stimulus. Heating, cooling, pressing, and so on always lead to depolarization: the inside of the cell becomes more positive than when at rest (see definitions in Chapter 1). This change in potential is excitatory and causes impulses to be initiated

[20]Fuortes, M. G. F., and Poggio, G. F. 1963. *J. Gen. Physiol.* *46*:435–452.

in the nerve fibers. When these impulses reach the end of an axon, they in turn cause the liberation of a chemical transmitter from the endings, and this substance influences the next cell in line. Much is now known about these processes; in particular, it is clear that the release of transmitter is triggered by depolarization of the nerve endings and suppressed by hyperpolarization (increased internal negativity). Depending on the chemistry of the transmitter substance and the membrane of the second-order cell, its response may be depolarization or hyperpolarization; in other words, excitation or inhibition. In contrast, recording from a vertebrate photoreceptor (Figure 7*B*) shows that illumination causes a hyperpolarization whose size is graded with the intensity of the light flash.

This is why the hyperpolarization response of rods and cones to light was so surprising. By analogy with other receptors, they behaved as though dark was the stimulus, since that was the condition in which they were depolarized, and light apparently signaled the absence of a stimulus. Again by analogy, this would imply that illumination turned off the liberation of transmitter which was proceeding continuously in the dark. It now seems that the explanation of this apparent paradox is probably that photoreceptors continually release transmitter in the dark and stop releasing it on illumination.

Photoreceptors are influenced predominantly by light falling directly upon their outer segments. Receptors, however, can also activate each other; they do this in two ways. First, Baylor, Fuortes, and O'Bryan[19] have shown that in the turtle, activation of cones by light may also affect neighboring receptors through the intermediary of horizontal cells that, in turn, can act back onto cones. This feedback action can create a surround effect in a receptor. Second, direct electrical interactions have been observed physiologically between photoreceptors, a result that is reinforced by the finding of anatomical junctions characteristic of electrical coupling (Chapter 8).

HORIZONTAL CELLS, which receive synaptic inputs from many receptors, give a variety of maintained responses to illumination. In the carp or goldfish, for example, one type of horizontal cell responds with either depolarization or hyperpolarization, depending on the wavelength of light, while another type just hyperpolarizes in a graded manner according to light intensity;[21] neither kind of cell generates conducted impulses. Since the horizontal cells are supplied by receptors over a large area of retina, their fields are far larger than those of bipolar or ganglion cells. In addition, neighboring horizontal cells are coupled to each other with junctions that permit current to flow between them, and this further increases the size of their receptive fields.

BIPOLAR CELLS respond to light with a sustained depolarization or hyperpolarization and, like horizontal cells, they do not generate impulses. The bipolar cells in the goldfish and in the mudpuppy have a concentric receptive field organization, with an antagonistic center-surround arrangement. An example is shown in Figure 8: a spot of light

---

[21]Kaneko, A. 1971. *J. Physiol.* *213*:95–105.

in the center hyperpolarizes and an annular light stimulus depolarizes. Other bipolar cells are depolarized by light at the center. Diffuse illumination of the whole field is relatively ineffective.[16] Hence, already at this level some of the key features of the receptive fields of ganglion cells are evident (Figure 4).

The center of the receptive field is presumably supplied directly by the receptors themselves and the extensive surround by the horizontal cells, whose receptive fields spread laterally over a wide area. The horizontal cells apparently can drive bipolar cells and through them elaborate the surround part of receptive fields of ganglion cells.[18] The output of the bipolar cells reports mainly the difference of the illumination between periphery and center of the receptive field, thereby providing a clear contrast mechanism. It is assumed that in this entire scheme the graded responses in receptors, bipolar cells, and horizontal cells are reflected in a graded output of transmitter.

In the second retinal layer, the AMACRINE CELLS receive no connection from the receptors, but only from bipolar cells and other amacrine cells. The amacrine cells apparently also feed back onto the bipolar cells, besides making connections with ganglion cells. Unlike ganglion and bipolar cells, amacrine cells in *Necturus* (the mudpuppy) tend to give transient discharges on illumination.

In birds, another important level of control in the retina has been established. EFFERENT FIBERS from the brain terminate upon amacrine cells and MODULATE their RECEPTIVE FIELD PROPERTIES. This, in addition to the observed feedback control between cells, reminds one that although the main flow of information is from receptors to ganglion cells, there also exists signaling in the opposite centrifugal direction (Chapter 15.)[22,23] The ganglion cell discharges presumably reflect the balance between the

[22]Cowan, W. M., and Powell, T. P. S. 1963. *Proc. R. Soc. Lond.* B *158*:232–252.
[23]Miles, F. A. 1972. *Brain Res. 48*:115–129.

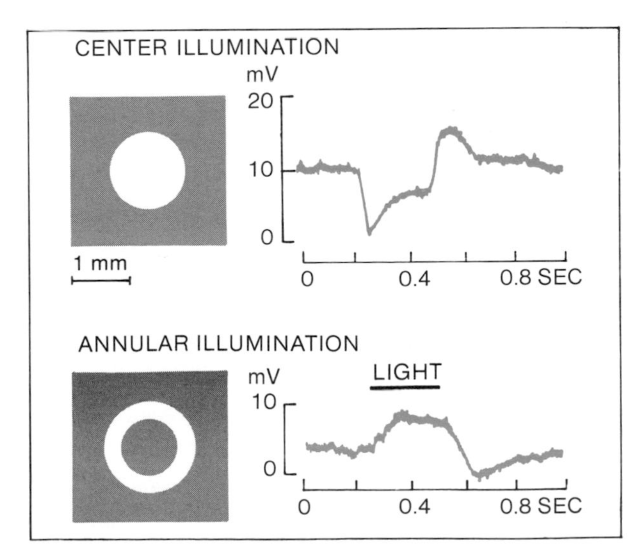

8
RECEPTIVE FIELD OF BIPOLAR CELL in a goldfish retina responding by hyperpolarization to illumination of its center and by depolarization to a ring of light. Other bipolar cells respond in the opposite way (depolarization with central illumination); neither type generates impulses. (From Kaneko, 1970)

bipolar and amacrine inputs, but the actual role of amacrine cells in producing the typical "on" or "off" center field is not clear.

In conclusion, the activity of ganglion cells is made up by the total contribution of four cell types. In the outer retinal layer the receptor and horizontal cells interact with each other and with the bipolar cells. These in turn, together with the amacrine cells, determine the signals that arise in the ganglion cells. The proviso of the preceding discussion is that the picture is a composite drawn from many lower vertebrates and only parts of the picture may apply to the mammalian retina.

The rich system of retinal interconnections is not yet understood in detail, and there remain wide gaps in understanding other aspects of retinal function. For example, little is known of synaptic mechanisms that link neurons, and in spite of promising chemical studies on isolated neurons,[24,25] knowledge about transmitters remains fragmentary. Nevertheless, in comparison with what was known 10 to 15 years ago, the advances seem dramatic, and a definitive outline of the functional organization of the retina has emerged.

What information do ganglion cells convey?

The most striking feature of ganglion cell signals is that they tell a different story from that of primary sensory receptors. They do not convey information about absolute levels of illumination because they behave in a similar fashion at different background levels of light. They ignore much of the information of the photoreceptors, which work more like a photographic plate or a light meter. Rather, they measure differences within their receptive fields by comparing the degree of illumination between the center and the surround. Apparently they are designed to notice simultaneous contrast, the transition from more to less or no light, while they are relatively insensitive to gradual changes in overall illumination. They are exquisitely tuned to detect such contrast as the edge of an image crossing the opposing regions of a receptive field.

What image of the world is presented to the brain by the retina? By comparison with our daily experience, it is a rather drab environment. The world is represented as a series of closely spaced dots and contours depending on DIFFERENCES in the level of illumination. Position on the retinal surface is indicated and the picture is livened by color. On the other hand, by comparison with information obtained from primary receptors alone (Chapter 15), the information provided by the entire retina is relatively lively. It is apparent, therefore, that the three retinal layers extract and analyze a great deal of information about the outside world. By choosing some aspects of the information collected by the primary receptors, and not others, a start has been made in selecting features that are important for form vision while jettisoning the level of background illumination.

Lateral geniculate nucleus

The optic nerve fibers running to the cortex from the eye terminate on cells of the right and left lateral geniculate nucleus, a distinctively

---

[24]Lam, D. M. K. 1972. *Proc. Natl. Acad. Sci. U.S.A., 69*:1987–1991.
[25]Drujan, B. D., and Svaetichin, G. 1972. *Vision Res. 12*:1777–1784.

layered structure (GENICULATE means bent like a knee). In the lateral geniculate nucleus of the cat there are three obvious, well-defined layers of cells (A, A_1, B), one of which (B) has been further subdivided.[26] In the monkey the lateral geniculate nucleus has six layers of cells (Figure 9). Anatomical and physiological studies have shown that in both cat and monkey each of these layers is predominantly supplied by one or the other eye. In the cat the inputs to layers A and B originate in ganglion cells in the eye on the other side of the animal, the fibers having crossed at the chiasm; A_1 is supplied with fibers from the eye on the same side that do not cross. Because of this orderly and systematic separation of fibers at the chiasm, the receptive fields of the cells in the lateral geniculate nucleus are all situated in the VISUAL field on the opposite side of the animal.

A striking topographical feature is the highly ordered arrangement of receptive fields within individual layers of the geniculate. Neighboring regions of the retina make connections with neighboring geniculate cells, so that the receptive fields of adjacent neurons overlap over most of their area. The area centralis, the region of the cat retina with small receptive field centers, projects onto the greater portion of each geniculate layer, while there are relatively few cells devoted to the peripheral retina. This heavy representation (the same as in the cortex) presumably reflects the use of these regions for high-acuity vision and the need for fine-grained resolution. Although there are probably an equal number of optic nerve fibers and geniculate cells, each geniculate cell receives its input from several optic fibers, which in turn split up to make synapses with several geniculate neurons.

[26]Guillery, R. W. 1970. *J. Comp. Neurol.* *138*:339–368.

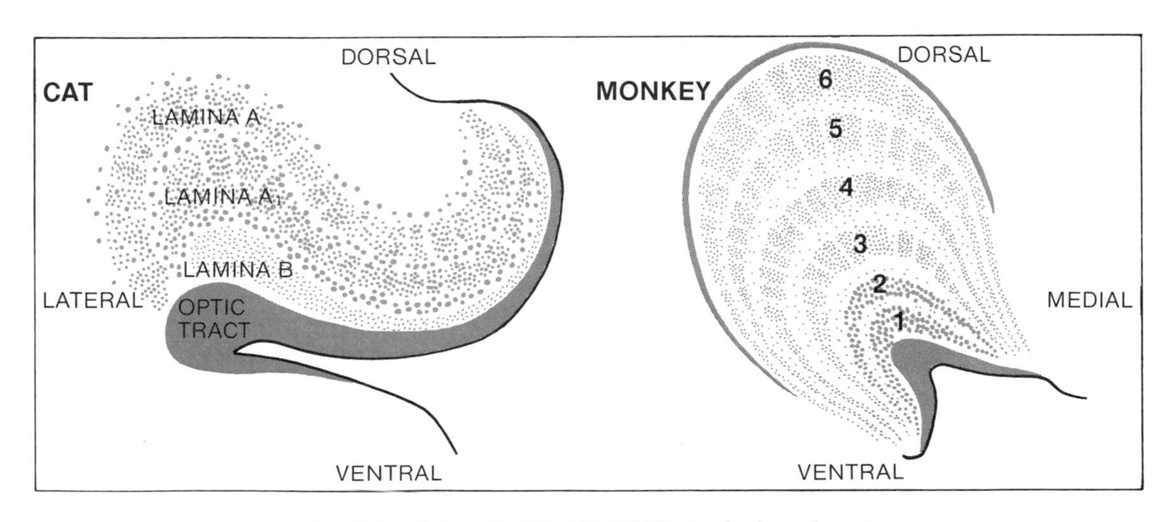

9

LATERAL GENICULATE NUCLEUS. In the lateral geniculate nucleus of the cat (A) there are three layers of cells. In the monkey (B) the lateral geniculate nucleus has six layers. In both animals each layer is supplied by only one eye. (From Szentagothai, 1973)

The topographically ordered sequence of connections is not confined to one individual layer; even cells in the different layers are in register. Thus, as a microelectrode passes from one layer of the geniculate to the next, which is supplied by optic nerve fibers from the other eye, one records from successive cells driven by first one eye and then the other; the positions of the receptive fields remain in corresponding positions on the two retinas representing the same area in the visual field.[27] There seems to be no intermediate area between the layers where extensive mixing of information or interaction between the eyes occurs, since binocularly excited cells (neurons with receptive fields in both eyes) are rare. This relatively small population of neurons has now been studied by Bishop and his colleagues, who have used computer techniques to detect them.[28]

Surprisingly, perhaps, the responses from geniculate cells do not differ drastically from those of retinal ganglion cells (Figure 10). Geniculate neurons also have concentrically arranged antagonistic receptive fields, either of the "off" center or the "on" center type, but the contrast mechanism is more finely tuned by a closer matching of the inhibitory and excitatory areas. The geniculate neurons, like retinal ganglion cells, require contrast for optimal stimulation but give weaker responses to diffuse illumination. Both X (sustained) and Y (transient) types are also seen.[8]

Studies of receptive fields of the lateral geniculate nucleus are still incomplete. For example, there are interneurons whose contribution has not been established. Knowledge of the synaptic organization is likely to gain in clarity as soon as the activity can be monitored by intracellular electrodes. Such a technique seems essential for sorting out the excitatory and inhibitory synaptic contributions of various optic nerve fibers.

In Chapter 18 the lateral geniculate nucleus is again discussed in connection with the changes that occur there in Siamese cats and albino animals of various species in which fibers cross abnormally at the chiasm as a result of a genetic defect. Also discussed are the structural and physiological effects of use and disuse in immature animals.

THE VISUAL CORTEX

General problems and question of numbers

Proceeding from the retina and lateral geniculate nucleus to the cerebral cortex raises questions that go beyond simple matters of technique. It has long been acknowledged that understanding the workings of any part of the nervous system requires knowledge of the cellular properties of its elementary units, the neurons—how they conduct and carry information and how they transmit that information from one cell to the next at synapses or specialized points of contact. Yet it was not

[27] Hubel, D. H., and Wiesel, T. N. 1961. *J. Physiol.* *155*:385–398.

[28] Sanderson, K. J., Bishop, P. O., and Darian-Smith, I. 1971. *Exp. Brain Res.* *13*:178–207.

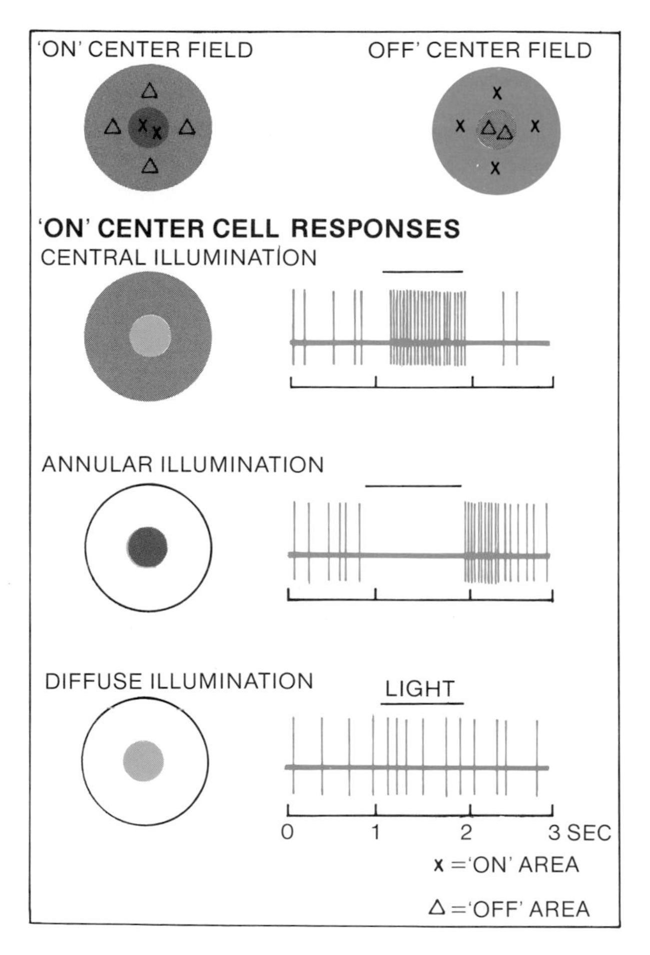

10
RECEPTIVE FIELDS OF LATERAL GENICULATE nucleus cells. The concentric receptive fields of cells in the lateral geniculate nucleus resemble those of ganglion cells in the retina, consisting of "on" center and "off" center types. The illustration is from an "on" center cell in the lateral geniculate nucleus of a cat. Bar above records indicates illumination. The central and surround areas antagonize each other's effects, so that diffuse illumination of the entire receptive field gives only weak responses (bottom record), less pronounced than in retinal ganglion cells. (After Hubel and Wiesel, 1961)

obvious that monitoring activity in single neurons could reveal many of the processes underlying higher functions. The argument usually took (and still does, at times) the following form. The brain contains some 10^{10} or more cells. Even the simplest task or event, like a movement, looking at a line or a square, engages hundreds of thousands of nerve cells in various parts of the nervous system. What chance does a physiologist have of gaining insight into complex actions within the brain if he samples only one or a few of those units, a hopelessly small fraction of the total number? This pessimistic view was in vogue until fairly recently.

On closer scrutiny, the logic of the argument about basic difficulties introduced by large numbers and complex higher functions is perhaps not so impeccable as it seems. As so frequently happens, some SIMPLIFYING PRINCIPLE turns up, opening a new and clarifying view. In the case of the retina, after all, well over 100 million cells deal with the infinite variety of the world. What simplifies the situation are the FEW CELL TYPES that are laid out in an apparently well-ordered manner, as REPEATING UNITS. Receptive field studies now indicate that this maze of over

100 million cells and many hundred million cross connections results in only a few stereotyped kinds of responses. What now holds back the investigator of the retina is not the question of numbers, but rather the technical problems associated with determining the patterns of connections and their integrative mechanisms at the synaptic sites. In some respects the situation is similar in the cerebellum, which contains many more neurons than the retina, but again, apart from neuroglia, has only five principal cell types (Chapter 1, Figure 2).

The retina analyzes visual events in a stepwise manner. Hubel and Wiesel acted on the assumption that the visual centers would perform their processing according to similar principles, but at a more advanced level. At the outset, however, they faced completely unanswered questions; for example, how signals are processed after they arrive by way of the optic radiation at various cortical cells. As a start, they took their clue from the simple anatomical observation that neurons in the visual system tend to be arranged in ordered layers. This layering is also present in the cortex, but is less striking than in the retina or the lateral geniculate nucleus. Therefore, it seemed reasonable to assume, as a working hypothesis, that information was handed on according to hierarchically ordered levels, progressing from one layer to the next.

The experimental procedure consisted of recording from neurons and finding the light stimulus best suited to arouse impulses. In practice, it was necessary to find the appropriate configuration of the receptive fields for various cells. The search resulted in the discovery of a marked increase in specialization of neurons: they respond selectively to visual patterns of progressively greater complexity at ascending levels in the hierarchy of the cortex. From the transformation of the signal patterns at different stages, one can derive much information about the general wiring diagram of connections within the visual cortex.

Structure of the cortex

Visual information passes to the cortex from the lateral geniculate nucleus through the optic radiation. In the cat the part of the visual cortex where the optic radiation ends consists of a folded plate of cells about 1 to 2 mm thick (Figure 11). This region of the brain, area 17 (also called the striate cortex or visual area I), lies posteriorly in the occipital lobe and can be recognized in cross section by its character-

David H. Hubel (left) and Torsten N. Wiesel during an experiment, about 1969. The cat, not shown, also faces the screen (see Figure 3).

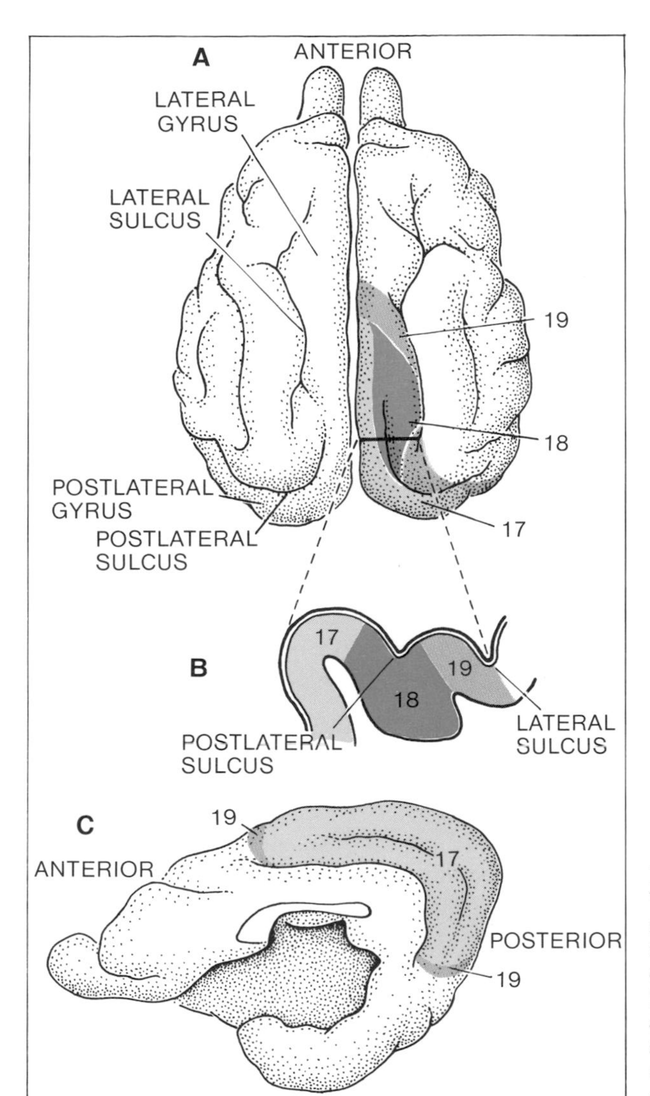

11
VISUAL AREAS IN CAT CORTEX. As in primates (Figure 1), three distinct areas can be recognized by several criteria. Area 17 (striate cortex, visual area I) projects to areas 18 and 19 (visual areas II and III). Parts of the areas are relatively inaccessible, lying on the medial surface of the hemisphere or within the folds of the cortex. A. View from above. B. Coronal section. C. Medial view.

istic appearance. Incoming bundles of fibers form a clear stripe in this area that can be seen by the naked eye—hence the name "striate." The areas of cortex adjacent to the striate cortex are also concerned with vision. They are called areas 18 and 19 (visual areas II and III) and receive inputs from area 17 (Figure 11). Their exact boundaries cannot be defined by simple inspection of the surface of the brain, but a number of structural criteria exist. For example, in area 18 the striate appearance is lost; large characteristic cells are found in the third layer; and coarse, obliquely running, myelinated fibers are seen in the deeper layers. From area to area there is much variation in the types of cells seen and the relative thickness of the different layers, and this provides additional criteria for demarcating boundaries.[29] In addition, physiologi-

[29]Otsuka, R., and Hassler, R. 1962. *Arch. Psychiatr. Nervenkr.* *203*:212–234.

cal experiments reveal functional differences in the types of processing carried out by the structurally different areas.[30,31]

The wealth of cell types and connections in the cortex immediately suggests that, in contrast to the geniculate, one can expect much greater integrating activity and transformation of signals to occur here. In monkey and man, the number of cells is even greater than in the cat, and histological examination shows the various layers and zones to be more clearly demarcated. A general feature of the mammalian cortex is that the cells are arranged in layers within the gray matter. Characteristically, the processes of the cells run for the most part in a radial direction, up and down through the thickness of the cortex (at right angles to the surface). In contrast, the great majority of their lateral processes are short. The main lateral connections between areas are made by axons that dip down and run in bundles through the white matter to surface again elsewhere (Figure 12).

Morphological features of cortical neurons

Cortical cells are classified on the basis of structural criteria, in the same way as are the cerebellar cells mentioned in Chapter 1. The two principal groups of neurons are STELLATE and PYRAMIDAL CELLS. Examples are shown in Figure 12. The main differences are in the length of the axons and the shape of the cell body. The axons of pyramidal cells are longer, dip down into white matter, and leave the cortex; those of stellate cells terminate locally. These two groups of cells exhibit many other variations, such as the presence or absence of spines on the dendrites. In addition, there are other fancifully named neurons (double bouquet cells, crescent cells) and the neuroglial cells (Chapter 13).[32]

Mapping visual fields in the striate cortex

As in the geniculate layers, visual fields from the eyes are represented in a strict topographical projection in each cortex. This was shown before the era of single-cell analyses by shining light onto small parts of the retina and recording with gross electrodes the potentials evoked on the surface of the cortex. Such potentials represent the summed activity of large numbers of neurons. Projection maps made in this manner by Talbot and Marshall[33] and by Daniel and Whitteridge[34] demonstrate that many times more cortical area is devoted to representation of the fovea than of the rest of the retina, as expected, since form vision is principally confined to foveal and parafoveal areas. Clinical tests involving destruction of these areas in the visual cortex and ablation studies in animals support these conclusions. But for an explanation of the cellular mechanisms on which perception or form vision is based, an analysis with higher resolution is needed; this is provided by recording from single cells. Chapter 3 discusses the finer grain of the architecture of the cortex and how cells are organized in columns of functionally related groups for the analysis of visual information.

---

[30]Hubel, D. H., and Wiesel, T. N. 1965. *J. Neurophysiol.* *28*:229–289.
[31]Zeki, S. M. 1974. *J. Physiol.* *236*:549–573.
[32]Ramón y Cajal, S. 1955. *Histologie du Système Nerveux.* II, C.S.I.C., Madrid.
[33]Talbot, S. A., and Marshall, W. H. 1941. *Am. J. Ophthalmol.* *24*:1255–1264.
[34]Daniel, P. M., and Whitteridge, D. 1961. *J. Physiol.* *159*:203–221.

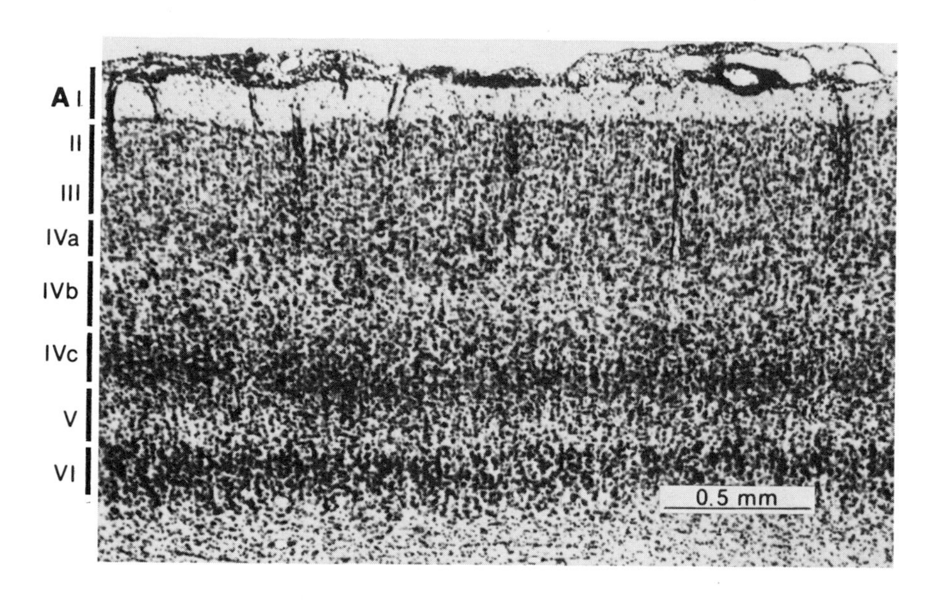

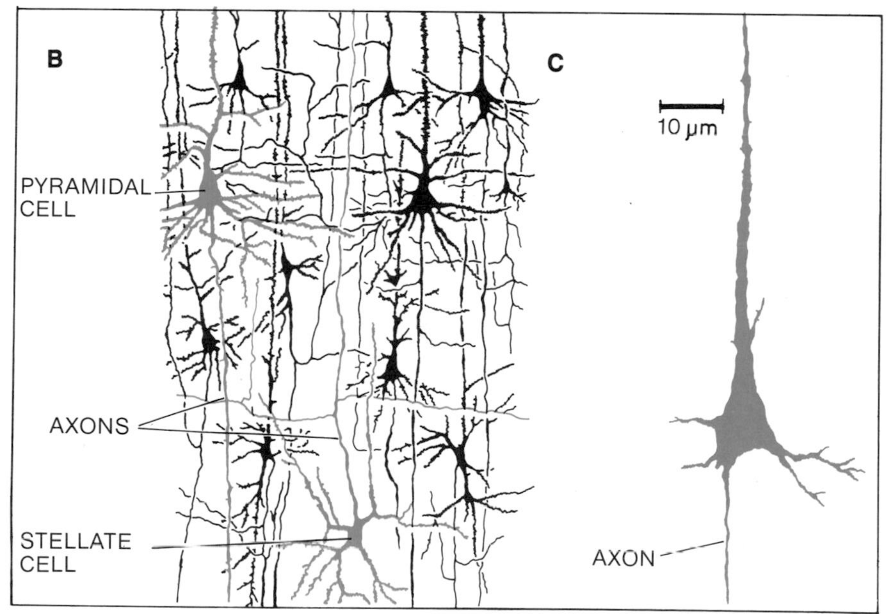

12

ARCHITECTURE OF VISUAL CORTEX. A. Distinct layering of cells in a section of striate cortex of the macaque monkey, stained to show cell bodies (Nissl stain). Fibers arriving from the lateral geniculate nucleus end in layers IVa, IVb, and IVc. B. Drawing a pyramidal and stellate cells (Golgi stain) in the visual cortex of the cat. The connections for the most part run radially through the thickness of the cortex and extend for relatively short distances laterally. C. Drawing from a photograph of a portion of a pyramidal cell in the cat cortex which had been injected with a dye (procion yellow) after its activity was recorded. This cell had a complex receptive field organization. (A from Hubel and Wiesel, 1972; B after Ramón y Cajal; C from Kelly and Van Essen, 1974)

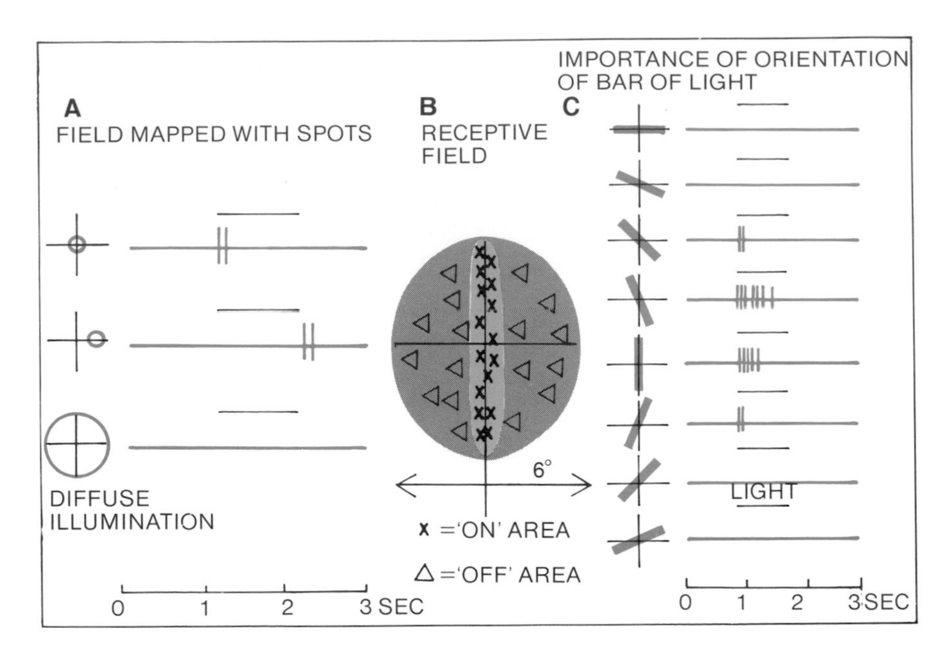

13 **RESPONSES OF A SIMPLE CELL in cat striate cortex to spots of light (A) and bars (C). The receptive field (B) has a narrow central "on" area flanked by symmetrical antagonistic "off" areas. The best stimulus for this cell is a vertically oriented light bar (1° × 8°) in the center of its receptive field (fifth record from top in C). Other orientations are less effective or ineffective. Diffuse light (third record from top in A) does not stimulate. Illumination indicated by bar. (After Hubel and Wiesel, 1959)**

Cortical receptive fields

Responses of cortical neurons, like those of the retinal ganglion and geniculate cells, tend to occur on a background of maintained activity. One of the most consistent observations is that discharges of CORTICAL NEURONS ARE NOT SIGNIFICANTLY INFLUENCED BY DIFFUSE ILLUMINATION of the retina. Bright flashes of light onto the entire retina, presented in the dark or at various levels of background illumination, do not noticeably change the irregular background discharge rate of neurons in the visual cortex. The almost complete insensitivity to diffuse light is an intensification of the process already noted in the retina and the lateral geniculate nucleus; it results from an equally matched antagonistic action of the inhibitory and excitatory regions in the receptive fields of cortical cells. Neuronal firing rate is altered only under special conditions when certain demands about the position and form of the stimulus on the retina are met. In other words, cortical neurons have receptive fields whose configurations differ from those of retinal or geniculate cells, so that spots of light often have little or no effect.

By following a progression of clues Hubel and Wiesel have worked out the appropriate light stimuli for various cortical cells whose receptive fields they have named according to their relative simplicity of construction. For want of more descriptive terms, the various field types are now known as SIMPLE, COMPLEX, HYPERCOMPLEX, and HIGHER ORDER HYPERCOMPLEX.[35]

Simple cells

The receptive fields of simple cells, like those of ganglion cells, are restricted in extent, but their internal geometry is different.[36,37] There are several varieties of this simple type. One that is closest to the ganglionic field consists of an extended narrow central portion flanked by two antagonistic areas. The center may be either excitatory or inhibitory. Figure 13*A* shows a receptive field of a simple cell in the striate cortex mapped out with spots of light that excited weakly, because they covered only a small part of the central area (marked with crosses), and inhibited in the surround (triangles).

The requirements of such a cell with a simple field are exacting. For optimal activation it needs a bar of light of not more than a certain width, entirely filling the central area, and oriented at a certain angle. This is shown in Figure 13*C* with the records taken from the same cell. As one might expect for this cell, a vertically oriented slit of light (fifth record from top) is most effective. With a deviation from this requirement the response deteriorates.

Different cells have receptive fields appropriate for a wide range of different orientations and positions. A new population of simple cells is therefore brought in by rotating the stimulus or by shifting its position in the visual field. The distribution of inhibitory-excitatory flanks in various simple receptive fields may not be symmetrical, or the field may consist of two longitudinal regions facing each other—one excitatory, the other inhibitory. Figure 14 shows examples of four receptive fields, all with a common axis of orientation, but with differences in the distribution of areas within the field. For receptive field A, a narrow light

[35]Hubel, D. H., and Wiesel, T. N. 1959. *J. Physiol.* *148*:574–591.
[36]Hubel, D. H., and Wiesel, T. N. 1962. *J. Physiol.* *160*:106–154.
[37]Hubel, D. H., and Wiesel, T. N. 1968. *J. Physiol.* *195*:215–243.

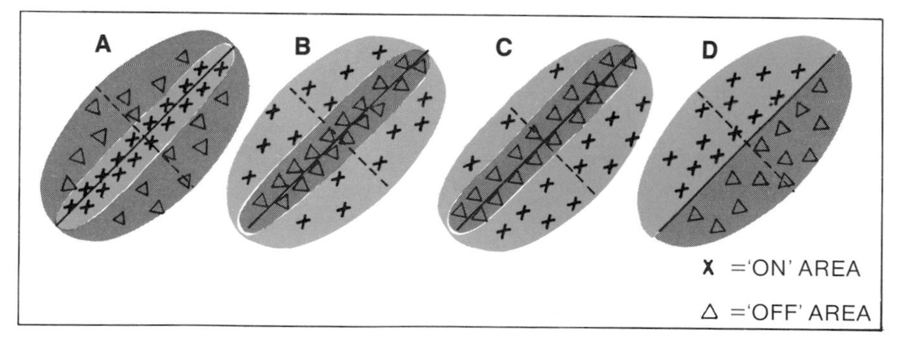

14

RECEPTIVE FIELDS OF SIMPLE CELLS in cat striate cortex. In practice all possible orientations are observed for each type of field. The optimal stimuli are for A, a narrow slit (or bar) of light in the center; for B and C, a dark bar; and for D, an edge with dark on the right. Considerable asymmetry can be present, as in C. (After Hubel and Wiesel, 1962)

slit with appropriate orientation on the central rectangle elicits the best response; a dark bar in the same place with light flanks suppresses the background discharge. Cells with field shapes B and C produce the opposite responses. For cell D an edge with light on the left and darkness on the right is most effective for "on" responses, while reversing the dark-light areas is best for "off" discharges.

In practice, one observes any number of asymmetrical receptive fields in which the portions flanking the central strip are of unequal strength. The common property of all simple cells is that they respond best to a properly oriented stimulus positioned at the border between the antagonistic zones. Another constant and remarkable feature is that in spite of all the different proportions of inhibitory and excitatory area relations, the sum of the two contributions match exactly and cancel each other's effectiveness, so that diffuse illumination of the entire receptive field produces a feeble response at best. The "off" areas in cortical fields are not always able to initiate impulses in response to dark bars. Frequently (particularly in the more elaborate fields to be described shortly) illumination of the "off" area can only be detected as reduction in an ongoing discharge that had been evoked from the "on" area or that represents "spontaneous" background activity.

In these examples the width of the narrow light or dark bar is comparable to the various diameters of the "on" or "off" center regions in the doughnut-shaped receptive field of ganglion or lateral geniculate body cells (Figures 4 and 10). Once again there is a specialization for detecting differences, but the spotlike contrast representation of ganglion cells has been transformed and extended into a line or an edge. Resolution has not been lost, but incorporated into a more complex pattern. In agreement with the above assumption, cortical cells whose fields are derived from the fovea are best excited by narrower bars than those arising in the peripheral area of the eye.

Responses of simple cells to moving stimuli

In initial studies on simple cells it was noted that moving edges or bars of the appropriate orientation were highly effective in initiating impulses. Indeed, some of the higher order cells to be discussed shortly respond only to moving stimuli (see Figure 16). As might be predicted from the diverse arrangement of "on" and "off" areas, different cells may be aroused by different types of movement. Some, like cell D of Figure 14, respond preferentially to a movement of an edge in one direction, from right to left, when light moves from an "off" area onto an "on" area; for others with symmetrical flanks (cells A and B), movement in either direction is equally effective. One would predict that cells with asymmetrical fields would be best designed for detecting movement in one direction and when light moves from an "off" area onto an "on" area. In practice, however, it is often not possible to explain the directional sensitivity of a cortical simple cell from the design of its field. In an extensive series of studies Bishop and his colleagues have examined the responses of cortical cells to moving edges or bars.[38] Their work has recorded a more detailed organization of the

[38]Bishop, P. O., Coombs, J. S., and Henry, G. H. 1971. *J. Physiol.* *219*:625–657.

receptive field center and surround and has enabled them to put forward a comprehensive scheme to account for movement sensitivity. Previously, Barlow and Levick had proposed cellular mechanisms to explain the directional sensitivity of certain ganglion cells in the retina of the rabbit.[39] In the cat cortex, X and Y neurons of the lateral geniculate nucleus with their sustained and transient responses presumably contribute to the movement sensitivity.[40] A fuller explanation will probably have to await information about the synaptic mechanisms involved in the formation of cortical receptive fields.

There is an interesting parallel between movement sensitivity in visual and somatosensory systems. Werner and his colleagues[41] have recorded in the somatosensory cortex of monkeys from cells that respond selectively to stroking the skin in one direction but not another. Again, this direction sensitivity is a property of higher order neurons and not of primary receptors.

Complex cells

In recordings made from individual neurons in the visual cortex, one finds, in addition to simple cells, other neurons whose behavior is quite different. These complex cells have one important property in common with simple cells: they require a specific field axis orientation of a dark-light boundary, while illumination of the entire field is ineffective. The demand, however, for precise positioning of the stimulus, observed in simple cells, is relaxed in complex cells. In addition, there are no longer distinct "on" and "off" areas. As long as a properly oriented stimulus falls within the boundary of the receptive field, most complex cells will respond, as in the examples illustrated in Figure 15. The meaning of the signals arising from complex cells, therefore, differs significantly from that of simple cells. The simple cell localizes the orientation of a bar of light (Figure 13) to a particular position within the receptive field, while the complex cell signals the abstract concept of ORIENTATION WITHOUT STRICT REFERENCE TO POSITION.

Hypercomplex cells

The hypercomplex cells' requirements for stimulation are still more refined than those for complex neurons. The responses of hypercomplex cells may be somewhat harder to understand at first. The best stimulus for them requires a certain orientation but in addition entails some discontinuity, such as a line that stops, an angle, or a corner. Figure 16 shows a cell that responds best to an edge with dark below and light above at an angle of about 45°. Diffuse illumination, other axis orientations, or spots are without effect. At first glance one might classify this as a complex cell similar to that shown in Figure 15. The difference, however, becomes clear in the fourth and fifth records from the top in Figure 16, when the dark edge is extended; this depresses the response of the hypercomplex cell. Interestingly, diffuse illumination of the right-hand field does not diminish the response (last record). A

---

[39]Barlow, H. B. and Levick, W. R. 1965. *J. Physiol.* *178*:477–504.

[40]Hoffmann, K. P., and Stone, J. 1971. *Brain Res.* *32*:460–466.

[41]Whitsel, B. L., Roppolo, J. R., and Werner, G. 1972. *J. Neurophysiol.* *35*: 691–717.

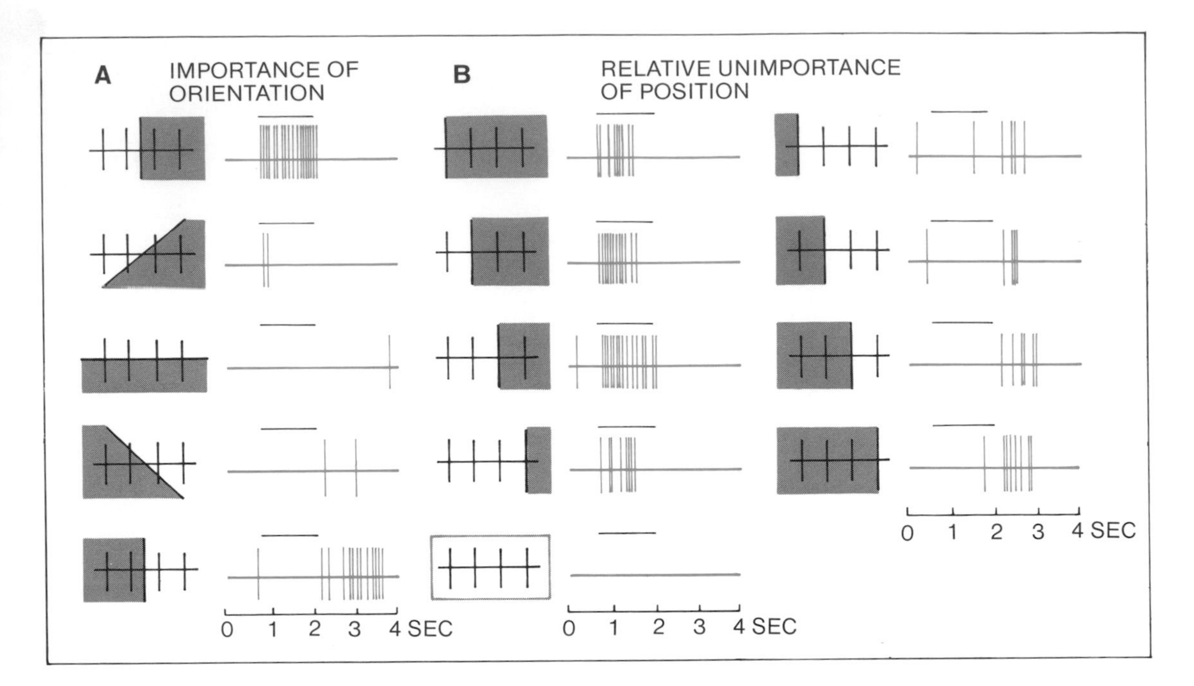

15 **RESPONSES OF A COMPLEX CELL in the striate cortex of the cat. Cell responds best to a vertical edge. A. With light on the left and dark on the right (first record), there is an "on" response. With light on the right (fifth record), there is an "off" response. Orientation other than vertical is less effective. B. Position of border within field is not important. Illumination of entire receptive field (bottom record) gives no response. (After Hubel and Wiesel, 1962)**

simple description of the best stimulus for this cell is a corner; moreover, the stimulus must move in one direction.

One other variant of hypercomplex cells may be mentioned. Its field is not difficult to understand if considered as an extension of the field of the hypercomplex cell just discussed. An extra inhibitory area is added on the left side (Figure 17), as though the field were composed of three components, two inhibitory and one excitatory. The best stimulus for this hypercomplex cell is a moving edge that covers the middle region; the stimulus, however, must not extend into EITHER of the two inhibitory lateral portions of the receptive field. If widened in either direction, the stimulus is weakened or ineffective. This second hypercomplex cell, therefore, is even more specific in its requirements. It signals that a narrow dark line is stopping in this part of the retina in a subdivision of its receptive field. It does not tell exactly where the line is in terms of position of the plane of movement (up or down), but it does indicate that the line is not wider than the amount specified by the position of the inhibitory portion of the receptive field. Hypercomplex cells reflect an increase in specificity with a reintroduction of lateral inhibition: POSITION WITHIN THE RECEPTIVE FIELD IS IMPORTANT.

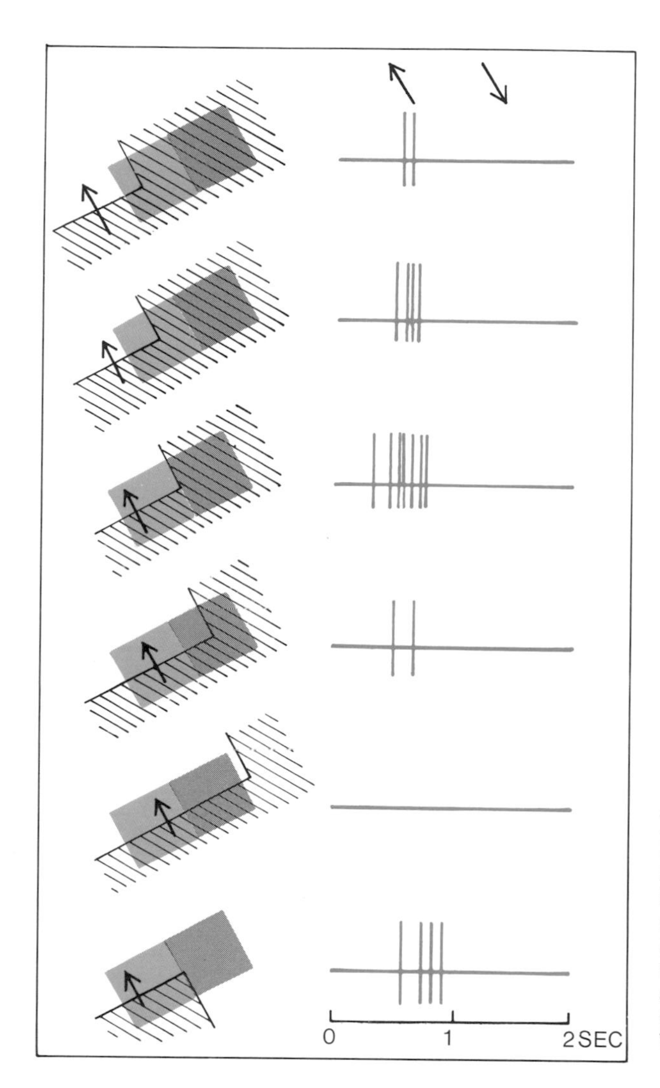

16
RESPONSES OF A HYPERCOMPLEX CELL in area 18 of the cat cortex. The best stimulus for this cell is a moving (arrows), oriented edge (a corner) that does not encroach on the antagonistic right-hand portion of the receptive field (third record from top). The records also show the selective sensitivity of the cell to upward movement. (After Hubel and Wiesel, 1965)

Often it takes several hours of trying out different patterns on the screen in front of the animal before the most effective stimulus can be identified, and at this stage the following question is frequently asked: How can you be sure that the best stimuli for the simple, complex, and hypercomplex cells are in fact straight lines rather than curves, the letter *E*, or perhaps the complex shape of a mouse or a hand? It is in practice not possible, of course, to try all combinations on each cell. In general, however, it has been found in many laboratories that curves, circles, and complex patterns do not stimulate cells as effectively as straight bars or edges. Naturally, "straight" can only be considered in relation to the size of receptive fields, so that a segment of a large circle on the retina is virtually a straight line. Just as a line can be constructed out of closely spaced dots, any desired shape can be formed out of straight lines of the proper length.

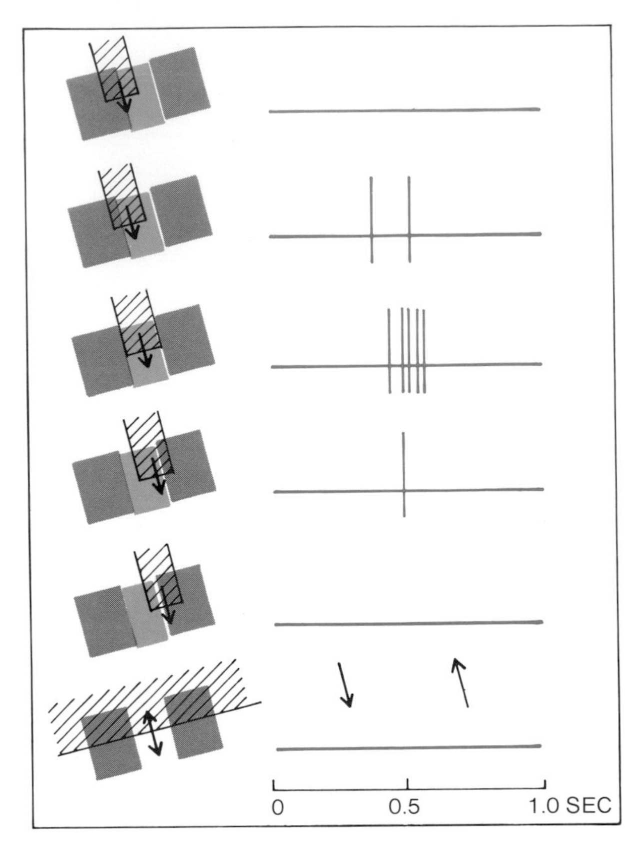

17

RESPONSES OF A HYPERCOMPLEX CELL in area 19 of cat cortex. The best stimulus is a narrow dark tongue in the middle moving downward. Shifting the dark edge sideways or widening it encroaches on antagonistic flanking areas and diminishes the response. (After Hubel and Wiesel, 1965)

Possible role of hypercomplex cells in perception

The various hypercomplex neurons are specialized to convey information about a discontinuity, such as when a line stops or changes its direction. This provides a clue to the explanation of a clinical observation in man—the COMPLETION PHENOMENON, which occurs when small retinal or cortical lesions cause blind areas or scotomas. In this condition, forms or shapes projected onto the retina appear to contain empty areas in the visual field corresponding to the site of the lesion. Yet, if the patient looks at striped wallpaper, a zebra, or a simple straight-line pattern, he sees the pattern continue through the blind area.

An interesting self-observation made by Lashley, a keen and perceptive reporter of psychophysical phenomena, during a migraine attack is worth quoting in this context:[42]

> Talking with a friend I glanced just to the right of his face whereon his head disappeared. His shoulders and necktie were still visible but the vertical stripes on the wall paper behind him seemed to extend right down to the necktie. Quick mapping revealed an area of total blindness covering about 30° just off the macula. It was quite impossible to see this as a blank area when projected on the striped wall paper or uniformly patterned surface although any intervening object failed to be seen.

[42]Lashley, K. S. 1941. *Arch. Neurol. Psychiat.* *46*:331–339.

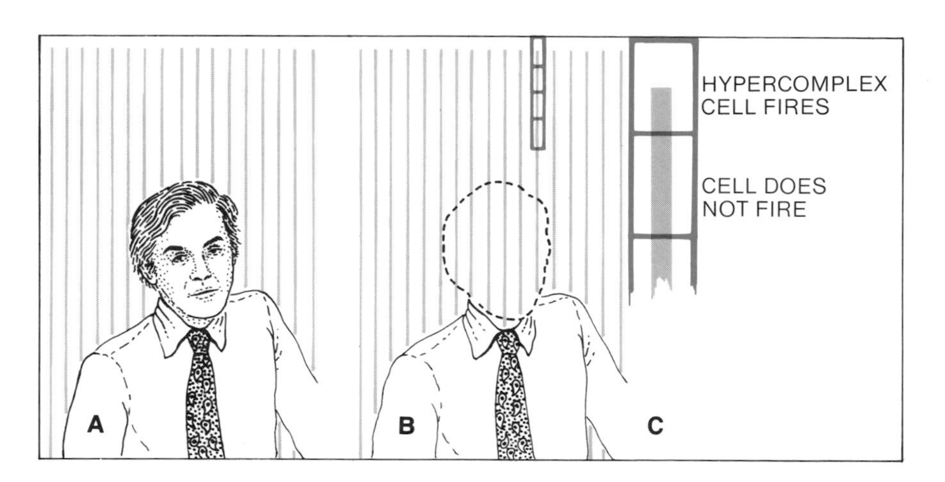

18

THE COMPLETION PHENOMENON. During a migraine attack (B) a small area of complete blindness occurs in the visual field occupied by the head of Lashley's friend. Yet the stripes on the wallpaper continue through the area. A possible explanation is that hypercomplex cells of the type shown in Figures 16 and 17 respond best to corners or lines that stop; hence, a stripe on the wallpaper acts as a good stimulus for a hypercomplex cell only at its end (C). A hypercomplex cell with its receptive field within the blind area would no longer be activated by the lines of the face; however, its silence could be interpreted by higher centers as an indication that the gray stripe had not stopped.

The events described by Lashley can be interpreted in terms of hypercomplex cells that signal information only about the terminations of lines (Figure 18). With an array of such cells, silence of members in the area of temporary (in Lashley's case) blindness produces little change in signaling since they are normally silent anyway, unless the line stops or changes direction. This type of reasoning provides a conceptual scheme that uses known properties of cells to explain complex phenomena. Some of the weaknesses and strengths of this approach are discussed later. It is ironic, incidentally, that this interpretation rests on cortical cells that have specific connections. This is contrary to the principles enunciated with clarity by Lashley himself, who stressed absence of specificity and localization in the cortex.

Receptive fields from both eyes converging on cortical neurons

Binocular interaction is introduced here because it provides another example of part of the design the brain uses to end up with perception of form. When we look at an object with one or both eyes, we see only one image, even if the size and the position of the object's projection are slightly different on the two retinas. Sherrington posed the problem:[43]

---

[43]Sherrington, C. S. 1947. *The Integrative Action of the Nervous System.* Yale University Press, New Haven.

> How habitually and unwittingly the self regards itself as one is instanced by binocular vision. Our binocular visual field is shown by analysis to presuppose outlook from the body by a single eye centered at a point in the midvertical of the forehead at the level of the root of the nose. It, unconsciously, takes for granted that its seeing is done by a cyclopean eye having a centre of rotation at the point of intersection just mentioned. In this visual field it obtains visual depth by unknowingly combining . . . crossed images of not too great lateral disparation. . . . Oneness is obtained by compromise between differences, if not *too* great, offered to the perceiving "self." There are other perceptual instances. The brightness of a binocular field differs hardly sensibly from that of either of two equally illuminated uniocular fields composing it. But the quantity of stimulus received by the eyes is roughly double in the binocular observation than that which it is in the uniocular.

Interestingly, well over 100 years ago Johannes Müller suggested that individual nerve fibers from the two eyes might fuse or become connected to the same cells in the brain. Thereby, he almost exactly anticipated Hubel and Wiesel's results. They found that about 80 percent of all cortical neurons in the visual areas of the cat can be driven from both eyes. Since the neurons in the various layers in the lateral geniculate nucleus are predominantly innervated from one eye or the other, the first opportunity for significant interaction between the eyes must occur in the cortex. In the cat it occurs in the fourth layer of area 17, the striate cortex, where geniculate axons make initial contacts with cortical cells. In the monkey the situation is somewhat different. The separation tends to be maintained in the fourth layer of area 17. Each simple cell is driven by only one eye, the other being without effect, a point referred to again in Chapter 3 in relation to the columnar organization of the monkey cortex. Mixing between the two eyes occurs in the subsequent relay stations, that is, in layers deeper toward the white matter and in more superficial cortical layers (toward the roots of the hair). A further difference from the cat is that some of the units recorded from in layer IV of the monkey cortex do not require oriented light patterns.

Examination of the receptive fields of a binocularly driven cell shows that (1) they are usually in exactly corresponding positions in the visual field of the two eyes, (2) their preferred orientation is the same, and (3) the corresponding areas in the two receptive fields add to each other's effect. An example of synergistic action between the two eyes is shown for a simple cell in Figure 19. Shining light onto an "on" region in the left eye sums with illumination onto the "on" area of the right eye. Simultaneous illumination of antagonistic areas in the two eyes reduces or suppresses an ongoing response and increases "off" discharges. Such cells would be useful for unifying images from the two eyes.

For DEPTH PERCEPTION there exists another binocular specialization of receptive fields.[44] Such fields are not necessarily in exactly correspond-

---

[44]Hubel, D. H., and Wiesel, T. N. 1970. *Nature 225*:41–42.

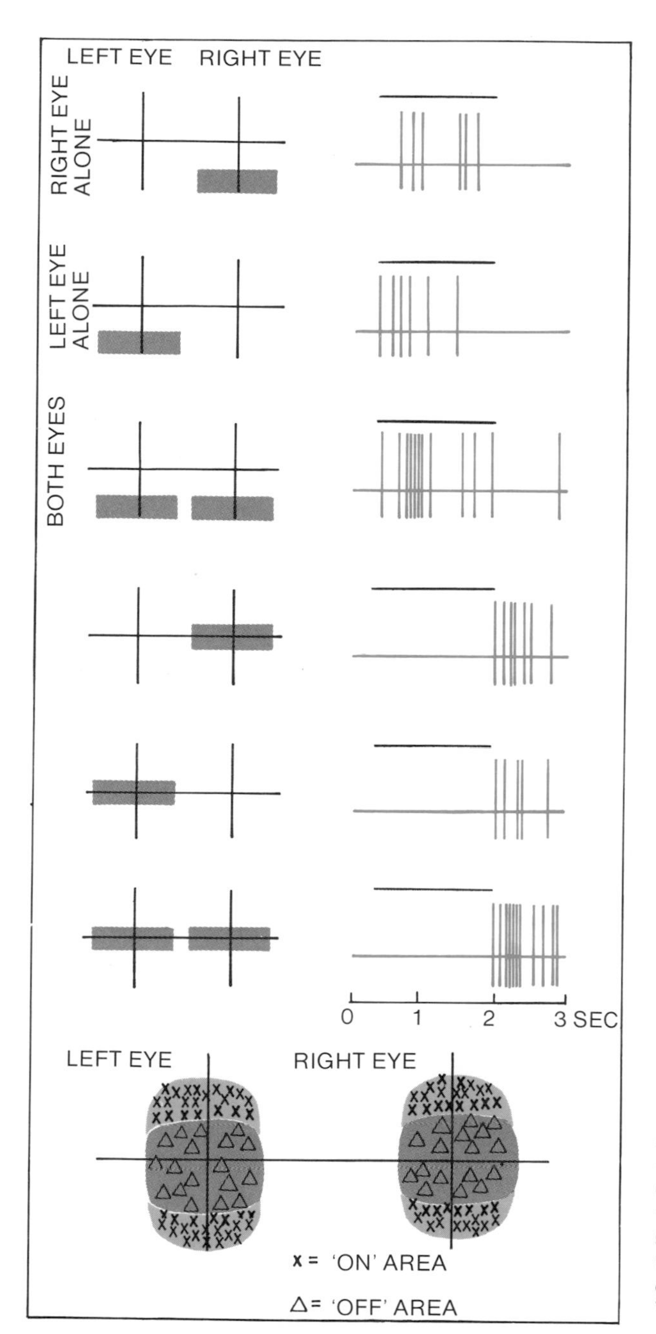

19
BINOCULAR ACTIVATION of a simple cortical neuron that has identical receptive fields in both eyes. Simultaneous illumination of corresponding "on" areas of right and left receptive fields is more effective than stimulation of one alone (upper three records). In the same way, stimulation of "off" areas in the two eyes reinforces each other's "off" discharges (lower records). (After Hubel and Wiesel, 1959)

ing points in the two retinas. Neurons whose performance fits them for depth perception do not respond to stimulation of one eye alone, but require a binocular stimulus either in front of, behind, or in the plane of fixation. Further, to be aroused they demand of a stimulus a particular size, shape, and orientation. This seems a good example of a high degree of specialization, in that a whole specific series of conditions must be met to make a stimulus suitable.

Connections for combining right and left visual fields

Each hemicortex is wired to perceive one half of the external world but not the other. This is equally true for sensations of touch and position and constitutes a general situation in relation to perception. It is natural to wonder what happens in the midline. How are the two cortices knitted together to produce a single image of the body and the world?

The obvious way to preserve continuity is for the two sides to be fused in register at the midline. This would allow a complete picture to be formed with a minimum number of connections between the two hemispheres. On the other hand, there would be little purpose in linking fields seen out of the corners of the two eyes that look on quite different parts of the world and observe, for example, a house and a frog.

Corpus callosum

The general question of transfer of information between the hemispheres has been studied in a most rewarding manner in man and monkey by Sperry, Myers, Gazzaniga, and their colleagues.[45,46] Concentrating on the coordinating role of the corpus callosum, a bundle of fibers that runs between the two hemispheres, they have shown that the fibers are actually involved in the transfer of information and learning from one hemisphere to the other. To cite one example, a normal right-handed person can name an object, such as a coin or a key, when it is placed in either hand (stereognosis). After section of the corpus callosum, however, the object can be named only if it is placed in the right hand, the information from which crosses before reaching the cortex and projects to the left hemisphere. This is because the main area responsible for language lies in the LEFT hemisphere. What happens if the object is placed in the left hand, which projects to the right hemisphere? Even though information reaches consciousness when the key is placed in the left hand, there is no way in which the concept "key" can be verbally expressed; this is because the center for language is situated on the left side of the brain, which cannot be reached without the corpus callosum. Thus a person without a corpus callosum can recognize the object with his left hand and may be able to use it, but cannot say the word KEY. Figure 20 expresses some of these ideas. Other experiments on the corpus callosum provide surprising insight into higher functions. For example, if deprived of cross connections, the two hemispheres can lead virtually separate existences. One small aspect of this important area of work is noted here because it bears directly on visual perception. It suggests that the fusing, or knitting together, of the two fields of vision is mediated by fibers in the corpus callosum. Cells with receptive fields that straddle the midline provide information about both sides of the visual world. There is good evidence that such cells combine inputs from both hemispheres and that the corpus callosum bundle provides the fibers responsible for extending the receptive field across the midline. Cutting or cooling (which blocks conduction) the corpus callosum causes the cell's receptive

---

[45]Sperry, R. W. 1970. *Proc. Res. Assoc. Nerv. Ment. Dis. 48*:123–138.
[46]Gazzaniga, M. S. 1967. *Sci. Am. 217*:24–29.

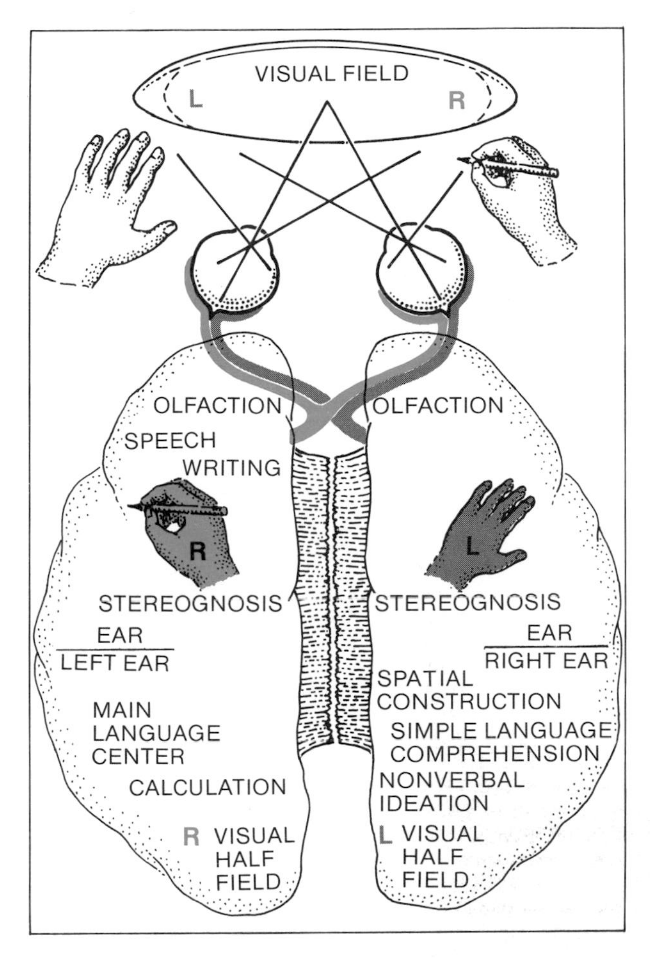

20
THE CORPUS CALLOSUM. Transection of the corpus callosum interrupts connections between the cerebral hemispheres, including those of receptive fields that straddle the midline. (After Sperry, 1970)

field to shrink and become confined to just one side of the midline (the usual arrangement with cortical cells). Furthermore, recording from single fibers in the callosum that are concerned with vision shows that their fields lie close to the midline, not in the periphery.[47–49] The role of callosal fibers is clearly demonstrated in an experiment of Berlucchi and Rizzolatti.[50] They made a longitudinal cut through the optic chiasm, thereby severing all direct connections to the cortex from the contralateral eye. Yet, provided the corpus callosum was intact, some cells in the cortex with fields close to the midline could still respond to appropriate illumination of the contralateral eye.

The general conclusion to be drawn from these demonstrations of binocular interactions on individual neurons is that relatively simple neuroanatomical concepts can explain how images from the two eyes are

[47]Choudhury, B. P., Whitteridge, D., and Wilson, M. E. 1965. *Q. J. Exp. Physiol. 50*:214–219.

[48]Hubel, D. H., and Wiesel, T. N. 1967. *J. Neurophysiol. 30*:1561–1573.

[49]Berlucchi, G., Gazzaniga, M. S., and Rizzolatti, G. 1967. *Arch. Nat. Biol. 105*: 583–596.

[50]Berlucchi, G., and Rizzolatti, G. 1968. *Science 159*:308–310.

Cortical organization for receptive fields

fused and how depth is assessed, an achievement that has simplified thinking about two complex problems related to perception.

In the retina, with its layering and sequence of connections, it is obvious that the performance of ganglion cells, and of their receptive fields, is determined by the neurons in the two preceding strata. Yet, attempts at synthesis are still only partially successful; the problems are due largely to incomplete knowledge of synaptic connections, their performance and mechanisms. Even less is known about the cortex, and only some initial but important approaches can be discussed.

The scheme of organization is such that more elaborate receptive fields are constructed by an ordered synthesis of simpler ones originating at a lower level. In the cortex simple cells behave as if they were built up of large numbers of geniculate fields. This idea is illustrated in Figure 21*A*, where the centers of concentric fields of geniculate neurons are lined up in such a way that a bar of light through their centers would excite them strongly, while a parallel shift of the bar into the inhibitory surround would reduce or stop the excitatory output of the cells. This scheme results in a central excitatory area and its opposing surround. The value of such a scheme is not diminished by several obvious shortcomings. For example, a simple excitatory input to cortical cells does not account for their silence to diffuse stimulation. Geniculate cells tend to respond, although not vigorously, with "on" as well as "off" discharges when diffuse light is flashed onto them or when it is turned off. The diagram of connections could be expanded to include inhibitory pathways. Nevertheless, the scheme of Figure 21*A* provides a working model for further experimental approaches. In the same way, one can tentatively construct complex receptive fields by lining up appropriate rows of simple fields.

The scheme illustrated in Figure 21*B*, which again presents just three sample components out of a great number that are needed, satisfies the requirements of a complex cell that is excited by a vertical edge stimulus that falls anywhere within the area of the receptive field. This is so because wherever the edge falls, one of the simple fields is traversed at its vertical inhibitory-excitatory boundary. None of the other components respond because they are uniformly covered by light or darkness. Diffuse illumination of the entire field covers all component fields equally and therefore none fires. One can postulate that only one or a few of the simple cells fire at any one position of the stimulus to evoke a response in a complex cell. Similarly, one can devise a way of combining complex cell fields to form hypercomplex fields as shown in Figure 21*C*.

Other ways for elaborating the fields of cortical cells from geniculate and intracortical connections have been proposed by Bishop and his colleagues.[51] The point to be emphasized is that the various schemes provide a conceptual framework, and their details should not be taken too literally. One can expect that when cortical cells are studied with intracellular electrodes, so that the synaptic potentials that excite or

---

[51]Bishop, P. O., Coombs, J. S., and Henry, G. H. 1973. *J. Physiol.* *231*:31–60.

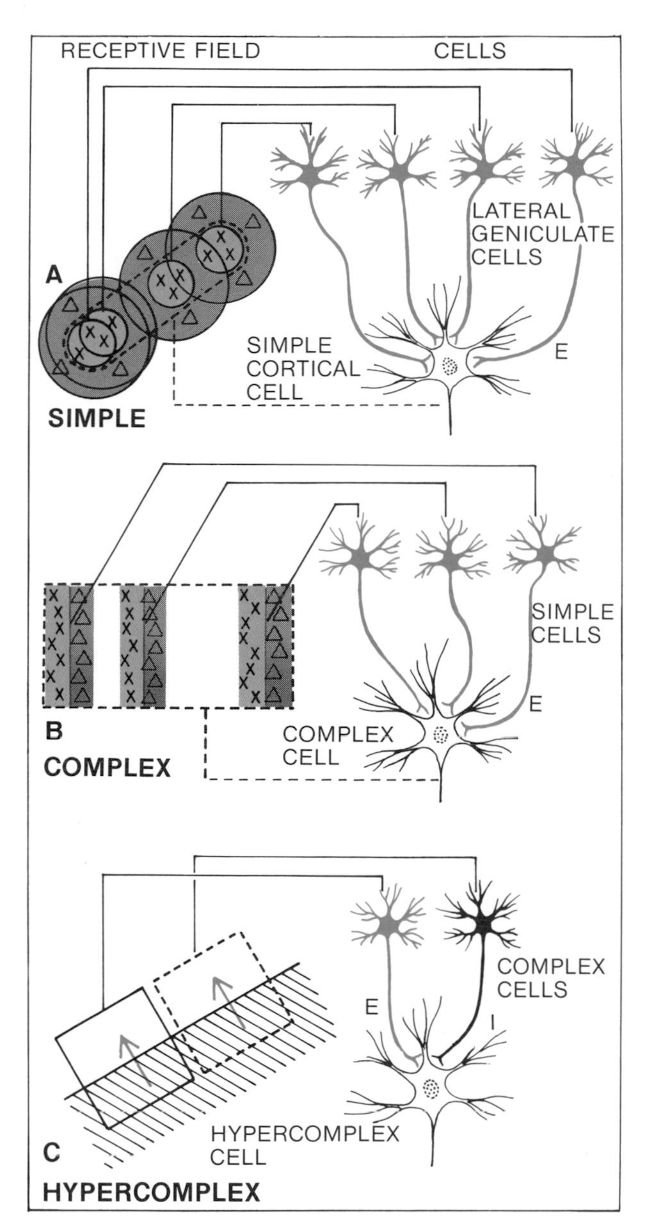

21

SYNTHESIS OF RECEPTIVE FIELDS. Hypothesis devised by Hubel and Wiesel to explain the synthesis of simple, complex, and hypercomplex receptive fields. In each case lower order cells converge to form receptive fields of higher order neurons. A. Fields of simple cells are elaborated by the convergence of many geniculate neurons with concentric fields (only four appear in the sketch). They must be arranged in a straight line on the retina according to the axis orientation of simple receptive fields. B. Simple cells responding best to a vertically oriented edge at slightly different positions could bring about the behavior of a complex cell which responds well to a vertically oriented edge situated anywhere within its field. C. Each of the two complex cells responds best to an obliquely oriented edge. But one cell is excitatory and the other is inhibitory to the hypercomplex cell. Hence an edge that covers both fields, as in the sketch, is ineffective, while a corner restricted to the left field would excite. (Hubel and Wiesel, 1962, 1965a)

inhibit them can be seen, the relative strength and distribution of the excitatory and inhibitory influences can be better estimated.

RECEPTIVE FIELDS: UNITS FOR FORM PERCEPTION

Table 1 lists some of the characteristics of receptive fields at successive levels of the visual system. Each eye conveys to the brain information collected from areas of varying size on the retinal surface. The emphasis is not on diffuse illumination or the energy absorbed by the photoreceptors, but on contrast. The ganglion cells prefer spots that fill the centers of receptive fields, but they also respond well to narrow bars of light or darkness, provided the bars traverse the center; POSITION is important. A small movement from the central position greatly alters the discharge by crossing over the border into the surround. Therefore, one would expect receptive fields with small centers to be best for RESOLUTION as well as CONTRAST. This is exactly what is seen as receptive fields become progressively smaller toward the foveal region. The process of contrast becomes further emphasized in the geniculate, where the antagonism between center and surround is stronger.

As the cortex is approached, the demands made on a stimulus for activation of neurons rise greatly. To the importance of position on the retina is added ORIENTATION of the stimulus (light or dark), which must be a line or bar rather than a spot. Directional sensitivities for a moving stimulus appear first in simple cortical cells; and at this level, in the cat, arises also the first opportunity for interaction between the two eyes. In complex cells the requirement of position is relaxed and

Table 1. **CHARACTERISTICS OF RECEPTIVE FIELDS AT SUCCESSIVE LEVELS OF THE VISUAL SYSTEM**

Type of cell	Shape of field	What is best stimulus?	How good is diffuse light as a stimulus?	Is orientation of stimulus important?
Receptor	⊗	Light	Good	No
Ganglion	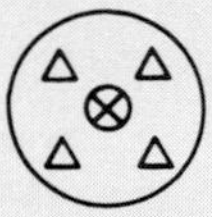	Small spot or narrow bar over center	Moderate	No
Geniculate	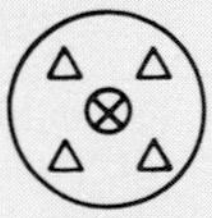	Small spot or narrow bar over center	Poor	No
Simple	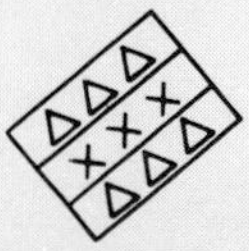	Narrow bar or edge	Ineffective	Yes
Complex	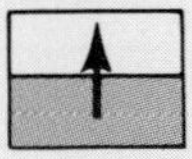	Bar or edge	Ineffective	Yes
Hypercomplex	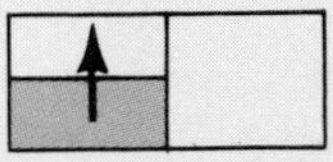	Line or edge that stops; corner or angle	Ineffective	Yes

orientation becomes generalized over a considerable area (up to 2 sq mm in the retina of the cat).

What types of signals are generated if a square patch of light such as that in Figure 22 (or a shadow of the same shape) is presented to the retina? The receptors in the retina seeing the bright central area absorb more light than those in the surround. Ignoring the receptors and bipolar cells and starting with the optic nerve fibers, the following types of signaling occur: the "on" center ganglion cells within the square increase their discharge (at least for the initial period), while the "off" center ganglion cells are suppressed. The best stimulated ganglion cells are those subjected to the maximum of contrast, that is, those whose centers lie immediately adjacent to the boundary between the light and dark areas. At the geniculate this contrast is even more pronounced because the cells respond only poorly to diffuse light; hence those whose receptive fields lie entirely within the dark or the light areas fire weakly, whether they are "on" center or "off" center cells. As is true of retinal ganglion cells, the geniculate cells that fire best

TABLE 1 (Continued)

Is position of stimulus important?	Are there distinct "on" and "off" areas within receptor fields?	Are cells driven by both eyes?	Can cells respond selectively to movement in one direction?
Yes	No	No	No
Yes	Yes	No	No
Yes	Yes	No	No
Yes	Yes	Yes (except in monkey layer IV)	Some can
No	No	Yes	Some can
Yes	Yes	Yes	Some can

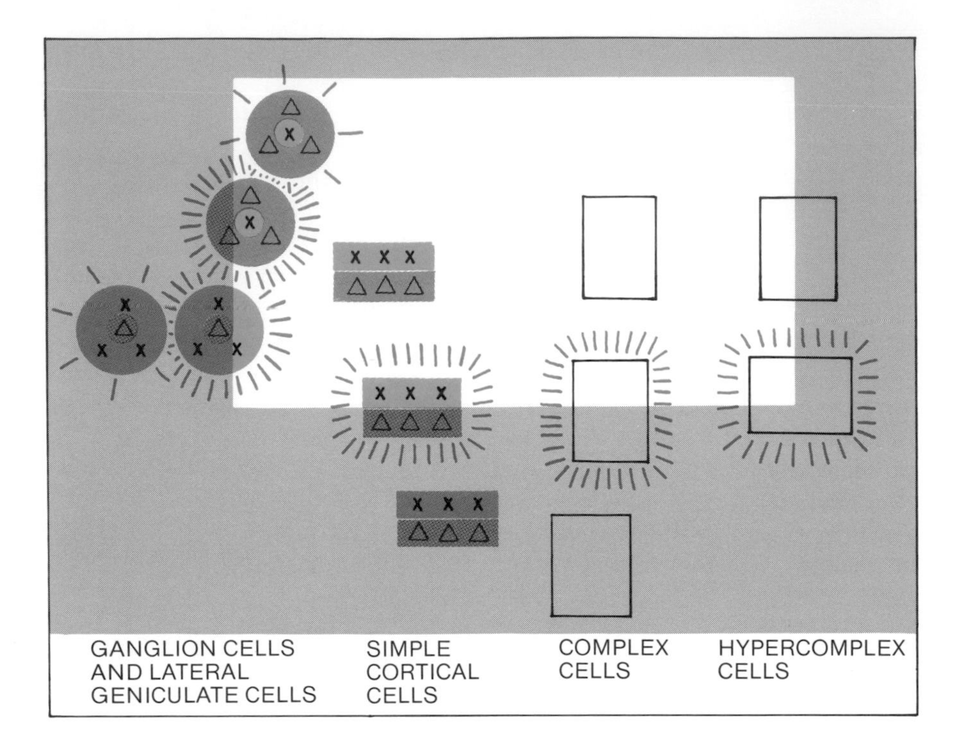

22 **RESPONSES OF NEURONS TO A PATTERN. When a square patch of light is presented to the retina, signals arise predominantly from receptive fields close to the borders of the square. The ganglion cells and lateral geniculate cells whose receptive fields are situated close to the border fire better than those subjected to uniform light or darkness. The only simple, complex, and hypercomplex cells that fire are those with appropriate receptive fields situated along a border or corner with the correct orientation.**

have their centers close to the border. This process is still further enhanced in the cortex. Cortical cells whose receptive fields lie completely either within the square or outside it send no signals (except for their usual maintained discharge), because diffuse illumination is not an effective stimulus. Only those simple cells with receptive fields whose axis orientation coincides with the horizontal or vertical boundaries of the square can be stimulated. Similar considerations apply to the stimulation of complex cells, which also require properly oriented bars or edges. There is an important difference, however, which depends upon the fact that at rest, small rapid saccadic movements are made continually by the eyes. They are essential for vision to be preserved while the eyes fixate, but they are not perceived as motion. Each movement causes a new population of simple cells with exactly the same orientation to be thrown into action.

For those complex cells that "see" the square, however, a boundary of appropriate orientation anywhere within the field is the only requirement. Many of the same complex cells therefore continue to fire even during eye movements, as long as the displacement is small and does not pass outside the receptive field. For such cells the position of the square on the retina does not appear to change.

In the lower right corner of the square is inserted a hypercomplex cell that responds best to a right angle—a corner.

If the above considerations are valid, the surprising conclusion is that the brain receives little information about the absolute level of uniform illumination within the square. Signals arrive only from the cells with receptive fields situated close to the border. This hypothesis is supported by a well-known psychophysical experiment. A square that appears white when surrounded by a black border can be made to appear dark merely by increasing the brightness of the surround. In other words, we perceive the difference or contrast at the boundary and it is by that standard that the brightness in the uniformly illuminated central area is judged.

The eye does have an index of the maintained level of brightness, expressed in the constancy of the pupillary size according to the absolute strength of ambient light over a wide range. Pupillary size is adjusted by a feedback mechanism (Chapter 15), the incoming loop of which leaves the eye through the optic nerve. However, the type of discharge in the afferent pathway of the pupillary reflex is not known, nor is there evidence of cortical cells that are supplied with information about absolute brightness.

Hierarchical synthesis of receptive fields

The orderliness, repetition, and progression of receptive field organization constitute convincing evidence that the connections between neurons are specific and laid out according to a functionally related ground plan. Anatomical evidence, presented in Chapter 3, supports this basic idea. There is much less certainty about the detailed proposals invoked to explain the actual synthesis of receptive fields.

One may assume that as information ascends from the eye some of it is also used (or perhaps diverted) at each stage for various purposes, such as eye movements, regulation of the pupil, depth perception, and other functions. Additional types of cells and input from other parts of the brain may intervene in the synthesis of fields, and some of the projections may miss a step, as Garey and Powell[52] have found with certain geniculate axons that in the cat project directly to area 18. In connection with Figure 21 examples have been cited in which the known properties of visual neurons could not satisfactorily be explained. These include the absence of inhibition in the scheme for simple cells, the directional sensitivity of cortical cells to moving stimuli, and the progressively increasing demands for movement in complex and hypercomplex cells. The scheme also fails to take into account the color coding of cells in the monkey cortex, which has been discovered by Zeki.[31,53]

The hierarchical scheme is not complete, correct in detail, or the

[52]Garey, L. J., and Powell, T. P. S. 1971. *Proc. R. Soc. Lond.* B *179*:41–63.
[53]Zeki, S. M. 1973. *Brain Res.* *53*:422–427.

only possible explanation. Rather it represents a useful hypothesis and an effective way of describing how the complex behavior of cells in the visual system could be brought about. While there remain huge gaps in knowledge, the description of receptive fields has supplied for the first time an inkling of how such concepts as verticality, length, thickness, depth, and squareness can be derived from properties and connections of individual cortical neurons.

Where do we go from here?

How much progress has been made in the last dozen years in explaining perception in neurophysiological terms? Earlier, the topographical representation of the retina on the cortex and ablation or stimulation experiments showed that certain specific areas are active during visual perception. But it was difficult to put forward clear-cut hypotheses in cellular terms. For example, in 1943, Craik[54] discussed the kind of device that might be employed in the brain to recognize the three angles of a triangle in the visual field. He suggested that there might be an "electronic scanning device in which the scanning beam could be made to move in a straight line and would then continue to move in that direction until the line suddenly changed its course at an angle." Craik pointed out that "the essential thing is that the scanning beam should be made to acquire the habit, so to speak, of moving along a line so the system shall be disturbed by an abrupt change of direction." This explanation showed great insight and was formally correct. However, it is hard to see how such a postulate could predict actual neural mechanisms that might be involved in detecting a line that stops at a corner and then changes its direction. These are, however, just the properties ascribed to the hypercomplex cell by virtue of the connections it has with lower order cells. The single-cell analysis, therefore, provides an idea of how the brain puts together and collates information used for perception.

The responses certain neurons will give when they "see" various configurations of stimuli can be predicted. Yet, this is only the beginning, since even the most complex receptive fields, which provide sophisticated abstract information, still cover only a small part of what goes on in the large visual field each eye surveys. An extension of the hypothesis of hierarchical organization predicts that in the future cells should be discovered that bring together larger and larger parts of the information that appears in the field of vision.[55] But how far can this concentration of information go? Will there be a small group of cells or, ideally, one pontifical cell that synthesizes and combines into a whole picture all the features perceived? The neurophysiologist obviously faces a dilemma. On the one hand, we know that visual information is synthesized into diverse meaningful units by various receptive fields. These are distributed among an incalculable number of cells; in this sense, perception is scattered and diffuse. On the other hand, there

---

[54]Craik, K. 1943. *The Nature of Explanation.* Cambridge University Press, London.

[55]Mountcastle, V. B. 1975. *The Johns Hopkins Med. Journal 136*:109–131.

should be cells that see the "big picture," since without them we end up stating that each group of cells looks at the next group and vice versa. The most encouraging development is that perception can be thought of in terms of known cell properties after traversing only six or seven synapses, and there are many more to come before the cells run out.

The main lesson of the discussion so far is recognition of the importance of the layout of connections. It seems that the machinery of the brain will perform properly and provide the required information if its parts are correctly assembled, just as a radio will emit music or noise, depending on how its circuits are wired. This chapter can appropriately close with the following quotation from Sherrington, written long before visual fields were mapped for single cells:[56]

> The chief wonder of all we have not touched on yet. Wonder of wonders, though familiar even to boredom. So much with us that we forget it all our time. The eye sends, as we saw, into the cell-and-fibre forest of the brain throughout the waking day continual rhythmic streams of tiny, individually evanescent, electrical potentials. This throbbing streaming crowd of electrified shifting points in the spongework of the brain bears no obvious semblance in space-pattern, and even in temporal relation resembles but a little remotely the tiny two-dimensional upside-down picture of the outside world which the eye-ball paints on the beginnings of its nerve-fibres to electrical storm. And that electrical storm so set up is one which affects a whole population of brain-cells. Electrical charges having in themselves not the faintest elements of the visual—having, for instance, nothing of "distance," "right-side-upness," nor "vertical," nor "horizontal," nor "colour," nor "brightness," nor "shadow," nor "roundness," nor "squareness," nor "contour," nor "transparency," nor "opacity," nor "near," nor "far," nor visual anything—yet conjure up all these. A shower of little electrical leaks conjures up for me, when I look, the landscape; the castle on the height, or, when I look at him, my friend's face and how distant he is from me they tell me. Taking their word for it, I go forward and my other senses confirm that he is there.

SUGGESTED READING

Papers marked with an asterisk (*) are reprinted in Cooke, I., and Lipkin, M. 1972. *Cellular Neurophysiology*, Holt, New York.

Retina and Lateral Geniculate Nucleus

Baylor, D. A., Fuortes, M. G. F., and O'Bryan, P. M. 1971. Receptive fields of cones in the retina of the turtle. *J. Physiol. 214*:265–294.

Baylor, D. A., and Hodgkin, A. L. 1973. Detection and resolution of visual stimuli by turtle photoreceptors. *J. Physiol. 234*:163–198.

Dowling, J. E., and Boycott, B. B. 1966. Organization of the primate retina: electron microscopy. *Proc. R. Soc. Lond.* B *166*:80–111.

[56]Sherrington, C. S. 1951. *Man on His Nature*. Cambridge University Press, London.

Hubel, D. A., and Wiesel, T. N. 1961. Integrative action in the cat's lateral geniculate body. *J. Physiol. 155*:385–398.

Kaneko, A. 1970. Physiological and morphological identification of horizontal, bipolar and amacrine cells in goldfish. *J. Physiol. 207*:623–633.

*Kuffler, S. W. 1953. Discharge patterns and functional organization of the mammalian retina. *J. Neurophysiol. 16*:37–68.

Kuffler, S. W. 1973. The single-cell approach in the visual system and the study of receptive fields. *Invest. Ophthalmol. 12*:794–813.

Rodieck, R. W. 1973. *The Vertebrate Retina: Principles of Structure and Function.* W. H. Freeman, San Francisco. (A detailed and comprehensive book that deals with both anatomy and physiology and contains abundant references.)

*Werblin, F. S., and Dowling, J. E. 1969. Organization of the retina of the mudpuppy *Necturus maculatus*: II. Intracellular recording. *J. Neurophysiol. 32*:339–355.

Visual Cortex

*Hubel, D. H., and Wiesel, T. N. 1962. Receptive fields, binocular interaction and functional architecture in the cat's visual cortex. *J. Physiol. 160*:106–154.

Hubel, D. H., and Wiesel, T. N. 1965. Receptive fields and functional architecture in two non-striate visual areas (18 and 19) of the cat. *J. Neurophysiol. 28*:229–289.

Hubel, D. H., and Wiesel, T. N. 1968. Receptive fields and functional architecture of monkey striate cortex. *J. Physiol. 195*:215–243.

Zeki, S. M. 1974. Cells responding to changing image size and disparity in the cortex of the rhesus monkey. *J. Physiol. 242*:827–841.

Color Vision

Daw, N. W. 1973. Neurophysiology of color vision. *Physiol. Rev. 53*:571–611. (A comprehensive review article.)

Zeki, S. M. 1973. Color coding in rhesus monkey prestriate cortex. *Brain Res. 53*:422–427.

Corpus Callosum

Gazzaniga, M. S. 1970. *The Bisected Brain.* Appleton, New York.

CHAPTER THREE

COLUMNAR ORGANIZATION OF THE CORTEX

Throughout the visual system neighboring groups of neurons respond to information from neighboring areas on the retinal surface. This principle has been well established at the cellular level in other sensory systems as well. Neighboring neurons in the visual cortex share other common functional properties; such cells are stacked in the form of columns or slabs which run at right angles to the cortical surface. For example, some columns of cells are best stimulated by vertical bars shone onto a small region of a particular area of the retina. Other columns of neurons in nearby areas of the cortex respond preferentially to a horizontal orientation, and so on for different angles. A comparable segregation into columns is also known for cells that are influenced preferentially by one or the other eye. These ocular dominance columns have been demonstrated by anatomical as well as physiological techniques.

The aggregation of neurons with related receptive field positions and functions is best interpreted by assuming that cortical cells are arranged together in a way that makes it easier for them to interconnect so that they can perform the type of analysis required of them.

The retina is not simply represented as a map on the cortex; instead, each retinal area is analyzed over and over again in neighboring cortical regions with respect to a number of different variables such as position, orientation, color, and so on. This type of cortical architecture provides some insight into the reason why so many components are needed for visual analysis. Cortical structure and functional organization go hand in hand. This emphasizes the great need for identifying the basic design behind the precise assembly of connections in the nervous system.

Interconnection of functionally related cells

This chapter presents evidence to show how functionally related nerve cells are aggregated and connected to each other at successive stages in the central nervous system. The basic principle of organization was first elaborated by Mountcastle and his colleagues,[1,2] who studied the somatosensory cortex. They discovered that sensory neurons, such as those serving light touch of skin, are grouped together, segregated from those neurons that respond to other stimuli—for example, deep pressure or rotation of joints. The grouping occurs in columns that run radially from the cortical surface to the white matter, and connections between cells are principally up and down along the columnar axis. The columnar organization has now been worked out in the visual system as well, by Hubel and Wiesel, and, as in Chapter 2, their findings form the basis of the following discussion.

In the visual system the pattern starts with the retina, where the functional unit consists of a group of photoreceptor cells connecting successively to bipolar and ganglion cells, with horizontal and amacrine cells feeding into the system. This unit, through its connections, elaborates the first integrated stage of visual information, which is handed on to the higher centers by the optic nerve fibers. As the impulses combine in the various cell stations in the cortex, the abstract significance of their message becomes transformed. It was suggested earlier, for example, that the field of a hypercomplex cell, which recognizes a corner, can be thought of as the convergence of fibers from the chain of lower order cells—complex, simple, geniculate, ganglion, bipolar, and primary receptor cells.

The formal scheme of building up receptive fields in a hierarchical fashion from layer to layer, even if correct, does not convey a structural picture of how cells are actually laid out to elaborate all the intricate receptive fields. Fortunately, analyses of receptive fields have clearly revealed the helpful simplifying principle that neurons performing similar tasks tend to be grouped together, and this has recently been confirmed histologically. Thus, a circumscribed area in the striate cortex receives its principal input from a group of geniculate axons with receptive fields in the same region of the retina. Within this cortical area are contained the very cells required for the synthesis of simple, complex, and hypercomplex fields with the same axis orientation, position, and ocular dominance. The outgoing information is then sent on to another area of cortex where another group of cells performs further processing, and so on.

The concept of grouping according to function adds new physiological insight to anatomical concepts. There is not simply a complete map of the retinal surface projecting onto the visual cortex, but a series of cell clusters, each of which performs its own special analysis and synthesis of the information coming from the retina. Further, circumscribed areas of the retina are represented over and over again in the various cortical regions.

Methods for tracing interconnections of cells

The general layout of neuronal pathways can be discerned by a

[1]Mountcastle, V. B. 1957. *J. Neurophysiol 20*:408–434.

[2]Powell, T. P. S., and Mountcastle, V. B. 1959. *Bull. Johns Hopkins Hosp. 105:* 133–162.

number of physiological and anatomical techniques. Mentioned earlier are the physiological studies in which evoked potentials were used (Chapter 2). Ideally, one would like to be able to trace an axon from its origin to its destination by simple visual inspection. This is almost never possible in the brain because of the complexity of the histological picture and the long distances involved. For example, axons arising in one layer of the lateral geniculate nucleus do not course to the cortex as a discrete, identifiable bundle, but are inextricably mixed with fibers from other layers. There is, however, a reliable way of sorting out the terminals that originate in a particular distant group of nerve cells. When the cell bodies of neurons are damaged, their axon terminals undergo characteristic degenerative changes over the next few days. The staining technique devised by Nauta and colleagues[3] allows ready identification of the degenerating nerve endings. The procedure, then, is to make a discrete lesion in a localized site—for example, in one layer of the geniculate—and search in different regions of the cortex for degenerating terminals (see below). For many years this was the only method of tracing the destinations of neurons.

Recently, two new methods have been developed which promise to be very useful. In one, an enzyme, horseradish peroxidase, is injected into a region of the brain. The enzyme is taken up by nerve terminals and transported backward along the axons to the cell bodies of origin, a process known as RETROGRADE TRANSPORT. The cells are then distinctively labeled by allowing the enzyme to react with a substrate, which causes a dense endproduct to be deposited. Cells stained in this way appear black in light micrographs; the endproducts of the reaction can also be seen by electron microscopy.[4] A second technique, devised by Grafstein[5] to explore visual pathways in fish, is similar in principle but makes use of radioactive amino acids. The amino acid is injected into a structure, such as the vitreous of the eye, from which it is taken up by nerve cell bodies of the retina and incorporated into protein. The labeled protein is then transported from ganglion cells through the optic nerve fibers to their terminals within the brain. There it can be detected by autoradiography, a method that consists of cutting a thin section of the tissue and putting it on a photoemulsion. The radioactive disintegrations in the terminals or geniculate neurons cause silver grains to be reduced in the photoemulsion. The technique detects only insoluble molecules like proteins precipitated by the fixative; water-soluble amino acids tend to be washed out of the tissue during dehydration and embedding. An unforeseen feature was that the label is also transferred from neuron to neuron across synapses. This finding has been put to good use in the mammalian visual system, where it permits entire pathways to be traced.

The information provided by the methods described above is a

---

[3]Nauta, W. J. H., and Gygax, P. A. 1954. *Stain Technol.* *29*:91–93.
[4]La Vail, J. H., and La Vail, M. M. 1974. *J. Comp. Neurol.* *157*:303–358.
[5]Specht, S., and Grafstein, B. 1973. *Exp. Neurol.* *41*:705–722.

prerequisite for studying functional architecture and the overall representation, or destination in the brain, of sensory pathways. These techniques, however, do not yet provide the resolution to determine the fine grain of neuronal interconnections. To gain such a resolution the interconnections must be analyzed at the level of individual cells. This can be done anatomically by the Golgi stain, which picks out individual cells and stains all their processes, or by injecting a fluorescent dye into a cell.[6] The latter method has the advantage of allowing the investigator to establish first the physiological performance of a neuron and then its geometry; for example, whether simple and complex cells are stellate or pyramidal neurons. The technique is difficult, particularly if the cells are small.

At present, electrical recording from individual neurons with external electrodes remains the simplest and most effective way of classifying individual cells according to their function. Their location in the cortex can be established by marking the position of the electrode tip at the end of a penetration by an electrolytic lesion. In this way the entire path of the electrode can later be reconstructed after fixation.

COLUMNS IN THE VISUAL CORTEX

Receptive field axis orientation columns

To study the functional architecture of the visual cortex, Hubel and Wiesel made a systematic survey of cells by penetrating adjacent areas of cortex with microelectrodes and listing the properties of the various cells.[7–10] When the results found on repeated cortical penetrations were combined, a clear, unambiguous pattern emerged. With the electrode moving at right angles to the surface through the thickness of the gray matter, the cells that were encountered in sequence had the same field axis orientation. A sample experiment is shown in Figure 1. A microelectrode is inserted normal to the surface of the cortex in area 17 of the cat. Each bar indicates one cell and its preferred orientation in the progression through the cortex. The circle at the end marks the site of a lesion and shows the final position of the electrode tip. From this end point and the electrode track observed in the fixed brain after the end of the experiment, the following sequence is seen: at first, all the cells are optimally driven by bars or edges, at about 90° to the vertical, at one position in the visual field. After penetration of about 0.6 mm, the axis of the receptive field orientation changes to about 45°. A second track to the right shows other cells, with slightly different receptive field positions and field axis orientations. Oblique penetrations show that the field axis changes repeatedly with only small movements of the electrode tip, as though a series of columns with different axis orientations is being traversed.

---

[6]Kelly, J. P., and Van Essen, D. C. 1974. *J. Physiol.* *238*:515–547.
[7]Hubel, D. H., and Wiesel, T. N. 1962. *J. Physiol.* *160*:106–154.
[8]Hubel, D. H., and Wiesel, T. N. 1963. *J. Physiol.* *165*:559–568.
[9]Hubel, D. H., and Wiesel, T. N. 1968. *J. Physiol.* *195*:215–243.
[10]Hubel, D. H., and Wiesel, T. N. 1974. *J. Comp. Neurol.* *158*:267–294.

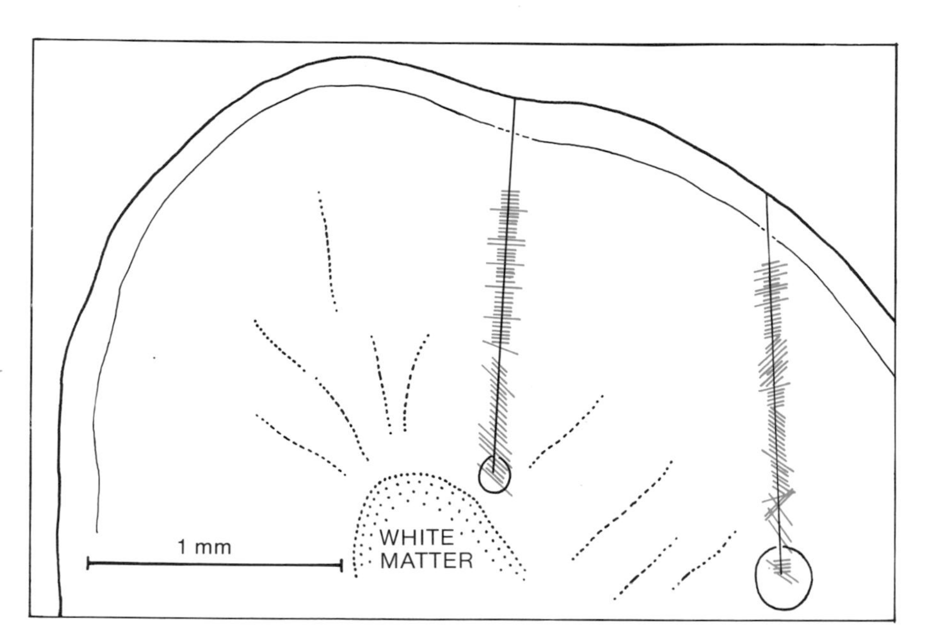

1

AXIS ORIENTATION of receptive fields of neurons encountered as an electrode traverses the cortex normal to its surface. Cell after cell tends to have the same axis orientation, indicated by the angle of the bar to the electrode track. The penetration to the right is somewhat more oblique and the axis orientations change more frequently. The electrode track is reconstructed by making a lesion at the end of the penetration (circle) and cutting serial sections through the brain. Such experiments have established that cat and monkey cells with similar axis orientation are stacked in columns running at right angles to the cortical surface. (After Hubel and Wiesel 1962)

From a large series of such experiments in cat and monkey it became apparent that in area 17 simple, complex, and hypercomplex NEURONS WITH SIMILAR RECEPTIVE FIELD AXIS ORIENTATION ARE NEATLY STACKED ON TOP OF EACH OTHER IN DISCRETE COLUMNS that run perpendicular to the cortical surface. The columns receive their input from largely overlapping receptive fields on the retinal surface. Separate columns exist for each axis orientation.

Other functional variables are also grouped in columnar aggregates of cortical cells. Zeki has shown that in the monkey there exist columns of cells with well-defined color sensitivity and other columns where the direction of movement of the visual stimulus is important.[11,12]

Ocular dominance columns

There are also COLUMNS FOR EYE PREFERENCE, a close grouping of binocular cortical neurons that are dominated, or more strongly influenced, by one eye or the other. Figure 2 illustrates these characteristics of neurons in the striate cortex of the monkey. The cells (total 1116) are subdivided into seven groups. Groups 1 and 7 are driven exclu-

[11]Zeki, S. M. 1973. *Brain Res. 53*:422–427.
[12]Zeki, S. M. 1974. *J. Physiol. 236*:549–573.

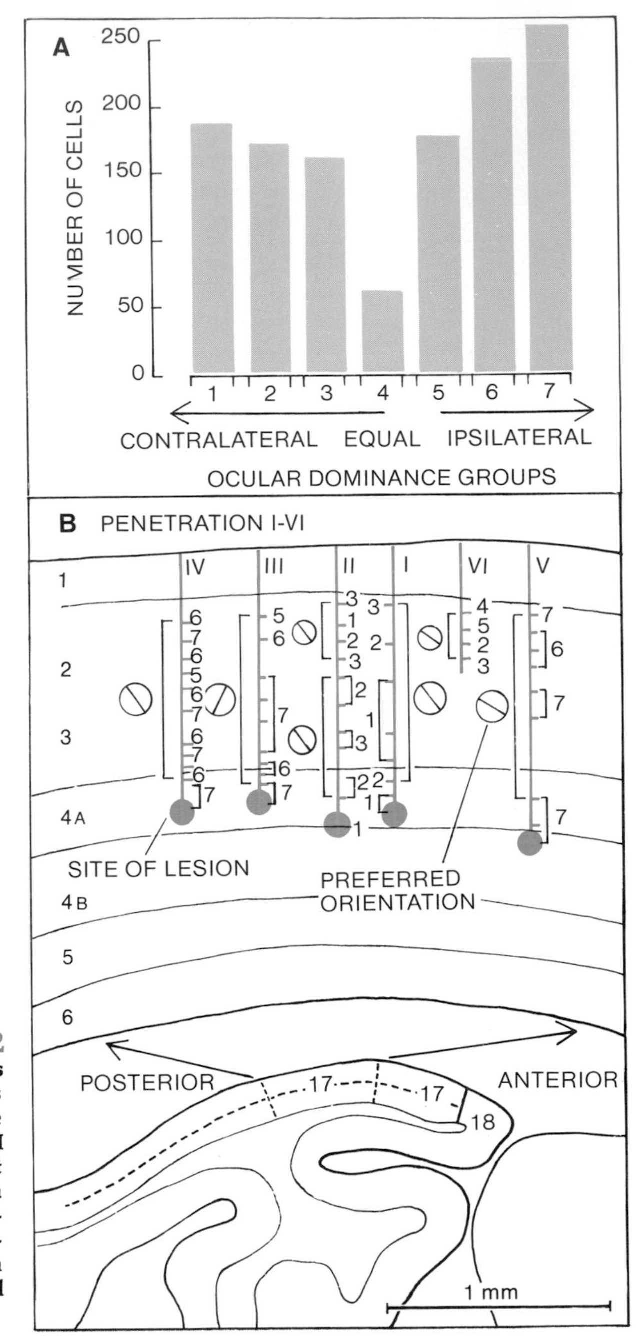

2

OCULAR DOMINANCE COLUMNS. A. shows the eye preference of 1116 cells in 28 Rhesus monkeys. Most cells (groups 2 through 6) are driven by both eyes. B. When penetrations (I to VI) are made through the cortex at right angles to the surface the cells encountered in one track have similar eye preference, indicated by numbers, as well as similar orientation specificity, indicated by lines within open circles (after Hubel and Wiesel, 1968; Wiesel and Hubel, 1974)

sively by one of the two eyes; in groups 2,3 and 5,6 the effect of one eye is stronger than that of the other, while the cells in the middle, group 4, are equally influenced. The majority respond to both eyes (Chapter 2).

In the striate cortex of the monkey, in contrast to that of the cat, the inputs from the two eyes are segregated at first (as in the geniculate) before being combined in an orderly manner. The incoming geniculate fibers end in layer IV of area 17. In this clearly defined region only simple cells are encountered, and they are stimulated by one eye only. Outside layer IV the simple, complex, and hypercomplex cells are driven for the most part by both eyes. For example, in some of the penetrations of Figure 2*B* all the cells are better driven by the left eye, while in others the cells are better driven by the right eye. These groupings are called OCULAR DOMINANCE COLUMNS.

The idea that cells are grouped as if to elaborate hierarchically ordered fields is reinforced by the characteristic distribution of the cell types observed during penetration through the thickness of the monkey cortex. As the electrode moves away from layer IV, either deeper down or more superficially, complex and hypercomplex cells are found, all with the same field axis orientation, ocular dominance, and position in the visual field. The arrangement of cells supports this scheme of organization through the thickness of the cortex. Thus, the processes of cortical neurons pass chiefly up and down at right angles to the surface of the cortex, but spread laterally only over short distances (see Figure 12, Chapter 2).

Shape and distribution of columns in the cortex

The discovery of the columnar arrangement through electrical recording methods was followed, some years later, by an anatomical demonstration of ocular dominance columns. This development reinforced confidence in the reliability of the physiological approach and provided new opportunities for a glimpse into the layout of functional synaptic organization. An opening was provided by the observation just mentioned, that in the monkey the simple cells in layer IV of area 17 are driven exclusively by one eye. As a result it has become possible to use a variety of procedures for demonstrating alternating groups of cells in layer IV supplied by one eye or the other.

One procedure is to make a small lesion in one layer of the lateral geniculate nucleus; degenerating terminals subsequently appear in layer IV. Degenerating endings of geniculate axons are distributed in a characteristic pattern of alternating strips or slabs (Figure 3*C*). These correspond to the areas driven by the one eye in whose line of connection the lesion is made. If a lesion is made so that axons from two layers of the geniculate are affected, carrying inputs from both eyes, a more evenly distributed pattern of degeneration is produced.[13]

Recently, a further striking demonstration of the ocular dominance columns was provided by the transport of radioactive fucose or proline from one eye, as described above.[14] It clearly shows the radioactivity around the terminals of geniculate neurons supplied by the injected eye; zones that get their input from the other eye remain clear. The columnar arrangement demonstrated in this way is similar to that seen after lesions to the lateral geniculate nucleus.

Finally, by staining the cortex with a modified silver stain it is

---

[13]Hubel, D. H., and Wiesel, T. N. 1972. *J. Comp. Neurol.* *146*:421–450.
[14]Wiesel, T. N., Hubel, D. H., and Lam, D. M. K. 1974. *Brain Res.* *79*:273–279.

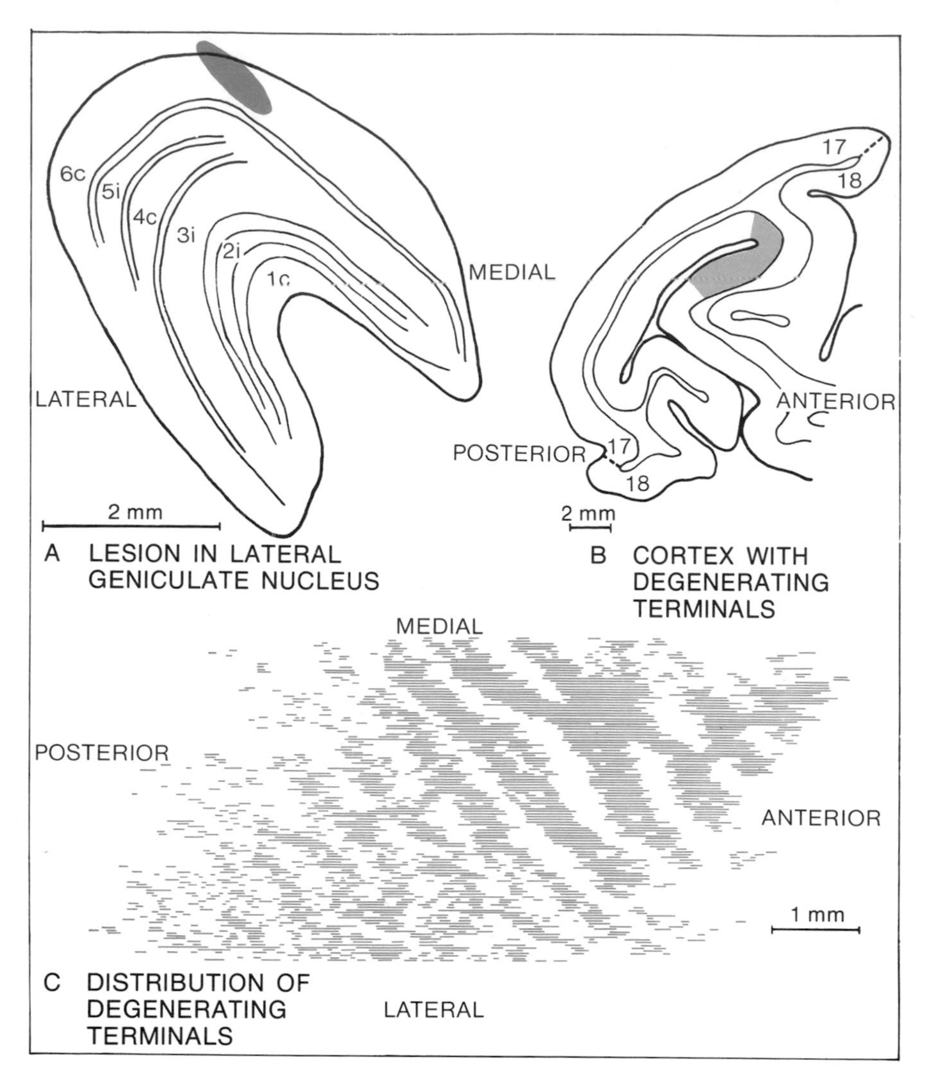

3 **OCULAR DOMINANCE COLUMNS are demonstrated in the monkey by detection of degenerating axonal endings in layer IV of the cortex after lesions in the lateral geniculate nucleus. In C these columns have been reconstructed from serial sections, 30μm thick, cut through the cortex. Each short line represents the region over which degenerating terminals were found in layer IVc of a section. The overview in C was produced by aligning all the drawings appropriately according to their relative positions in adjacent sections. The columns actually have the shapes of slabs. (From Hubel and Wiesel, 1972)**

possible to observe the boundaries of the columns in monkey cortex both histologically and physiologically.[15] Figure 4*A* shows a tangential section chiefly through layer IV in which the light bands delineate

[15]Le Vay, S. Hubel, D. H., and Wiesel, T. N. 1975. *J. Comp. Neurol. 159*:559–576.

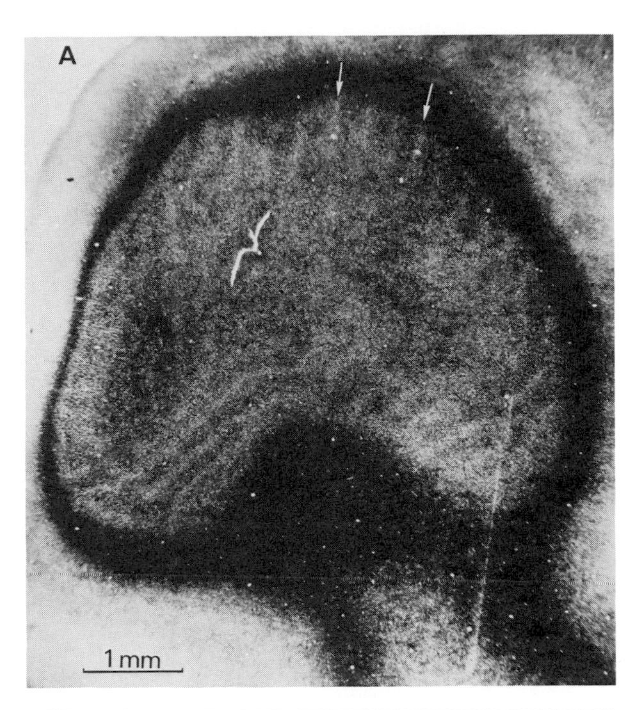

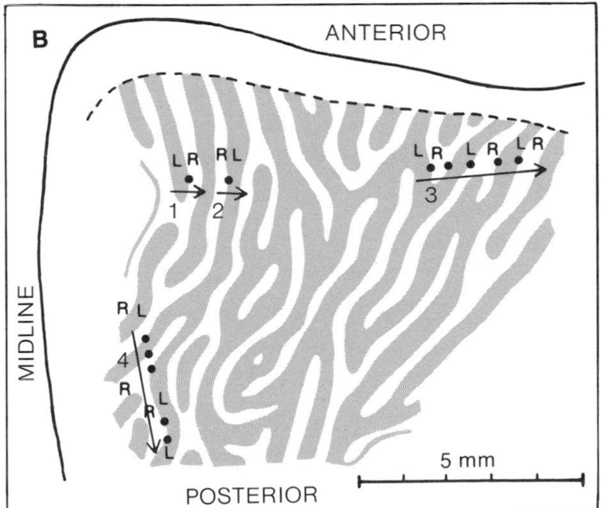

4
STAINING METHOD for demonstrating ocular dominance columns. A. The light bands delineate the borders of ocular dominance columns in monkey striate cortex, revealed by a reduced silver staining method. The section passes for the most part horizontally through layer IV. Two pale bands are marked by lesions (below arrows), which appear as round white dots (the other dots are blood vessels). B. Shows a reconstruction of the columnar patterns. The areas for the left (L) eye are shaded. Four electrode tracks are indicated by arrows. At each point when the eye preference changed (from L to R or R to L), a lesion was made (shown by dark circles). There is good agreement between the boundaries established physiologically and subsequently by histology. (See also Figure 6, Chapter 19) (After LeVay, Hubel, and Wiesel, 1975)

boundaries between columns. This area of cortex was investigated previously by electrical recording in the animal. As the electrode moves horizontally along layer IV, the cells are driven first by one eye and then by the other. The transition points are marked at the circles in Figure 4*B* by making small lesions with the electrode. Clearly there is excellent agreement between the physiological and the histological definitions of the margins of columns. All the techniques described above indicate that the slabs for each eye dominance are about 0.25 to 0.5 mm wide and run horizontally for long distances (Figures 4 and 6). Strictly

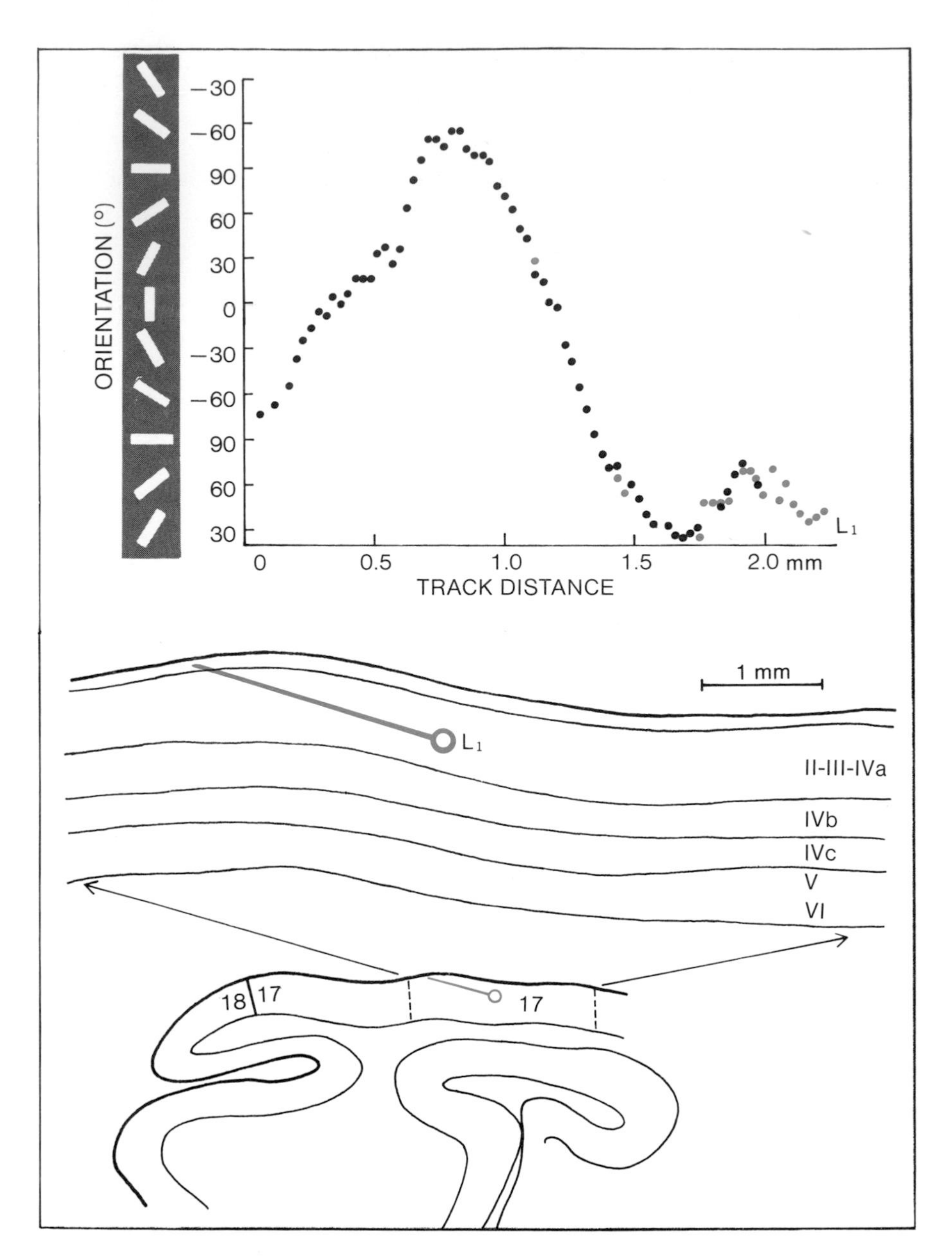

5 **ARRANGEMENT OF ORIENTATION COLUMNS. As the electrode moves obliquely through the cortex, the orientation specificity of the cells encountered shifts systematically. The shift is about 10° for each 50μm, as though a series of columns or slabs were being traversed in a regular sequence. (After Hubel and Wiesel, 1974)**

speaking, they are not shaped like columns, but the term has become established and is generally retained.

There is no simple anatomical means available for displaying orientation columns. However, a more detailed physiological examination of the system of cells with similar field axis orientation has revealed that the orientation columns, like the ocular dominance columns, are shaped like slabs. This information about their arrangement in the visual cortex of monkeys and cats was obtained by making oblique or tangential electrode penetrations through the cortex. An example is shown in Figure 5, which reveals once again the orderliness of the arrangement of cells. Each 20-μm shift of the electrode along the cortex is accompanied by a change in field axis orientation of about 10° in a regular sequence through 180°. The field axis orientation columns are narrower than those for ocular dominance, 20 to 50 μm compared with 0.25 to 0.5 mm.

A scheme for the way in which the two sets of columns might be arranged in the cortex is shown in Figure 6. For simplicity in elaborating fields, one would expect the orientation and ocular dominance columns to run at right angles to each other. If they did, inputs from the two eyes could be combined to produce coherent binocular fields with similar orientation for the synthesis of simple, complex, or hypercomplex

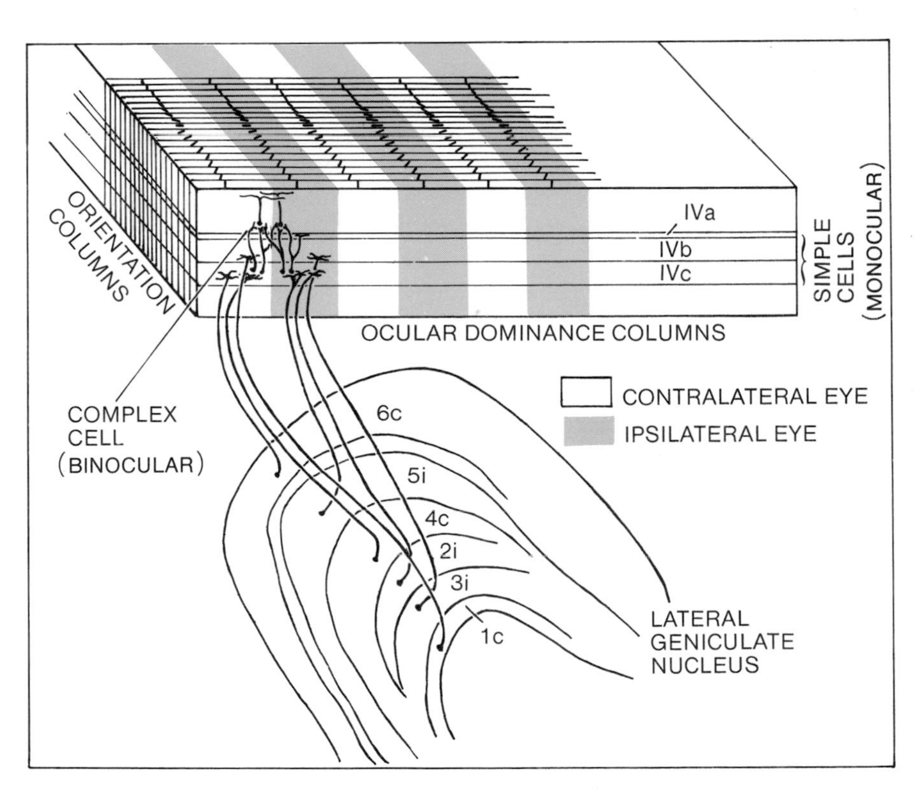

6

RELATION BETWEEN OCULAR DOMINANCE AND ORIENTATION COLUMNS. Scheme in which the ocular dominance and orientation columns run at right angles to each other. An example of a complex cell is shown in an upper layer, receiving its inputs from two simple cells that lie in two neighboring ocular dominance columns, but share the same orientation column. (From Hubel and Wiesel, 1972)

fields positioned in the same part of the visual field. The illustration provides a further clue to cortical organization and the relation of the orientation to the ocular dominance columns. The functional unit of the visual cortex appears to be a roughly square area in which all the possible orientations are represented for a receptive field area in the same place in each of the two eyes. An adjacent square of cortex would analyze information in the same way for an adjacent part of the visual field, and so on. In parts of the cortex that deal with peripheral fields of vision the receptive fields of individual eyes become larger. At the same time, a small area of cortex deals with a relatively large area of peripheral retina. But the basic organization of the cortex remains similar in that orientation and ocular dominance columns have the same width as their counterparts in cortical areas dealing with the fovea, where receptive fields are small.

Significance of cell groupings

The existence of separate columns for all the diverse orientations and their presumed use in providing components for processing information suggest why the cortex needs such a staggering number of cells. Look, for example, at just a very small segment of the primary visual cortex. Such a segment consists of columns with their rows of cells concerned exclusively with a small part of the visual field that is analyzed for movement, orientation, color, and other stimulus parameters. Each part of the primary visual cortex consists of basically similar stereotyped repeating arrays of cells. And each small region of the retina is represented not once, but over and over again, in column after column, first for one receptive field axis orientation and then for another.

The columnar system enables the cortex to analyze the visual world in terms of many different variables. As expected, columns are not entirely independent units; through their connections they hand on the result of their processing by linking up with other cortical areas where further analysis takes place along a similar repeating pattern. There is direct anatomical evidence to show that fibers leave the primary visual cortex (area 17) to supply the adjacent areas (18 and 19), which contain more elaborate and higher order cells.[16]

At this stage it is natural to wonder what other features of the perceptual world are organized in columns and how the rest of the cortex is arranged. The first demonstration of columns was made, as mentioned earlier, by Mountcastle[1] in the somatosensory cortex, where mechanosensory cells are grouped according to modality. Within a column all the cells respond either to touch or to deep pressure and joint position. Similarly, cells in the motor cortex are arranged in columns according to the muscles they innervate.[17] Other types of groupings may occur in different areas. For example, in the face area of mouse and rat somatosensory cortex there are barrel-shaped arrange-

---

[16]Hubel, D. H., and Wiesel, T. N. 1965. *J. Neurophysiol.* *28*:229–289.
[17]Asanuma, H. 1975. *Physiol. Rev.* *55*:143–156.

ments of neurons aligned in well-defined rows.[18] Each group of cells corresponds to one of the hairs, the vibrissae, on the animal's face. Here, however, the functional organization of the cortical cells is not understood; yet the principle remains the same—cells with similar functions are aggregated together.

An encouraging aspect of the recent studies of the functional organization of the cortex is that a few methodologies have already yielded so many advances.

SUGGESTED READING

Columnar Organization of Visual Cortex

Hubel, D. H., and Wiesel, T. N. 1962. Receptive fields, binocular interaction and functional architecture in the cat's visual cortex. *J. Physiol. 160:* 106–154. (The original description of orientation and ocular dominance columns in cat cortex.)

Hubel, D. H., and Wiesel, T. N. 1968. Receptive fields and functional architecture of monkey striate cortex. *J. Physiol. 195:*215–243.

Hubel, D. H., and Wiesel, T. N. 1972. Laminar and columnar distribution of geniculo-cortical fibers in the macaque monkey. *J. Comp. Neurol. 146:* 421–450.

Hubel, D. H., and Wiesel, T. N. 1974. Sequence regularity and geometry of orientation columns in the monkey striate cortex. *J. Comp. Neurol. 158:267–294.* (A number of basic concepts about columnar organization, receptive fields, and cortical function are incorporated in this paper.)

LeVay, S., Hubel, D. H., and Wiesel, T. N. 1975. The pattern of ocular dominance columns in macaque visual cortex revealed by reduced silver stain. *J. Comp. Neurol. 159:559–576.*

Zeki, S. M. 1973. Color coding in rhesus monkey prestriate cortex. *Brain Res. 53:*422–427.

Zeki, S. M. 1974. Cells responding to changing image size and disparity in the cortex of the rhesus monkey. *J. Physiol. 242:*827–841.

Somatosensory and Motor Cortex

Asanuma, H. 1975. Recent developments in the study of the columnar arrangement of neurons within the motor cortex. *Physiol. Rev. 55:*143–156.

Mountcastle, V. B. 1957. Modality and topographic properties of single neurons of cat's somatic sensory cortex. *J. Neurophysiol. 20:*408–434. (This is the first description of columnar organization.)

---

[18]Walker, C., and Woolsey, T. A. 1974. *J. Comp. Neurol. 158:*437–454.

PART TWO

MECHANISMS FOR NEURONAL SIGNALING

CHAPTER FOUR

ELECTRICAL SIGNALING

The signals used by nerve cells to transmit information consist of electrical currents generated across their surface membranes. The currents flow through the intracellular and external fluids and result principally from the movements of charges carried by sodium, potassium, calcium, and chloride ions. Compared with insulated metallic materials, such as cables, nerves are very poor conductors of electricity. The interior of a neuron is separated from the outside by the cell membrane, which is an imperfect insulator and permits some leakage of ions in both directions. The membrane has also the capacity to store and separate electrical charges. These properties impose restrictions on the way in which electrical signals can be conducted.

Neurons carry only two types of signals: localized potentials and action potentials. The localized, graded potentials are used over short distances, usually limited to 1 to 2 mm. They play an essential role at special regions, such as at sensory endings, where they are called generator potentials, or at junctions between cells, where they are named synaptic potentials. The localized potentials enable individual nerve cells to perform their integrative functions and are also an essential part of the mechanism for the conduction of impulses. The action potentials are conducted regenerative impulses that do not decrease over distances.

These two types of signals are the universal language of nerve cells in all animals that have been studied.

Current flow in nerve cells

How do the materials that the body uses to conduct electrical signals compare with those that are in everyday use, such as metal wires? A nerve fiber can be considered as a tube containing axoplasm, a watery solution of salts and proteins. The cell membrane restricts the diffusion of ions and separates the axoplasm from the outside fluid, which has roughly the same ionic strength but a different composition (Table 1, Chapter 5). Axoplasm as a conductor of electricity is about 10^7 times worse than metal wire because the number of charge carriers is smaller and their mobility is less: ions rather than electrons are used to carry current. These factors inevitably restrict the amount of current a nerve fiber can carry. Thus, in a solution of sodium chloride current flows through the movement of positively charged sodium ions (cations) toward the negative pole of a battery, called the cathode, while negatively charged chloride ions (anions) move to the positive pole, or anode. Conduction of electrical currents along nerves is further hampered by the fact that the surface membrane is an imperfect insulator and, therefore, some leakage occurs between the inside and the outside medium. In addition to these purely electrical defects, the nerve fibers are small, and this further limits the amount of current. Hodgkin provided a dramatic illustration of the consequences these factors have on the spread of electrical signals:[1]

> If an electrical engineer were to look at the nervous system he would see at once that signalling electrical information along the nerve fibers is a formidable problem. In our nerves the diameter of the axis cylinder varies between about 0.1μ and 10μ. The inside of the fibre contains ions and is a reasonably good conductor of electricity. However, the fibre is so small that its longitudinal resistance is exceedingly high. A simple calculation shows that in a $1\ \mu$ fibre containing axoplasm with a resistivity of 100 ohm cm, the resistance per unit length is about 10^{10} ohms per cm. This means that the electrical resistance of a metre's length of small nerve fibre is about the same as that of 10^{10} miles of 22 gauge copper wire, the distance being roughly ten times that between the earth and the planet Saturn. An electrical engineer would find himself in great difficulties if he were asked to wire up the solar system using ordinary cables.

A further property of the nerve cell membrane that tends to distort signals is its electrical capacity, which enables electrical charges to be stored and separated. As a result, signals are not only severely attenuated in amplitude as they spread over a distance along a nerve fiber (due to leakage of current) but in addition their time course becomes distorted (due to membrane capacity).

These considerations are discussed more fully in Chapter 7; for now, the conduction of heat along an iron rod serves as a familiar analogy. Everyday experience, such as holding an iron poker in a fire, confirms the statements that follow. The material out of which the rod is made determines the distance over which heat applied to one end spreads. If the rod is immersed in conducting fluid, but surrounded by an insulating membrane, then the insulator determines how far the increase in temperature spreads. If the insulator is leaky, the temperature rises less at the far end than it does if the insulator is perfect. Finally, if the bar or the insulator stores heat, there is a delay in the rate at which heat can spread from end to end (because each part of the rod has to be heated up before the next part can become warmer); a difference

---

[1]Hodgkin, A. L. 1964. *The Conduction of the Nervous Impulse.* Liverpool University Press, Liverpool, p. 15.

therefore appears in the rate at which the temperature rises at the site of heating and at a distance. The equations that formally describe heat conduction also apply to the spread of electricity in an undersea cable and in a nerve fiber. In summary, the physical nature of the matter that makes up the nervous system imposes severe restraints upon the mechanisms by which information can be carried from point to point without drastic attenuation.

Types of signals

The electrical signals generated by nerve cells overcome and make use of the "defects" described above. The general signals can be classified into two groups: (1) graded, passive, or LOCALIZED POTENTIALS that depend on the cablelike properties of the cell, and (2) the impulses or ACTION POTENTIALS that travel rapidly, and without distortion, from one end of a nerve to the other.

The main characteristics of passive potentials are that they are continuously graded and do not spread for distances greater than 1 mm or so without severe attenuation. They occur at specialized regions of nerve cells, where responses are initiated or where they are inhibited. At sensory nerve endings the passive, localized potentials are called GENERATOR POTENTIALS or RECEPTOR POTENTIALS; at synapses they are called EXCITATORY POSTSYNAPTIC POTENTIALS if they tend to give rise to impulses and INHIBITORY POSTSYNAPTIC POTENTIALS if they counteract or suppress excitation. Localized potentials also play an essential part in the process of impulse conduction. (For comments on nomenclature, see page 81.)

In contrast to the passive localized potentials, the nerve impulse is not attenuated with distance and is an ALL-OR-NONE explosive event which has much the same amplitude in all nerve cells of the body and varies mainly in the velocity with which it propagates. Like standard dots or dashes in Morse code, the signals are all just about the same in amplitude and time course in an optic nerve, an auditory nerve, or a sensory fiber carrying a message from the skin of the big toe to the spinal cord. Initiation of impulses and their mechanism of conduction are essentially similar in neurons of a wide range of invertebrates and vertebrates. It is unnecessary to refer each time to specific animals or nerves when discussing general features or impulse mechanisms. ELECTRICAL IMPULSES ARE THE UNIVERSAL LANGUAGE OF ALL KNOWN NERVOUS SYSTEMS.

Measurement of membrane potentials

It is convenient to begin by examining the types of electrical potentials that are generated in a "standard" nerve fiber. In practice, it is easiest to use muscle fibers for an experiment of the type to be described, but the principles are similar for nerves.

Two electrodes are used to measure the potential difference across the membrane, one making contact with the axoplasm and the other with the outside fluid. The difference in potential between them is amplified and fed into an oscilloscope. The usual procedure is to take a fine glass capillary with a tip of less than 1μm and insert it, with the help of a micromanipulator, through the cell membrane to make connection with the intracellular fluid. Capillary microelectrodes are generally used when the voltages generated by a nerve or muscle cell are

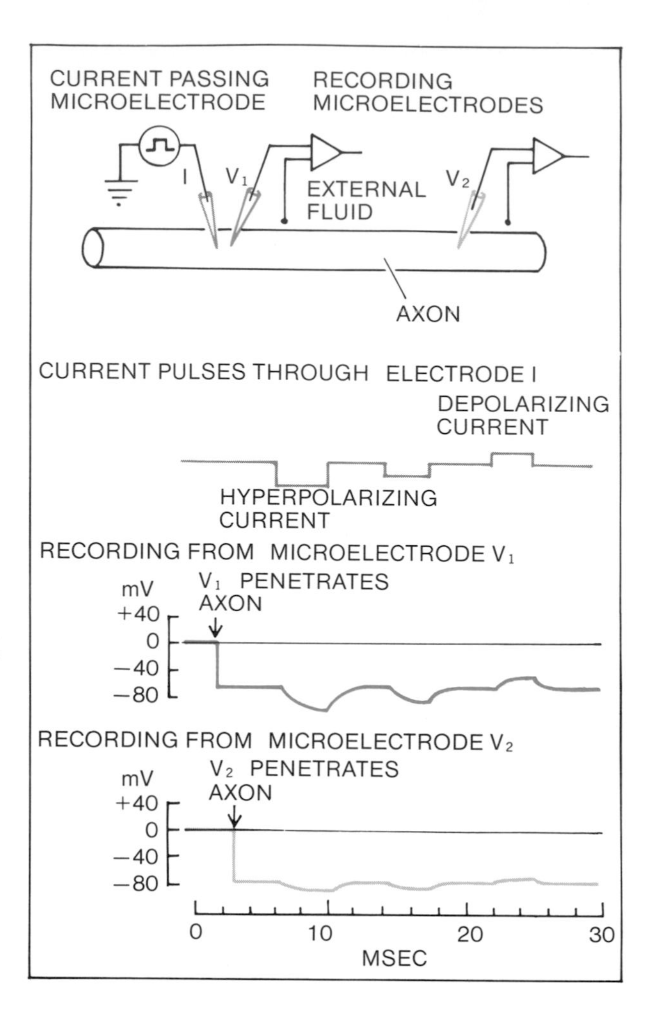

1

INTRACELLULAR RECORDING from a large axon with microelectrodes. One electrode (V_1) is inserted into the axon and records a resting potential of −70 mV (inside negative) with respect to the outside. A second electrode (I), next to V_1, is used to pass pulses of current that produce localized graded potentials. The first two are hyperpolarizations and the third is a depolarization. Electrode V_2, about 1 mm away from V_1, also measures a resting potential of −70 mV when it penetrates the axon. But the localized graded potentials are smaller and slower than at V_1, owing to the passive electrical properties of the axon.

to be measured precisely. The construction, manipulation, and electrical properties of the electrodes pose numerous difficulties for the investigator and the electronics engineer. The tip must be small enough to penetrate the cell membrane without causing damage, which inevitably means that the electrical resistance is high, 200 MΩ or so in the case of fine micropipettes with an outer diameter of close to 0.1μm. Special equipment is required to draw the electrode from molten glass, to move it into the cell, and to ensure that the potentials displayed on the oscilloscope are not distorted through the electrical characteristics of the electrode itself.

With both electrodes in fluid outside the cell no potential difference is recorded; as the microelectrode is pushed through the membrane, a sudden jump of about 0.1 V is registered, the inside of the fiber being negative (Figure 1). This voltage is called the RESTING POTENTIAL and indicates that the membrane is electrically polarized at rest; any reduction in its magnitude (bringing the voltage toward zero) is a DEPOLARIZATION of the membrane, and any increase in the membrane potential is a HYPERPOLARIZATION. In most of the figures in this section, voltage is plotted against time. The negative and positive signs refer to the potential at the tip of the microelectrode with respect to the indifferent electrode—that is, the one outside the cell. Increased negativity is shown in the downward direction.

A second microelectrode can be used to pass current through the membrane and determine its electrical characteristics. Figure 1 shows the effects of passing current pulses that hyperpolarize or depolarize by injecting negative or positive charges into the axon. The potentials have the following properties:

1. They are graded. The amplitude of the voltage change varies with the current and approximately obeys Ohm's law. Such potentials can therefore sum.
2. The duration of the potential change also varies with the duration of the current, but it rises and falls more slowly because of the electrical capacity of the membrane (Chapter 7). One important consequence of this is that the change in potential outlasts the current flow that gave rise to it.
3. The change in potential decreases and becomes further slowed with distance. Little or no change in membrane potential can be discerned several millimeters away from the region in the fiber where current is being injected.

Localized, graded potentials of this type, which depend on the passive cable properties of the membrane, are therefore useless for

A NOTE ON NOMENCLATURE

The localized changes in membrane potential that depend on the passive electrical properties of the membrane are all basically similar, but they have been given a variety of names depending on the mechanism that generates the potential change, its effect, and the site at which it occurs. Defined here are some commonly used terms.

Potential Change	Designation
Brought about by passing current through electrodes	ELECTROTONIC POTENTIAL
Brought about by chemical transmitters at synapses	SYNAPTIC POTENTIAL, excitatory or inhibitory; often abbreviated as epsp and ipsp (excitatory and inhibitory postsynaptic potentials); the excitatory postsynaptic potential in muscle is called the END PLATE POTENTIAL (epp)
At electrical synapses, caused by current flow	COUPLING POTENTIAL
In sensory end organs (or sensory receptors), brought about by the appropriate stimulus	GENERATOR or RECEPTOR POTENTIAL

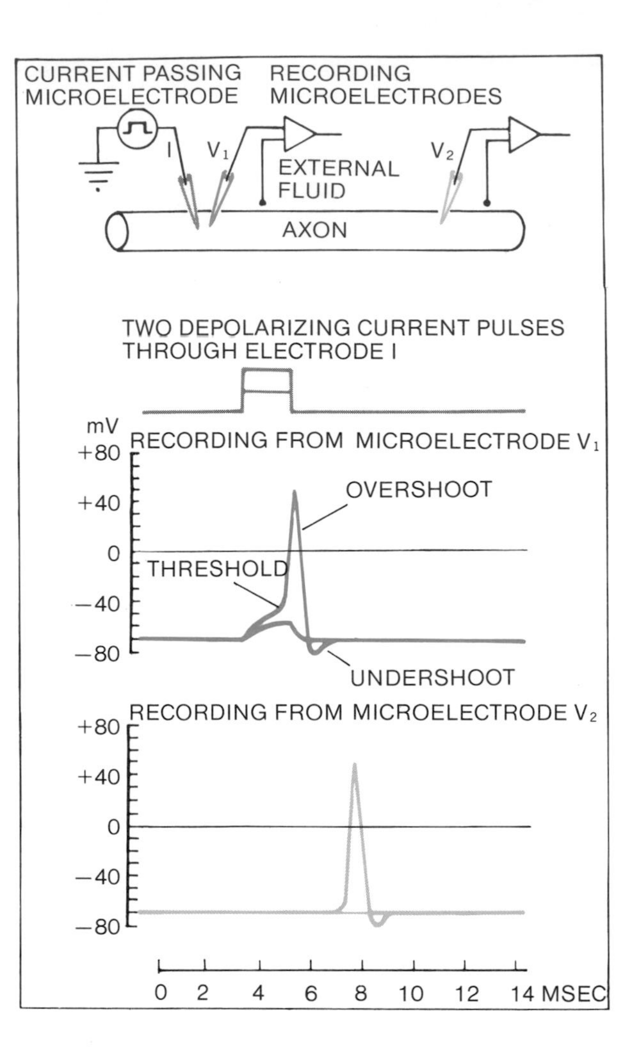

2

ACTION POTENTIALS recorded from a large axon by means of intracellular microelectrodes V_1 and V_2. The resting potential is −70 mV (inside negative); each trace shows two successive sweeps of the oscilloscope beam. During one sweep a relatively small depolarizing current pulse is passed through microelectrode I, causing a small localized potential (recorded in V_1) that does not reach threshold. The second current pulse is larger, giving rise to a depolarization that reaches threshold and initiates an action potential that propagates rapidly along the axon and reaches V_2 at a distance of 2 cm. Unlike the graded localized potential which cannot spread more than 1 to 2 mm, and is therefore not recorded at V_2, the action potential is the same size all along the axon.

long-range signaling. They are, however, essential for the workings of the nervous system and provide a clue to understanding many basic neural processes, such as the mechanisms that operate in neurons during integration, during the initiation of sensory signals, and when impulses propagate along axons.

Nerve impulses

Within the body, impulses are normally initiated at terminals of sensory fibers or at synapses. For the present purpose we can mimic the excitatory event by artificially injecting current into the nerve fibers. Current is passed in such a direction as to make the inside of the membrane relatively more positive (depolarized). If such currents are small, the resulting changes in membrane potential are passive, and they decay in amplitude over a distance (Figure 1). However, once the membrane becomes depolarized to a certain critical level (the THRESHOLD), a totally new type of event appears. The membrane potential undergoes a series of changes that bear no relation to the original stimulus. The depolarization goes rapidly through 0 mV (OVERSHOOT), so the voltage becomes transiently reversed (negative outside, positive inside).

After reaching a peak value of about +40 mV (inside) the membrane potential returns to its original value; in many nerve cells there is a transient hyperpolarization (the UNDERSHOOT). This sequence of potential changes, called an action potential or impulse, lasts 1 to 2 msec and is illustrated in Figure 2. The impulse or action potential has the following features:

1. It is not graded, but has a distinct threshold. Once started, its amplitude does not depend upon the amplitude of the stimulus; larger currents do not give rise to larger action potentials, nor do stimuli of longer duration prolong nerve impulses. Hence the action potential is a triggered explosive all-or-nothing event. Its fixed size arises from the fact that the stimulus, a depolarization, merely initiates events that are determined by ionic concentration gradients and the selective permeability of the membrane (Chapters 5 and 6).
2. The entire sequence of potential changes must occur before another action potential can be initiated. After each action potential there is a period of enforced silence (the REFRACTORY PERIOD) during which a second impulse cannot be initiated. This and the all-or-nothing nature of the impulse preclude the possibility of summation.
3. The action potential does not decline with distance. Whether recorded 1 mm or 1 meter from the stimulating electrode, it is much the same size and shape. In man and other mammals the fastest impulses in the largest fibers travel at a speed of over 120 meters/sec (432 km/hr or 270 miles/hr) and are therefore capable of conveying information rapidly over long distances. The action potential is not distorted and attenuated by the passive electrical properties of the nerve because it is renewed in each succeeding patch of membrane. The impulse in one patch acts as the stimulus for an impulse in the next, and in this way the signal is carried unfailingly over long distances (Chapter 7).

SIGNALS USED IN A SIMPLE REFLEX

The characteristic features of short- and long-distance signals can be illustrated in the simple STRETCH REFLEX, which exemplifies many of the components of signaling and integration. A familiar demonstration of this reflex is the knee jerk, in which the patellar tendon below the knee is tapped with a hammer to stretch a group of muscles that extend the leg. The reflex is used in the body in regulating movements and in maintaining posture under control of the nervous system. Since only two types of neurons are involved in this reflex, one can examine in a simple form some of the mechanisms by which the central nervous system performs its tasks. The following account provides a framework for discussing the various aspects of signaling that are taken up in detail later.

Neurons involved in stretch reflex

Figure 3 shows the structural elements that subserve the stretch reflex. First to be involved is a sensory nerve fiber, whose endings form an intimate attachment with a specialized structure called the MUSCLE SPINDLE. The muscle spindles lie within the main mass of the muscle fibers that move the limb; they provide the nervous system with information about stretch. (In Figure 3 the muscle spindle is drawn outside the muscle.) The spindle acts as a mechanoelectrical transducer—sensitive to one form of energy, transforming it into another (stretch into nerve impulses). Other sensory nerve fibers are sensitive to other forms of energy and resemble different types of transducers, such as microphones, photoelectric cells, or thermocouples; in each instance, one specific form of energy (sound, light, or heat) gives rise to a stereotyped electrical signal (Chapter 15).

The sensory nerve fiber enters the spinal cord, where it branches extensively. For the present we need consider only the synaptic connections with a particular group of cells in the cord, the motoneurons. These motoneurons send their fibers back to the same muscle from which the stretch-sensitive sensory axons came. There the motor axons form excitatory synapses and thereby are directly responsible for the muscle's contraction.

This is one more example of the remarkable specificity in the design of the neural connections (discussed in Chapters 2 and 3 in rela-

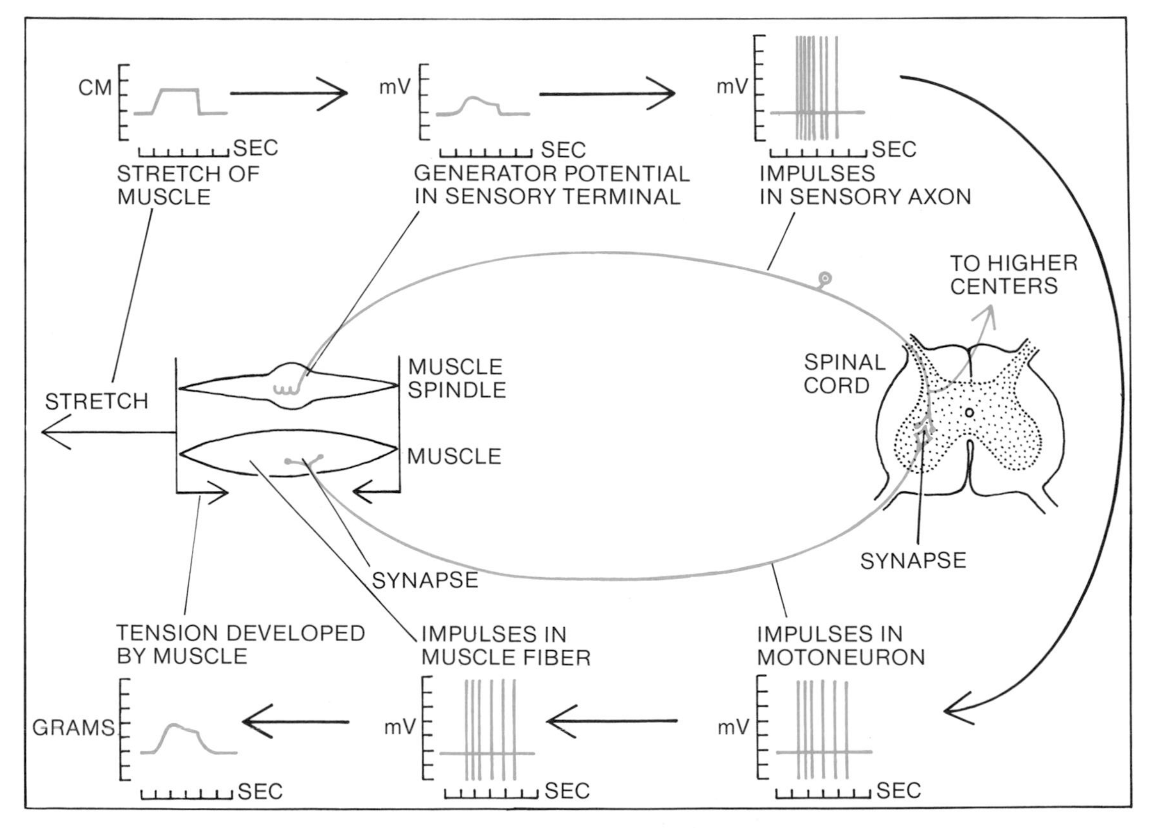

3 **STRUCTURES AND SIGNALS INVOLVED IN THE STRETCH REFLEX. For simplicity, the drawing does not show important structural and functional features that are described in detail in Chapters 8 and 15.**

tion to the visual system). The sensory fiber selects not any motoneuron but one that supplies the same muscle. Other branches of the sensory nerve ascend for long distances in the spinal cord to reach preappointed groups of cells in the next relay station, responsible for higher levels of motor coordination. This orderly anatomy is, of course, the first prerequisite for the reflex to function.

The anatomical discontinuity between the sensory and motor nerve cells constitutes the first synapse in the reflex arc. The membranes here are separated by a gap of about 30 to 50 nm. There is a second synapse between the peripheral termination of the motor nerve fibers and the muscle fibers.

In their least complicated form, the functional components of the two-neuron reflex work in the following manner: to maintain posture—for example, during standing—the muscles contract in such a way as to maintain a constant length. When a person starts to lose his balance, certain groups of muscles are stretched, activating the stretch reflex, which restores the muscles to their former length. If the postural mechanisms work well, he does not even notice these small correcting fluctuations, because the deviations from the desired posture are small and the correction is rapid (see Chapter 15 for additional control of the stretch reflex).

Figure 3 also depicts sequentially the main electrical events during the stretch reflex. Stretch gives rise to a change in membrane potential in the sensory terminal, called the generator potential or receptor potential. It is confined to about 1 mm of the terminal portion of the sensory neuron, and its size and duration reflect the amount and duration of stretch. If this were all that happened, the spinal cord, which is many inches away, would never receive news of the stretch. However, when the stretch (and the potential change) reaches a critical level, all-or-none propagating impulses are started in the sensory nerve. Stronger stretches give rise to larger generator potentials and higher frequency of impulses. These long-range signals travel rapidly at about 100 meters/sec (about 10 times faster than a very good sprinter) to the spinal cord. Here the impulses reach the gap that separates the sensory fiber from the motoneuron. Transmission between the cells is effected by means of a chemical liberated from the terminals of the sensory nerve. As a result, the traveling all-or-none sensory impulse of fixed size becomes translated into a depolarizing synaptic potential localized in the motoneuron. The synaptic potentials resemble in some respects generator potentials and start a train of impulses in the motoneuron. These are similar in time course and size to the sensory impulses and travel rapidly down to the muscle. Here they initiate in muscle fibers excitatory synaptic potentials that in turn lead to conducted muscle impulses and contraction. This whole cycle of events (generator potential → sensory nerve impulses → synaptic potentials in a motoneuron → motor nerve impulses → synaptic potentials in muscle → muscle impulses → contraction) is rapid and takes a small fraction of a second in man. It is therefore adequate for making the speedy adjust-

How does a neuron take account of different converging influences?

ments in muscle length that are necessary for maintaining a desired posture.

The monosynaptic stretch reflex is a PRIORITY REFLEX in that it overrules other events when the primary task is maintenance of posture. However, many thousands of synapses are formed on each mammalian motoneuron, resulting from a convergence of neurons, some excitatory, others inhibitory (cf. Figure 9, Chapter 16); and the stretch reflex can itself be overruled by more pressing stimuli. For example, a pin stuck into the toe tends to cause the knee to bend—the opposite reaction to the stretch reflex. This is because a sufficiently painful stimulus gives rise to inhibitory synaptic potentials that prevent impulses from being initiated in the motoneuron. The process that sorts out these various influences, some acting in concert, others opposing each other, was called by Sherrington[2] the INTEGRATING ACTION OF NEURONS. Integration at the cellular level is simply the way in which impulses (long-range signals) that converge on a cell become translated into postsynaptic potentials (local or short-range signals). These then determine the firing of the neuron, that is, the generation of a new impulse that contains the synthesis of all the various inputs.

In Chapters 15 and 16 integration is discussed more fully and the elegant biasing mechanism is described whereby the central nervous system can set the sensitivity of the muscle spindle to stretch. However, in all these activities for transferring information only two types of signals are used, and they convey all the abstractions of the surrounding world. Integration by the motoneuron, which adds up excitation and inhibition and then fires an impulse or remains silent, is strikingly similar to integration by the nervous system as a whole. The cell and the brain both "decide" whether or not to act on the basis of information derived from a wide variety of sources.

Many of these general principles we owe to Sherrington, who discovered them through recording the tension generated by skeletal muscles involved in the stretch reflex before electrical recordings were possible from individual nerve cells. The following quotation is still a useful, concise description of the different neural signals:[3]

> The nerve-nets are patterned networks of threads. The human brain is a vast example, offering immense numbers of determinate paths, and immense numbers of junctional points. At these latter the travelling signal so to speak hesitates and sets up a local gradable state which may have to accumulate before transmitting further, or indeed may there subside and fail. These junctional points are often convergence points for lines from several directions. Arrived there signals convergent from several lines may coalesce and thus reinforce each other's exciting power.
>
> At such points too appears a process which instead of exciting, quells and precludes excitation. This inhibition, like its opposite process, excita-

---

[2]Sherrington, C. S. 1947. *Integrative Action of the Nervous System.* Yale University Press, New Haven.

[3]Sherrington, C. S. 1933. *The Brain and Its Mechanism.* Cambridge University Press, London.

tion, does not travel. It is evoked, however, by travelling signals not distinguishable from those which call forth excitement. The travelling signals calling up excitement and those calling up inhibition never, however, reach the nodal point by the same path, never have paths in common.

The two are relatively antagonistic. Each can be neutralized gradually by a dosage of the other. The inhibition may be a temporary stabilization of the membrane at the nodal point, which is potentially a relay station. The inhibitory stabilization produced by a travelling signal is evanescent; a train of signals is required to maintain it. While it lasts, the nodal point is blocked to signals, or only transmits them slowly.

These two opposed processes, excitation and inhibition, cooperate at nodal point after nodal point in the nerve-circuits. Their joint operation at any moment settles what will be the conduction pattern, and so the motor outcome, of the signalling going forward in the brain.

C. S. Sherrington with one of his pupils (J. C. Eccles) in the mid 1930's.

CHAPTER FIVE

IONIC BASIS OF RESTING AND ACTION POTENTIALS

The electrical potential difference between the inside and the outside of the membrane of an excitable cell depends on the ionic concentration gradients and the selective permeability of the membrane. The details have been worked out best in the giant axon of the squid, which has served as a good model for other neurons and for excitable tissues in general. Neglecting the relatively small contribution of other ions, one can state that changes in the permeability of the membrane to potassium, to sodium, or to both ions can shift the potential to any value within a range of about 130 mV, from −75 mV (inside negative) to +55 mV (inside positive).

The inside fluid in nerve cells contains relatively high concentrations of potassium, but low concentrations of sodium and chloride; in the outside fluid the ratio is reversed (low potassium, high sodium and chloride). The resting membrane is much more permeable to potassium than to sodium. Potassium ions, therefore, tend to leak out of the cell along their concentration gradient and give rise to an electrical potential that is inside negative. This negative potential across the membrane balances the concentration gradient, driving potassium out. The relation between membrane potential and potassium concentrations is described quantitatively by the Nernst equation.

The concentration and electrical gradients for sodium across the membrane are directed inward. When the membrane becomes permeable to sodium, sodium enters the cell, thereby making the inside more positive, until at 55 mV (inside positive) the electrical and chemical gradients for sodium ions are at equilibrium.

The general conclusion is that electrical signals can be generated by the membrane without the direct intervention of metabolism, by selectively changing the ionic permeabilities and allowing ions to flow downhill along their electrochemical gradients. With each action potential the cell gains sodium and loses potassium, but for a single impulse the amounts are small enough for the resultant concentration changes to be ignored. Metabolic energy is expended to restore the resting (original) concentrations.

The membrane can be represented by a simple circuit in which the permeability channel for each ion is described in electrical terms as a conductance. This scheme has the great advantage of allowing permeability to be estimated by electrical measurements, which confers a high degree of resolution.

Chapter 4 utilizes a simple reflex arc to demonstrate that nerve cells and muscle fibers generate localized and conducted electrical signals. The mechanisms by which such universal signals arise are now to a large extent understood; it is known what forces move charges across the membrane and at what stage of the process work is done. This chapter shows that metabolic energy is expended only to maintain the resting state and that during both the depolarizing and the repolarizing stages of the action potential ions run downhill along their concentration and electrical gradients. The keys to controlling the membrane potential are the chemical gradients and electrical potentials acting on certain ions and the selective permeability of the membrane. The permeability itself is controlled by the level of the membrane potential. An early review by Hodgkin[1] first brought together many of the concepts described in this chapter.

Ionic basis of resting potential

The development of ideas concerning the role of ions in signaling is interesting enough to warrant a brief historical sketch. It has long been known that electricity is associated with transmission of nerve impulses and muscular contraction. In the last two decades of the eighteenth century, the intellectual community was "galvanized" into new activity as the result of the experiments of Galvani, who noted that frog muscles contracted when a metal hook piercing the spinal medulla made contact with an iron railing.[2] He argued that the muscles had a reservoir of electricity that somehow caused the contractions. This led to an acrimonious debate with Volta, whose basic claim was that electricity had to be applied to a muscle for it to contract. He had little trouble convincing most of his contemporaries, because stimulating nerves or muscles with "piles of Volta" or with "forceps" made of dissimilar metals was easier than recording from them. Galvani's proposal that muscles produce electricity ("animal electricity") had to wait many decades for a clear demonstration.

Julius Bernstein[3] in 1902 was the first person to propose a satisfactory hypothesis for the origin of the resting potential in nerve and muscle fibers. From chemical analyses he knew that the interior of nerve fibers and muscle fibers was rich in potassium but contained little sodium or chloride. He further knew that there were anions within the cell to which the resting membrane was not permeable; and the evidence available at the time suggested that the membrane was

[1]Hodgkin, A. L. 1951. *Biol. Rev.* *26*:339–409.
[2]de Santillana, G. 1965. *Sci. Am.* *212*:82–91.
[3]Bernstein, J. 1902. *Pflügers Arch.* *92*:521–562.

impermeable to sodium ions (but see below). He therefore proposed that the voltage across the membrane came about as a result of unequal distribution of potassium ions as predicted by the NERNST EQUATION. This relation is derived from physical chemistry and predicts what voltage will be developed across a living or nonliving membrane permeable to one ion species only (here potassium) when the concentrations of that ion are not equal on the two sides of the membrane (see below). For potassium:

$$E_K = \frac{RT}{zF} \log_e \frac{K_o}{K_i}$$

where E_K is the potassium equilibrium potential, which is the potential across the membrane at which there is no net flow of potassium ions from one side to the other, that is, when the system is in equilibrium; K_o is the outside and K_i is the inside potassium concentration; R is the gas constant; T is the absolute temperature; z is the valency of the ion; and F is the faraday (see below).

At 20° C, this equation reduces to:

$$E_K = 58 \log_{10} \frac{K_o}{K_i}$$

E_K is inside negative when K_o and K_i are in the physiological range.

Bernstein could not test the hypothesis satisfactorily because the resting potential across the cell membrane could not be measured accurately with the instruments available. One approximation was to cut muscle fibers and place an electrode onto the cut area, thereby effectively making contact with the fluid within the muscle fibers and recording an "injury potential." By varying the absolute temperature, Bernstein tried to assess whether the resting potential conformed to the Nernst equation. Within the error of his measurements, he found good agreement between the predicted behavior of the membrane potential and what he actually observed.

In more recent years thorough tests have been made by varying the external and internal potassium concentrations to see whether the membrane potential does actually behave as one would expect from the Nernst equation. Many of the basic ideas were formulated by Boyle and Conway,[4] who, like Bernstein, worked on muscles, but the experiments discussed below were done for the most part on large nerve fibers (up to 1 mm in diameter) that innervate the mantle of the squid and are used to initiate rapid movements. These giant nerves are particularly useful because their size makes them suitable for a number of tests that could not be performed on smaller axons.[5] Many of the pioneering investigations on signaling were done on these nerve fibers and the results have turned out to be relevant to the function of many different tissues—for example, heart muscle, vertebrate muscle, and

[4] Boyle, P. J., and Conway, E. J. 1941. *J. Physiol.* *100*:1–63.
[5] Young, J. Z. 1936. *J. Micr. Sci.* *78*:367–386.

vertebrate nerve fibers. A. L. Hodgkin, who together with A. F. Huxley initiated many experiments on the squid axon, has said:[6]

> It is arguable that the introduction of the squid giant nerve fibre by J. Z. Young in 1936 did more for axonology than any other single advance during the last forty years. Indeed a distinguished neurophysiologist remarked recently at a congress dinner (not, I thought, with the utmost tact), "It's the squid that really ought to be given the Nobel Prize."

It is instructive to consider first the ionic contents of the blood and the axoplasm (Table 1) and the ratio of the two across the membrane of squid nerves. From the ratio of potassium concentrations in the blood and inside the cell (1:20), the resting potential should be −75 mV according to the Nernst equation, provided that (1) the potassium ions inside the cell are not bound, but are free to move (Chapter 6), and (2) the membrane is permeable exclusively to potassium ions. Instead of the predicted value of −75 mV, the voltage generally recorded is slightly less (−70 mV) for a nerve in situ. (The reasons for this discrepancy will become apparent later.) After the nerve has been dissected out of the body, the voltage is usually lower still, about −60 to −65 mV, probably as a result of minor damage. The majority of membrane potentials in squid nerves discussed in this and the next chapters have been measured with capillary tubes about 0.1 mm in diameter, filled with salt solution. The tube can be inserted longitudinally through the cut end of an axon and pushed for various distances (several centimeters) along its axis without causing noticeable injury (Figure 1).

A more thorough test of the Bernstein hypothesis is to see how the potassium concentration ratio affects the membrane potential. From the Nernst equation one would expect that when the external or internal potassium concentration is changed by a factor of 10, the membrane potential should change by 58 mV at a temperature of about 20° C. In Figure 2, the external potassium concentration is plotted on a log scale

[6]Hodgkin, A.L. 1973. *Proc. R. Soc. Lond.* B *183*:1–19.

Table 1.

CONCENTRATIONS OF IONS INSIDE AND OUTSIDE FRESHLY ISOLATED AXONS OF SQUID

	Concentration (mM)		
ION	AXOPLASM	BLOOD	SEAWATER
Potassium	400	20	10
Sodium	50	440	460
Chloride	40–150	560	540
Calcium	0.3×10^{-3}*	10	10

*The precise value of ionized intracellular calcium is not known. Data from Hodgkin (1964) and Baker, Hodgkin, and Ridgway (1971).

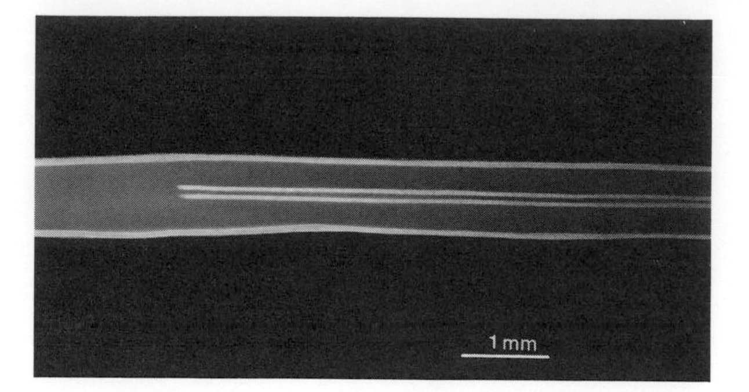

1

RECORDING ELECTRODE inside an isolated giant axon of the squid. (From Hodgkin and Keynes, 1956)

and the membrane potential is shown along the ordinate. The experimental points lie close to the theoretical line except in the region of low external potassium concentrations.

If the resting potential arises from potassium gradients across the membrane separating the interior and the outside, a critical test is to change not merely the outside concentration but also that inside the cell. This improbable sounding experiment was actually performed successfully on the squid axon by Baker, Hodgkin, and Shaw.[7] The axon is large enough for its axoplasm to be squeezed out and the nerve cannulated and perfused with selected solutions without destroying the properties of the surface membrane. With internal solutions containing the normal potassium concentration (but not the usual proteins), such axons continue to give many thousands of normal action potentials when stimulated, provided the membrane is not damaged. Figure 3 depicts the impulses in a normal (*B*) and a perfused (*A*) axon. This result dramatically emphasizes the importance of the membrane in signaling; in the short run, no special role in the conduction of impulses need be assigned the axoplasm. A wide range of internal potassium concentrations has been tested in perfused axons, and these results are also in agreement with those predicted by the Nernst equation. For example, when the

[7]Baker, P. F., Hodgkin, A. L., and Shaw, T. I. 1962. *J. Physiol.* *164*:330–354, 355–374.

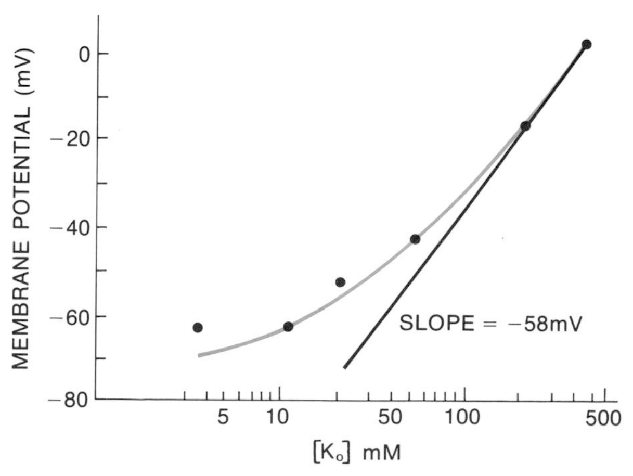

2

MEMBRANE POTENTIAL AND EXTERNAL POTASSIUM concentration in a squid, axon plotted on a log scale. The solid straight line is drawn with a slope of −58 mV, according to the Nernst equation. Because the membrane is also slightly permeable to sodium and chloride, the points deviate from the slope of −58 mV especially at low external potassium concentrations (see text). (After Hodgkin and Keynes, 1955)

external potassium concentration is high (540 mM) and the internal is low (100 mM) (in other words when the K_o/K_i ratio is 5.4:1 instead of 1:20), the membrane potential is reversed and the outside becomes negative with respect to the inside. Further, from the Nernst equation one would expect the membrane potential to be zero when the external and internal potassium concentrations are equal; experimentally, it was found that the resting potential is abolished when the inside and outside potassium concentrations are equal over the range between 10 and 540 mM.

Alterations in the ratios of the sodium or chloride concentration do not produce comparable changes in the resting potential (see below). From these results we can infer that the membrane behaves as though it were permeable to potassium but only slightly permeable to sodium.

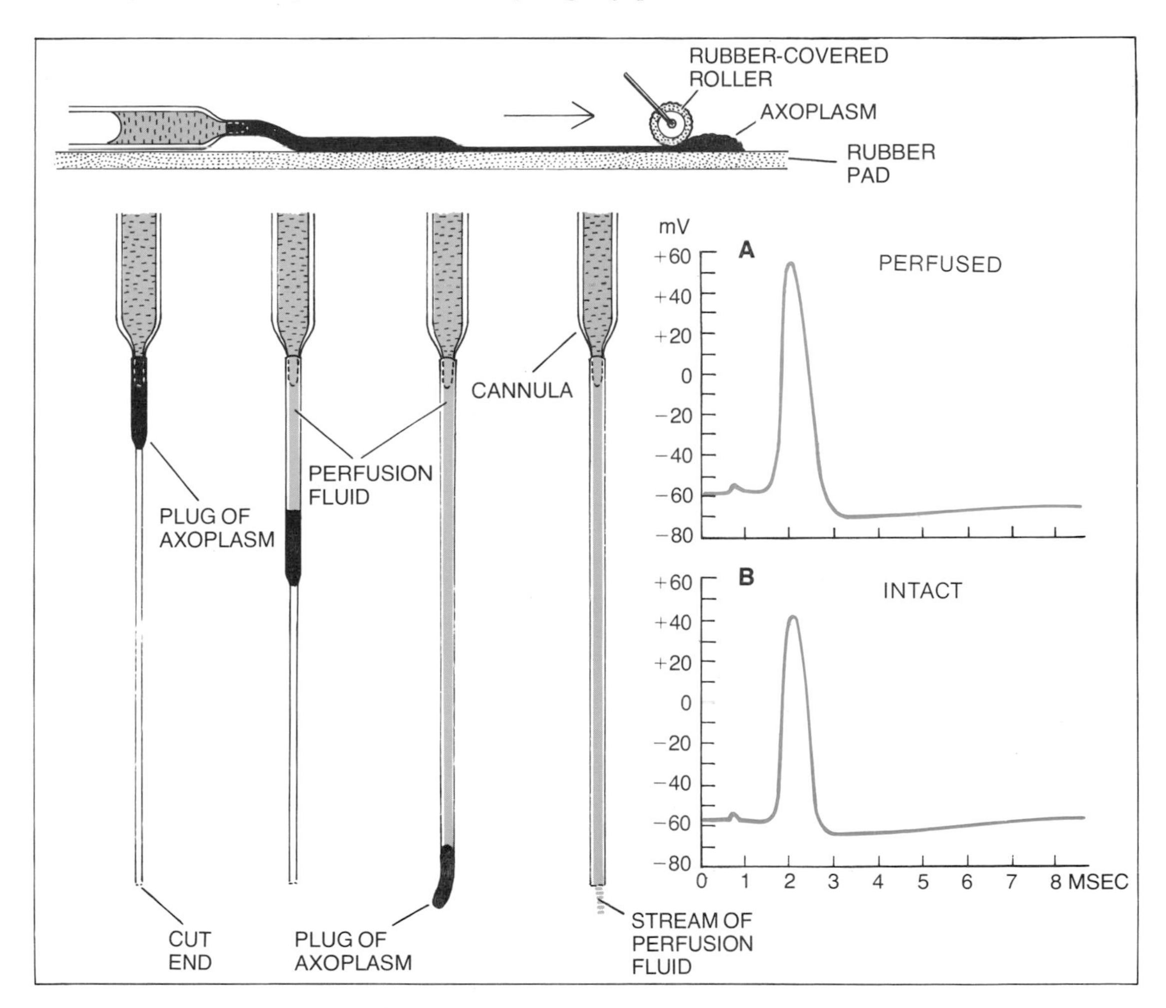

AXOPLASM BEING EXTRUDED from a squid axon, which is then cannulated and internally perfused. Upper record (A) is an impulse from a perfused axon; lower record (B) is from a normal axon. (After Baker, Hodgkin, and Shaw, 1962) 3

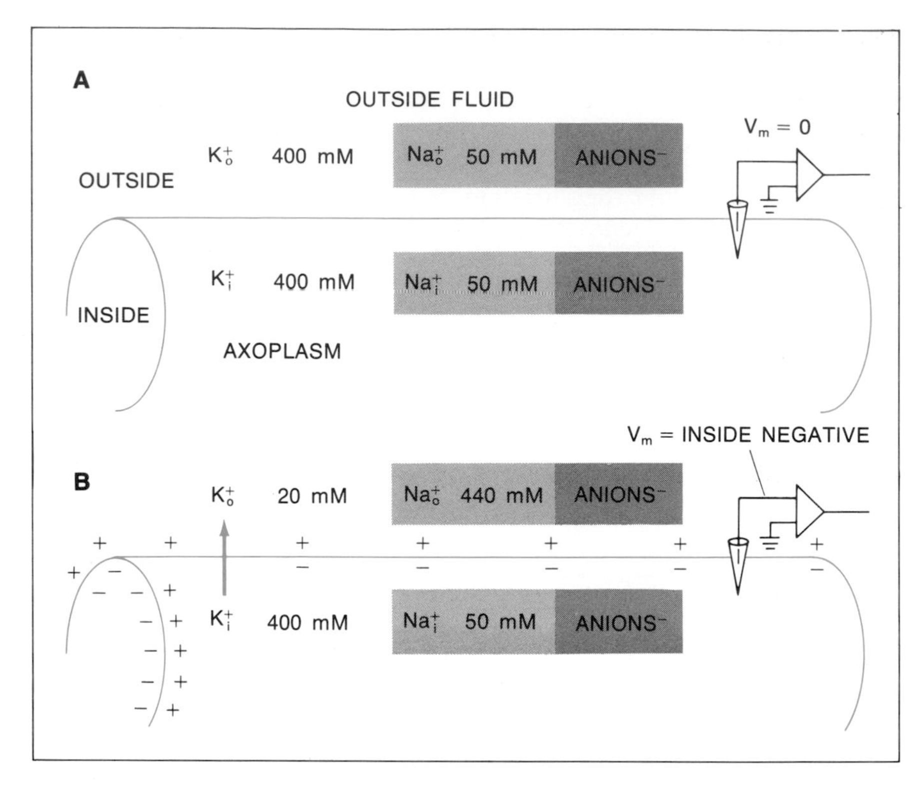

4 **ORIGIN OF RESTING POTENTIAL for a cell membrane permeable to potassium (K^+) but not to sodium (Na^+) or anions− (for simplicity these are not specified). A. The resting potential is zero for the cell bathed in fluid resembling axoplasm. B. When the axon is transferred to seawater, positively charged potassium ions move out along their concentration gradient. Positively charged sodium ions and anions cannot, however, move across the membrane, and as a result the inside fluid becomes negative with respect to the outside.**

The contribution of sodium and chloride ions to the resting potential is discussed later.

Origin of electrochemical potentials

How does the distribution of potassium ions actually cause an electrical potential? Full descriptions of the Nernst equation are given in textbooks of physical chemistry; for a simple intuitive explanation, it is useful to consider an idealized experiment and assume (1) that the axoplasm contains a high concentration of potassium, a low concentration of sodium and chloride, and a high concentration of internal anions (Table 1, Figure 4), and (2) that the membrane is completely impermeable to sodium and to all the internal anions, ignoring the contribution of chloride ions. (In fact, there is a small but significant permeability to sodium and chloride, as will be shown.) If the nerve is placed into a solution with the same ionic composition as its axoplasm (Figure 4*A*), the resting potential is, as expected, zero. If the nerve is next, put into seawater containing sodium chloride and a low concentration of potassium, the ions arrange themselves in a predictable manner, according to the electrical and chemical forces acting upon them. Sodium and the internal anions are not able to move across the membrane, but potassium ions tend to leave the cell since the concentration gradient is directed outward (20 times more potassium inside the cell than outside). However, as potassium ions move out they cannot be accompanied by the internal anions (Figure 4*B*). The migration of positively charged potassium ions constitutes an outward current ($I_{K\ out}$). (The term OUTWARD implies that positive charges are leaving the cell or negative charges entering it.) This current influences the distribution of charge across the membrane, making the inside solution more negative with respect to the outside. Once this electrical potential is sufficiently large, it stops any further net outflow of potassium ions. At this point the potential, E_K, exactly opposes the outwardly directed concentration gradient.

This scheme is shown graphically (Figure 5) by two vectors of the same size pointing in opposite directions. One represents the concentration gradient tending to push positively charged potassium ions out; the other represents the opposing electrical potential, pushing positively charged ions in. This is a verbal description of the Nernst equation in which E_K stands for the electrical potential and $(RT/zF)\ \log_e\ (K_o/K_i)$ for the concentration gradient converted to electrical terms. The resemblance to the gas laws is not fortuitous: one equation refers to the work required to move a charged particle along a concentration gradient, the other to the movement of a molecule of gas along a pressure gradient.

An important consequence of the Nernst equation is now clear: when the membrane potential is exactly at the equilibrium potential (E_K), the electrical potential exactly balances the concentration gradient, and there is no net movement of potassium either into or out of the cell (Figure 5*A*). At this value of the membrane potential an individual potassium ion might move in or out, but the net potassium movements are exactly balanced ($I_K = 0$). This then is the equilibrium potential for an ion when it carries no *net* charge in either direction. On the other hand, when the concentration gradient is not exactly balanced by the electrical potential, potassium ions move in or out and carry a current across the membrane ($I_{K\ in}$, $I_{K\ out}$). No work is required to keep the membrane potential steady at the equilibrium potential; it stays there unless the potential is displaced from this value or the concentration gradient is changed.

Continuing such an idealized experiment, one can examine the equilibrium situation in more detail by imagining what happens if a pulse of sodium ions is injected into the cell through a microelectrode (Chapter 12). The entry of positively charged impermeant ions reduces the membrane potential. But E_K is defined as the exact electrical potential required to balance the 20:1 potassium concentration gradient. Consequently, when the membrane potential falls ($V_m < E_K$), the concentration gradient that pushes potassium ions out of the cell is no

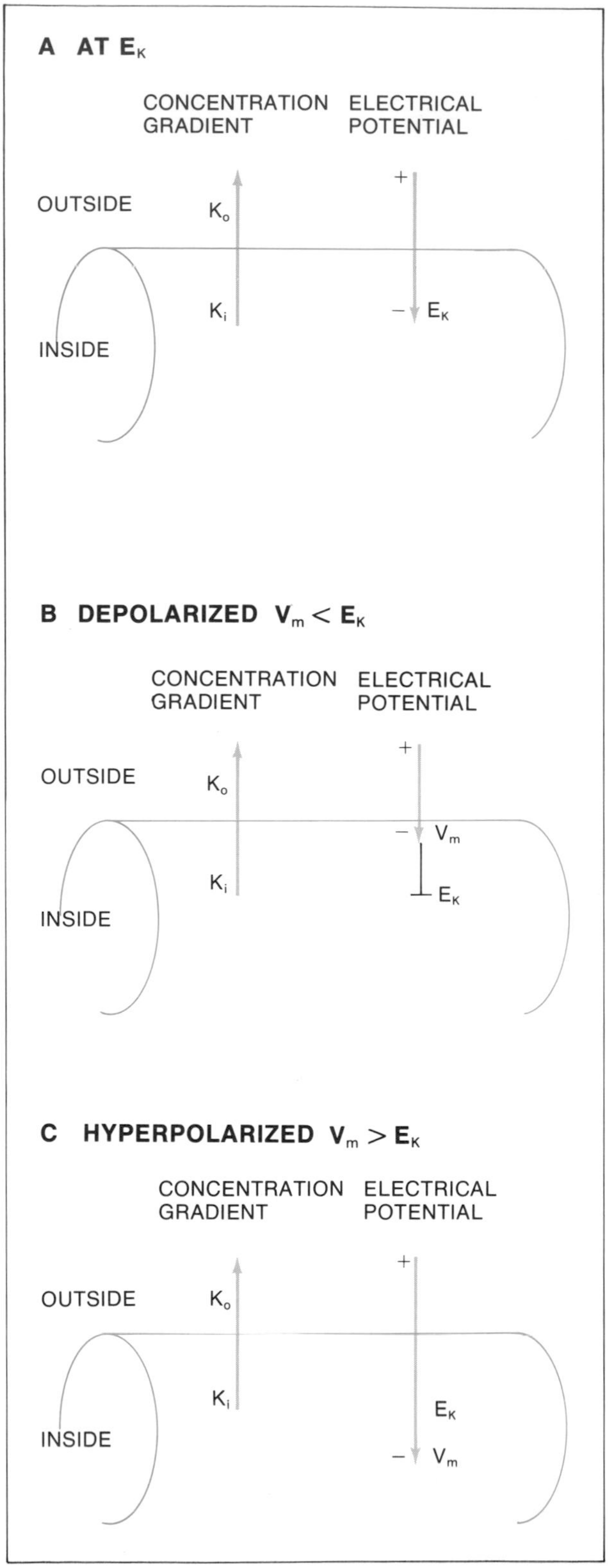

5

CONCENTRATION AND ELECTRICAL GRADIENTS for potassium. Direction of potassium movement resulting from changes in membrane potential. In A the electrical and concentration gradients are opposite and equal, indicated by length of arrows. In B the membrane has been depolarized ($V_m < E_K$) and potassium is therefore able to leak out along the concentration gradient ($I_{K\ out}$). In C the membrane has been hyperpolarized ($V_m > E_K$), and potassium ions move into the cell ($I_{K\ in}$).

longer opposed by an equivalent electrical potential (Figure 5B). The result is an outward movement of potassium ions, that is, an outward current ($I_{K\ out}$). As the positively charged potassium ions leak out, the inside of the cell again becomes more negative relative to the outside; this continues until the membrane potential once more reaches the equilibrium value E_K, where the electrical potential and the concentration gradient are exactly balanced. Conversely, if the membrane potential is displaced in the opposite direction—hyperpolarized through the injection of some impermeant anion—the potential becomes larger than the equilibrium potential for potassium ions ($V_m > E_K$). The effect of the electrical potential on potassium, pushing it inward, is now larger than the concentration gradient in the opposite direction; potassium ions therefore enter the cell (an *inward* current) against the concentration gradient (Figure 5C). This movement continues until V_m is once more equal to E_K, the equilibrium potential for potassium ions. It is, therefore, the difference between E_K and V_m that determines how potassium ions move across the cell membrane—inward, outward, or not at all. The term $V_m - E_K$ is known as the DRIVING FORCE.

Contribution of sodium ions to resting potential

For simplicity, the previous section discusses the nerve membrane according to the Bernstein hypothesis as if it were completely impermeable to sodium. This, however, is not so; the membrane is slightly permeable to sodium. Both the concentration gradient and the membrane potential (inside negative, outside positive) tend to drive sodium into the cell (Figure 6); the reason relatively little can enter at rest is that the sodium permeability is low. Nevertheless, the consequences of sodium ions trickling slowly into the cell at a steady rate are important. As the positive charges enter they constitute an inward sodium current ($I_{Na\ in}$) that reduces the membrane potential. In other words, the potential is less than E_K, and the concentration gradient is able to drive potassium ions out at exactly the same rate that sodium leaks in ($I_K = -I_{Na}$, where positive and negative indicate opposite directions of current). Under these conditions the membrane potential (V_m) reaches a steady state where there is no net current flow. It is not a true equilibrium potential, however, because as sodium enters and potassium leaves, the internal concentrations gradually change. Nevertheless, the membrane potential continues at its steady level for seconds or minutes because the leakage rate for sodium is so low that it takes a relatively long time for the internal concentrations to change appreciably. To keep the system from running down and to maintain concentrations constant in the long run, there must be a mechanism that transports sodium out as fast as it enters. This is achieved by the sodium pump, discussed in Chapter 12. Moreover, as in several other systems, this pump is coupled, carrying potassium into and sodium out of the cell.

Sodium ions and the action potential

Bernstein's original hypothesis comes close to explaining the ionic basis of the resting potential. But since a small resting permeability for sodium ions does exist, the hypothesis clearly needs modification. Furthermore, a satisfactory explanation is needed for the action potential,

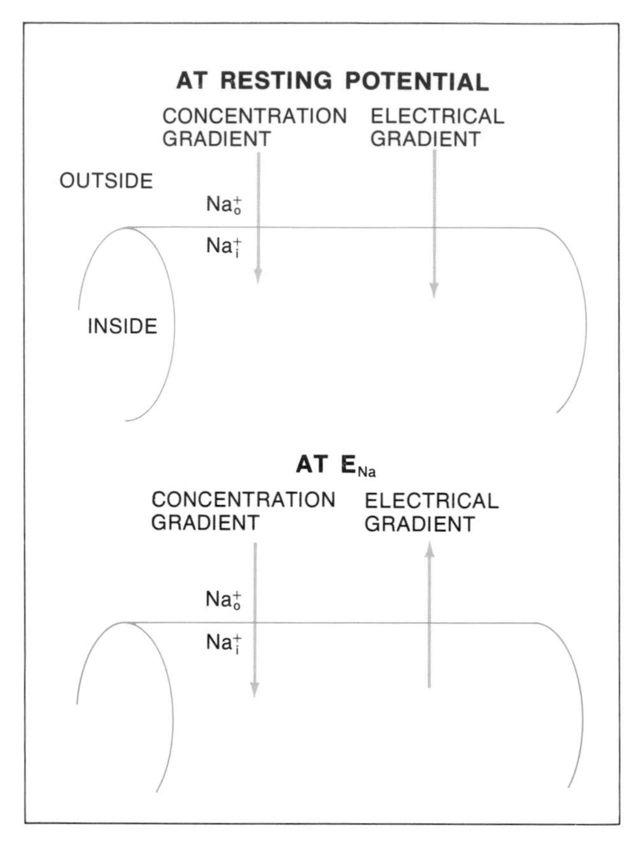

6

CONCENTRATION AND ELECTRICAL GRADIENTS for sodium. At resting potential both electrical and concentration gradients for sodium ions are directed inward. At E_{Na} the potential across the membrane exactly balances the concentration gradient.

which Bernstein attributed to a breakdown of the selective permeability of the membrane.

At the same time as Bernstein proposed his scheme in 1902, Overton[8] made the important discovery that sodium ions are necessary for nerve and muscle cells to maintain their electrical excitability and that lithium is the only ion that can substitute for sodium. Overton suggested that the action potential might come about through sodium ions running into the cell, but he pointed out that excitable cells conducted literally millions of action potentials during the life of the animal without becoming richer in intracellular sodium.

The role of sodium ions for impulse conduction could not be clarified until two important observations were made: (1) in 1939 Curtis and Cole[9] and Hodgkin and Huxley[10] showed that the action potential is actually larger than the resting potential; there occurs an OVERSHOOT that makes the inside of the nerve cell transiently positive with respect to the outside; (2) in 1949 Hodgkin and Katz[11] showed that changes in external sodium concentration affect the amplitude of the overshoot of

---

[8]Overton, E. 1902. *Pflügers Arch. 92*:346–386.
[9]Curtis, H. J., and Cole, K. S. 1940. *J. Cell. Comp. Physiol. 15*:147–157.
[10]Hodgkin, A. L., and Huxley, A. F. 1939. *Nature 144*:710–711.
[11]Hodgkin, A. L., and Katz, B. 1949. *J. Physiol. 108*:37–77.

A

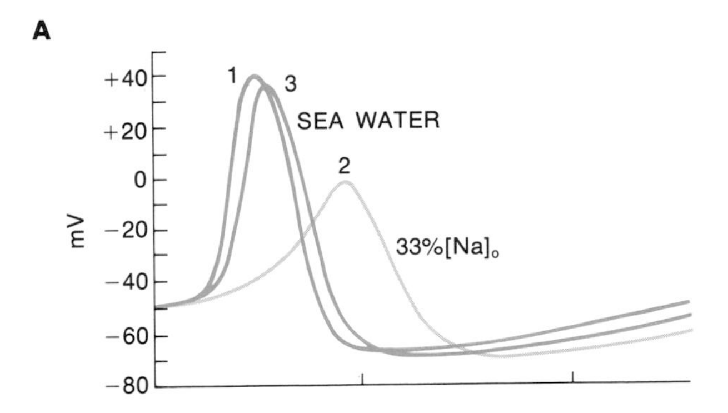

B

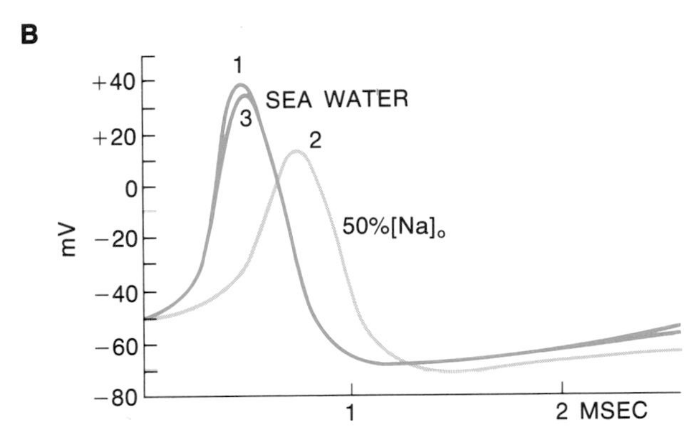

7

ROLE OF SODIUM in conduction of an action potential in a squid axon. In record 2 of A, the external sodium concentration is reduced to one third of the normal and in B it is reduced to half. Records 1 and 3 in each case are the control values in normal seawater before and after testing low sodium solution. (From Hodgkin and Katz, 1949)

the action potential (Figure 7). Over a wide range the relation agrees well with that predicted from the Nernst equation for sodium:

$$E_{Na} = 58 \log_{10} \frac{Na_o}{Na_i} = +55 \text{ mV}$$

Similar results have also been obtained by Baker, Hodgkin, and Shaw,[7] who varied the INTERNAL sodium concentration of perfused squid axons. Together these results show that during the action potential the selective permeability of the membrane changes so that sodium ions dominate the membrane potential.

It is now possible to predict the ionic movements and changes in potential that follow from changes in selective permeability. These are shown diagrammatically in Figure 8. If a membrane that is permeable only to potassium suddenly becomes (by an undefined mechanism) permeable only to sodium ions, sodium dominates the membrane potential (Figure 8*B*). Starting at −75 mV (E_K), both the concentration gradient and the membrane potential tend to push sodium ions into the cell. Notice that once the permeability barrier is removed the cell need use no metabolic energy to allow sodium ions to enter. They enter by flowing downhill along their electrochemical gradient. However, as more sodium runs in, the inside becomes more positive until eventually the electrical potential is large enough to prevent further inward sodium movement. At this voltage of +55 mV (inside positive) (E_{Na}), the concentration gradient is exactly balanced by the electrical potential, which remains constant unless other changes are brought about.

What happens if the membrane somehow reverts to its original state by a sudden DECREASE in sodium permeability and an INCREASE in permeability to potassium (Figure 8C)? The electrochemical forces on potassium at E_{Na} are such that both the concentration gradient and electrical potential are in the same direction and push potassium ions out of the cell (the inside of the cell being positive to the outside). As potassium moves out, however, the inside becomes less positive and eventually becomes negative; the leakage continues until the potential again exactly balances the concentration gradient with the membrane potential at E_K. The important conclusion is that THE POTENTIAL ACROSS THE MEMBRANE CAN BE DRIVEN TO E_{Na} OR TO E_K simply by changing the relative permeabilities to sodium or to potassium ions without the expenditure of metabolic energy. Use of such a mechanism can shift the membrane potential anywhere between −75 mV (E_K) and +55 mV (E_{Na}).

Changes in intracellular concentration resulting from ionic movements

If the interior of the nerve gains sodium during the rising phase of the action potential and loses potassium during its falling phase, one might expect the intracellular potassium and sodium concentrations to change, as well as E_K and E_{Na}. If this were true, after a nerve impulse the membrane potential should not return exactly to its original value. The amount of concentration change depends on the number of ions required for a change in membrane potential of ±130 mV. The number is so small that for most purposes intracellular concentration changes can be ignored in the short run. This can be shown in two ways: by a simple calculation and by direct measurements.

The calculation depends on the fact that the nerve cell membrane has an appreciable electrical capacity to store and separate charges (Appendix and Chapter 7). For a capacitor, be it in a radio set or in a living membrane, the potential difference between the two plates (V) depends

8

INFLUENCE OF POTASSIUM AND SODIUM PERMEABILITY on the membrane potential. At E_K (−75 mV) potassium permeability dominates the membrane potential; at E_{Na} (+55 mV) membrane potential is determined by the sodium permeability. The values of −75 mV and +55 mV are derived from the Nernst equation for the concentrations of potassium and sodium on the two sides of the membrane. At B an unspecified mechanism increases the sodium permeability relative to potassium. Sodium ions flow in until the potential reaches E_{Na}. No further ion flow occurs until C, when the original membrane properties are restored (relative potassium permeability high) and potassium ions flow out until the potential reaches E_K again. Depending on the permeability ratio for potassium and sodium, the potential can be shifted to any value between E_{Na} and E_K (see text).

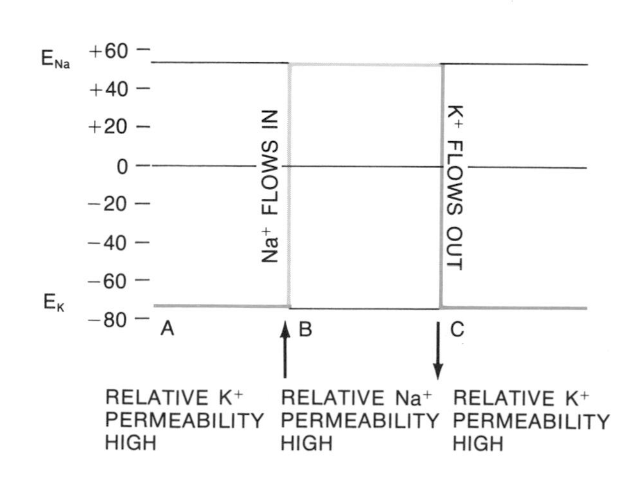

on the amount of charge (Q) held by the capacitance (C). A simple equation relates the three variables: $V = Q/C$. Every change in the potential across the membrane must therefore be brought about by the movement of charges onto or off the capacitor. This process is not instantaneous; it takes a finite time and therefore causes the slowing down of electrical signals in nerves, as mentioned earlier.

The membrane capacitance meaured in squid nerve (1 μF/sq cm) is similar to the value found in most other nerve fibers within the body. Knowing this, one can calculate how many charges or ions have to be displaced across 1 sq cm of membrane to change its potential by 130 mV from E_K (−75 mV) to E_{Na} (+55 mV), the size of an action potential. Solving the equation for charges $Q = CV = 10^{-6}$ F $\times$ 0.13 V = 1.3×10^{-7} coulombs for each square centimeter of membrane for each impulse. The electrical charges can be converted to the number of ions by means of the faraday (F) (used earlier in the Nernst equation), a constant which gives the proportionality between ions and charge as 96,516 coulombs per mole of univalent ion. To produce a potential change of 130 mV in either direction, the number of sodium or potassium ions that must move across 1 sq cm of membrane is therefore about 1.3×10^{-12} moles.

Does this constitute a large or a small proportion of the ions already present inside the cell? From the diameter of the squid nerve fiber (0.5 to 1.0 mm) and the internal concentration of potassium ions, one can calculate that 1 sq cm of membrane encloses about 10^{-5} moles of potassium. It is clear that to produce the potential change associated with an action potential about 1 potassium ion in 10^7 must be replaced by sodium. Only negligible changes in intracellular potassium are therefore produced by a large number of action potentials. For a smaller nerve fiber the surface/volume ratio is less favorable but the number of potassium ions exchanged for sodium is still only about 1 in 3,000 in a nerve fiber with a diameter of 1μm for a single impulse.

These calculations are satisfactorily supported by measurements of the amount of radioactive potassium and sodium leaving and entering single squid fibers during a nerve impulse. The value of 3×10^{-12} to 4×10^{-12} moles/sq cm of potassium or sodium[12] is somewhat larger than the value predicted theoretically. This is not surprising because in order to make the calculation it was assumed arbitrarily that the membrane became uniquely permeable either to sodium or to potassium, and that the two ions could not move simultaneously in opposite directions. If the permeability changes were not completely separated in time, more ions would have to move to produce the same potential change.

Again the important conclusion is emphasized: only very small changes occur in internal sodium or potassium concentrations with single impulses; for many purposes E_K and E_{Na} can be considered constant (but see Chapter 13 for a discussion of changes in external potassium).

---

[12]Keynes, R. D., and Lewis, P. R. 1951. *J. Physiol.* *114*:151–182.

Contribution of chloride ions

The distribution of chloride ions has so far been neglected. At the resting potential, close to E_K, the concentration gradient of chloride ions is directed inward, while the electrical potential tends to drive these ions outward (chloride being negatively charged). If the membrane were permeable to chloride and if it were not pumped, the ion would be distributed passively according to the membrane potential. At the value of −65 mV set by the movements of sodium and potassium, the internal chloride would reach a concentration that satisfies the Nernst equation:

$$E_{Cl} = V_m = -58 \log_{10} \frac{Cl_o}{Cl_i}$$

If the membrane potential changed for any reason, chloride would move into or out of the cell so that it was once again in equilibrium at the new potential. At present chloride movements need not be considered in detail because the effects are small and, more important, the chloride permeability does not change during the action potential. The changes in chloride permeability that occur at inhibitory synapses are discussed more fully in Chapter 8.

Relation of permeability and membrane potential

Up to this point it has been shown that the distribution of potassium ions is mainly responsible for determining the resting potential, while an influx of sodium ions gives rise to the action potential. The key to controlling the membrane potential is the selective permeability of the membrane. It is not, however, immediately obvious how one can express permeability in electrical terms so that it can be related to membrane potential (V_m) or ionic current (I). A convenient measure of membrane permeability is the conductance (g) of the membrane to an ion; this is the reciprocal of electrical resistance (ohms) and is measured in reciprocal ohms (mhos); conductance is an index of the ease with which charges can flow. In a simple circuit with a battery and a resistor, the current $I = V/R$, where V is potential and R is resistance. Using conductance, Ohm's law can be rewritten $I = gV$. Thus, the conductance of the membrane to a particular ion can be used to relate current and the electrical driving force with the general formula:

$$I_{ion} = g_{ion} (V_m - E_{ion})$$

When the membrane potential (V_m) is at the equilibrium potential for a particular ion (E_{ion}), the driving force is zero and that ion carries no net current across the membrane no matter how great the conductance. For sodium and potassium ions the equations are:

$$I_{Na} = g_{Na} (V_m - E_{Na})$$

$$I_K = g_K (V_m - E_K)$$

These equations have been of considerable use for a number of reasons. First, at any one instant the relation between the membrane current and V_m is linear—that is, the conductance does not change instantaneously (Chapter 6). This means that the conductance can be derived by measuring the current carried by an ion through the membrane and the membrane potential. Second, one can calculate the relative conductances to sodium and potassium at any steady value of membrane potential (again ignoring the small contribution of chloride). For example, at a resting potential lower than E_K (say −60 mV), a small amount of sodium slowly leaks into the cell, and this inward current is exactly matched by an outward potassium current. In this case, or whenever the potential is steady, the net current flow across the membrane is zero. Consequently:

$$I_{Na} = -I_K$$

and

$$g_{Na}\,(V_m - E_{Na}) = -g_K\,(V_m - E_K)$$

By simple algebra:

$$V_m = \frac{(g_{Na}/g_K)(E_{Na}) + E_K}{g_{Na}/g_K + 1}$$

where E_{Na} and E_K are effectively constants. The only variable in the equation is the ratio g_{Na}/g_K. What would happen to V_m if the sodium and potassium conductances became equal? Starting at E_K, sodium ions move into and potassium out of the cell, so that the membrane potential ends up at a value midway between E_{Na} and E_K —that is, $(E_{Na} + E_K)/2$. This is similar to the situation at the neuromuscular junction, where acetylcholine (ACh) makes the membrane simultaneously more permeable to both sodium and potassium ions, and the membrane potential comes to lie between E_{Na} and E_K (Chapter 8). Using these equations, any value of membrane potential can be expressed in terms of the permeability of the membrane to two ions and their equilibrium potentials.

In this and other discussions we use the terms ionic permeability and ionic conductance interchangeably. This is a convenient simplification that has proved useful experimentally under many but not all conditions. Thus, if the membrane were permeable to an ion that was not present in either the internal or external fluids, then the conductance would, of course, be zero. The permeability—that is, the ease with which the substance could cross—would still be high (see Katz[13] for a fuller discussion).

Ionic currents and conductances

An electrical model can now be drawn of a patch of the membrane.[14] In Figure 9, E_{Na} represents a sodium battery; its positive pole faces inward because E_{Na} from the Nernst equation is 55 mV (inside positive). E_K and E_{Cl} are the potassium and the chloride batteries with their positive poles facing outward. The conductances for sodium and potassium have arrows which indicate that they can be changed, but the chloride conductance is fixed. The electrical capacity of the membrane is shown

[13] Katz, B. 1966. *Nerve, Muscle, and Synapse.* McGraw-Hill, New York, p. 72.
[14] Hodgkin, A. L. 1964. *The Conduction of the Nervous Impulse.* Liverpool University Press, Liverpool.

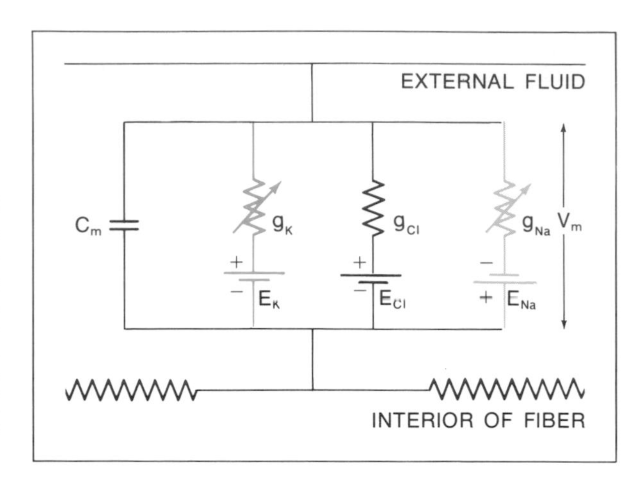

9

EQUIVALENT CIRCUIT for a patch of excitable membrane. E_K, E_{Cl}, and E_{Na} are the potentials determined by the Nernst equation. The resistors represent 1/g for each of the ions, and the arrows indicate that they are variable for potassium and sodium. C_m is the membrane capacity and V_m the membrane potential. The external fluid and the internal fluid connect this patch of membrane with the neighboring elements.

by C_m. Capacitative currents flow through it only while the membrane potential is actually changing.

With this diagram one can readily determine in which direction ions will move at any value of membrane potential (V_m). It was already shown that whenever the potential across the membrane is equal to the equilibrium potential for an ion, that particular ion cannot carry net current across the membrane. For example, when the membrane potential is at E_K (−75 mV inside), then the two ends of the g_K conductance channel are at the same potential ($E_K = V_m$); no current flows through this part of the circuit. Similarly, if V_m is less than −75 mV, a positive potassium current flows outward; that is, potassium ions leave the cell ($V_m < E_K$). If, on the other hand, the membrane potential is larger than E_K, potassium ions flow into the cell as an inward current ($V_m > E_K$). Similar considerations apply to the movement of sodium and chloride ions through the membrane. When the membrane potential is at E_K, some sodium ions tend to enter the cell, but the low sodium permeability (the small value of g_{Na}) limits the amount of current that can actually flow through this channel.

In cells other than the squid axon similar principles operate, but the resting permeabilities to certain ions are different. For example, chloride ions make a significant contribution to the resting potential of skeletal muscle fibers; in many invertebrate neurons, the resting potentials are lower, about −45 mV, owing to a high resting sodium conductance. On the other hand, glial cells (the satellite cells in the brain; Chapter 13) have resting potentials at E_K and behave as though their membranes were permeable to potassium only.

SUGGESTED READING

Papers marked with an asterisk (*) are reprinted in Cooke, I., and Lipkin, M. 1972. *Cellular Neurophysiology*. Holt, New York.

General

Hodgkin, A. L. 1951. The ionic basis of electrical activity in nerve and muscle. *Biol. Rev. 26*:339–409.

Katz, B. 1966. *Nerve, Muscle, and Synapse.* McGraw-Hill, New York, Chap. 4.
(Both provide a comprehensive description of the Nernst equation and contain many key references to older literature.)

Selected Original Papers

Baker, P. F., Hodgkin, A. L., and Shaw, T. I. 1962. Replacement of the axoplasm of giant nerve fibres with artificial solution. *J. Physiol. 164*:330–354.

*Baker, P. F., Hodgkin, A. L., and Shaw, T. I. 1962. The effects of changes in internal ionic concentrations on the electrical properties of perfused giant axons. *J. Physiol. 164*:355–374.

Hodgkin, A. L., and Horowicz, P. 1959. The influence of potassium and chloride ions on the membrane potential of single muscle fibres. *J. Physiol. 148*:127–160. (Especially useful for considering the part played by chloride.)

*Hodgkin, A. L., and Katz, B. 1949. The effect of sodium ions on the electrical activity of the giant axon of the squid. *J. Physiol. 108*:37–77. (This paper not only describes the effect of sodium on the action potential but also deals quantitatively with the influence of permeability on membrane potential.)

CHAPTER SIX

CONTROL OF MEMBRANE PERMEABILITY

The ionic mechanisms responsible for generating the nerve impulse have been quantitatively described in squid axons, largely through the use of the voltage clamp method, which sets and maintains the membrane potential at a particular value. As a result one can determine the current flow carried by each ion across the membrane after the potential has been changed to a new steady level. From this one can estimate the magnitude and time course of the permeability of the membrane to ions as a function of the electrical potential.

The selective permeabilities of the membrane to sodium and potassium, which determine the membrane potential, are influenced by the potential across the membrane. Depolarization increases sodium permeability, and also, more slowly, potassium permeability. The increase in permeability to sodium as a function of depolarization and time is transient and turns itself off by inactivation. The sodium and potassium permeability mechanisms in the membrane behave independently. This conclusion is reinforced by the action of a poison, tetrodotoxin, which selectively blocks the increase in sodium conductance produced by depolarization. Similarly, tetraethylammonium selectively blocks the potassium conductance mechanism.

The sequential timing and magnitude of the permeability changes account quantitatively for the rising and falling phases of the action potential. The dimensions of the permeability channels for sodium and their distribution in the membrane have been estimated. The opening of the ionic gates in response to depolarization appears to be brought about through the movement of charged particles within the membrane.

The findings on squid axons have helped to explain many of the basic properties of excitable membranes. Similar general principles also apply to other cells, such as vertebrate nerves, muscle fibers, and heart muscle.

Permeability and the membrane potential

Chapter 5 has shown that the potential of the neuronal membrane is determined by its permeability; this chapter presents evidence to show that the principal factor controlling the permeability is the membrane potential itself. The discussion is based chiefly on the initial experiments of Hodgkin and Huxley and on the more recent work of Armstrong, Chandler, Hille, Keynes, Moore, and their colleagues.

One clue to the regulation of permeability is that a depolarization of the membrane is required to trigger the action potential. Consider what happens if depolarization increases the sodium conductance (g_{Na}). This increases the rate of inward movement of sodium (I_{Na}), which further reduces the membrane potential, which in turn produces a further increase in g_{Na}, and so on (Figure 1). This explosive process resembles a gunpowder explosion in which heat accelerates the chemical reaction, which in turn produces more heat, which in turn accelerates the chemical reaction (positive feedback). If g_{Na} is not turned off, the membrane potential approaches and then stays at E_{Na}, the sodium equilibrium potential.

What happens if depolarization increases the permeability for potassium ions (g_K) alone? As the membrane potential becomes displaced from the equilibrium potential (E_K), potassium ions move outward ($I_{K\ out}$ increases), which leads to a repolarization—that is, an increase in membrane potential. In other words, the effect tends to restore the previous state and drive the voltage back to its original value, E_K; this restitution process is therefore a form of negative feedback. In theory, then, the rising and falling phases of the action potential can be explained by assuming that depolarization causes an initial increase in the permeability to sodium, which then is turned off and followed by an increase

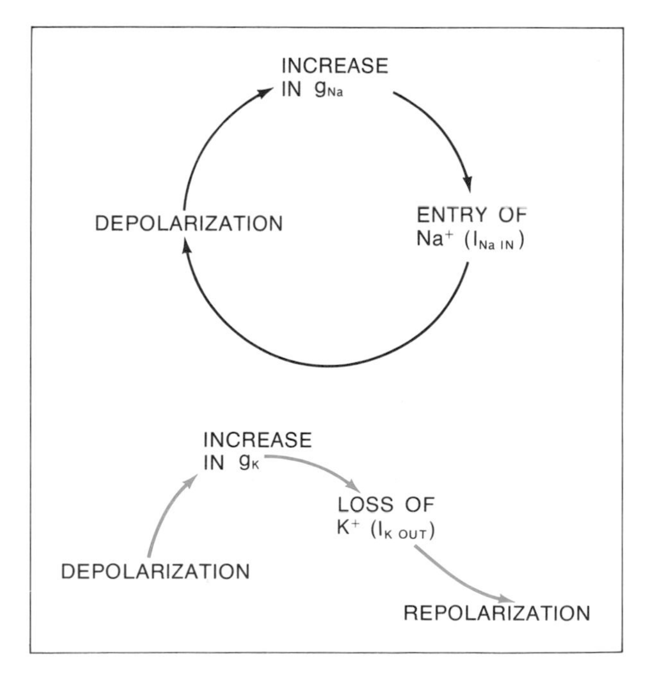

1
EFFECTS OF INCREASING g_{Na} and g_K on membrane potential. Sodium entry reinforces depolarization, while potassium efflux leads to restitution (repolarization).

in the permeability to potassium, which restores the membrane potential. This constitutes a possible mechanism that can shift the membrane potential anywhere between E_{Na} and E_K, for example, between +55 mV and −75 mV, depending on the concentration gradients discussed in Chapter 5.

The following discussion demonstrates that changes in sodium and potassium conductances account quantitatively for the action potential and that they are large enough and correctly timed. A simultaneous increase in permeability to both sodium and potassium could not give rise to the action potential, since the membrane potential would be driven to a value somewhere between E_K and E_{Na} (Chapter 5).

The principal problem was solved by Hodgkin and Huxley (see Suggested Reading at the end of this chapter), who determined how g_{Na} and g_K vary with membrane potential, current, and time. After they had obtained a full knowledge of these variables, they were able to reconstruct the action potential from first principles and to explain many other properties of excitable membranes, such as threshold, refractory period, undershoot, and propagation (see later).

At first it might seem that to measure conductance one need only determine the amount of current flowing into and out of the membrane at various levels of potential, since for each ion:

$$I_{ion} = g_{ion} (V_m - E_{ion})$$

It is clear, however, that three major problems must be solved experimentally:

1. How can one measure separately the currents carried by sodium and by potassium at each instant, since the total current across the membrane is the sum of all the individual ionic currents?
2. The current (I) changes the membrane potential (V_m), which in turn changes the conductance (g). How, therefore, can one hold the system steady and determine systematically the way in which conductance depends on potential? The difficulty is severe, as the action potential

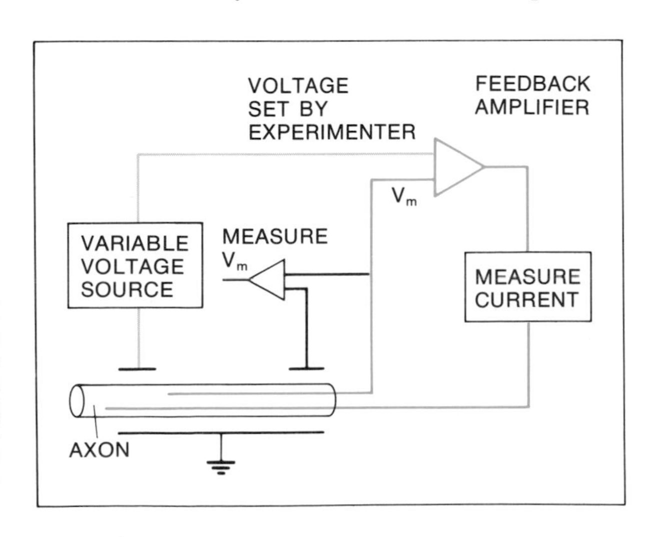

2
VOLTAGE CLAMP TECHNIQUE for squid axon. One longitudinal electrode measures V_m and is connected to the feedback amplifier, which passes current into the axon through a second electrode. With this circuit the membrane potential can be displaced abruptly and held constant at a new value while the current flowing across the membrane is measured (see text and Figure 4).

rises at a rate of several hundred volts per second; accordingly a system that operates in microseconds is needed. Of the variables that can be controlled, the membrane potential seems the most suitable, since current can be passed within the required microsecond range so as to keep the membrane at a particular value of potential; it is difficult to clamp conductance, and time is a variable that cannot be altered. As is shown later, the current itself does not influence the conductance, which is affected by membrane potential and time.

3. A variable fraction of the current during an impulse flows through the membrane capacity and not through the ionic permeability channels. How can this current be eliminated or subtracted from the total so that conductance can be measured? The way this can be done is discussed below.

Principle of the voltage clamp

The technique for holding the membrane potential at a constant value, the VOLTAGE CLAMP, was devised by Cole and his colleagues[1,2] and further developed by Hodgkin, Huxley, and Katz.[3] A fine silver wire electrode connected to an oscilloscope for the measurement of membrane potential is inserted longitudinally into a squid axon. The signal leads to a special differential amplifier that delivers current to the inside of the cell through a second longitudinal electrode. The essential feature of this circuit is that current flows through its output when its two inputs are not at the same potential (Figure 2). With one input from the membrane potential and the other from a potential source with a variable setting, the system works as follows. When the setting on the variable voltage source is the same as the resting potential (for example, −65 mV), no current flows in the external circuit. When the setting of the variable potential is changed from −65 mV to −15 mV (the new desired "clamped" potential), a difference of potential appears between the two inputs to the amplifier, which therefore passes a current in such a direction as to drive the inside of the nerve fiber to the new potential of −15 mV. This is a process of negative feedback. However, in the course of depolarizing the nerve fiber to −15 mV under normal circumstances, an action potential would be initiated. Sodium ions run in and make the inside more positive. The response of the electrical system to this is once more to hold the potential at −15 mV, but to do so it passes current in the opposite direction (preventing the potential from drifting toward −14 mV). Thus, the feedback amplifier sets the membrane potential to a particular voltage and keeps it there by compensating for ionic currents through the membrane.

In principle, this system is like a thermostat. With a thermostatically regulated bath, a variable control is adjusted to the desired value, while a thermometer records the actual temperature. When the temperature in the bath falls below the temperature setting on the control, heat is applied to the bath; if the temperature rises, the bath is cooled.

---

[1]Marmont, G. 1949. *J. Cell. Comp. Physiol.* *34*:351–382.

[2]Cole, K. S. 1968. *Membranes, Ions and Impulses.* University of California Press, Berkeley.

[3]Hodgkin, A. L., Huxley, A. F., and Katz, B. 1952. *J. Physiol.* *116*:424–448.

The voltage clamp measures the current the amplifier puts out to keep the membrane potential constant. This current is exactly equal and opposite to the current carried by ions flowing across the membrane. The voltage clamp, therefore, measures the total current flowing through the membrane and allows one to calculate g, since V_m, E_{Na}, and E_K are known. In the thermostat analogy, the heat produced by an explosion of gunpowder can be determined by measuring the heat that must be drawn off to keep the temperature of the bath constant.

Note that the experimental setup of longitudinal wire electrodes abolishes longitudinal current flow, which normally accompanies conduction from point to point along the nerve fiber (Chapter 7). As a result of the introduction of a silver wire, the interior of the axon becomes isopotential, and no current can flow in the longitudinal direction. All current is directed radially through the membrane. When the feedback system is not used with an intracellular wire in place, the axon still gives a standard action potential. The potential change, however, occurs across the whole area of membrane surrounding the wire. During the course of the impulse, no net current can flow into or out of the cell; therefore, the only charges that actually move are those involved in altering the voltage across the two sides of the membrane capacity. There is no circuit for current flow along the axon, into it, or out of it, since the inside fluid, like the outside fluid, is isopotential. Under these conditions, $I_C = -I_{\text{ionic}}$, where I_C is the current flowing through the capacitor and I_{ionic} is that flowing through the ionic conductance channels (see below). In effect this experiment, in which the membrane potential is allowed to change (that is, is not clamped), is a CURRENT CLAMP, because the net current flowing through the membrane is kept at zero once the initial stimulus has been delivered. The experiment demonstrates two important principles: (1) the cell can give impulses in the absence of longitudinal current, and (2) the sequence of changes observed during an impluse can occur in the absence of net current through the membrane.[3]

The voltage clamp and capacitative currents

At first the presence of the membrane capacity appears to pose a serious problem. A circuit diagram of the membrane (Figure 3) shows two pathways for current flow across the membrane: (1) the permeability channels through which ions move and (2) the membrane capacity. How can one use the current flowing through the membrane to measure ionic movements and permeability if at the same time some current is flowing through the capacity? What is the relative importance of these two currents at different times and voltages? The difficulty is overcome by the voltage clamp technique. With this technique the membrane potential is displaced from one value to another within a few microseconds and then held constant. As a result, only for this very brief period does current flow through the capacitance. The current (I_C) flowing through a capacitor (C) is a function of changes in potential (see Appendix). Thus:

$$I_C = C\frac{dV}{dt}$$

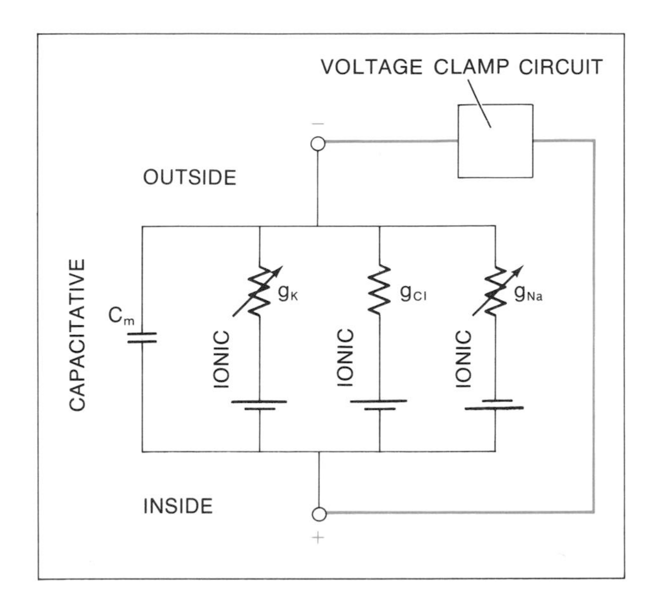

3

PATHWAYS FOR CURRENT FLOW through membrane. When the potential of the membrane is displaced from one value to another (for example, inside positive), capacitative current (I_C) flows only while the potential is changing. In particular, this means that capacitative currents last for only a few microseconds, whereas ionic currents begin after a delay and continue for milliseconds (see also Figure 4).

Therefore, once the potential has reached its steady value, *all* the current flowing through the membrane is carried by ions moving through permeability channels, since $I_C \doteq 0$. Only an abrupt change in membrane potential achieves this effect of diminishing I_C after a few microseconds. Analysis of membrane currents is far more complex if the voltage is changed gradually.

Ionic current as a function of membrane potential

How does current flow across the membrane of a squid axon when it is suddenly depolarized (within a few microseconds) from −65 mV to −9 mV, at which it is clamped (Figure 4)? The current consists of three phases: (1) first a brief outward surge lasting only a few microseconds while the membrane potential is being changed, (2) then a transient inward current, and (3) finally a delayed outward current that is maintained for as long as the membrane potential is clamped at −9 mV. The large, brief outward surge of current at the beginning, during the rising phase of the potential step, is a capacitative current produced by changing the charge on the membrane capacitance (Figure 4*B*; see also Chapter 7 and Appendix). Plainly, the next inward phase of the current must be caused by positive ions moving in or negative ions moving out of the axon, and the opposite holds for the outward phase of current that follows. In addition to these currents, there is a small constant LEAKAGE CURRENT that persists while the potential is kept at either depolarized or hyperpolarized values. This current represents nonspecific ion movements (for example, of chloride) and is linearly related to the change in potential.

Do these results fit with what is known about the action potential in an unclamped axon, particularly that the depolarization is caused by a transient increase in sodium conductance which is followed by an increase in potassium conductance? The observed currents seem to be appropriate, but by themselves do not reveal the identity of the ions

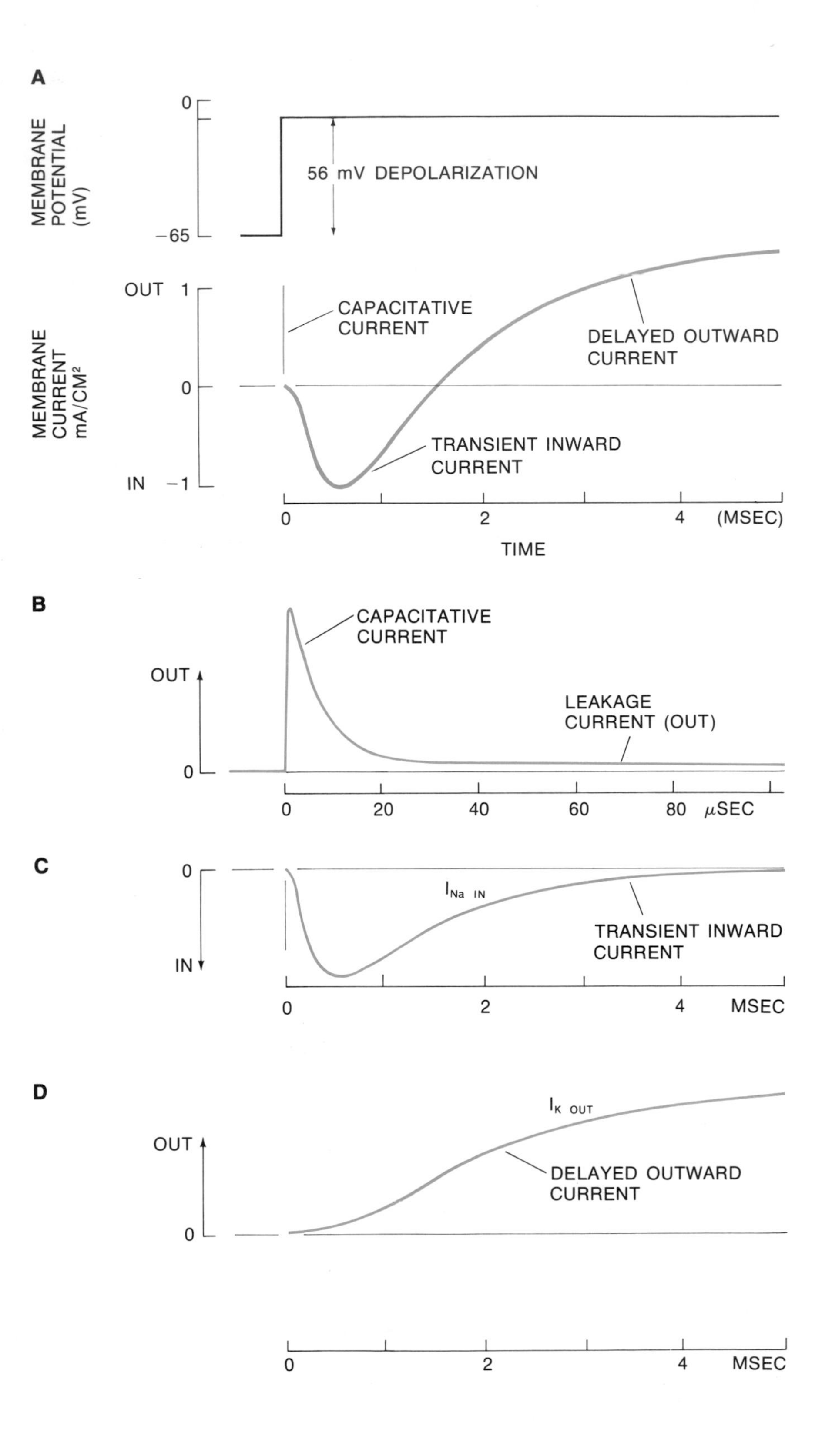
A
MEMBRANE POTENTIAL (mV)
0
−65
56 mV DEPOLARIZATION
MEMBRANE CURRENT mA/CM²
OUT 1
0
IN −1
CAPACITATIVE CURRENT
DELAYED OUTWARD CURRENT
TRANSIENT INWARD CURRENT
0
2
4
(MSEC)
TIME
B
CAPACITATIVE CURRENT
OUT
LEAKAGE CURRENT (OUT)
0
0
20
40
60
80
μSEC
C
0
$I_{Na\ IN}$
TRANSIENT INWARD CURRENT
IN
0
2
4
MSEC
D
$I_{K\ OUT}$
OUT
DELAYED OUTWARD CURRENT
0
0
2
4
MSEC

responsible. As a first step, one can show that alterations in chloride concentration have no effect on either the transient current or the prolonged outward current, although they do affect leakage current. Therefore, the chloride and the leakage currents can be ignored for the moment.

Currents carried by sodium and potassium ions

Hodgkin and Huxley made a number of tests to determine which ions were carrying the current. The experiments described below show that the early inward current is due to sodium entering the axon (Figure 4C) and the delayed outward current to potassium leaving it (Figure 4D).

One of their first procedures was to vary the external sodium concentration and determine how this influenced the initial inward phase of current flow. When the external sodium concentration was reduced to zero (by substituting choline), the initial phase of inward current disappeared and was replaced by a small transient outward current (Figure 5). The second maintained phase of outward current was unchanged. This result fits well with the hypothesis that the initial current results from an increase in sodium conductance. When the sodium concentration outside is zero, the driving force tending to push sodium ions out of the cell becomes large; consequently, an increase in sodium conductance actually allows sodium ions to move out instead of in.

The second experimental test of the ion species involved in the currents was to displace the membrane potential toward E_{Na} (Figure 6). At E_{Na} sodium ions cannot carry current across the membrane and, as expected, no inward current flows. Instead of the two phases of current flow, there is now a single, maintained S-shaped curve of outward current. At E_{Na} it is therefore possible to obtain the time course of the change of the late potassium current (I_K) in isolation and thereby to calculate potassium conductance (g_K).

Reconstructing the action potential requires knowledge of how g_{Na} and g_K vary with time at all values of membrane potential. Hodgkin and Huxley did this by varying the outside sodium concentration and thereby changing E_{Na} to a new level of membrane potential. For example, Figure 7 shows that when 90 percent of the external sodium

4
CURRENT FLOW ACROSS MEMBRANE during depolarization. A. Membrane Currents measured by voltage clamp during a 56-mV depolarization of membrane potential of a squid axon. The resulting currents (lower tracing) consist of a brief outward capacitative current, a transient phase of inward current, and a delayed, maintained outward current. These are shown separately in B, C, and D. The capacitative current (B) lasts for only a few microseconds. The small outward leakage current is due largely to movement of chloride. The transient inward current (C) is due to sodium entry (see Figure 5) and the prolonged outward current (D) to potassium movement out of the fiber. In this and other voltage clamp records from squid axons the membrane potential is assumed to be 60 to 65 mV. (After Hodgkin and Huxley, 1952a)

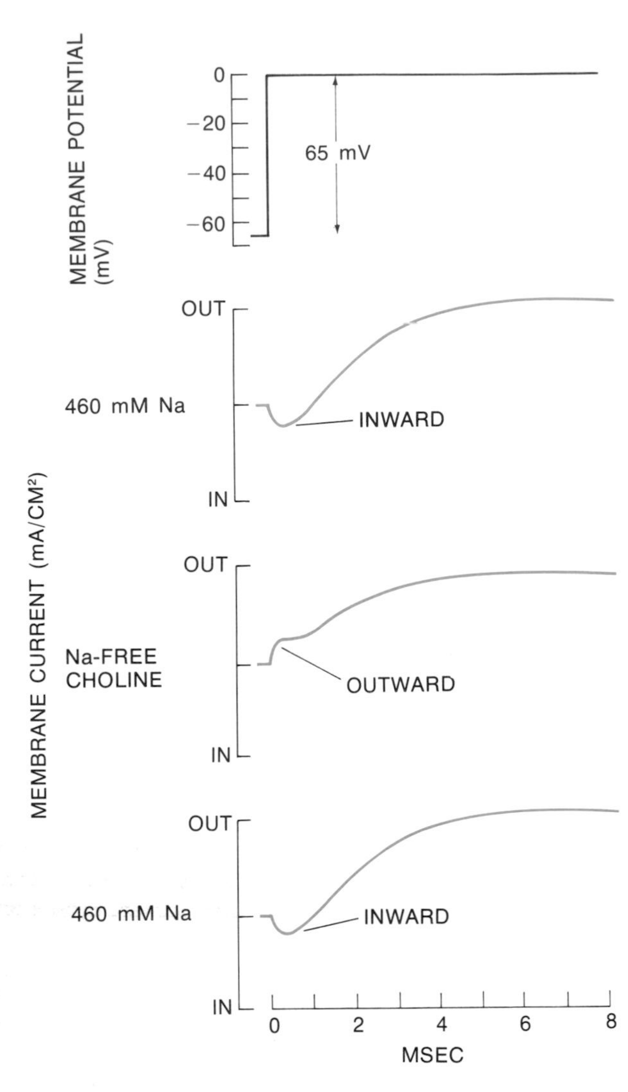

5

MEMBRANE CURRENT contributed by inward flow of sodium when the membrane potential of a squid axon is depolarized by 65 mV. In sodium-free solution the inward current disappears. It is replaced by a small transient outward current as sodium moves out of the fiber. The inward current is restored (last record) when sodium is reintroduced. (After Hodgkin and Huxley, 1952a)

concentration is removed, E_{Na} is correspondingly reduced, according to the Nernst equation. In this way, two patterns of current flow are recorded at one value of V_m: (1) with the normal sodium concentration present in the bathing fluid, an inward movement of current is followed by an outward movement (Figure 7, curve A); (2) with the external sodium concentration reduced so that the membrane potential corresponds to E_{Na}, sodium can make no contribution (Figure 7, curve B). The current can be carried only by potassium ions, and this gives rise to the smooth outward current that is maintained. By subtracting one current from the other, the inward sodium current can be determined (Figure 7, curve C). Similarly, if the external potassium concentration is increased and V_m is clamped at the new value of E_K, the sodium current is left, uncomplicated by potassium movements.

Hodgkin and Huxley directly tested the hypothesis that potassium movement causes the outward maintained current. They loaded a nerve

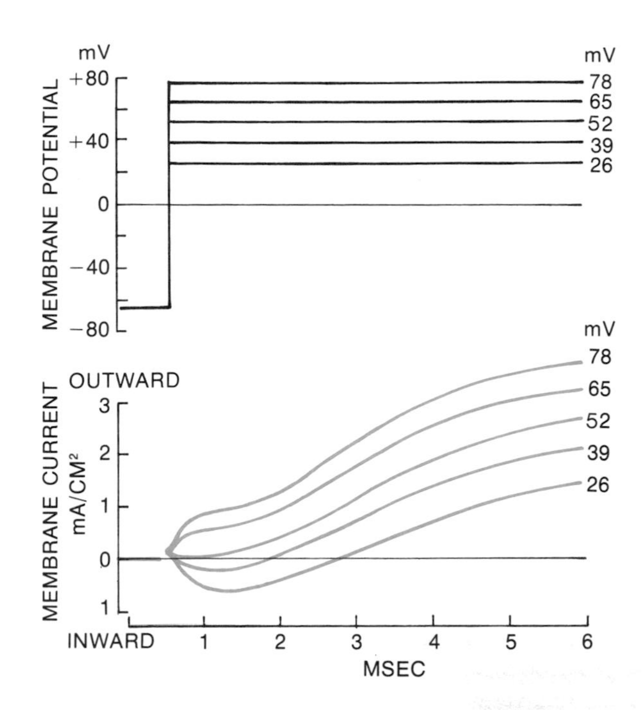

6
EFFECT OF MEMBRANE POTENTIAL on membrane current. Membrane currents in squid axon for five different displacements of the membrane potential (upper traces). At +52 mV, near E_{Na}, the inward sodium current (lower traces) has disappeared, the slower potassium component remaining. (After Hodgkin, Huxley and Katz, 1952)

fiber with radioactive potassium and clamped the membrane at a depolarized membrane potential. In this way the number of positively charged ions that had moved across the membrane to produce the maintained outward current was compared with the number of radioactive potassium ions that appeared in the solution. The agreement was excellent and provided direct evidence that the outward current was caused by movement of potassium ions.

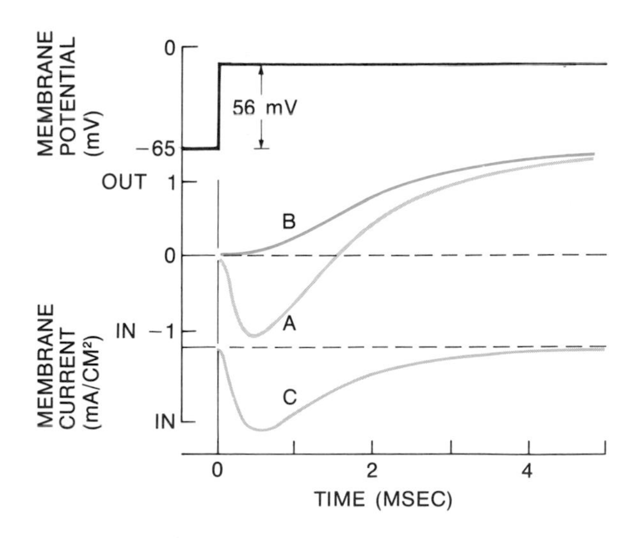

7
SODIUM AND POTASSIUM COMPONENTS of membrane current. With a displacement of 56 mV of the membrane potential in a squid axon in normal sea water one sees the normal sodium and potassium membrane current components in A. After replacing most of the sodium, so that E_{Na} is now at —9mV, the pure potassium current (B) remains. From the difference between A and B one obtains the sodium current (C). (After Hodgkin and Huxley, 1952a)

Selective poisons for sodium and potassium permeability: tetrodotoxin and tetraethylammonium

In the years since Hodgkin and Huxley's original experiments, new and convenient pharmacological methods for dissecting the sodium and potassium currents have been found. Tetrodotoxin (TTX), in particular, has turned out to be useful for a wide range of additional problems in neurobiology. TTX is a virulent poison, concentrated in the ovaries of certain fish, whose potent effects have given rise to the Chinese proverb, "To throw away life eat blowfish" (puffer fish). Kao[4] has reviewed the fascinating history of TTX, beginning with the discovery of its effects by the legendary emperor of China, Shun Nung (2838–2698 B.C.). He personally tasted 365 drugs while compiling a pharmacopoeia and lived (for an amazingly long time) to tell the tale.

One great advantage of TTX for neurophysiological studies is that its action is highly specific. Working with squid axons, Moore, Narahashi, and their colleagues[5] have shown that it selectively blocks the voltage-sensitive sodium permeability mechanism. When an axon poisoned by TTX is clamped to a depolarizing potential, the inward sodium current disappears and only the delayed outward potassium current remains, unchanged in amplitude and time course. Internal perfusion of the squid axon by TTX does not block the sodium current, indicating that the site of action is on the outside of the membrane (see below). Many other cells that produce regenerative sodium action potentials are affected by TTX in a similar manner, including vertebrate myelinated and nonmyelinated axons and skeletal muscle fibers. Figure 8*B* shows the effect of TTX on the sodium current in a myelinated nerve fiber.[6]

On the basis of results from many tests, one can now infer that a conductance mechanism involves sodium if it is blocked by TTX. The action of TTX has also been exploited for other purposes, described later. They include (1) estimating the number and molecular dimensions of sodium permeability channels in a unit area of membrane and (2) analysis of synaptic transmission without impulses.

In squid axons and in frog myelinated axons, Armstrong, Hille, and their colleagues[7] have shown that the potassium permeability mechanism is selectively blocked by tetraethylammonium (TEA; Figure 8*D*). Widespread applications have also been found for TEA in studying permeability and synaptic mechanisms. This MOLECULAR APPROACH, using various substances to bind specifically to membrane components, is one of the major advances of the last few years.

Sodium and potassium permeability as functions of membrane potential and time

By the methods outlined above, I_{Na} and I_K can be determined as functions of V_m and time. One can, therefore, estimate g_{Na} and g_K (since E_{Na}, E_K, and V_m are known) from the equations:

$$I_{Na} = g_{Na}\,(V_m - E_{Na})$$

$$I_K = g_K\,(V_m - E_K)$$

---

[4]Kao, C. T. 1966. *Pharmacol. Rev. 18*:977–1049.

[5]Moore, J. W., Blaustein, M. P., Anderson, N. C., and Narahashi, T. 1967. *J. Gen. Physiol. 50*:1401–1411.

[6]Hille, B. 1970. *Prog. Biophys. Mol. Biol. 21*:1–32.

[7]Armstrong, C. M., and Hille, B. 1972. *J. Gen. Physiol. 59*:388–400.

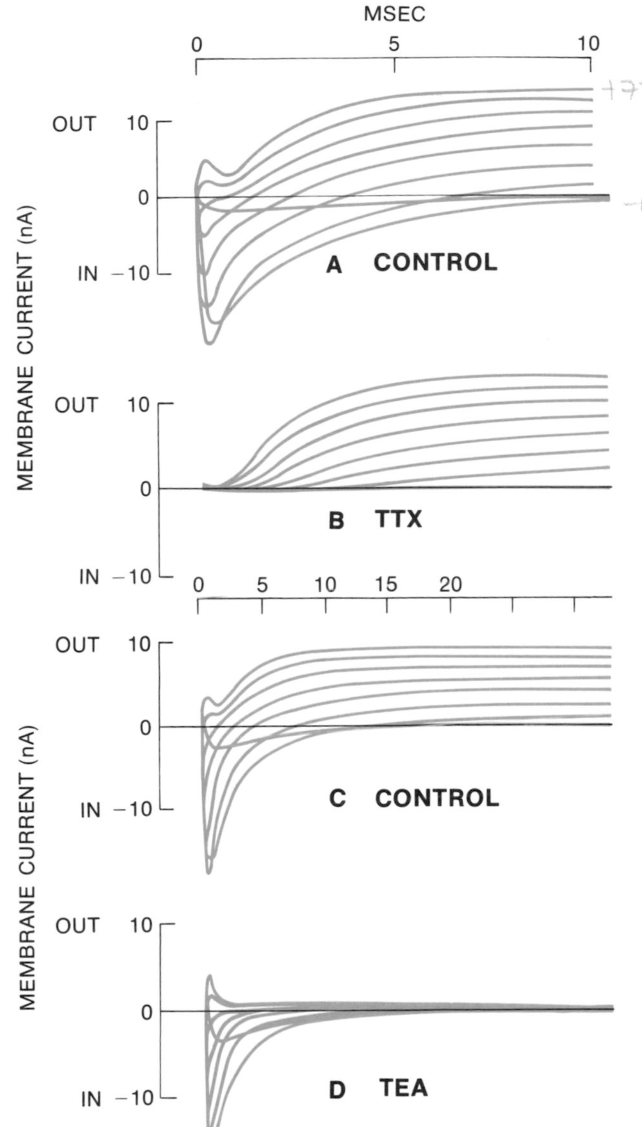

8
PHARMACOLOGICAL SEPARATION of membrane current into sodium and potassium components. The membrane potential of a myelinated nerve fiber is displaced to various intermediate levels between —60 mV and +75 mV. A and C are control records in normal fluid. In B, addition of 300 nM of tetrodotoxin (TTX) causes the sodium current to disappear, while the potassium current remains. In D, addition of tetraethylammonium (TEA) blocks the potassium current, revealing the sodium contribution. (After Hille, 1970)

The results show clearly that g_{Na} and g_K are both increased by depolarizing the membrane.

From what has been said already it is clear that the current itself has no effect on the sodium or potassium conductance. Thus, the time course and amplitude of I_K at any one value of V_m are the same, whether or not sodium current is flowing, for example, in the presence of TTX or when the potential is at E_{Na} in the experiments mentioned earlier. Similarly, I_{Na} is not affected by TEA. Finally, a brief shock can trigger an action potential in which no net current flows across the membrane.

The preceding discussion presents several lines of evidence showing

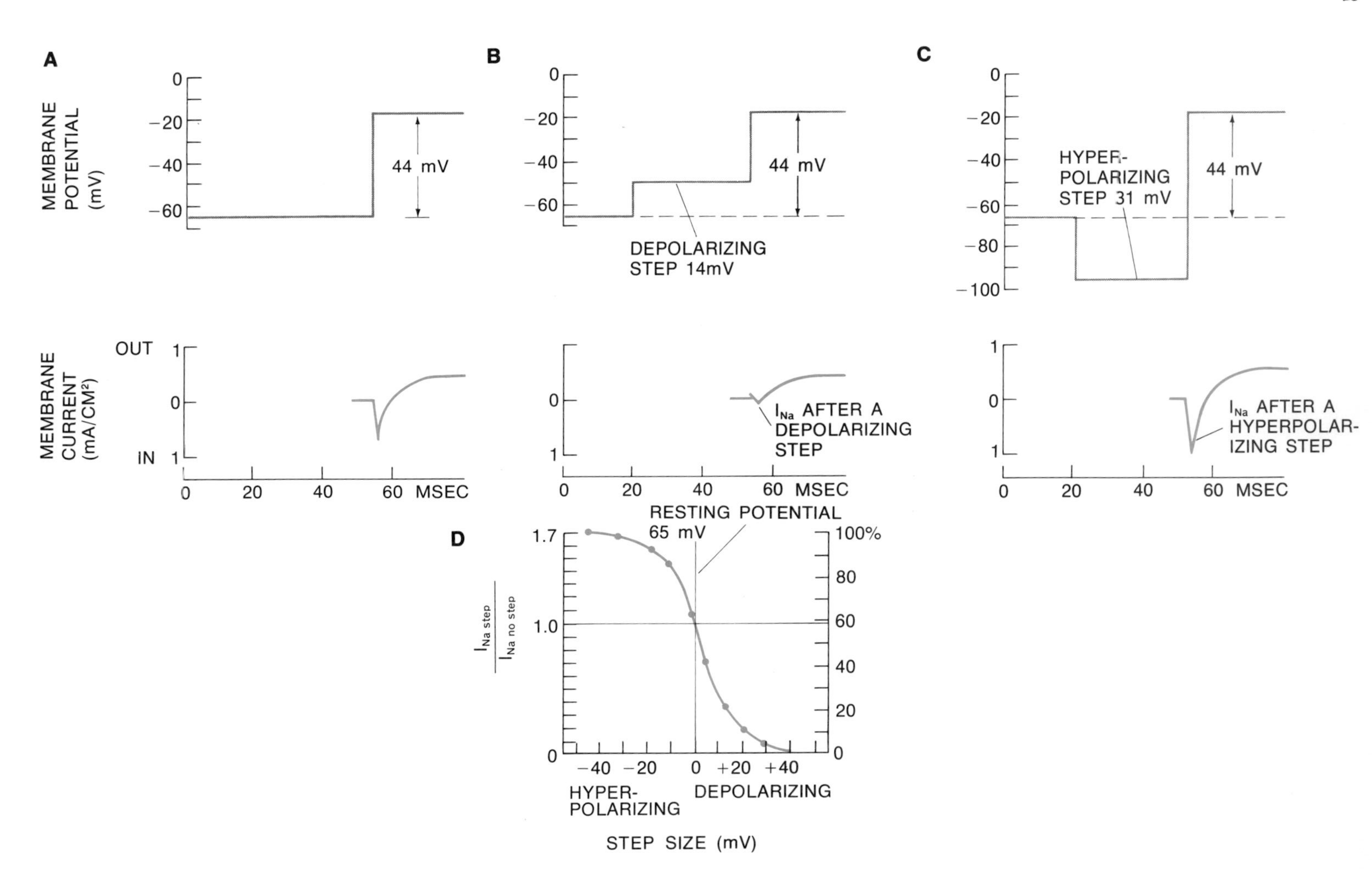
A
B
C
MEMBRANE
POTENTIAL
(mV)
44 mV
DEPOLARIZING
STEP 14mV
HYPER-
POLARIZING
STEP 31 mV
MEMBRANE
CURRENT
(mA/CM²)
OUT
IN
I_Na AFTER A
DEPOLARIZING
STEP
I_Na AFTER A
HYPERPOLAR-
IZING STEP
MSEC
D
RESTING POTENTIAL
65 mV
I_Na step / I_Na no step
100%
HYPER-
POLARIZING
DEPOLARIZING
STEP SIZE (mV)

that the sodium current rises rapidly and then gradually declines. On the other hand, the potassium current rises slowly and is maintained for the duration of the depolarization. This poses an interesting problem: What causes the decline of the sodium conductance? This INACTIVATION PROCESS appears to be of general significance, since it is likely to bear on many other properties of the nerve, such as the time course of repolarization and the refractory period.

Inactivation of sodium permeability

Hodgkin and Huxley analyzed the properties of the inactivation process that decreases sodium permeability in time by assuming that depolarization has a dual effect: that depolarization first produces a rapid increase of permeability to sodium and then a relatively slow decrease. They measured these opposite parameters by testing the effect of depolarizing pulses of varying amplitude and duration on the response to a subsequent standard depolarization. This experiment is illustrated in Figure 9. The membrane potential is first displaced abruptly from −65 mV to −21 mV, and the inward sodium current is measured (Figure 9*A*). Next, the membrane potential is depolarized by 14 mV for about 35 msec, followed by a further depolarization to −21 mV (Figure 9*B*). The small initial step of depolarization (by 14 mV) reduces the inward sodium current that occurs subsequently with the larger depolarizing step. Two factors influence the process: the size of the initial depolarization and its duration. When the depolarizing step preceding the test is short, about 1 msec, almost no effect is seen. With longer steps, a maximal effect appears, after which maintaining the initial step longer makes no further difference. Similarly, small steps of depolarization, about 1 mV, are less effective than larger ones. In contrast, a hyperpolarizing step increases the subsequent initial phase of sodium current when the membrane is once again depolarized to −21 mV (Figure 9*C*).

These results can be expressed quantitatively by calculating the ratio of sodium currents that occur with and without a preceding step of depolarization or hyperpolarization. When the voltage step is over about 30 mV in the depolarizing direction and lasts for a few milliseconds, inactivation is complete:

$$\frac{I_{Na}\ \text{step}}{I_{Na}\ \text{no step}} = 0$$

No matter how large the second depolarization, it is not accompanied by a phase of inward sodium current. Similarly, a ceiling is reached

9

EFFECT OF DEPOLARIZATION AND HYPERPOLARIZATION on sodium current in a squid axon. If a depolarizing step between —65 mV and —21 mV (A) is preceded by a smaller conditioning step (B), the sodium current component is greatly reduced. However, an initial hyperpolarizing step (C) increases the subsequent sodium current. D depicts the effects of various conditioning hyperpolarizing and depolarizing steps on the amplitude of the sodium current. After Hodgkin and Huxley, 1952c)

with steps hyperpolarizing the membrane by more than about 30 mV beyond the resting potential. Larger hyperpolarizations produce no more increase in sodium current. These results are shown in Figure 9*D*, in which depolarization and hyperpolarization are plotted along the abscissa and the value of $I_{Na\ step}/I_{Na\ no\ step}$ is plotted along the ordinate. The curve is S-shaped, reaching a maximum at about 30 mV of hyperpolarization and zero at 30 mV of depolarization. The curve can be represented by a single variable (h). When $h = 100$ percent, at a hyperpolarized membrane potential, sodium conductance is not inactivated; when $h = 0$, inactivation is complete.

In addition to TTX and TEA, there is a third agent that can be used to influence permeability characteristics in a selective manner. When a proteolytic enzyme, pronase, is perfused through the inside of a squid axon, the inactivation of sodium conductance is reduced and then abolished.[8] The activation of g_{Na} and g_K is not affected. Interestingly, the molecules responsible for the sodium-inactivating mechanism are accessible to pronase only from the inside surface of the membrane.

Reconstruction of the action potential

At this stage, the way in which g_{Na} and g_K increase with time at every value of membrane potential is known, as is the way in which g_{Na} decreases with time at a particular membrane potential. Permeability is determined not simply by the instantaneous value of the membrane potential, but also by its history—how it got there and where it came from.

Knowing V_m and how g_{Na} and g_K vary with time, Hodgkin and Huxley were able to account for the rising and falling phases of the action potential. But to estimate how the potential across the membrane would change in time after a superthreshold shock, they had to be able to work out (at every instant) exactly how the conductance would change, how the current would flow, and how the voltage would change. For this they had to fit their experimental results with curves that could be described by simple equations. The results are shown in Figure 10. The time course for g_K activation, with its S-shaped inflection, could conveniently be fitted by an equation using a fourth-power variable (n^4). The activation of g_{Na} was fitted by a third-power variable (m^3), while the inactivation variable was a single power (h). Starting with a depolarizing step sufficient to bring V_m above threshold, they calculated what the subsequent potential changes would be at successive intervals of 0.01 msec. Thus, during the first 0.01 msec after the membrane had been depolarized to, say, −15 mV, they calculated how g_{Na} and g_K would change and what values of I_{Na} and I_K would result. Knowing this, they estimated the change in V_m, g_{Na}, g_K, I_{Na}, and I_K during the next 0.01 msec. Then they once more estimated how V_m changed, and so on all through the rising and falling phases of the action potential, using only their curves based on experimental results.

The end result is a truly outstanding intellectual achievement. Using

---

[8] Armstrong, C. M., Bezanilla, F., and Rojas, E. 1973. *J. Gen. Physiol.* 62:375–391.

g_{Na} and g_K as functions of V_m and time, Hodgkin and Huxley reconstructed perfectly almost the entire course of the action potential (Figure 10*B*, *C*). The figures they used for the calculation were based on current measurements made under completely artificial conditions, without longitudinal currents or propagation, with the membrane potential held first at one value and then another, with low and high sodium or potassium concentrations in the external fluid. In addition to describing conductance changes during the impulse, they were able to explain in terms of ionic permeability the propagation of the action potential and many properties of axons that had remained mysterious (such as the refractory period, the threshold, anode break excitation, and the undershoot, some of which are discussed below). Further, their findings have been found to apply to a wide variety of tissues and have influenced thinking about other excitable cells in which the same experimental evidence has not yet been obtained.

Refractory period and threshold

How do the findings of Hodgkin and Huxley explain the refractory period? Clearly, two changes develop that make it impossible for the nerve fiber to give another impulse during the falling phase of the action potential: (1) g_K is now far larger than g_{Na}, and (2) inactivation, which prevents the further increase in g_{Na}, is also maximal. As a result a second depolarization cannot cause g_{Na} to increase to the point where I_{Na} exceeds I_K and initiates another impulse (see below).

And how do their findings explain the threshold membrane potential at which the impulse takes off, especially when it might seem that a discontinuity, like threshold, would require a discontinuity in the curves of g_{Na} and g_K? The explanation is that when a patch of membrane is depolarized, threshold is that value of V_m at which the outward current carried through potassium and leakage channels is exactly equal and opposite to the inward current carried by sodium. (This is also the condition at rest.) The threshold potential of the membrane represents an unstable equilibrium at which one of two events may occur. If an extra sodium ion enters the cell, the depolarization is increased, g_{Na} increases, more sodium enters, and so on. If, on the other hand, a potassium ion leaves the cell, the membrane potential is driven away from threshold, g_{Na} decreases, and further potassium ions leave until the potential returns to its resting value. Depolarizing currents that bring the membrane potential to beyond threshold allow g_{Na} to increase sufficiently for inward movement of sodium to swamp outward movement of potassium. In contrast, subthreshold depolarization fails to increase g_{Na} enough to enable the regenerative sodium mechanism to become effective.

A convenient analogy for threshold is the critical flash point of an explosive reaction. An increase in temperature accelerates the reaction, which produces more heat, raises the temperature further, and so on. At the critical point, if the temperature falls, the whole reaction is damped down to quiescence.

Calcium ions and excitability

An ion of considerable importance, not yet discussed, is calcium, which plays a key role in a variety of processes. For example, calcium

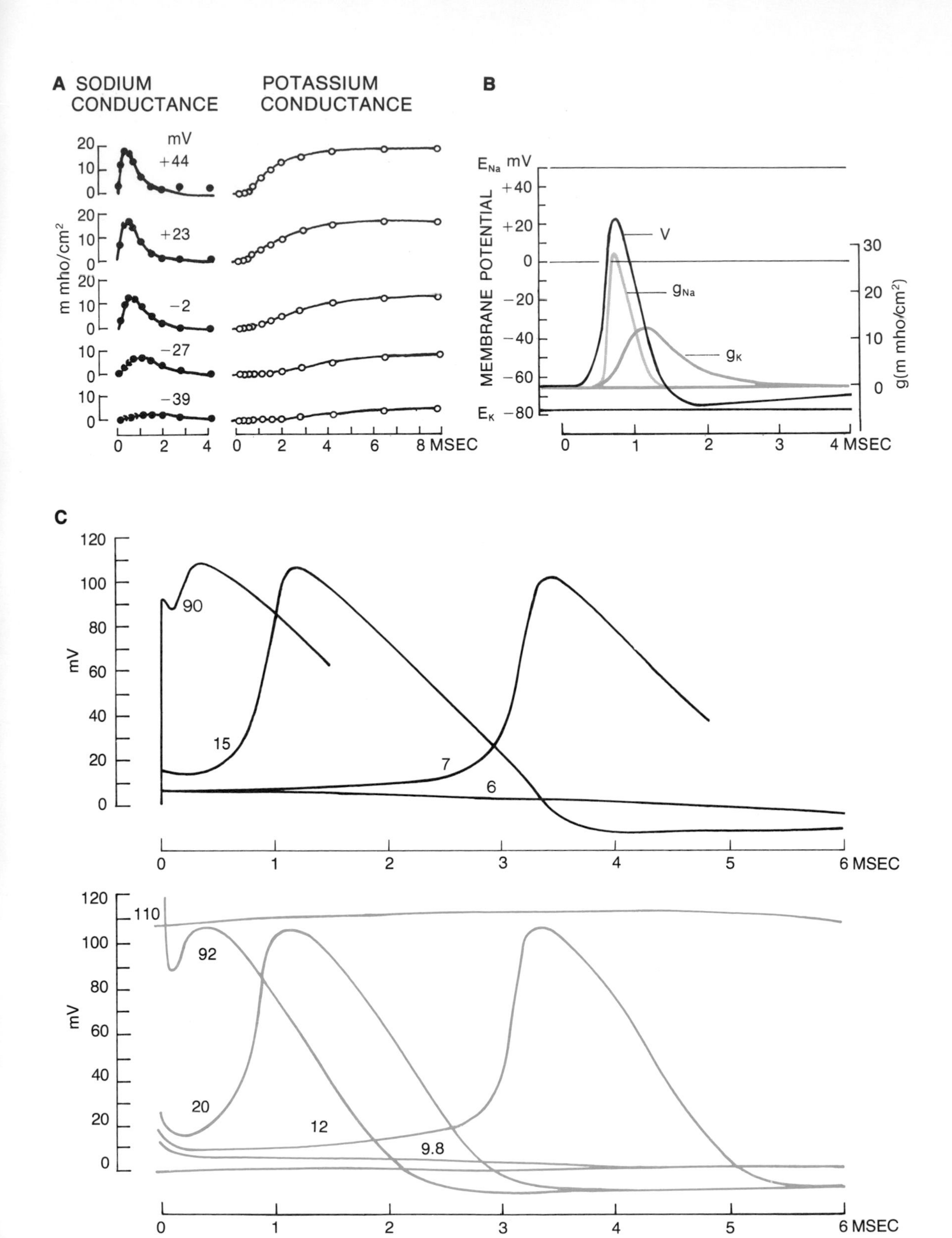
A SODIUM CONDUCTANCE
POTASSIUM CONDUCTANCE
B
m mho/cm²
mV
+44
+23
−2
−27
−39
8 MSEC
E_Na mV
MEMBRANE POTENTIAL
V
g_Na
g_K
E_K
g(m mho/cm²)
4 MSEC
C
mV
90
15
7
6
6 MSEC
110
92
20
12
9.8
6 MSEC

ions are an essential link in the coupling between changes in membrane potential and secretion of chemical transmitters by nerve terminals, and in the initiation of contraction in muscle fibers. The first of these actions is discussed more fully in later chapters. Here we consider the effects of calcium on excitability and the entry of calcium into the axon during the nerve impulse.

It has long been known that a reduction in external calcium concentration lowers the threshold for the initiation of impulses in certain nerves and skeletal muscles. An increase in external calcium acts in the opposite way, raising the threshold and decreasing excitability. Using the voltage clamp technique on squid axons, Frankenhaeuser and Hodgkin[9] measured the effects of different concentrations of calcium on membrane currents. When the external calcium concentration is lowered, a large and reversible increase occurs in the sodium and potassium currents produced by a depolarizing step. Quantitatively, the effect of a fivefold reduction in calcium is equivalent to a membrane depolarization of 10 to 15 mV. Thus, extracellular calcium ions influence the relation between the membrane potential and the conductance mechanism for sodium and potassium and tend to stabilize the membrane by increasing the amount of depolarization needed to reach threshold. Recently, it has been shown that increases in intracellular calcium can give rise to a pronounced increase in g_K.[10]

In addition to these actions, calcium itself enters squid axons and other nerve fibers during impulses. This was first demonstrated with radioactive tracer measurements and subsequently with an elegant optical technique.[11] A protein, aequorin, obtained from certain jellyfish (aequoria) is injected into the interior of squid axons. This substance emits light in the presence of ionized calcium and is a reliable, sensitive indicator of the calcium activity. Accordingly, one can measure the background level and small changes in concentration with very high time resolution. In the squid axon, as in other neurons, the internal calcium concentration is very low, about 0.3μM; it is kept at this low level partly by transport out of the cell. A transient regulation is also brought

[9]Frankenhaeuser, B., and Hodgkin, A. L. 1957. *J. Physiol.* *137*:218–244.

[10]Meech, R. W. 1974. *J. Physiol.* *237*:259–277.

[11]Baker, P. F., Hodgkin, A. L., and Ridgway, E. B. 1971. *J. Physiol.* *218*:709–755.

10

RECONSTRUCTION OF ACTION POTENTIAL. A. Time course of sodium and potassium conductances in a squid axon clamped at different membrane potentials, indicated by numbers. Circles are experimental values; the solid lines are predicted from equations. B. Theoretical reconstruction of an action potential (curve V) and the sodium and potassium conductances, using results from voltage clamp experiments. C. Observed action potentials following brief shocks at three different intensities (lower tracings) and the theoretical solution for such depolarizations (upper tracings). (After Hodgkin and Huxley, 1952d)

about by mitochondria, which can take up calcium and unload it again, thereby exerting a buffering action (Chapter 9). In the presence of aequorin, trains of impulses or steps of depolarization lead to an increased emission of light, indicating an increase in intracellular concentration.

Calcium entry occurs in two phases. The early phase is blocked by tetrodotoxin, becomes inactivated with time, and has a time course similar to that of the movement of sodium. In this phase calcium probably enters through the sodium channel (see below), but the membrane is about 100 times less permeable to calcium than to sodium. In the late delayed phase calcium entry behaves differently: it is inactivated only slowly, depends simply on the outside calcium concentration and membrane potential, and is not blocked by TTX or TEA. These findings suggest the existence of a separate voltage-sensitive pathway for calcium ions resembling those for sodium and potassium.

Calcium movements have at least two important consequences for signaling: (1) the delayed phase of calcium entry is probably related to the secretion of neurotransmitters by nerve endings (Chapter 9), and (2) calcium entry in some neurons and muscle fibers can carry sufficient current to contribute to or even be responsible for action potentials. Examples are mammalian cardiac muscle fibers, skeletal muscle fibers in crustaceans, and certain neurons in invertebrates.[12] This behavior would be expected for cell membranes with a large voltage-sensitive calcium conductance. For calcium, as for sodium, the electrical driving force and concentration gradient are directed inward. Thus, if depolarization increases g_{Ca}, calcium ions enter and produce a regenerative impulse in a manner analogous to the behavior of sodium. In squid axons and most neurons the calcium contribution to the total charge is considerably less than that of sodium, because of the relatively small value of g_{Ca}.

The fertilized eggs of certain starfish and tunicates provide an interesting illustration of the role of calcium currents. These egg cells give full-sized overshooting impulses in response to electrical stimulation. At early stages both sodium and calcium contribute to the potential. Takahashi, Miyazaki, Hagiwara, and their colleagues have followed the subsequent changes in permeability to sodium and calcium in the developing embryos.[13] It is natural to wonder what role these impulses play in an egg. Are they merely an outward expression of a process to be repressed later? Or do the currents and ionic movements perform a function related to development?

Channels for sodium and potassium movement: mechanism of permeability changes

In their comprehensive series of papers in 1952, Hodgkin and Huxley discussed the molecular mechanisms that allow ions to move across the membrane under the influence of changes in electrical potential. Among the questions that arose from their experiments were the fol-

---

[12]Hagiwara, S. 1974. In G. Eisenman (ed.). *Membranes: A Series of Advances*, Vol. 3. Dekker, New York.

[13]Miyazaki, S., Takahashi, K., and Tsuda, K. 1974. *J. Physiol.* *238*:37–54.

lowing: (1) Are there separate pathways for sodium and potassium movement? (2) Do sodium and potassium ions cross the membrane as ions or in combination with a lipid-soluble carrier? (3) How many sites for ion flow are there in each unit area of membrane? (4) How does depolarization open the permeability gates for sodium and potassium?

The analysis of such problems by Hodgkin and Huxley laid the groundwork for many of the experiments that were still to come, and their tentative predictions have in specific instances proved accurate.

One example, mentioned earlier, is the subsequent confirmation by means of TTX and TEA that separate and distinct pathways are used by sodium and potassium to cross the membrane. Similarly, experiments with pronase[8] have confirmed that a distinct molecular mechanism is associated with the inactivation of sodium conductance, equivalent to Hodgkin and Huxley's h variable.

Many lines of evidence have now come together to provide a more detailed picture of permeability mechanisms in nerves than was possible in 1952. Studies with natural and artificial membranes have revealed some of the general principles of ionic permeability. Also, it is now possible to estimate the dimensions of the channels in the membrane for sodium and potassium.[6] (For convenience, the term CHANNEL is often used to denote the structure or pathway through which an ion moves through the membrane.) Using the voltage clamp on single frog axons, Hille tested a variety of substances for their ability to carry current through sodium and potassium conductance channels. Some ions were excluded entirely; others were more or less permeant than sodium or potassium. From the dimensions of the various substances, their configurations and charges, and the ease with which they passed through the membrane, Hille[14,15] estimated that the sodium channel is about 0.3×0.5 nm wide, while the potassium channel is narrower and shorter with a diameter of about 0.3 nm. At the narrowest part are negative charges, provided by a circle of oxygen atoms, which exclude anions and attract cations. As expected, the sodium channel is somewhat permeable to potassium ions, but the value is so much smaller than for sodium ions that the contribution of potassium may in practice be neglected. In the squid axon the sodium channel is 12 times more permeable to sodium than to potassium.[16] At the same time, several lines of evidence (some of which have already been mentioned) clearly show that sodium and potassium channels are indeed separate and that one does not become converted to the other in the course of depolarization.

Two questions arise from these considerations. What is the conductance of a single open channel and how many are there in a unit area of membrane? Two independent approaches have been used to tackle these problems, with remarkably similar results. One method is

---

[14]Hille, B. 1971. *J. Gen. Physiol.* *58*:599–619.

[15]Hille, B. 1973. *J. Gen. Physiol.* *61*:669–686.

[16]Chandler, W. K., and Meves, H. 1965. *J. Physiol.* *180*:788–820.

essentially electrical. From the dimensions of a single sodium channel Hille[6] estimated the number of ions that would move through it in 1 msec and the conductance change it would contribute to the membrane. The conductance per channel is approximately 10^{-10} mho in normal Ringer's fluid and is thus estimated to be about 4×10^{-10} mho in seawater (for squid axon). For a unit area of membrane the maximum value of the sodium conductance is known, and so the number of sodium channels can be calculated. The other method of counting the sodium channels in a nerve employs TTX or another compound (saxitoxin) that selectively blocks g_{Na}. In one of the first experiments of this type lobster nerves were bathed in a 300-nM solution of TTX until conduction was blocked;[17] then the toxin remaining in the bath was measured. From this, the amount of membrane-bound TTX was estimated after making allowance for the extracellular space. Ritchie and his colleagues[18,19] performed similar experiments on a variety of nerves from lobster, rabbit, and garfish. In addition, they measured directly the amounts of radioactive TTX and saxitoxin that are bound.

From these diverse approaches a consistent picture has emerged: there are surprisingly few permeability channels for sodium in the membrane. The numbers estimated per square micron of membrane are of the order of tens or at the most hundreds. For garfish, rabbit, crab, lobster, and squid unmyelinated fibers the number of sites ranges from 20-500/sq μm. Assuming that the channels are arranged in a square array, distances of the order of tenths of a micron rather than angstroms separate one from the next. For garfish nerve this is of the same order as the fiber diameter.

Inevitably, the calculations involve a number of assumptions, such as (1) that one molecule of toxin blocks one sodium channel, (2) that sodium channels are either fully open or completely closed, and (3) that the channels constitute a homogeneous population.

Activation of permeability: gating mechanisms for ionic channels

Another important question concerns the nature of the permeability changes and how they are brought about by depolarization. In the discussion of this problem by Hodgkin and Huxley their aim was to distinguish between the theories that were excluded by the results of their own experiments and those that were consistent with them. Their first conclusion, mentioned earlier, was that the changes in permeability depend on membrane potential. They went on to suggest that the dependence of g_{Na} and g_K on membrane potential indicates that the permeability changes arise from the effect of the electrical field on the distribution or orientation of molecules with a charge or dipole moment (Figure 11*B*). But how can this affect the ease with which ions cross the membrane? After considering alternative possibilities, they suggested that within the membrane are charged ACTIVATING OR GATING

---

[17]Moore, J. W., Narahashi, T., and Shaw, T. I. 1967. *J. Physiol.* *188*:99–105.

[18]Colquhoun, D., Henderson, R., and Ritchie, J. M. 1972. *J. Physiol.* *227*:95–126.

[19]Henderson, R., Ritchie, J. M., and Strichartz, G. R. 1973. *J. Physiol.* *235*:783–804.

PARTICLES. These do not serve as carriers (for example, of sodium) in the usual sense, but act as though they open gates at the entrance to the sodium channel. When the particles occupy particular sites in the membrane, sodium ions can pass along the appropriate pathway. Similarly, the decline of sodium conductance with inactivation can be attributed to the relatively slow movement of another particle, which blocks the sodium permeability channel. For potassium, Hodgkin and Huxley suggested the existence of other, completely separate charged particles that move relatively slowly under the influence of an electric field to open the potassium gates. According to this scheme, specific charged activating and inactivating particles move within the membrane to open or to close the gates of the sodium and potassium channels. Figure 11*B* illustrates this diagrammatically.

Next, Hodgkin and Huxley considered how many charges each gating particle was likely to bear. A clue to thinking about this problem is to consider the steepness of the slope of the curve that expresses the relation between the conductance of the membrane and the potential. In any system of this type, living or inert, the slope depends upon the valence of the particles that move under the influence of an electrical field. (The relation is given in the Boltzmann equation.) Thus, for a particle with a single charge, one would expect an *e*-fold change in conductance to be produced by a 25 mV potential change. (This is the situation for the movement of electrons in a conventional vacuum tube.) In the squid axon, however, about six times less depolarization is needed for an *e*-fold change in g_{Na} (4 mV instead of 25 mV). Thus the gating particle must bear either six charges at one end or three positive charges at one end and three negative at the other. Alternatively, the opening of the sodium permeability gate may require the arrival at the same place of six particles, each with a unit charge. A consequence of a mechanism of this type is that in the physiological recordings the movement of the charged particles within the membrane should appear as a capacitative current directed outward in response to depolarization (see Appendix). Hodgkin and Huxley estimated that the gating currents within the membrane must be very small in comparison with the ionic currents carried by sodium ions, perhaps only a few percent.

An obvious next step was to try to measure directly these gating currents associated with the turning on of the sodium permeability mechanism. Knowing the amplitude, time course, and dependence on potential, one might then be in a position to make inferences about the gating molecules, their charges and kinetics of movement.

A number of technical difficulties stood in the way, and it was not until 1973 that two groups succeeded, almost simultaneously, in measuring the gating currents for sodium channels.[20,21] One problem is that the currents are small; another is that they tend to be swamped by other capacitative currents flowing at the same time. The principle of the

---

[20]Armstrong, C. M., and Bezanilla, F. 1974. *J. Gen. Physiol.* *63*:533–552.

[21]Keynes, R. D., and Rojas, E. 1974. *J. Physiol.* *239*:393–434.

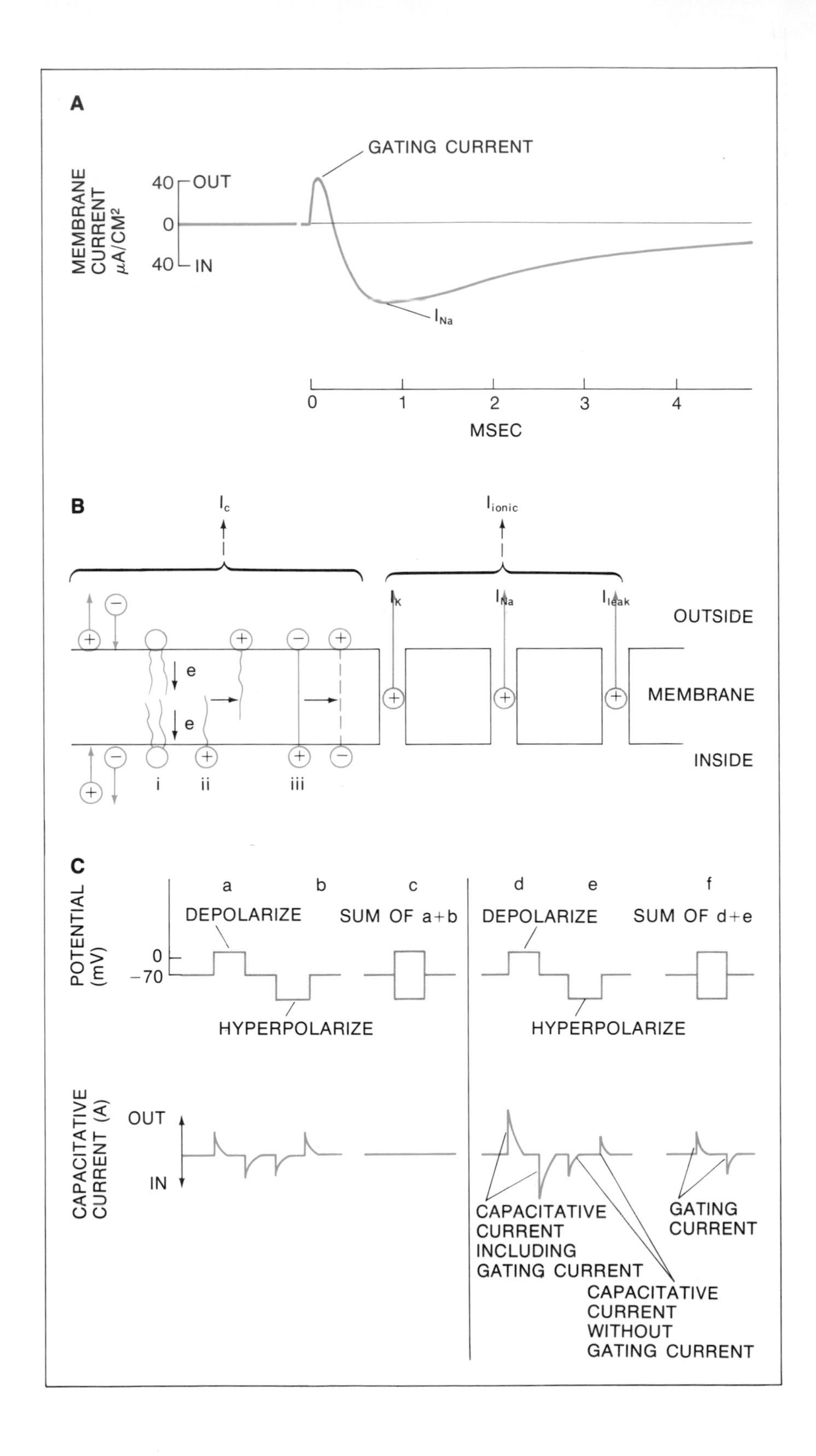
A
GATING CURRENT
MEMBRANE CURRENT μA/CM²
40 OUT
0
40 IN
I_{Na}
0
1
2
3
4
MSEC
B
I_c
I_{ionic}
I_K
I_{Na}
I_{leak}
OUTSIDE
e
e
MEMBRANE
INSIDE
i
ii
iii
C
a
b
c
d
e
f
DEPOLARIZE
SUM OF a+b
DEPOLARIZE
SUM OF d+e
POTENTIAL (mV)
0
−70
HYPERPOLARIZE
HYPERPOLARIZE
CAPACITATIVE CURRENT (A)
OUT
IN
CAPACITATIVE CURRENT INCLUDING GATING CURRENT
GATING CURRENT
CAPACITATIVE CURRENT WITHOUT GATING CURRENT

measurements is illustrated diagrammatically in Figure 11. Using internally perfused squid axons and a voltage clamp, the membrane potential was first abruptly depolarized and then hyperpolarized by an equal amount. Over a number of trials the currents flowing as a result of the two steps in opposite directions were summed. In this way linear symmetrical membrane currents were eliminated, revealing nonlinear responses associated with gating. Implicit in this reasoning is that the bulk of the capacitative current in the membrane behaves passively, a depolarization and a hyperpolarization producing identical but opposite surges of current as charges flow from one side to the other and back again (Figure 11*B*, *C*). In contrast, if the gating current is to account for the opening of sodium channels, it must be larger for depolarization than for hyperpolarization, which cannot shut channels that are already closed. The gating current should therefore stand out, particularly if a number of trials are averaged (Figure 11*C*, *d–f*).

An example of the gating current followed by inward sodium ionic current is shown in Figure 11*A*. To reduce the ionic currents in these experiments, the sodium concentration in the bathing fluid was only 5 percent of the normal and the inside of the axon was perfused with cesium fluoride.

The gating current is isolated even more effectively if TTX is added to the outside fluid, abolishing the ionic g_{Na} component. While it may appear surprising that gating current is still present under these conditions, there is evidence that TTX acts by physically preventing sodium ions from entering the sodium channel. An action of that type would not suppress the movement of particles within the membrane.

The experiments on gating currents have opened up the possibility of obtaining more detailed information about the events associated with the activation of sodium permeability.

SUGGESTED READING

Papers marked with an asterisk (*) are reprinted in Cooke, I., and Lipkin, M. 1972. *Cellular Neurophysiology*, Holt, New York.

11
GATING CURRENT IN SQUID AXON. A. The brief outward gating current precedes the inward sodium current. The gating current is a small component of the total capacitative intramembranous current produced by depolarizing the axon. Within the membrane (B) electrons are redistributed (i), and charged (ii) or dipolar (iii) molecules become reoriented under the influence of the electrical field. To separate out the gating components from the rest of the capacitative current, the procedure shown in C is used. Symmetrical depolarizing (a) and hyperpolarizing (b) pulses are applied. For a perfect capacitor, the sum of the outward and inward currents should be zero (c). If, however, a nonlinear gating component is produced by depolarization (d) but not by hyperpolarization (e), it should appear in isolation when the inward and outward currents are summed (f). (After Armstrong and Bezanilla, 1974)

General reviews

Hille, B. 1976. Ionic Basis of Resting and Action Potentials. In E. Kandel (ed.). *Handbook of the Nervous System,* Vol. I. American Physiological Society, Bethesda, Md., Chap. 3.

Hodgkin, A. L. 1964. *The Conduction of the Nervous Impulse.* Liverpool University Press, Liverpool.

Katz, B. 1966. *Nerve, Muscle, and Synapse.* McGraw-Hill, New York, Chap. 5.

Voltage clamp technique

Moore, J. W. 1971. Voltage Clamp Methods. In W. J. Adelman (ed.). *Biophysics and Physiology of Excitable Membranes.* Van Nostrand Reinhold, New York, pp. 143–167.

Voltage clamp analysis of squid axons

*Hodgkin, A. L., Huxley, A. F., and Katz, B. 1952. Measurement of current-voltage relations in the membrane of the giant axon of *Loligo. J. Physiol. 116*:424–448.

*Hodgkin, A. L., and Huxley, A. F. 1952. Currents carried by sodium and potassium ions through the membrane of the giant axon of *Loligo. J. Physiol. 116*:449–472.

*Hodgkin, A. L., and Huxley, A. F. 1952. The components of membrane conductance in the giant axon of *Loligo. J. Physiol. 116*:473–496.

*Hodgkin, A. L., and Huxley, A. F. 1952. The dual effect of membrane potential on sodium conductance in the giant axon of *Loligo. J. Physiol. 116*:497–506.

*Hodgkin, A. L., and Huxley, A. F. 1952. A quantitative description of membrane current and its application to conduction and excitation in nerve. *J. Physiol. 117*:500–544.

*Hodgkin, A. L., and Huxley, A. F. 1953. Movement of radioactive potassium and membrane current in a giant axon. *J. Physiol. 121*:403–414.

Properties of sodium and potassium channels

GENERAL REVIEWS

Hille, B. 1970. Ionic channels in nerve membranes. *Prog. Biophys. Mol. Biol. 21*:1–32.

Landowne, D., Potter, L. T., and Terrar, D. A. 1975. Structure-function relationship in excitable membranes. *Annu. Rev. Physiol. 37*:485–508.

Narahashi, T. 1974. Chemicals as tools in the study of excitable membranes. *Physiol. Rev. 54*:812–889.

ORIGINAL PAPERS

Armstrong, C. M., and Hille, B. 1972. The inner quaternary ammonium ion receptor in potassium channels of the node of Ranvier. *J. Gen. Physiol. 59*:388–400.

Henderson, R., Ritchie, J. M., and Strichartz, G. R. 1973. The binding of labelled saxitoxin to the sodium channels in nerve membranes. *J. Physiol. 235*:783–804.

Gating currents

Armstrong, C. M., and Bezanilla, F. 1974. Charge movement associated with the opening and closing of the activation gates of Na channels. *J. Gen. Physiol. 63*:533–552.

Role of calcium

GENERAL REVIEW

Baker, P. F. 1972. Transport and metabolism of calcium ions in nerve. *Prog. Biophys. Mol. Biol. 24*:177–223.

ORIGINAL PAPERS

Baker, P. F., Hodgkin, A. L., and Ridgway, E. B. 1971. Depolarization and calcium entry in squid giant axons. *J. Physiol. 218*:709–755.

Frankenhaeuser, B., and Hodgkin, A. L. 1957. The action of calcium on the electrical properties of squid axons. *J. Physiol. 137*:218–244.

A. L. Hodgkin, 1949

A. F. Huxley, 1974

CHAPTER SEVEN

NEURONS AS CONDUCTORS OF ELECTRICITY

Impulses propagate along axons by the longitudinal spread of current. As each region of membrane generates the all-or-nothing action potential, it depolarizes and excites the adjacent not-yet-active region and gives rise to a new regenerative impulse. For an understanding of impulse propagation, as well as synaptic transmission and integration, one must know how an electric current spreads passively along a nerve.

As current spreads along a nerve fiber, it becomes attenuated with distance. This depends on a number of factors, including the diameter of the fiber and its membrane properties. A longitudinal current spreads farther along a fiber with a large diameter and a high membrane resistance. The electrical capacitance of the membrane influences the time course of electrical signals, but not necessarily their longitudinal distribution. To estimate how far a subthreshold potential change will spread, one needs to know the nerve's dimensions, geometry, and membrane characteristics.

In many vertebrate nerve cells, segments of the axon are covered by a high-resistance, low-capacitance myelin sheath. This acts as an effective insulator and forces currents associated with the nerve impulse to flow through the membrane at the intervals where the insulating myelin wrapping is interrupted. The impulse jumps from one myelin-free area of the membrane (a node of Ranvier) to the next, and thereby the conduction velocity is increased. Myelinated nerves are used by the nervous system for signaling in pathways where speed is essential.

Passive electrical properties of nerve and muscle membranes

Preceding chapters discuss the permeability change and ionic currents that flow over small patches of membrane under voltage-clamped conditions. This chapter examines more fully the way currents spread along nerves and how these currents produce localized, graded potentials. Local potentials behave in many respects like the signals in undersea cables, in which the principles were first worked out.

The study of the electrical properties of cells is not simply a preoccupation of physiologists who like to use electrical circuits as conceptual models for neuronal signals. The cable properties of nerves are essential for signaling in the nervous system. At sensory end organs, they are the link between the stimulus and the production of impulses; along axons, they allow the impulse to spread and propagate; at synapses, they enable the postsynaptic neuron to add and subtract the synaptic potentials that arise in it from numerous converging inputs. It will be shown in Chapter 16 that the synapses on neurons are usually situated close enough together to allow spatial integration of the subthreshold potentials that spread passively over the cell surface.

The pathways through which currents flow have been studied quantitatively in crab nerves by Hodgkin and Rushton,[1] who passed subthreshold current pulses through the cell membrane with external electrodes. Later on, intracellular electrodes were used for similar studies. The following account of local circuits and the way they influence signaling in various types of neurons is mainly descriptive.

Figure 1*B* shows the analogous circuit for a length of axon membrane in which each component represents a well-defined electrical property for a unit length. For a start, only resistors are needed. The three pathways (resistive elements) for current flow are (1) the axoplasm (or myoplasm) in the interior of cells, (2) the external fluid, and (3) the cell membrane that separates the two media. Under most conditions the resistance of the extracellular fluid (r_o) is insignificant and can be ignored. The membrane capacity, which slows the time course of signals, is discussed later.

In Figure 1 the direction of current flow is shown by the arrows, r_i is the resistance of intracellular fluid for 1 cm of nerve, and r_m is the membrane resistance over the same length. Provided the potential changes used to study the membrane properties are small and well below threshold, current and potential across the membrane are related by Ohm's law. Accordingly, the membrane resistance (r_m) represents predominantly the conductances to potassium and chloride at or near the resting potential. (With larger potentials, approaching threshold, the resistance, of course, decreases, owing to the increase in permeability to sodium and potassium.)

The principle of the analysis is to measure the change in potential produced by a square pulse of current. From the decrement that occurs with distance along the fiber, one can calculate the values of these circuit elements, r_i and r_m. At the same time, if the dimensions of the fiber are known, one can estimate the specific resistance of the fluids and of the membrane. While r_i and r_m refer to the values for a particular nerve fiber and, as described later, depend on its diameter, the SPECIFIC RESISTANCES (R_i and R_m) are measures of the conducting properties of the materials of which the nerve is made and the fluid that surrounds it.

[1]Hodgkin, A. L., and Rushton, W. A. H. 1946. *Proc. R. Soc. Lond.* B *133*: 444–479.

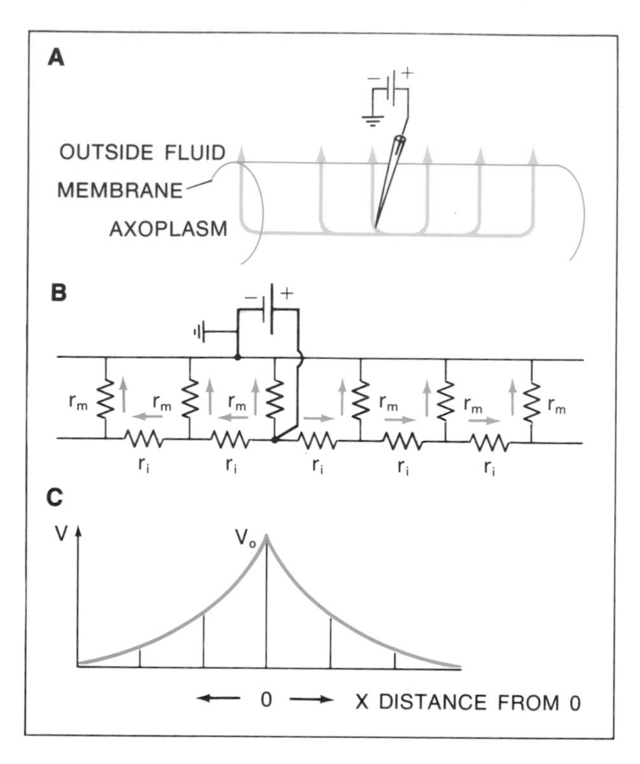

1

PATHWAYS FOR CURRENT FLOW. A. Current flow across the surface membrane generated by injection of charge through a microelectrode. B. Equivalent circuit, neglecting external resistance and capacity (see Figure 2). r_i, longitudinal resistance of intracellular fluid per unit length; r_m, membrane resistance for unit length. C. Decrement of potential (V) along axon from its original amplitude (V_o).

Knowledge of specific resistance is therefore important if one wishes to compare the leakiness of one membrane with that of another.

It is worth examining the units used to express the various electrical resistances. The resistance (r) in ohms of a cylindrical conductor is given by $R(L/\pi\rho^2)$, where L is the length, ρ is the radius, and R is the specific resistance for a unit volume of the material; r_i is the resistance for a unit length of nerve and is therefore expressed as ohms per centimeter, while the specific resistance (R_i) is expressed as ohms $\times$ centimeter. The membrane, however, is a flat sheet of material through which current flows transversely. Since its thickness (L) is not known, the true value of the specific resistance for a unit volume of membrane cannot be measured directly. A rough estimate is that for a nerve membrane 7 nm thick the specific resistance is approximately $10^9\ \Omega \cdot$cm. In practice it is more useful to deal with the membrane resistance for a unit length of axon (r_m) and unit area (R_m) whose dimensions are expressed as ohms $\times$ centimeter and ohms $\times$ square centimeter respectively,

The specific internal resistance (R_i) of nerve fibers or muscle fibers in an animal is similar from one cell to another, since the intracellular fluid has a uniform ionic composition. In squid nerves, R_i is about $30\Omega \cdot$cm, or about 10^7 times worse than copper. In frogs and mammals, where the ionic strength is weaker, the values are higher. The specific membrane resistance does, however, vary in different cells. A common figure is $1{,}000\Omega \cdot$sq cm; but in some cells it may be higher than $5{,}000\Omega \cdot$sq cm, while in others it may be considerably less. The leakiness to ions per unit area of membrane is therefore not constant in all nerve and muscle cells.

The decrement of subthreshold potential changes that occurs along the length of an axon depends upon the values of r_m and r_i for that particular fiber. From Ohm's law it is evident that the longitudinal spread of current is favored if r_m is high in relation to r_i, since less current leaks out through the membrane. One would therefore expect potentials to spread over greater distances in fibers that have a high membrane resistance and a low internal resistance. Conversely, the current spreads over a shorter distance and the potentials decrease more steeply in a fiber that has a low membrane resistance and a high internal resistance. In other words, as current leaks out of the fiber there is less current available farther along to change the potential across the membrane (Figure 1).

The relation between steady state potential and distance is given by the exponential equation:

$$\frac{V_x}{V_o} = e^{-x/\lambda}$$

where λ is $\sqrt{r_m/r_i}$, V_x is the potential at a distance (χ) from the site at which current is applied and where the potential is V_o. λ is known as the length constant and is the distance over which the potential declines to $1/e$, or roughly one third of its original amplitude (V_o).

In practice, the size of the fiber is one of the most important variables influencing λ. In any one animal the specific internal resistance (R_i) is constant. However, the internal resistance of a particular fiber (r_i) varies inversely with $\pi\rho^2$. As a result, λ may be as large as 5 mm in a giant fiber like a squid axon with a diameter of 500μm, while in a frog muscle fiber, 100μm in diameter, λ is about 2 mm. In many of the fibers and processes in the vertebrate nervous system that are only a few microns or less in diameter, λ is smaller still—a fraction of a millimeter.

It has been pointed out that a close parallel for current flow is the spread of heat in a metal rod surrounded by an insulator and in a conducting material. If one end of the rod is heated, heat is lost to the outside; at greater distances from the heated end, the temperature falls and the amount of heat loss decreases progressively because the rod itself becomes less hot. The fall in temperature with distance is an exponential function, like the fall of voltage with distance in a nerve. How far the heat spreads depends on (1) the material of which the rod is made, (2) the properties of the insulator, (3) the diameter, and (4) the heat-conducting properties of the material outside.

The flow of current in Figure 1 can now be described in ionic terms: positive charges flow into the cell from the tip of the microelectrode when the switch is closed. These positive charges repel other cations and attract anions. By far the most mobile and abundant ion in the axoplasm is potassium, which therefore carries most of the current and moves away from the electrode toward the membrane. However, the membrane is an insulator and acts as a barrier to the flow of current or ions. Potassium ions coming up against the membrane are therefore not able to leave immediately, but accumulate and spread longitudinally down the axoplasm. In reality, no one ion migrates far; the displacement of ions more closely resembles the collisions along a series of billiard balls arranged in a row. In ionic terms, the spread of the potential changes depends on the membrane resistance and diameter. In a cell with a low membrane resistance (r_m), the potassium conductance (g_K) is large and the ions can therefore leak out without spreading far along the nerve fiber. In contrast, in a cell with a high membrane resistance, g_K is low and a greater proportion of potassium is displaced laterally before leaking out. Along the nerve, less and less potassium accumulates and so the number of potassium ions leaving each adjacent unit area of membrane declines in an exponential manner with the distance away from the current-passing electrode. In a small fiber the internal resistance is higher; as a result the longitudinal fall of potential is steeper.

Membrane capacity and the time constant of the membrane

In the discussion of the steady state value of the membrane potential at various distances along the nerve from the point at which its voltage was changed, there was no need to consider the capacity of the membrane. This, as shown earlier, slows the rise and fall of localized potentials compared with an abrupt current pulse passed into the cell. The distortion in time course is caused by the ability of the membrane to store and separate charges.

General considerations of capacitors suggest that a cell membrane should have a relatively large electrical capacity. A capacitor consists of two conducting plates separated by a dielectric or a good insulator. The closer together the two plates are, the greater the capacity to store and separate charges. The intracellular and extracellular fluids, corresponding to two conducting plates, are separated by the lipoprotein membrane, which is only about 7 nm thick; its dielectric properties are dramatically shown by the potential gradient of the resting potential: 70 mV across 7 nm corresponds to about 10^5 V/cm, a value close to the breakdown point of a good insulator. Unlike the membrane, the axoplasm and the extracellular fluids cannot store charges without producing a change in voltage, because the ions are free to move.

As explained in Chapter 5, for a capacitor $C = Q/V$, where Q is the charge applied to the plates to produce a potential difference between them (V_C). The current (charges per second) flowing onto or off the plates of a capacitor is given by:

$$I_C = \frac{dQ}{dt} = C\,\frac{dV}{dt}$$

In the circuits of Figure 2 a delay in the rate at which the potential changes across a resistance (R) is introduced by the presence of a capacitor (C). Another way of describing the circuit is that the capacitor acts as a variable short circuit across the resistor. At the instant when the switch is closed or opened, charges flow onto or off the capacitor,

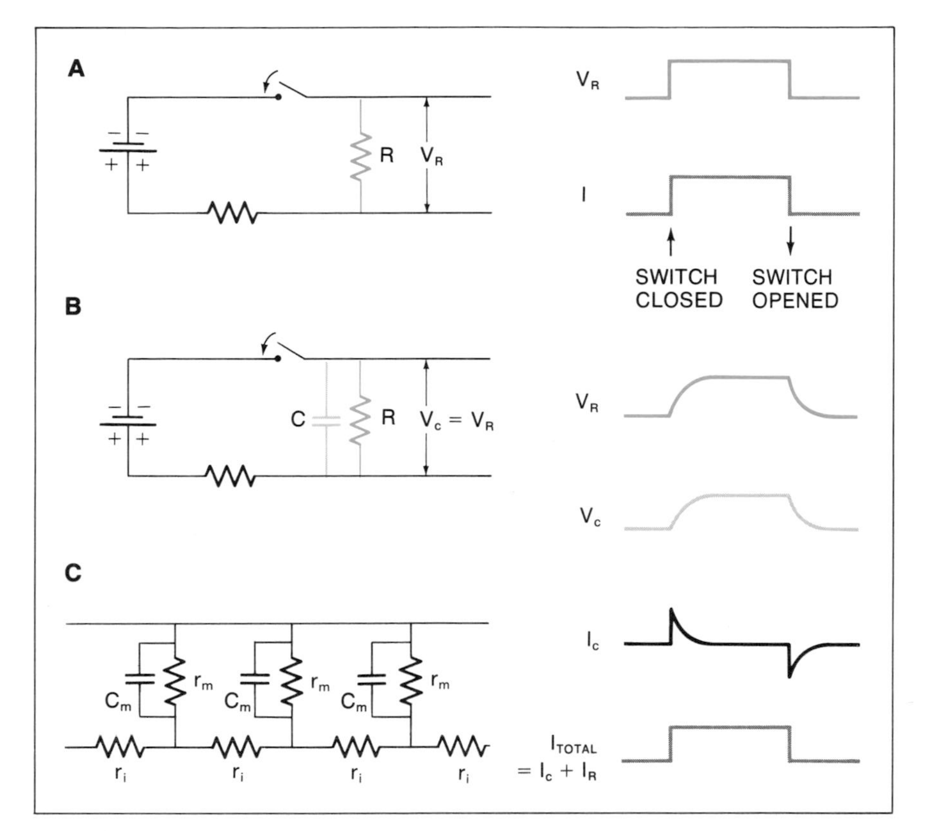

EFFECT OF CAPACITY on time course of potentials. A. Time course of potential (V_R) resulting from current (I) in a purely resistive network is not distorted. B. With the introduction of capacity, a surge of current (I_C) flows through the capacitor when the switch is closed and opened; this is responsible for the gradual buildup and decline of potential shown in V_C and V_R. Such a situation applies in a nerve membrane, whose equivalent circuit is shown in C.

2

giving rise to the initial surge of current (I_C). At the end, when the capacitor is fully charged, $I_C = 0$. Only then does all the current flow across the resistor with the potential at its final steady value. For a simple circuit with a resistor and a capacitor in series, the voltage change produced by a steady current switched on at $t = 0$ is given by the equation:

$$\frac{V_t}{V_o} = (1 - e^{-t/\tau})$$

where the time constant $\tau = RC$, V_o is the voltage at $t = 0$, and V_t is the voltage at time t.

The unit of capacitance is the farad (F). For capacitors arranged in series, the total capacitance is reduced; in parallel, the capacitance increases (see Appendix).

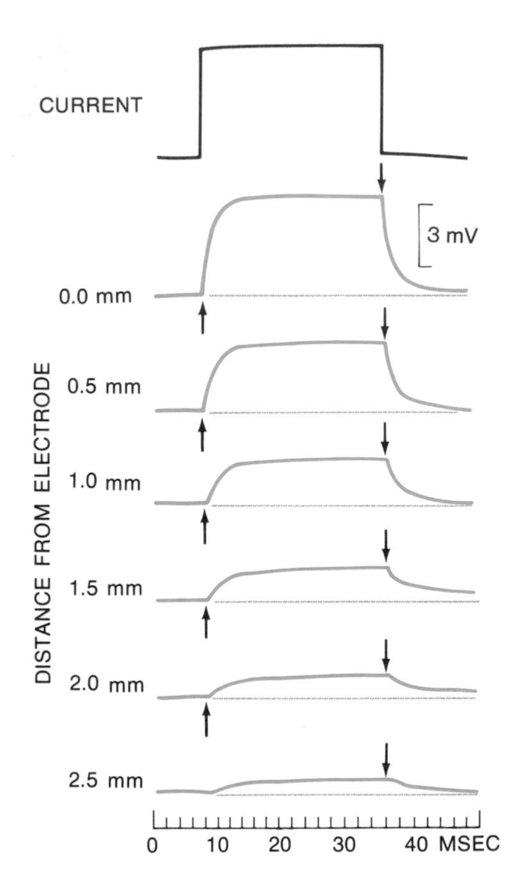

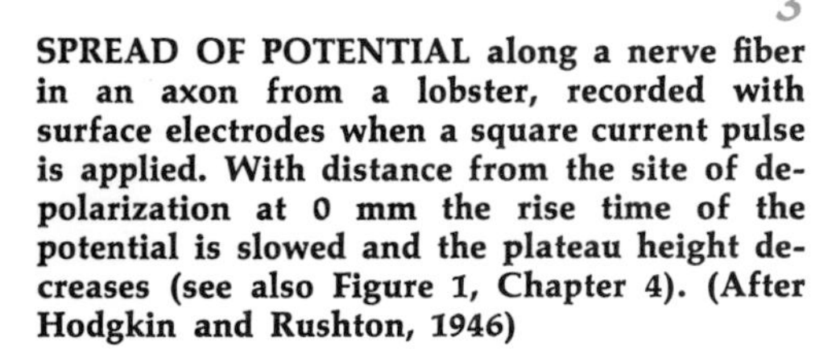
3

SPREAD OF POTENTIAL along a nerve fiber in an axon from a lobster, recorded with surface electrodes when a square current pulse is applied. With distance from the site of depolarization at 0 mm the rise time of the potential is slowed and the plateau height decreases (see also Figure 1, Chapter 4). (After Hodgkin and Rushton, 1946)

The effect of the membrane capacity can also be explained in ionic terms. When positive charges are injected into the cell, potassium ions on the inside move toward the membrane, which has the capacity to store charges and contains only scattered leakage points. For a change in potential (V_C) to occur, a certain number of ions (Q) must accumulate or be displaced from the capacitor (C_m). The movement of these ions takes time and constitutes a capacitative current that alters the charge distribution on the membrane without actual movement of ions through it (Figure 11, Chapter 6). In addition, potassium ions can carry current out of the cell by moving through the permeability channels. Eventually the membrane potential reaches its final value (V_C), whose magnitude is determined by $I \times r_m$. At this potential the capacitor is fully charged, and all the potassium movement occurs through permeability channels. The time required to reach this final value depends on c_m and r_m, that is, on the number of charges that must be siphoned off from the ionic current to charge or discharge the capacity.

The effect of membrane capacity is to distort the time course of rise and fall of the potential change. For neurons the time constant is usually of the order of 0.5 to 5.0 msec. The farther along an axon one records, the slower the potentials become (Figures 2C and 3). This is because the capacity in each successive region of the nerve adds its effect to the time course of the signal. The capacity does not reduce the plateau of a sustained potential. However, potentials that are brief in relation to the time constant, such as synaptic potentials, may be attenuated in their peak amplitude as well as slowed.

It has been mentioned that the change of temperature along a metal rod heated at one point is a good analogy for the longitudinal spread of potentials; it is also a good analogy for the time course. The temperature rises more slowly and with a delay at a distance from the part of the rod that is being heated, because the metal has a finite capacity to store heat and each piece must be heated up before the next. In the nerve the site of the capacitor is in the membrane, while in the rod it is within the metal. Nevertheless, the equations, originally designed to describe how heat is lost in a solid conductor, fit the distribution of electrical potentials in an undersea cable or a nerve fiber represented by a circuit diagram, as in Figure 2C. This is why the passive electrical properties are known as CABLE PROPERTIES.

Knowledge of membrane resistance and capacity is often of great value in interpreting signals in neurons. For example, with this information about a cell one can estimate how far a synaptic potential will spread along it. Similarly, one can work backward and estimate the current that gave rise to the synaptic potential. This in turn may set limits on the type of mechanism responsible for producing the potential.

Propagation of action potentials

Conduction of impulses in a nerve is considerably influenced by internal and external resistances, membrane resistance and capacitance, and the threshold. This is important because conduction velocity plays a significant role in the scheme of organization of the nervous system. It varies by a factor of more than 100 in nerve fibers that transmit messages of different information content. In general, those nerves that conduct most rapidly, at 100 m/sec or more, are involved in mediating rapid reflexes that need instant attention, such as those used for regulating posture; slower conduction frequently serves responses involving internal organs—heart, gut, glands, and blood vessels.

To illustrate how localized, graded potentials are involved in propagation, the action potential can be plotted along a nerve fiber in terms of its spatial distribution at an instant. Knowing the conduction velocity and duration of the action potential, one can estimate the spread of potential on either side of the peak. In Figure 4 the impulse is shown frozen in time. An instant later it would have moved onward to the right. The peak of the action potential corresponds to a transient reversal of the normal membrane potential, the inside being positively charged with respect to the outside. For regions of the membrane above threshold, the net movement of positive charges is inward, as sodium ions run into the fiber and depolarize the adjacent region of the membrane in advance of the impulse. The outward current in a distant region where the depolarizing voltage is still subthreshold is mainly capacitative; as is true of subthreshold currents considered earlier, positive charges accumulate on the inside surface and are removed from the outside. A short time later, as threshold is reached, the membrane current reverses its direction and inward sodium movement dominates.

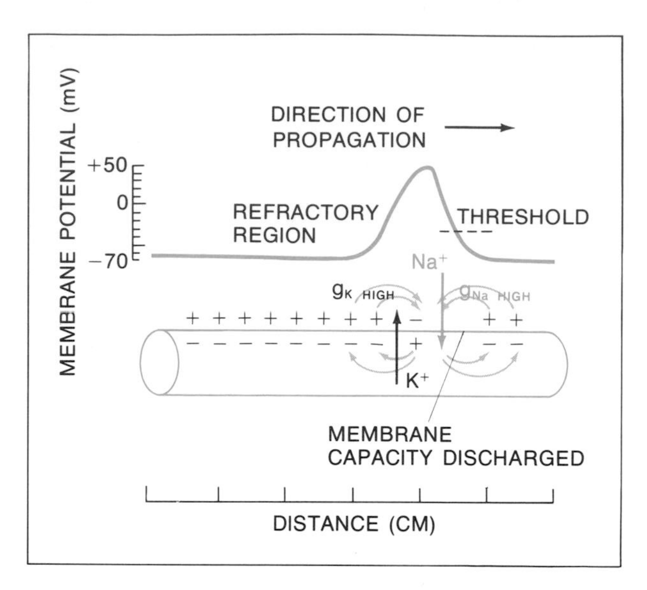

4

CURRENT FLOW during a nerve impulse at one instant in time. Positive charges move to the depolarized area from the outside surface of the membrane ahead of the impulse, discharging the membrane capacity. At threshold, sodium ions carry positive charges into the fiber, while in the region of nerve being repolarized positive charges are carried out by potassium.

If an action potential is set up in the middle of an axon, it propagates both ways. Normally in the body the impulses move in one direction only. The region through which an impulse has just passed cannot be reexcited because of the refractory period. In this area, indicated in Figure 4, g_K is high, and potassium ions move out to repolarize the membrane. The repolarizing potassium current far exceeds the depolarizing effect of current spread from the adjacent active region.

The cable properties of the nerve influence the conduction velocity. If the length constant (λ) is large, currents spread farther ahead of the "active" impulse region and the action potential propagates more rapidly. Conversely, the greater the internal resistance, the capacitance, and the external resistance, the slower the conduction velocity. As already noted, the internal resistance depends mainly on the cross-sectional area of the fiber, because the specific resistance of axoplasm is the same in all the nerves in a particular animal. The internal resistance, therefore, varies inversely with the square of the radius, which is one of the most important variables in determining conduction velocity. Except in the case of myelinated nerves (see below), the external resistance can be ignored.

Evidence of involvement of local circuits in the conduction of action potentials

Experiments by Hodgkin[2] in 1937 clearly demonstrate the role of local circuits in the propagation of action potentials. In the first series of experiments he blocked conduction by applying pressure or by cooling a small region of the frog sciatic nerve and showed that electrical potentials did spread through the blocked region. When impulses initiated some distance away arrived at the block, some of their currents spread through the blocked area and produced a depolarization for several millimeters beyond it. This is illustrated in Figure 5, in which the recording electrode was moved to five different positions past

[2]Hodgkin, A. L. 1937. *J. Physiol.* *90*:183–210, 211–232.

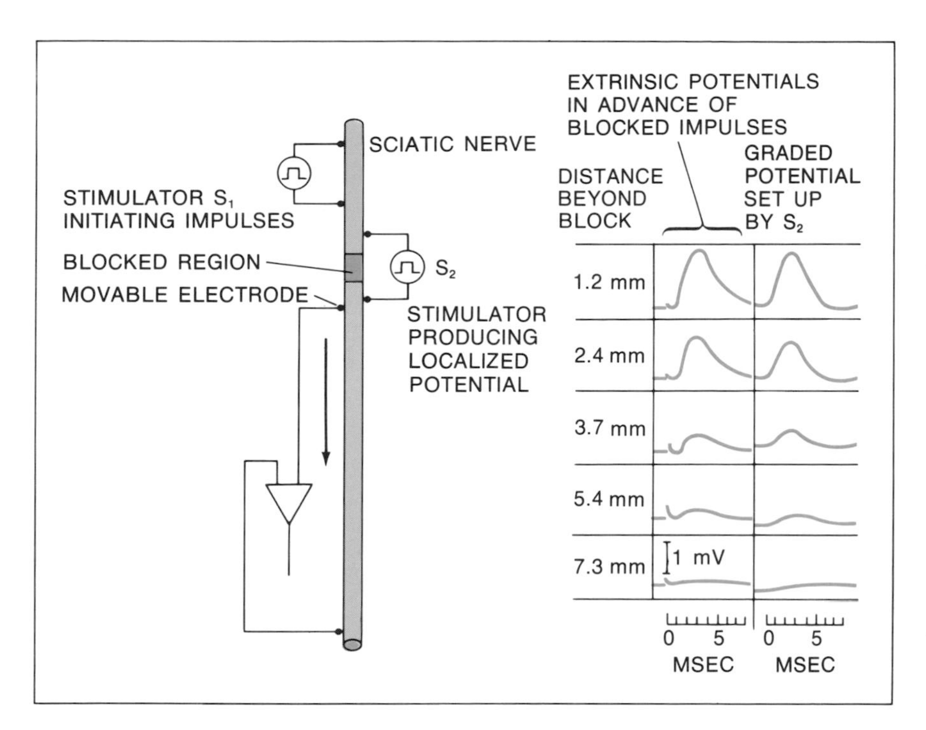

THE ROLE OF CURRENT FLOW in the propagation of impulses. Nerve impulses in a sciatic nerve generate currents that spread across a blocked region, where they cause depolarizing potential changes that decline with distance. Similar potentials that decrease beyond the block can be set up by passing a small current pulse through electrode S_2. (After Hodgkin, 1937)

5

the blocked region. The decline with distance was no different if the blocked nerve impulses were replaced with an electrode (S_2) that passed a small current through the area just before the block. Hodgkin also made tests which showed that the excitability at various points beyond the block changed in parallel with the spread of the potentials produced either by a blocked action potential or by artificial electrical current pulses. Thus, if the membrane had already lost some of its resting potential (was depolarized) it was nearer to threshold and a smaller current was needed for excitation.

In another series of experiments,[3] local circuit theory was further tested by showing that alterations in the external resistance affect the conduction velocity in the expected way. A large axon from a crab was stimulated at one end while electrical recordings of the action potential were made at the other. The experiment was done first in seawater, in which the external resistance (r_o) is low; then a high external resistance was produced by replacing the bathing solution with mineral oil, leaving only a thin film of seawater around the axon. As expected, the conduction velocity slowed when the nerve was pulled into the oil.

[3]Hodgkin, A. L. 1939. *J. Physiol.* 94:560–570.

Finally, in the converse experiment, the external resistance was reduced even below the normal r_o in Ringer's solution by laying the nerve along a series of separate silver plates that could be linked by throwing a switch. When the silver plates were separated, the action potential traveled at its normal velocity. With the silver plates linked together, however, the action potential increased in velocity, owing to the decreased external resistance.

Myelinated nerves and saltatory conduction

A good test for the local circuit theory of transmission of impulses has been obtained in myelinated nerves, the rapidly conducting axons in our bodies. Myelin is formed by Schwann cells which during development wrap themselves around axons. Since the fused membranes have a high resistance and a low capacity, the wrapping acts as an insulating material. At intervals of about 1 mm the myelin is interrupted, exposing patches of axonal membrane (the nodes of Ranvier). Since ions cannot flow in or out across the high-resistance myelinated portion of a nerve fiber, the ionic currents during an impulse take the relatively low-resistance path through the nodes of Ranvier. Only a small restricted portion of the axonal membrane becomes involved, and relatively few charges are needed to displace the membrane potential. As a result,

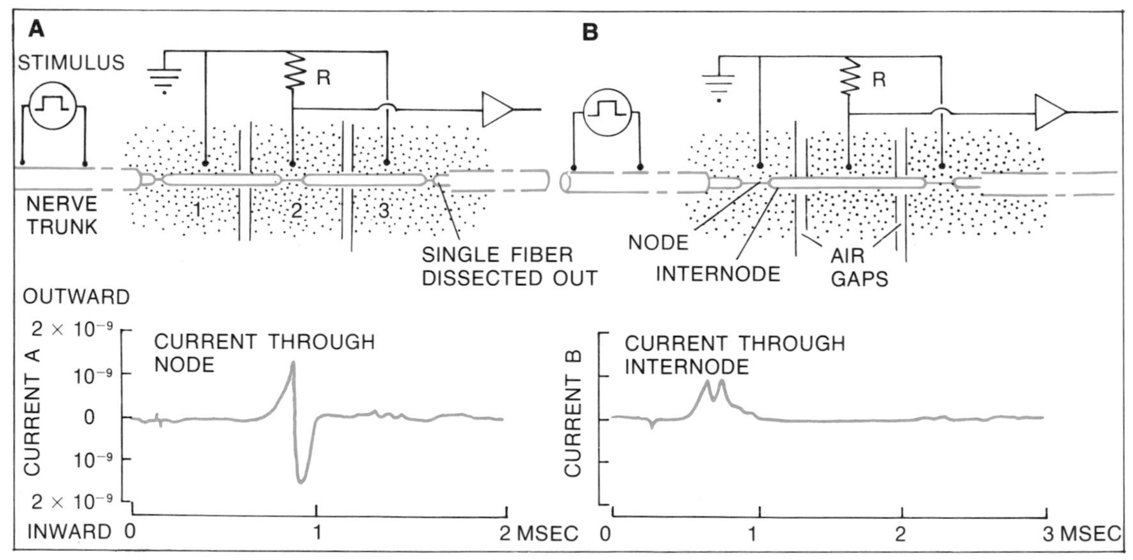

6 **CURRENT FLOW THROUGH A MYELINATED AXON. A single myelinated nerve fiber passes through two air gaps which create three compartments that are not linked by extracellular fluid. Radial or membrane current during the propagated action potential is recorded as a voltage drop across the resistor R. In A the node of Ranvier is in compartment 2. The first deflection represents the depolarizing outward current, followed at threshold by an inward current as sodium enters the node. In B an internodal region is in compartment 2. Only outward capacitative current is seen. (After Tasaki, 1959)**

the impulse propagates more rapidly and jumps from node to node (hence the name SALTATORY).

Experiments to demonstrate saltatory conduction were first made in 1941 by Tasaki[4] and later by Huxley and Stämpfli,[5] who recorded the current at the nodes and internodes. According to the saltatory hypothesis, only outward current should be recorded in the internodal regions during the time the small capacitance of the internode becomes discharged; but at the nodal regions there should be recorded first an outward current corresponding to the subthreshold depolarization of the membrane (as its capacity is discharged), followed by a large inward surge of current as the node becomes active and sodium enters the fiber. Records obtained from myelinated nerves confirm this scheme (Figure 6).

By restricting the area over which ionic movements take place to the small nodal region, myelin not only speeds conduction but also increases the metabolic efficiency of the axon. Fewer sodium ions enter the cell with each impulse and therefore fewer must be pumped back to restore the original concentrations (Chapter 12).

Another demonstration of the essential role of local circuits is provided by an experiment in which a single myelinated nerve fiber is placed in three pools of Ringer's fluid (Figure 5*B*), while the internode lies in a sucrose solution or in an air gap of infinite resistance. Under these conditions the impulse is blocked unless the external circuit is completed by a salt bridge connecting the pools of Ringer's fluid.[4]

Recording from nerves with external electrodes

For much physiological work in the central and peripheral nervous system extracellular electrodes are used to stimulate or record from neurons of various diameters. During extracellular recording, potentials are registered that are generated in the extracellular fluid by current flow. The amount of current that individual neurons can contribute depends on their diameter. The larger the fiber, the greater its cross-sectional area and surface, and the lower its effective resistance (which depends on r_i and r_m). As a result larger fibers produce larger currents during an impulse and larger signals at external electrodes. By the same token, when external electrodes are used to stimulate a nerve, the largest fibers usually require the least current and therefore have the lowest thresholds. This situation is frequently important for various experimental approaches, because it enables stimulation of large nerves selectively, without involving the groups of smaller axons.

A quite different method of registering impulse activity may, in time, offer considerable advantages because it dispenses with the conventional electrode system. The method being developed by Cohen and his colleagues[6] relies on dyes that fluoresce under the influence of a potential difference such as is generated by nerve impulses. Replacing electrical

---

[4]Tasaki, I. 1959. In H. W. Magoun (ed.). *Handbook of Physiology*, Vol. I. American Physiological Society, Bethesda, Md.

[5]Huxley, A. F., and Stämpfli, R. 1949. *J. Physiol.* *108*:315–339.

[6]Salzberg, B. M., Davila, H. V., and Cohen, L. B. 1973. *Nature* *246*:508–509.

with optical recording systems may make it easier to follow simultaneously the separate activity of many individual neurons or clusters of cells.

SUGGESTED READING

Papers marked with an asterisk (*) are reprinted in Cooke, I., and Lipkin, M. 1972. *Cellular Neurophysiology.* Holt, New York.

Full treatment of passive electrical spread

*Hodgkin, A. L., and Rushton, W. A. H. 1946. The electrical constants of a crustacean nerve fibre. *Proc. R. Soc. Lond.* B *133*:444–479.

Myelinated nerve

*Huxley, A. F., and Stämpfli, R. 1949. Evidence for saltatory conduction in peripheral myelinated nerve fibres. *J. Physiol. 108*:315–339.

Rushton, W. A. H. 1951. A theory of the effects of fibre size in medullated nerve. *J. Physiol. 115*:101–122.

Waxman, S. G. 1972. Regional differentiation of the axon: A review with special reference to the concept of the multiplex neuron. *Brain Res. 47*: 269–288. (Summarizes recent work that correlates structure with function in myelinated axons.)

CHAPTER EIGHT

SYNAPTIC TRANSMISSION

Synaptic transmission—the transfer of signals from one cell to another—can be treated as an extension of the principles that have been discussed for conduction of nerve impulses. Two distinct modes of transmission are known, one electrical and the other chemical.

At electrical synapses currents generated by an impulse in the presynaptic nerve terminal spread directly into the next neuron through a low-resistance pathway. The sites for electrical communication between cells have been identified in electron micrographs as gap junctions, in which the usual intercellular space of several tens of nanometers is reduced to about 2 nm.

At chemical synapses the fluid-filled gap between pre- and postsynaptic membranes prevents a direct spread of current. Instead, the nerve terminals secrete a specific substance, the transmitter. This diffuses across the synaptic gap to the postsynaptic cell, where it changes the permeability of the membrane to specific ions. Depending on the species of ions, either excitation or inhibition results. The direction of current and the resulting synaptic potentials depend on the particular ions, their concentrations on both sides of the membrane, and the value of the membrane potential at the time the transmitter acts. The membrane potential of the postsynaptic cell does not significantly affect the specificity or the magnitude of the permeability change produced by the transmitter. In the best studied case of chemical excitation, at vertebrate neuromuscular synapses, the chemical transmitter, acetylcholine (ACh), produces a simultaneous increase in the permeability to sodium and potassium, and this leads to a depolarizing synaptic potential.

At chemical inhibitory synapses the permeability to potassium or chloride, or both, is increased, counteracting depolarization and driving the postsynaptic membrane potential away from threshold. Nerve cells in the central nervous system receive both excitatory and inhibitory synapses whose effects are graded. The generation of impulses depends on the balance between the excitatory and inhibitory influences.

A second mode of inhibition is presynaptic; the inhibitory nerve terminal secretes its transmitter onto a neighboring excitatory nerve terminal. This leads to a reduction in the output of excitatory transmitter. It serves to eliminate selectively certain pathways converging on a cell without influencing the effectiveness of others.

Chemical synaptic transmission enables a small terminal to generate considerable amounts of current in a large postsynaptic cell. The transmitter opens ionic permeability gates and acts as a trigger for releasing stored energy.

Initial approaches

In the second half of the nineteenth century a vigorous discussion took place between proponents of the CELL THEORY, who considered that neurons were independent units, and those who believed that nerve cells were generally connected to each other by protoplasmic bridges. It was not until the early part of this century that the cell theory won general acceptance and most biologists started to think of nerve cells as being similar to other cells in the body. Although in retrospect the opponents of the cell theory may appear unnecessarily stubborn, one should realize that at the time it was difficult to obtain convincing histological evidence to show whether or not there was continuity between neurons. It remained for electron microscopy to show that each neuron is completely surrounded by its own membrane. As long as the issue of continuity versus contact between neurons remained unsettled, relatively little experimentation was done on how impulses were transmitted between cells. Even now, continuity must be considered, but in the form of tenuous connections of molecular dimensions that link the interiors of certain cells. There are known to exist direct routes that permit the flow of ions and small molecules between cells.

Since unexpected modes of synaptic transmission have still been found in the past dozen years, such as electrical synaptic excitation and inhibition, and since additional forms of interaction or specific exchanges between cells are likely to become established, it is instructive to review briefly the development of a few of the relevant ideas.

Concept of chemical transmission between cells

In 1843, Du Bois-Reymond showed that flow of electric current was involved not only in muscle contraction but also in nerve conduction, and it was but a small extension of this notion that transmission of excitation between nerve and muscle was brought about by current flow.[1] The idea of "animal electricity" had a potent hold on people's thinking. For almost 100 years, no contrary evidence could budge the attractive assumption of "electrical transmission" between cells; the idea was further extended to excitation between nerve cells in general, even though DuBois-Reymond himself had stated an alternative explanation—the secretion by nerve of an excitatory substance that then

[1]Du Bois-Reymond, E. 1848. *Untersuchungen über thierische Electricität.* Reimer, Berlin.

caused contraction. In fact, he favored this mechanism over electrical transmission.

The idea of chemical transmission was not attractive because it was known that between nerve and skeletal muscle, and especially between nerve cells in the central nervous system, transmission takes place in a small fraction of a second. No such difficulties existed in theory for autonomic nerves (sympathetic and parasympathetic) that innervate glands and smooth muscles lining viscera and blood vessels, because they act relatively slowly. Accordingly, evidence from experiments on such tissues carried little weight in the argument. For example, Elliot[2] in 1904 pointed out that an extract of the adrenal gland, adrenaline (also called epinephrine), mimicked the action of sympathetic nerves, and he cautiously suggested that it might be secreted by nerve terminals and act as a transmitter. He was discouraged from even fully publishing his results. Interesting sidelights on the men and their scientific ideas about this subject at the turn of the century and later are contained in the writings of Dale,[3] for several decades one of the leading figures in British physiology and pharmacology. Among his many contributions were the clarification of the action of acetylcholine (ACh) at synapses in autonomic ganglia, the establishment of its role in skeletal neuromuscular transmission, and the elucidation of the action of histamine and ergot alkaloids. As one of our colleagues has delicately put it: "He walked in a cow pasture and trod only on the daisies."

Henry Dale (left) and Otto Loewi, mid 1930's. (Courtesy of Lady Todd and W. Feldberg.)

In 1921, Otto Loewi did an experiment whose directness and simplicity made the idea of chemical transmission more attractive to many physiologists.[4] He perfused the heart of a frog and electrically stimulated the vagus nerve, causing stoppage or slowing of the heartbeat. When the fluid from the inhibited heart was transferred to a second unstimulated heart, this heart also began to beat more slowly. Apparently, the vagus nerve had released into the perfusing solution an inhibitory substance. Loewi and his colleagues later showed (by various bioassays) that this substance was indistinguishable from ACh.

It is an amusing sidelight that Loewi had the idea for the experiment in a dream, wrote it down in the middle of the night, but could not decipher the writing the next morning. Fortunately, the dream returned, and this time Loewi took no chances; he rushed to his laboratory and performed the experiment. In a personal account, he reflected:

---

[2]Elliot, T. R. 1904. *J. Physiol.* *31*:20.
[3]Dale, H. H. 1953. *Adventures in Physiology.* Pergamon Press, London.
[4]Loewi, O. 1921. *Pflügers Arch.* *189*:239–242.

> On mature consideration, in the cold light of the morning, I would not have done it. After all, it was an unlikely enough assumption that the vagus should secrete an inhibitory substance; it was still more unlikely that a chemical substance that was supposed to be effective at very close range between nerve terminal and muscle be secreted in such large amounts that it would spill over and, after being diluted by the perfusion fluid, still be able to inhibit another heart.

Subsequently, in the early 1930s support in favor of ACh as the chemical synaptic transmitter was extended to other autonomic synapses whose presynaptic neurons liberate ACh. In particular, the evidence obtained by Brown, Feldberg, Gaddum, MacIntosh, and their colleagues[5] was strong in sympathetic ganglia.

Pharmacological approaches were indispensable for such experiments. For example, the hydrolysis of ACh was prevented by a drug, eserine, that inhibits the enzyme cholinesterase. Another indispensable tool was curare (Figure 2*A*), an Indian arrow poison, at that time prepared as a rather crude extract of cinchona bark; this drug was used in the last century by Claude Bernard, who showed that it blocked neuromuscular transmission. Langley confirmed this observation in a number of studies made in the early 1900s.

A turning point came in 1936, when Dale and his colleagues demonstrated that stimulation of motor nerves to skeletal mammalian muscle caused the release of ACh.[6] In addition, injection of ACh into arteries supplying a muscle showed that it had the required stimulating action, since it caused a large synchronous twitch.

The evidence, therefore, for chemical transmission at the end of the 1930s was strong, not only at autonomic synapses, but also in the rapidly acting skeletal muscles. There remained a number of vexing problems; for example, it was not possible to demonstrate that ACh was actually released during the critical period after the arrival of the motor nerve impulse at the terminals and before the initiation of the muscle action potential at the neuromuscular junction.

Synaptic potentials at myoneural junctions

A new phase in the study of synaptic transmission started with the use of improved electrical recording techniques on mammalian and amphibian nerve-muscle preparations. Figure 1 illustrates diagrammatically a vertebrate neuromuscular synapse and a nerve-nerve synapse. The area of nerve-muscle contact is called the END PLATE, and the potential change recorded in that region was originally named "end plate potential" (epp).[7,8] It is now clear that there is no real difference between end plate potentials at myoneural junctions and synaptic potentials at nerve-nerve junctions. The distinction, which seemed advisable during the initial phases of the work, has lost its usefulness. In this chapter end

---

[5]Feldberg, W. 1945. *Physiol. Rev. 25*:596–642.
[6]Dale, H. H., Feldberg, W., and Vogt, M. 1936. *J. Physiol. 86*:353–380.
[7]Göpfert, H., and Schaefer, H. 1938. *Pflügers Arch. 239*:597–619.
[8]Eccles, J. C., and O'Connor, W. J. 1939. *J. Physiol. 97*:44–102.

plate potentials are called SYNAPTIC POTENTIALS, corresponding to the analogous potentials elsewhere in the nervous system. A synaptic potential that excites a postsynaptic cell is generally abbreviated as epsp and one that inhibits as ipsp.

The synaptic potentials are confined to the vicinity of nerve terminals in skeletal muscle and have turned out to be of particular importance. They provided the missing link in electrical events between the all-or-none motor nerve impulse and the analogous muscle impulse that triggers contraction. As a consequence, the study of the mode of transmission, either by a chemical transmitter or by electric current

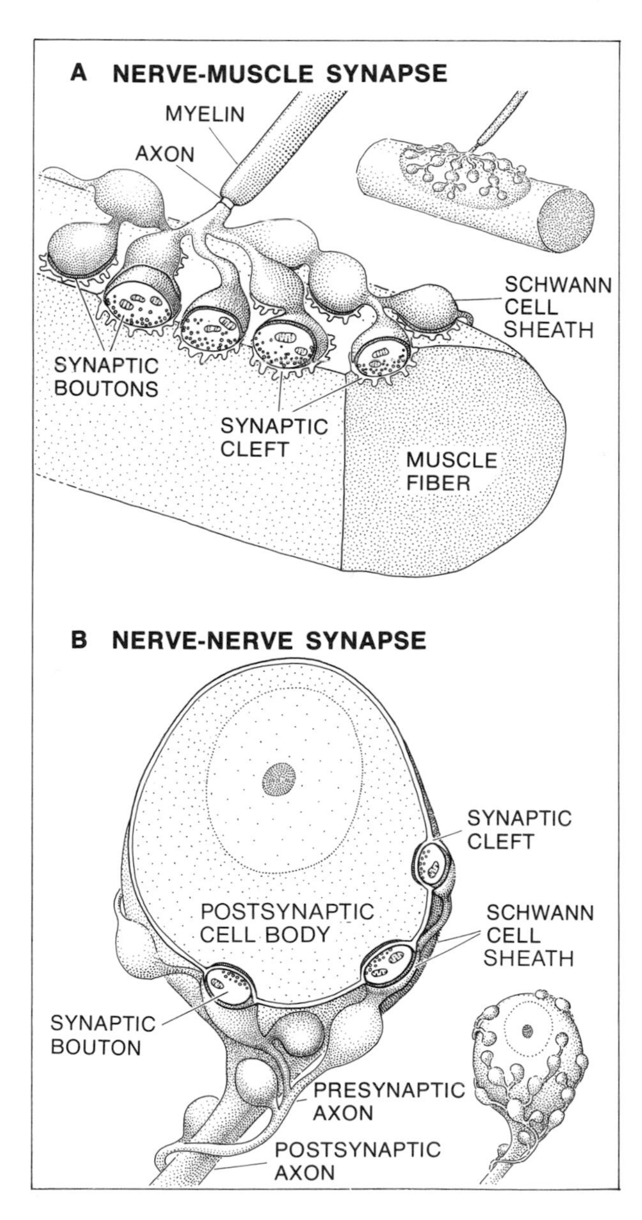

1

STRUCTURAL FEATURES OF CHEMICAL SYNAPSES. A. The end plate formed by a motor nerve on a skeletal muscle fiber of the snake consists of a collection of synaptic boutons. Boutons contain synaptic vesicles and mitochondria and are always separated from the postsynaptic cell by a cleft about 50 nm wide. They are covered by fine lamellae of Schwann cells. B. Similar synaptic boutons are distributed on a ganglion cell in the heart of the frog. For details see Chapter 10. (After McMahan and Kuffler, 1971)

spread, concentrated from 1939 onward on the question of the manner in which the nerve impulse gives rise to the synaptic potential at nerve-muscle junctions or at junctions between neurons in the nervous system.

According to the chemical hypothesis, ACh was the agent underlying the synaptic potential. Experiments soon tested this proposition. For example, in the presence of anticholinesterases, such as eserine, the hydrolysis of ACh is inhibited, allowing ACh to remain longer at the site of its liberation. As a result, the size and total duration of the synaptic potential is greatly prolonged (Figure 2*B*). This finding supported the proposition that ACh, released by nerve terminals, is responsible for the synaptic potential; that is, its release occurs at the required critical period.[9] Figure 2 also illustrates the usefulness of curare, which enables the synaptic potential to be observed on its own without the action potential. Several lines of evidence showed that curare acts postsynaptically to reduce the sensitivity to ACh without affecting release (see also Chapter 10). Subsequently, work on isolated synapses of single nerve-muscle fibers showed that no electrical event occurs prior to the synaptic potentials which are directly linked with the initiation of the postsynaptic impulse (Figure 2*A*). Moreover, the currents generated by the nerve impulses in motor nerve terminals do not spread effectively into the postsynaptic region and cannot be responsible for the epsp.[10]

Bernard Katz (right) John Eccles and friend in Australia, about 1941.

These and similar studies with extracellular electrodes gave conclusive support for chemical transmission at nerve-muscle junctions. The full evidence for the parallel between chemical transmission at various neural and neuromuscular synapses is discussed below and in Chapters 9 and 10.

The most rapid advances occurred after the introduction of intra-

---

[9]Eccles, J. C., Katz, B., and Kuffler, S. W. 1942. *J. Neurophysiol.* 5:211–230.
[10]Kuffler, S. W. 1948. *Fed. Proc.* 7:437–446.

cellular recording methods,[11] which from 1950 on were used extensively. Only then did the presynaptic and postsynaptic sites become available for the direct measurement of membrane potentials and ACh sensitivity.

An important step was the technique devised by Nastuk[12] for studying the excitatory action of ACh applied directly. One microelectrode is used to measure the membrane potential of a skeletal muscle fiber at the end plate, while a movable micropipette, filled with ACh, is held just outside the cell (Figure 3*A*). To apply ACh to a localized region of the membrane, a brief current is passed through the pipette, causing positively charged ACh ions to be carried out. In this way, by IONTOPHORESIS, it can be shown that ACh has a pronounced depolarizing effect in the region of the motor end plate (Chapter 10), provided it is applied to the outside of the membrane; intracellular injection is without effect. The iontophoretic method has also made possible detailed mapping of the distribution of ACh receptors in the postsynaptic membrane of certain nerve cells[13] as well as muscle fibers.[14] Figure 3 illus-

---

[11]Ling, G., and Gerard, R. W. 1949. *J. Cell Comp. Physiol. 34*:383–396.

[12]Nastuk, W. L. 1953. *Fed. Proc. 12*:102.

[13]Dennis, M. J., Harris, A. J., and Kuffler, S. W. 1971. *Proc. R. Soc. Lond.* B *177*:509–539.

[14]Miledi, R. 1960. *J. Physiol. 151*:24–30.

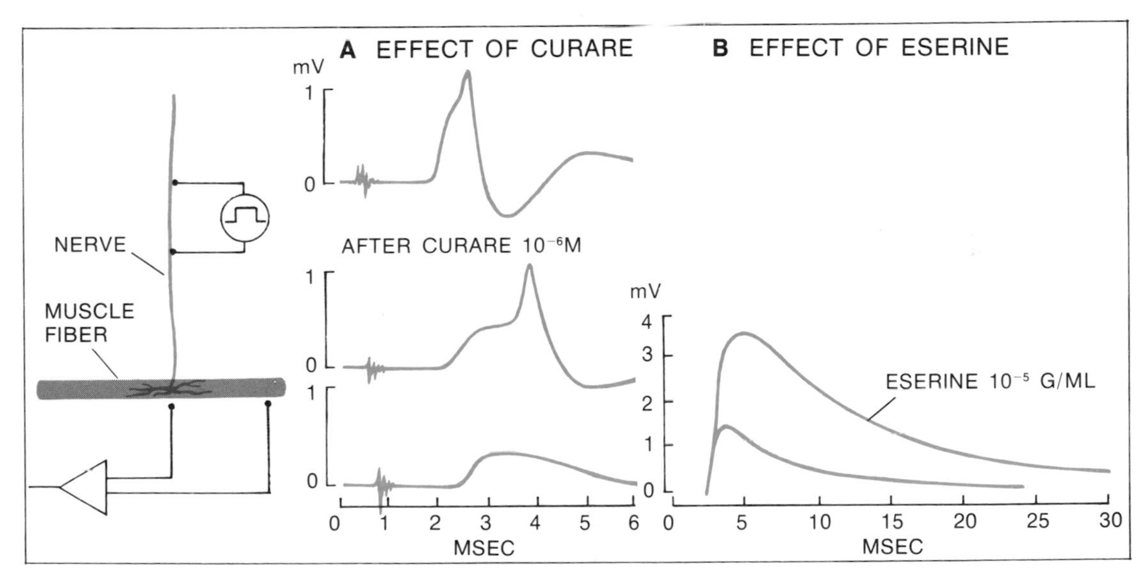

EFFECT OF CURARE AND ESERINE on synaptic potential. At a neuromuscular synapse nerve stimulation leads to a synaptic potential that initiates a muscle impulse (upper record in A). In the presence of curare the separation of these two events is more pronounced, and eventually only the synaptic potential remains (bottom record in A). In B the curarized synaptic potential (lower record) is increased in size and prolonged by eserine, which prevents the hydrolysis of ACh. The potentials were recorded with extracellular electrodes. (A after Kuffler, 1942; B after Eccles, Katz, and Kuffler, 1942) 2

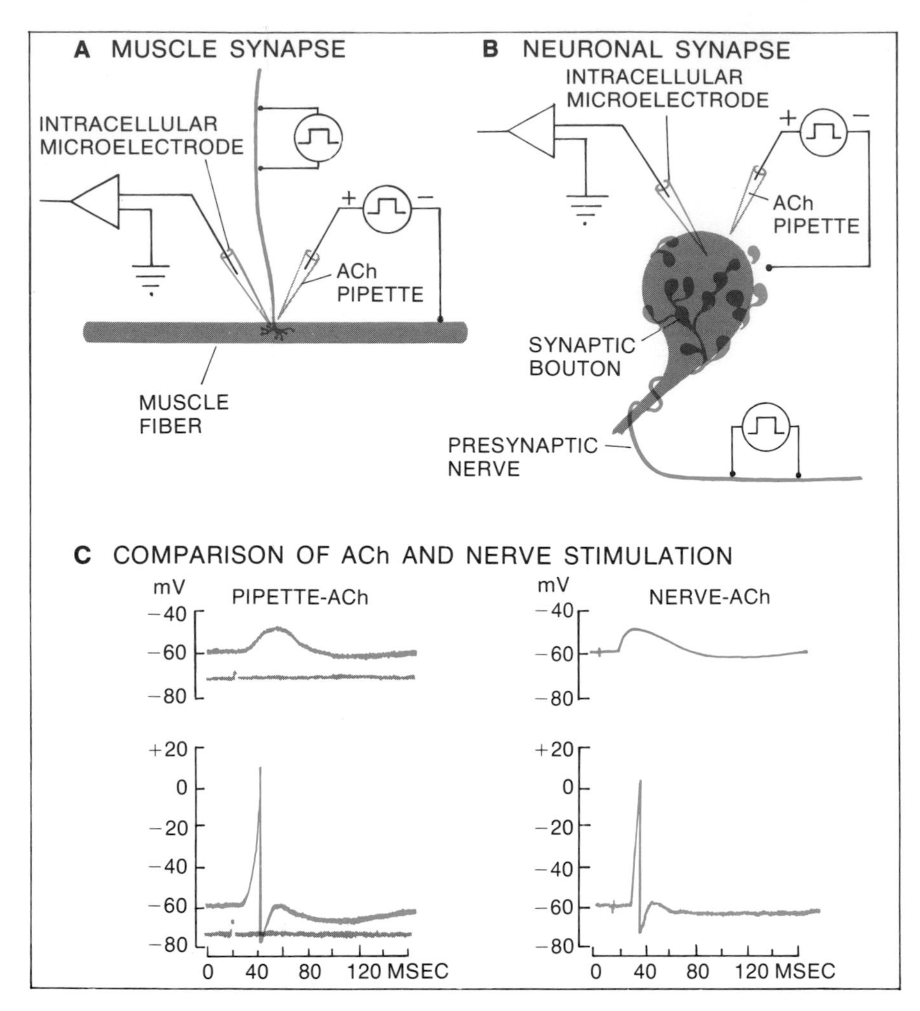

3 IONTOPHORESIS OF ACH. ACh applied directly by a pipette to a neuromuscular synapse (A) and to a synapse on a ganglion cell in the heart (B) mimics nerve stimulation. C shows a nerve-evoked synaptic potential in a ganglion cell (upper right) that in the lower record initiates an impulse. The records on the left were obtained from the same nerve cell with brief pulses through an ACh-filled micropipette. The first pulse is subthreshold (upper record). The potentials were recorded with an intracellular electrode. (After Dennis, Harris, and Kuffler, 1971)

trates the good correlation between the effects of ACh produced by a micropipette and by the nerve at a synapse between two neurons in the frog heart. The chemosensitivity is highest in the immediate vicinity of the nerve terminals, at the synaptic boutons where ACh is released. Release at a distance of several microns from the synaptic area is much less effective. One clear difference between ACh released from the nerve and from the pipette is that an appreciable delay, close to

0.5 msec, occurs between the arrival of a nerve impulse in the terminal and its postsynaptic action, a delay that is absent when ACh is applied directly. This is discussed in Chapter 9, since it is important for understanding the release mechanism of transmitter.

The development of microelectrodes also enabled more accurate examination of the spread of synaptic potentials along the muscle fiber. A new sequence of studies was initiated by Fatt and Katz,[15] who measured the time course and amplitude of the potentials at various distances from the end plate (Figure 4). They found that the decrease in amplitude with distance and the slowing of the rising and falling phases can be entirely accounted for by the resistance and capacity of the muscle membrane. This confirmed the end plate potential as a conventional, localized, graded synaptic potential whose spread depends upon the passive electrical properties of the muscle membrane. From these studies one can derive the amplitude and the duration of the ionic current that gives rise to the voltage change. In their estimates Fatt and Katz showed that the synaptic current produced by nerve stimulation is a brief surge that does not last as long as the potential change. They also provided an explanation of the mechanism whereby ACh produces ionic current flow. This is discussed later, together with voltage clamp studies. The essential conclusion was that the transmitter acts by increasing the permeability of the postsynaptic membrane to small ions.

Accordingly, by the 1950s chemical synaptic transmission was generally accepted, and alternate explanations had almost disappeared.

---

[15]Fatt, P., and Katz, B. 1951. *J. Physiol.* *115*:320–370.

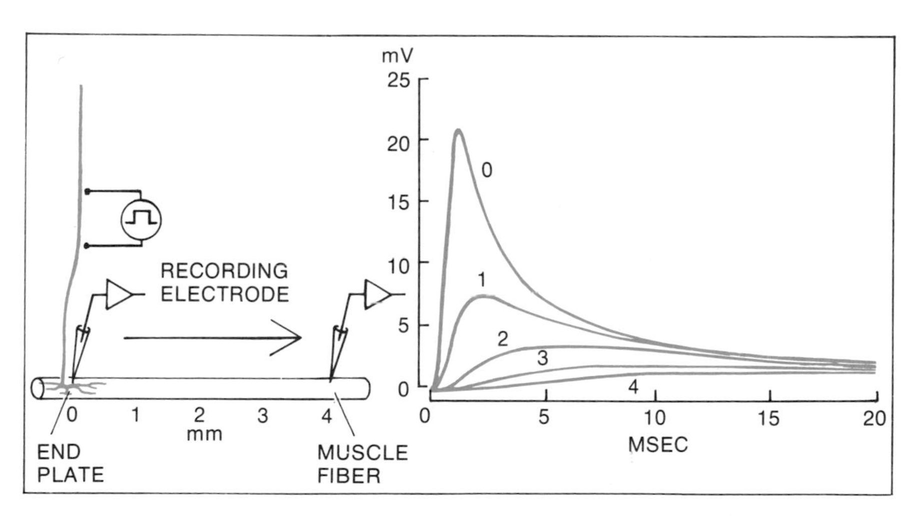

4

SYNAPTIC POTENTIALS at various distances from end plate, showing the decline in size and slowing of the rise time of subthreshold synaptic potentials with distance from the site of release of ACh. An intracellular electrode was inserted into the muscle fiber successively at distances of 1, 2, 3, and 4 mm from the end plate. (After Fatt and Katz, 1951)

Physiologists could pay full attention to exploring HOW chemical transmission operates.

The definite establishment of chemical transmission came just in time for the first direct demonstration of electrical synaptic transmission at an authentic synapse by Furshpan and Potter in 1959.[16] One witnessed an amusing and instructive situation. The attractive and long-championed electrical hypothesis had been almost universally accepted for about 100 years on inadequate evidence and at the wrong type of synapse, the nerve-muscle synapse. About the time it was finally forsaken by its supporters, for very good reasons, the principle of the hypothesis was shown to be valid after all, but at a different set of synapses.

ELECTRICAL SYNAPTIC TRANSMISSION

The electrical hypothesis had never been seriously challenged, especially for transmission of excitation at skeletal neuromuscular junctions. In its simplest form the hypothesis states that the passive current flow responsible for conduction along axons is also capable of transferring a nerve impulse directly across a synapse. If current is to pass from one cell to the next, electrical continuity must be maintained and the cable properties of the neurons must not be interrupted at the contact areas.

A simple scheme based on the original experiments by Furshpan and Potter[16] is shown in Figure 5. In the first model (Figure 5*A*), the adjoining cell membranes of two neurons, A and B, are closely apposed or fused. For the transmission to work, the resistance to current flow at this contact area (r_c) must be very low compared with that of the rest of the cell membrane. If positive charges are passed into cell A through a microelectrode, current flows out across the membrane of cell A, but some also crosses into cell B. If the low-resistance coupling membranes are replaced with a septum whose specific resistance is as high as that of the regular surface membrane (usually about 1000 Ω. sq cm), what sort of barrier will such a transverse membrane provide to the flow of ionic currents? For a fiber with a diameter of 5μm, the interposed membrane patch is equivalent to a resistance of 3000 MΩ in series with the current flow. The input impedance of a 5 μm axon is close to 20 MΩ, so the efficacy of the impulse in cell A for driving depolarizing current through cell B is reduced to less than 1 percent and therefore the conducted impulse is not transmitted.[17] The situation between most nerve cells appears even less favorable for electrical continuity, because the membranes are separated by a gap of several tens of nanometers which is filled with extracellular fluid having a low specific resistance (Figure 5*B*). This extracellular pathway provides a shunt for current to flow away without entering the next cell. The

[16]Furshpan, E. J., and Potter, D. D. 1959. *J. Physiol.* *145*:289–325.

[17]Katz, B. 1966. *Nerve, Muscle, and Synapse.* McGraw-Hill, New York, pp. 100–101.

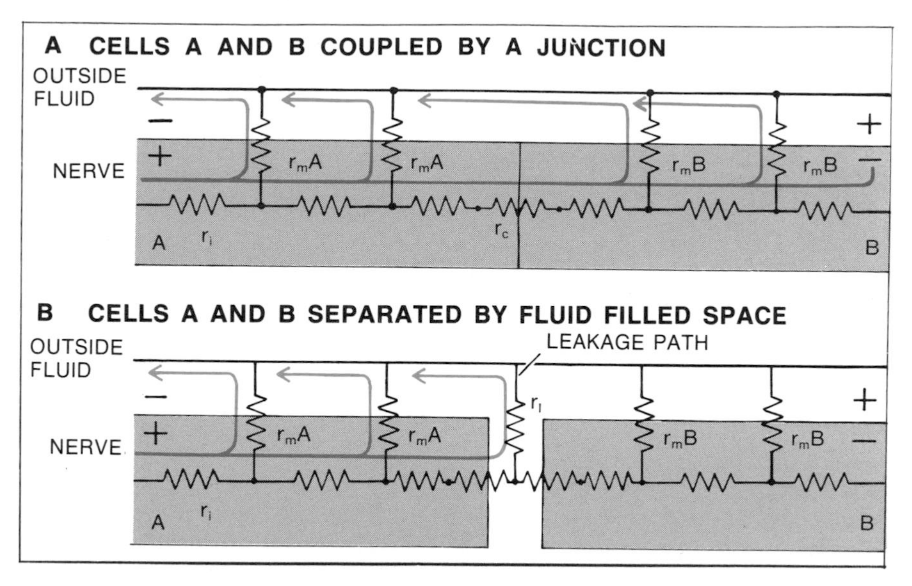

5 **PATHWAYS FOR CURRENT FLOW during electrical synaptic transmission. A. A model of two cells in which current can flow from cell A to cell B. The coupling resistance (r_c) must be low compared with the other membrane resistances (r_m). In B the cells are separated by a fluid-filled gap. Current flow between cells A and B is excluded.**

model in Figure 5*B* has in series with the input impedance of fiber A the high-resistance membrane at the synaptic terminal of cell A, the shunt resistance of the gap, and another high resistance at the other border of the gap in cell B. Electrical transmission is therefore excluded unless some special structure allows electrical continuity to be established between the interior of one neuron and the interior of another (see below).

In addition to the presence of specialized junctional membranes and a high resistance to reduce leakage to the outside fluid, there must be a good match between the sizes of the pre- and postsynaptic processes for effective electrical transmission to occur. A very small presynaptic terminal with a high resistance cannot deliver enough current to depolarize a large postsynaptic cell with a low input impedance even if the junctional membrane has a negligibly low resistance. To return to the earlier analogy for the conduction of heat (Chapter 7), a hot knitting needle cannot heat up a cannonball.

In their original demonstration of electrical transmission, Furshpan and Potter used the synapse between a large nerve in the abdominal nerve cord of the crayfish and a motor nerve that innervates fast-acting flexor muscles of the tail. Stimulating and recording electrodes were inserted into the pre- and postsynaptic fibers on each side of the synapse. Depolarizing current passed with little attenuation from the pre- to the postsynaptic fiber, showing that the cable properties be-

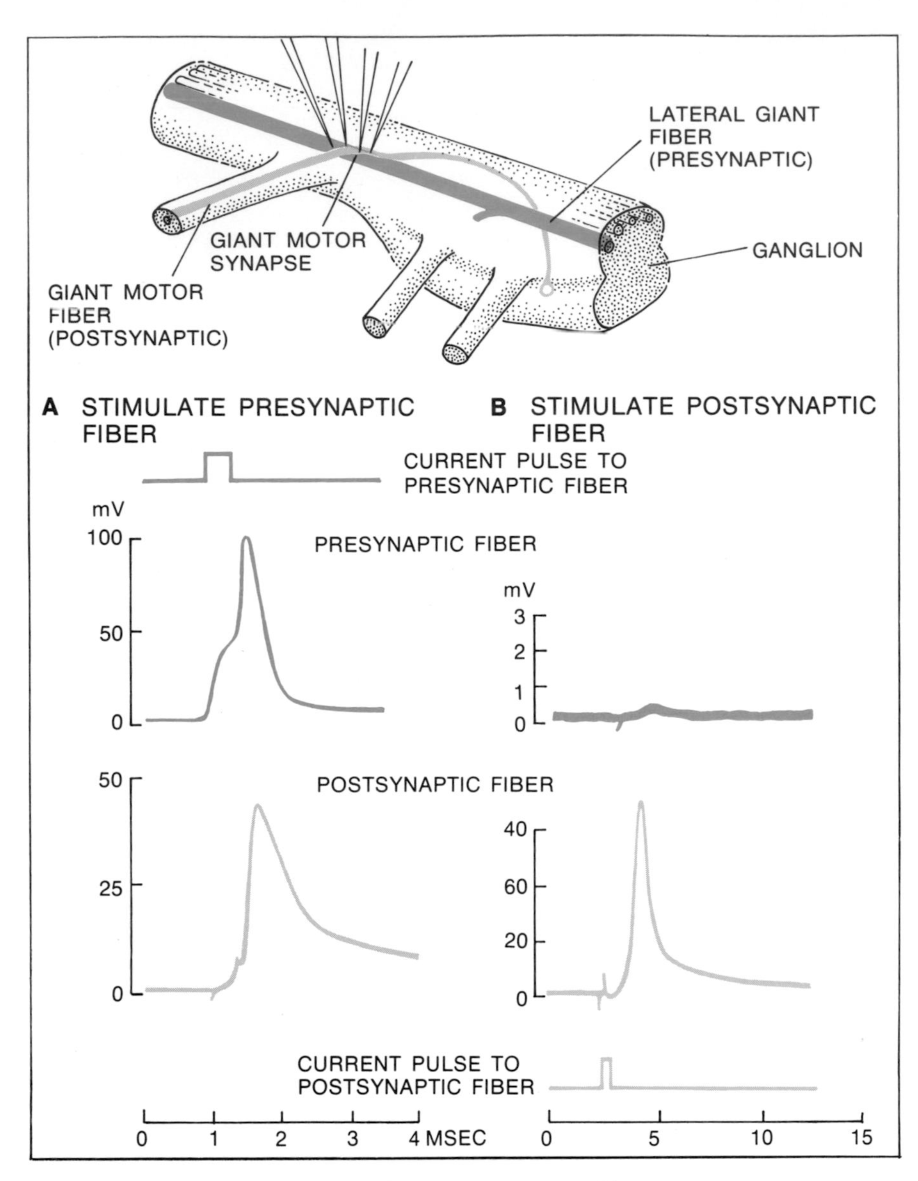

6 **ELECTRICAL TRANSMISSION between two nerve fibers in the nerve cord of the crayfish. Stimulating and recording electrodes are inserted into the presynaptic and the postsynaptic neurons. The current flow generated by an impulse in the presynaptic axon (A) spreads into the postsynaptic fiber. The cable continuity between the two neurons is maintained, and no synaptic delay is detected between the pre- and postsynaptic potentials in the upper and lower records of A. In contrast, depolarization of the postsynaptic cell by a nerve impulse in B does not spread effectively into the presynaptic cell. (After Furshpan and Potter, 1959)**

tween the neurons are maintained and that the leak to the outside fluid is relatively small. Since transmission depends only on the flow of electric current, signals conduct through the junction without the characteristic delay (Figure 6) that occurs at chemical synapses. Another interesting property of this synapse is that the synaptic contact area permits depolarizing current to flow from pre- to postsynaptic fibers, but not in the reverse direction. On the other hand, hyperpolarizing potentials pass readily in the opposite (antidromic) direction. Thus, the junction shows RECTIFYING properties that enable current to flow in only one direction. This allows impulse transmission to go forward but not antidromically; like a chemical synapse, it is polarized.

In addition to the electrically transmitting synapse in the crayfish abdominal nerve cord, electrical transmission between neurons has now been demonstrated in a number of sites in various vertebrate and invertebrate nervous systems. For example, cells electrically coupled to each other have been found in nervous systems of annelids, mollusks, and arthropods.[18] In the medulla of the puffer fish groups of large nerve cells are linked together.[19] In the spinal cord of the frog, recordings from motor neurons in the anterior horn indicate that current can spread directly between groups of cells,[20] and an appropriate structural specialization for the site of coupling between the neurons has been seen.[21] Evidence for electrical coupling of nerve cells has also been obtained in mammals.[22] Many of these synapses do not exhibit rectification, but conduct equally well both ways.

Another interesting example was discovered in the ciliary ganglion of the chick, whose neurons regulate the accommodation of the lens in the eye and the diameter of the pupil. Here electrical and chemical mechanisms are combined in one synapse.[23] It is perhaps more frequently observed that a postsynaptic neuron receives separate chemically and electrically mediated inputs from different fibers that converge on it. This dual arrangement was seen originally in neurons of the crayfish, later in the large Mauthner cells of the goldfish and in the central nervous system of the leech. Some of these cells are considered in detail in Chapters 16 and 17 to illustrate the diverse ways in which different converging influences can be integrated.

Structural basis of electrical transmission

The demonstration of electrical transmission stirred up an intensive search for some specialized relation of the membranes at electrical junctions. When cells known to be electrically coupled are scrutinized in detail, the common finding is that certain regions exist in which the cells come in close contact with each other. Instead of the usual 20 nm cleft, the outer lamellae of the unit membranes at electrical synapses

---

[18]Bennett, M. V. L. 1974. *Synaptic Transmission and Neuronal Interaction.* Raven Press, New York, pp. 153–178.

[19]Bennett, M. V. L. 1973. *Fed. Proc.* *32*:65–75.

[20]Grinnell, A. D. 1970. *J. Physiol.* *210*:17–43.

[21]Sotelo, C., and Taxi, J. 1970. *Brain Res.* *17*:137–141.

[22]Llinas, R., Baker, R., and Sotelo, C. 1974. *J. Neurophysiol.* *37*:560–571.

[23]Martin, A. R., and Pilar, G. 1963. *J. Physiol.* *168*:443–463.

are separated by a space of only about 2 nm. These sites of close apposition appear as round plaques and are called GAP JUNCTIONS. Such gap junctions have been seen in various parts of the mammalian brain—in monkey, cat, rat, and mouse.[24] Two examples from the cat and chicken are shown in Figure 7 (see also Figure 5, Chapter 16). Gap junctions as well as electrical transmission between receptor cells and between horizontal cells have been demonstrated in the retinas of frogs and turtles,[25] and the list of creatures examined continually lengthens.[26]

In contrast to the narrowed intercellular spaces at gap junctions, the two outer lamellae of the apposing membranes are fused in many nonneural cells. These areas of special contact have a pentalaminar appearance and are called TIGHT JUNCTIONS. They occlude the extracellular space and can prevent diffusion along the intercellular spaces if they form a complete seal around the circumference of the cells (Chapter 14).

In conclusion, there now is a large body of evidence in a variety of excitable tissues that associates low-resistance electrical contacts between the interiors of neurons with gap junctions. Many details of the fine structure of gap junctions have been revealed by recent electron microscopic studies. A common feature of gap junctions is the presence of small structures, closely packed together as a polygonal array on the surfaces of the membrane spanning the gap between the cells. It is suspected that these structures serve as channels between the interiors of the coupled cells. When the plasma membranes are split by the method of freeze fracture, which reveals broad areas of their interior,

---

[24]Pappas, G. D., and Waxman, S. G. 1972. *Structure and Function of Synapses.* Raven Press, New York, pp. 1–43.

[25]Lasansky, A. 1971. *Phil. Trans. R. Soc. Lond. (Biol. Sci.) 262*:365–381.

[26]McNutt, N., and Weinstein, R. S. 1973. *Prog. Biophys. Mol. Biol. 26*:45–101.

7

GAP JUNCTIONS between neurons. A. Two dendrites (D) in the inferior olivary nucleus of the cat are joined by a gap junction (arrow). To the left is an axon terminal (Ax) that contains numerous vesicles and makes chemical synapses on both dendrites. The inset shows a high magnification of a gap junction. The usual space between the cells is almost completely obliterated in the contact area which is traversed by cross bridges, better seen if the section is made in a slightly different plane (see Figure 5, Chapter 16). B. Freeze fracture through the presynaptic membrane of a nerve terminal that forms gap junctions with a neuron in the ciliary ganglion of a chicken. A broad area of the cytoplasmic fracture face is exposed, showing numerous specialized areas that correspond to sites of gap junctions (arrows). C. Higher magnification of cytoplasmic fracture face of a gap junction; a cluster of closely packed particles (about 9 nanometers in diameter) is seen within the membrane. The particles are assumed to form part of the channel system between the presynaptic terminal and the postsynaptic cell. For details about freeze-fracture consult Chapter 10 Figure 3. (A from Sotelo, Llinas, and Baker, 1974; B and C from Cantino and Mugnaini, 1975)

one sees particles corresponding to those observed on the surface. They are shown in Figure 7B and C. (For more about the method see Figure 3, Chapter 10.)

Still unknown is how the structure seen in association with gap junctions provides the channels either for current flow or for molecule passage between the interiors of adjoining cells. In recent studies a protein, called 'connexin,' has been extracted from a preparation of isolated gap junctions in liver cells. This opens the way to a chemical analysis of the constituents of gap junctions and the role of individual proteins in electrically linking one cell to another.[27,28]

[27]Goodenough, D. A. 1976. In *The Synapse. Cold Spring Harbor Symp. Quant. Biol.* 40:37–44.

[28]Gilula, D. 1974. In R. P. Cox (ed.). *Cell Communication.* John Wiley & Sons, New York, pp. 1–29.

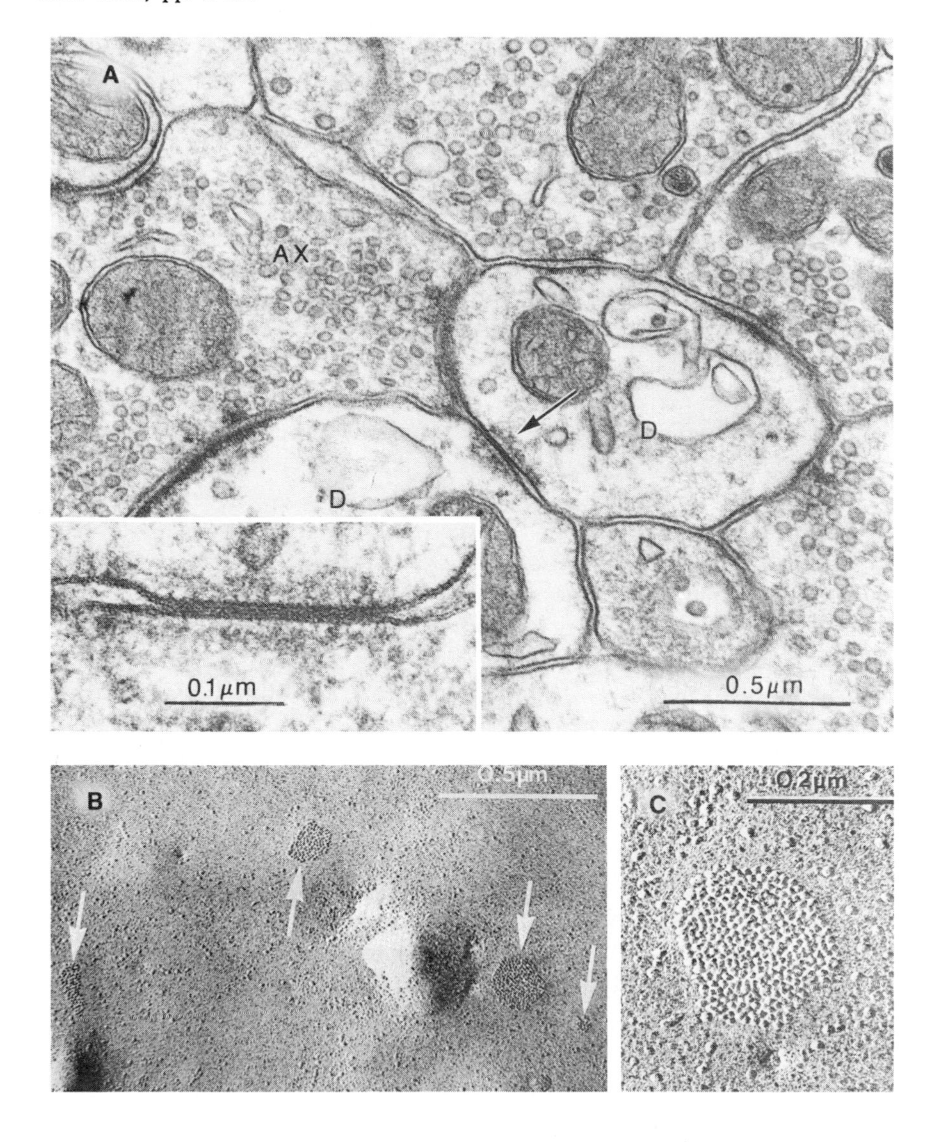

Significance of electrical synapses for signaling

What are the possible functions of electrical synapses? One role of some of the electrical synapses in the nervous system is probably related to the absence of the synaptic delay that in chemically transmitting synapses is approximately 0.5 msec, introduced mainly by the secretory process (Chapter 9). Electrical synapses provide greater SPEED as well as a high degree of certainty that an impulse in the presynaptic fiber will give rise to another in the postsynaptic cell. For example, certain electrical synapses in the goldfish brain probably mediate a rapid reflex reaction whereby the fish gives a strong tail flip, suitable for escape, when the surface of the water is disturbed, for example, by a bird. In such a case, even a fraction of a millisecond can make a vital difference. Chapter 16 describes an electrical synapse located in the Mauthner cell that may serve just such a purpose. Similarly, electrical synaptic connections in the medulla of the puffer fish mediate rapid synchronous movements.[18] In invertebrates—for example, the leech—electrical synapses in the ganglia connect homologous cells supplying the two sides of the body so that they act in concert (Chapter 17).

There is no theoretical reason for assuming that electrical transmission is used exclusively for rapid direct transmission of signals. Subthreshold or integrative actions (see later) in nerve cells can also be produced. Effects of electrical transmission vary widely, depending on the amount of current that can be delivered by a presynaptic impulse to the next cell. Effects are described in terms of COUPLING RATIOS, so that a ratio of 1:2 means that half the total presynaptic voltage change appears in the postsynaptic cell. Thus, stronger or weaker coupling between nerve cells may be one way to grade and control the amount of current flow between neurons. In addition, multiple electrical synapses converging on a neuron have simple additive effects. Such a mechanism of transmission most likely fluctuates very little compared with the chemical synaptic processes discussed later. These possibilities remain open. For the present, no well-defined role has been allocated to electrical synaptic transmission in schemes for the functional organization of the mammalian nervous system.

Electrical transmission between cells, like axonal conduction, is not readily modified by drugs such as curare or neostigmine that strongly influence chemical transmission. It follows that different experimental and therapeutic approaches are needed for dealing with electrical as contrasted to chemical synapses. Therefore, in any analysis of synaptic transmission, an essential first step is to determine whether the mechanism is chemical.

Electrical coupling in nonneural tissue and in embryos

The occurrence of electrical connections in the body is far more common than the examples given above suggest. Coupling is a general feature in epithelial cells in the skin and gut, where cells are interconnected, as are cells in glands and organs like the liver.[29] Low-resistance pathways that enable current to spread have also been demonstrated in glial cells, smooth muscle cells, and heart muscle fibers. In

[29]Loewenstein, W. R. 1966. *Ann. N.Y. Acad. Sci.* *137*:441–472.

all these tissues current flow has been demonstrated directly by recording electrically, and gap junctions have been identified in most instances by means of the electron microscope. In contrast, the vast majority of neurons and skeletal muscles are not coupled—a reasonable situation, since the nervous system could hardly function if everything were coupled to everything else.

An example of widespread coupling can be seen in various embryos, where all the cells may be in electrical continuity with each other and with the yolk sac. For example, in the squid embryo, in which the phenomenon was first described, generalized coupling persists even after organs have visibly differentiated.[30] Eventually, however, the nervous system splits off, and its cells become uncoupled while the remaining cells in different organs remain coupled to each other. The phenomenon has now been described in various vertebrate and invertebrate embryos (birds, amphibians, fish, lobsters, squid, and starfish).[19]

What is the significance of coupling between cells like epithelial or glial cells (Chapter 12) that are not able to give impulses? A clue is the fact that if current can spread between cells, ions must be able to move. And if ions can move, perhaps other molecules of interest, such as polypeptides or proteins, can also pass from cell to cell. Among substances that have been seen to spread between cells are procion yellow and fluorescein.[31,32] The implications are intriguing. One may speculate that intercellular transport across the junctions has a role in regulating cell functions during growth and embryonic development, or there may simply be a metabolic interaction among sheets of cells. There may be a relation between electrical coupling and growth of some cancer cells. Evidence from several tissues suggests that cancer cells are not coupled to each other or to normal cells, but the general rules are not yet clear. In a promising series of experiments Loewenstein[32] and his colleagues have investigated the genetic control of the formation of electrical coupling by fusing cell lines of different types in culture. Their results show that the capacity to form gap junctions behaves like a genetically dominant factor.

CHEMICAL SYNAPTIC TRANSMISSION

Some features of the mechanism of chemical transmission have already been described. Certain obvious questions arise when one considers the elaborate scheme for transmission that entails the secretion of a specific chemical by nerve terminals. How does the presynaptic nerve fiber liberate the chemical? Does the action potential with its characteristic sequence of permeability changes have some feature that is specialized to cause secretion of transmitter? Or is the release mechanism activated

---

[30]Furshpan, E. J., and Potter, D. D. 1968. *Curr. Top. Dev. Biol.* 3:95–127.

[31]Payton, B. W., Bennett, M. V. L., and Pappas, G. D. 1969. *Science* 166: 1641–1643.

[32]Loewenstein, W. R. 1974. *Membrane Transformations in Neoplasia.* Academic Press, New York.

by any depolarization of the presynaptic nerve fiber? Release of transmitter is considered in Chapter 9; the present discussion is concerned with the question of how the transmitter produces a potential change in the postsynaptic cell that leads to excitation or inhibition.

Many of the pioneering experiments were made on relatively simple preparations, such as the neuromuscular junction of the frog. This particular preparation has the great advantage that the transmitter has been definitely identified as acetylcholine. In addition, it allows detailed examination of the mechanism of action of ACh on the postsynaptic membrane and of the way in which the nerve impulse releases ACh. Most chemical synapses between neurons, particularly in the brain, are far less easy to study. In fact, in only a few specific examples can the transmitters be identified and rarely can the synaptic mechanisms be analyzed in detail. The best studied peripheral neuronal synapses are in the stellate ganglion of the squid and the parasympathetic ganglia in the septum of the frog heart.

The few junctions mentioned may seem poor models for the study of the properties of synapses in general. At these synapses, each impulse in the presynaptic fiber invariably gives rise to such a large synaptic potential that an impulse is produced in the muscle fiber or postsynaptic nerve cell and there is, therefore, no integration. Nevertheless, as Katz has so aptly pointed out, the neuromuscular synapse does contain in a concealed form the special synaptic properties that enable a nerve cell to take account of convergent signals arriving in a number of different pathways. It turns out that the neuromuscular junction is composed of a large number of discrete synapses operating in synchrony, and the operation of any of these is not unlike the operation of synapses on neurons (Chapter 10). Other neuromuscular synapses that have provided good models are found in crustaceans, which possess inhibitory as well as excitatory innervation.

General mechanism of action of ACh

How does the nerve impulse give rise to current flow in the postsynaptic membrane? One possibility, that can now be definitely ruled out, is that synaptic transmission between the motor nerve ending and muscle occurs through an electrical mechanism. According to this idea, even though ACh meets all the requirements to act as a transmitter, the nerve action potential is also fully capable of stimulating the muscle by the direct spread of current. The arguments against this are as follows: (1) there is a minimum delay of about 0.5 msec between the arrival of the presynaptic action potential at the terminal and the first sign of a voltage change in the postsynaptic membrane—too long a time to be accounted for by the passive spread of electric currents; (2) substances that prevent the release of ACh from the terminals (low calcium, high magnesium, botulinum toxin) block transmission without abolishing the action potential in the nerve; and (3) experimentally produced voltage changes in the presynaptic nerve terminal do not directly influence the membrane of the postsynaptic cell, but do cause synaptic potentials (see below). In fact, electrical transmission is quite unlikely on anatomical grounds alone, in view of the gross mismatch of size of pre- and postsynaptic structures and the width of the intercellular cleft, as discussed earlier in connection with electrical transmission.

Another possibility is that ACh acts by entering the membrane of the end plate. If a sufficient number of positively charged ACh ions move along their electrochemical gradients into the cell, the membrane will be depolarized. However, from the amount of current that actually flows through the end plate as a result of a nerve impulse, one can estimate how many positively charged ions would have to cross the membrane. Many thousand times too little ACh is liberated by each impulse to account for the size of the synaptic potential.[15] If ACh were released in adequate amounts, its store in the terminals would be rapidly exhausted. Thus, flow of ACh into the postsynaptic cell cannot be responsible for generating the synaptic potential. In addition, since each ACh molecule is responsible for the entry of thousands of ions into the postsynaptic membrane, chemical synapses provide a large amplification mechanism.

Ionic permeabilities and the concept of the reversal potential

The clue to the mode of action of ACh was provided by Fatt and Katz.[15] They reached the important conclusion that ACh produces a marked increase in the permeability of the synaptic membrane to small ions. One modification had subsequently to be made in this hypothesis, namely that the chloride permeability is not increased (see later). The idea in some respects resembles Bernstein's original hypothesis for the action potential, which suggested that the membrane becomes permeable to all ions at once.

What are the consequences of assuming that the synaptic membrane becomes equally permeable to sodium and potassium ions under the influence of ACh? Such an increase in permeability would tend to drive the voltage toward a level between E_K and E_{Na}, close to zero potential, if ACh acted when (1) the membrane was at its natural resting level (for example −90 mV inside) or (2) when the membrane potential was reversed, as during the peak of an action potential, so that the inside was now positive (for example +30 mV) as in Figure 8. Thus, the larger the potential across the membrane at the time the transmitter acts, the larger the voltage change it should produce. At one value, at the REVERSAL POTENTIAL, ACh should produce no change in potential. ACh can therefore give rise to (1) an inward current at the normal resting potential, causing a depolarization; (2) an outward current and repolarization during the peak of the action potential; or (3) no current at all. The reversal potential at skeletal neuromuscular junctions has been found to be near −15 mV (Figure 8).

Evidence that ACh increases both sodium and potassium conductances at the neuromuscular junction

After the initial analysis by Fatt and Katz, a number of questions remained concerning the action of ACh. What is the magnitude of the permeability change and which ions are involved? Is the action of ACh on permeability independent of the level of membrane potential? This sort of analysis is an essential prerequisite not only for understanding how ACh works, but also for understanding other synaptic actions, including inhibition.

Two techniques are available for assessing the permeability changes

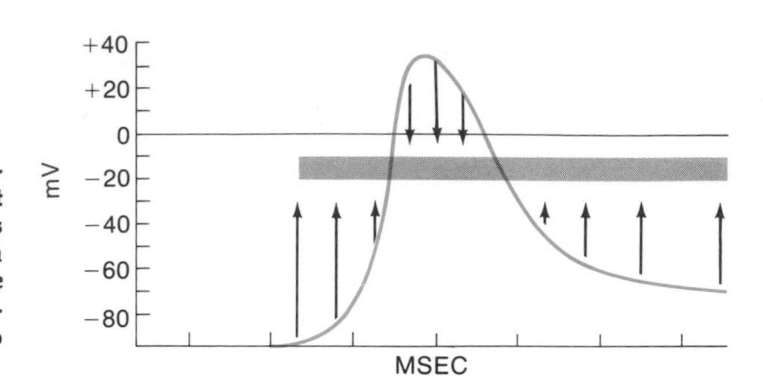

8

DIRECTION OF CHANGE in membrane potential produced by ACh when the cell is at rest or when it is at various other levels during an impulse. Arrows indicate direction and magnitude of change. The shaded range is the reversal potential at which ACh produces no potential change. (After del Castillo and Katz, 1956)

produced by ACh. The first involves use of radioactive tracers. This method has shown that the permeability of the postsynaptic membrane is increased to both sodium and potassium (and also calcium), but not chloride.[33] This experiment provides convincing evidence concerning the ion species involved, but does not reveal the timing of the conductance changes. For this, one must use the second technique—the voltage clamp. Experiments in which the membrane potential at the motor end plate was clamped were first performed by the Takeuchis.[34] They inserted two microelectrodes into the synaptic regions of a frog muscle fiber, using one to record the membrane potential (V_m) and the other to inject sufficient current to displace V_m to selected values, at which it was held. In an experiment of this type ionic currents generated by the membrane are measured by applying equal and opposite current from a feedback amplifier; this balances the changes and keeps the membrane potential constant after it has been displaced to the new value, as discussed in Chapter 6.

To study conductance changes produced by ACh, the membrane potential was clamped at various levels. Then the nerve was stimulated to release ACh, or alternatively, ACh was applied directly from a micropipette close to the synapse. The Takeuchis found that when the membrane is held near the normal resting potential, the release of ACh causes a brief surge of inward current, as though positively charged ions are moving into the postsynaptic cell. This additional surge of current at the synapse produced by ACh is called ΔI.

What happens to ΔI when the membrane potential at the end plate is kept constant at different values? The more the membrane is hyperpolarized, the greater the inward current through the end plate (Figure 9). On the other hand, if the membrane potential is held at +38 mV, ACh gives rise to a current flow in the outward direction. These measurements by Magleby and Stevens[35] are plotted in a more complete form in Figure 10, which shows the relation between the peak ACh current and the membrane potential. At one membrane potential near zero, nerve-released ACh causes no extra peak current flow. That is the level (V_r) at which the current changes direction.

---

[33]Jenkinson, D. H., and Nicholls, J. G. 1961. *J. Physiol.* *159*:111–127.
[34]Takeuchi, A., and Takeuchi, N. 1960. *J. Physiol.* *154*:52–67.
[35]Magleby, K. L., and Stevens, C. F. 1972. *J. Physiol.* *223*:151–171.

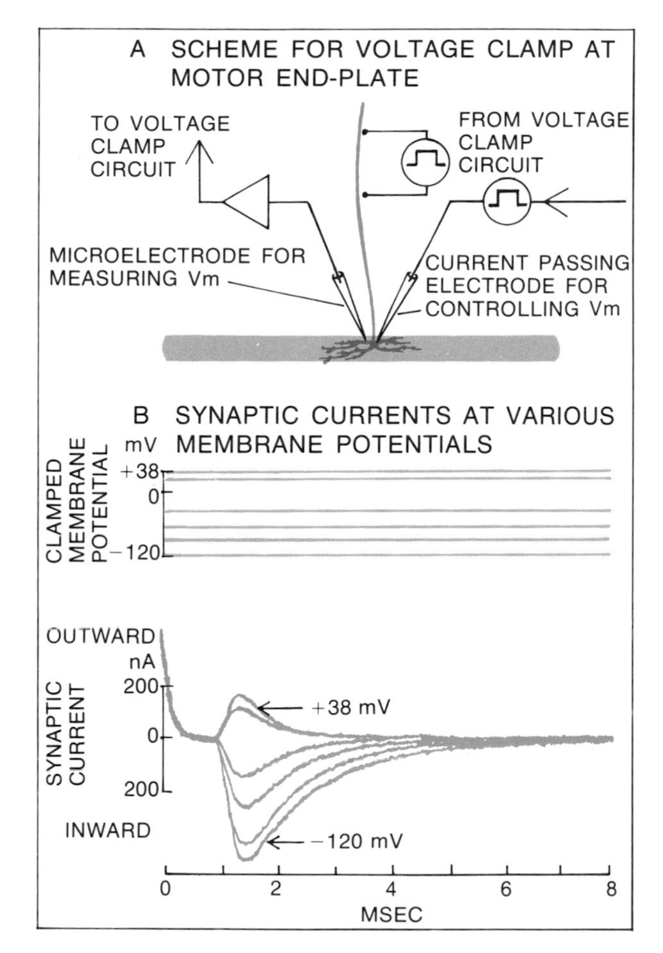

9

SYNAPTIC CURRENTS contributed by nerve-released ACh. The membrane potentials were clamped at values between —120 mV and +38 mV. The holding current has been subtracted. (After Magleby and Stevens, 1972)

The points in Figure 10 are almost in a straight line, so that ΔI is roughly proportional to the membrane potential. This is equivalent to saying that Ohm's law holds. Thus:

$$\Delta I = (V_m - V_r)/r_m = \Delta g\ (V_m - V_r)$$

The main conclusion to be drawn from this experiment is that since the

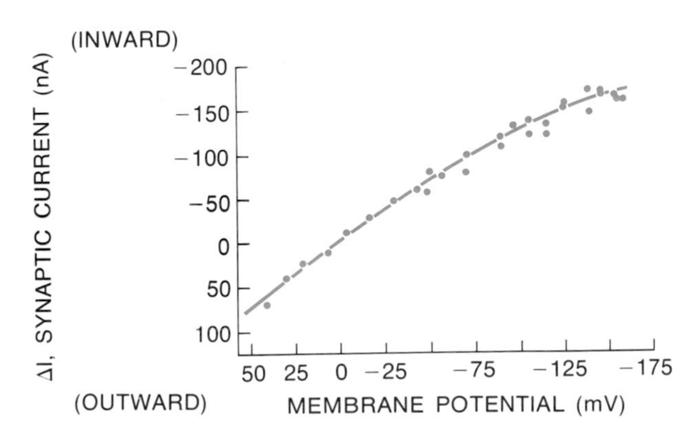

10

PEAK SYNAPTIC CURRENTS at various clamped membrane potentials (see Figure 9). Close to zero membrane potential the synaptic current changes direction. Negative values indicate inward currents. (After Magleby and Stevens, 1972)

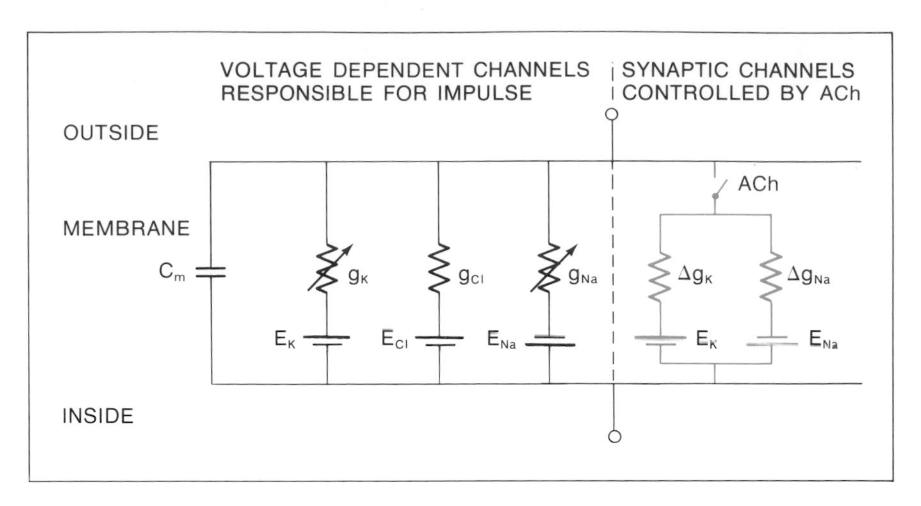

11 **ELECTRICAL MODEL of the synaptic membrane activated by ACh and of the rest of the membrane controlled by the change in membrane potential.**

slope is approximately constant, Δg (the permeability increase caused by ACh) is not significantly affected by the membrane potential.

While the peak ACh current is influenced only a little by the level of the membrane potential, the same does not apply to the duration of the permeability change. At both vertebrate and invertebrate synapses the conductance change can be either prolonged or shortened.[35,36] These results do not invalidate the general conclusion that the synaptic permeability change produced by ACh is relatively independent of the membrane potential, especially by comparison with the axonal permeability sequences that are governed by potential changes.

The next series of experiments demonstrates that ACh increases the permeability to both sodium (Δg_{Na}) and potassium (Δg_K), but not to chloride. It is convenient to draw once again an electrical model of the membrane, inserting the special channels influenced by ACh at the nerve-muscle synapse, in addition to the usual channels involved in the action potential (Figure 11).

At the patch of membrane where ACh acts, the net current flow produced by closing switches depends on the membrane potential (V_m). At the normal resting potential, sodium enters (ΔI_{Na} inward) and potassium leaves (ΔI_K outward). At one value of membrane potential, no net current flows through the membrane when the switches are closed. This occurs at the reversal potential (V_r), where the currents produced by sodium and potassium are equal and opposite ($\Delta I_{Na} = -\Delta I_K$), and ACh has no tendency to change the membrane potential. From the relation of current to conductance and driving force, it follows that:

$$\Delta I_{Na} = \Delta g_{Na} (V_m - E_{Na})$$
$$\Delta I_K = \Delta g_K (V_m - E_K)$$

[36]Dudel, J. 1974. *Pflügers Arch.* 352:227–241.

where ΔI_{Na} and ΔI_K are the currents that flow when ACh interacts with the receptors to produce permeability changes of Δg_{Na} and Δg_K. When the membrane potential (V_m) is at the reversal potential, $\Delta I_{Na} = -\Delta I_K$. At this value of V_m:

$$V_m = V_r = \frac{(\Delta g_{Na}/\Delta g_K)(E_{Na}) + E_K}{\Delta g_{Na}/\Delta g_K + 1} \quad \textbf{(1)}$$

From this equation it is clear that if ACh increases the permeability to sodium and potassium by an equal amount, then $\Delta g_{Na}/\Delta g_K = 1$, and $V_r = (E_{Na} + E_K)/2$, which at the frog neuromuscular junction is -25 mV. In fact, however, the reversal potential observed is closer to zero (nearer to E_{Na}). Therefore, ACh produces a simultaneous increase in sodium and potassium permeability that slightly favors sodium. From equation 1 it is also possible to estimate how changes in E_{Na} and E_K alter the reversal potential, if the permeability change produced by ACh remains constant. Over a wide range of sodium and potassium concentration changes, there is good agreement between the predicted and the observed changes in reversal potential, indicating that the permeability change produced by ACh ($\Delta g_{Na}/\Delta g_K$) remains constant. In contrast, in tests in which the external chloride concentrations are changed no significant influence on the reversal potential is seen.[33,37] Tests with calcium show that ACh also increases the permeability to calcium.[33,38] While this change may have important physiological consequences, it has little effect on the reversal potential.

The channels opened by ACh behave differently from those involved in the action potential mechanism. Clearly, the changes in Δg_{Na} and Δg_K produced by ACh must be simultaneous, since at the reversal potential no additional current flows as the result of ACh action. If the two channels were opened at different rates, one would expect to see a biphasic synaptic current at this potential rather than no effect at all. Some of the special properties that distinguish synaptic and extrasynaptic membranes are summarized in Table 1.

Similar principles hold at other chemical excitatory synapses. The ratio of Δg_{Na} to Δg_K need not, however, be the same in every case. At the squid giant synapse and at crustacean neuromuscular junctions, for example, V_r is approximately 20 mV inside POSITIVE.[39,40] What matters for excitation is that the reversal potential be beyond threshold.

Chapter 10 will show that considerable progress has been made in isolating the specific membrane protein that acts as the ACh receptor. Once again, a problem that began with the analysis of electrical phenomena is reduced to the molecular level. One key to this approach was use of α-bungarotoxin, a paralyzing substance obtained from snake venom which binds strongly and practically irreversibly to the receptor. Another major advance has been to measure the conductance change

[37]Takeuchi, N. 1963. *J. Physiol.* *167*:141–155.
[38]Katz, B., and Miledi, R. 1969. *J. Physiol.* *203*:689–706.
[39]Takeuchi, A., and Onodera, K. 1973. *Nature (New Biol.)* *242*:124–126.
[40]Miledi, R. 1969. *Nature* *223*:1284–1286.

and ionic flow produced by the opening of a single ACh channel (Chapter 10).

SYNAPTIC INHIBITION

In the past few years it has become possible to discuss synaptic inhibition (and excitation) as part of the general problem of conduction and excitation. In the process of inhibition the same principles are involved, with minor but important variants. Both inhibitory and excitatory transmitters act by changing the permeability of the membrane, but excitation is achieved by driving the membrane potential toward threshold, while inhibition involves keeping the membrane potential below threshold—at or near the resting value. If the inhibitory transmitter were to increase permeability to an ion whose equilibrium potential was about −90 mV, then its effect would be to hyperpolarize if the resting membrane potential was less than −90 mV. On the other hand, if the membrane were hyperpolarized to beyond −90 mV, the transmitter would produce a depolarization (Figure 12). Two ions that could in theory be involved in bringing about inhibition are potassium or chloride (or both) since their equilibrium potentials are below the threshold, close to the resting potential.

The skeletal muscles of the frog have no inhibitory innervation. However, crustaceans (lobsters, crayfish) possess some relatively simple

Table 1. **PROPERTIES OF IMPULSE AND SYNAPTIC POTENTIAL AT NERVE MUSCLE SYNAPSE**

	Impulse	Synaptic potential
Initiated by	Depolarization	ACh
Changes of membrance conductance		
During rising phase	Specific increase in g_{Na}	Simultaneous increase of g_{Na} and g_K
During falling phase	Specific increase in g_K	Passive decay
Equilibrium potential of active membrane	E_{Na} approx. +50 mV	Reversal potential close to 0 mV
Other features	Regenerative ascent, followed by refractory period	No evidence for regenerative action or refractoriness
Pharmacology	Blocked by TTX, not influenced by curare	Blocked by curare, not influenced by TTX

Modified from del Castillo and Katz (1956).

inhibitory synapses that are among the few whose inhibitory transmitter has been identified. The presynaptic terminals at these synapses release γ-aminobutyric acid (GABA)[41,42] (Chapter 11). Another useful preparation is the spinal motoneuron in the cat in which stimulation of the presynaptic inhibitory nerves usually leads to a small hyperpolarization in the postsynaptic cell.[43] Figure 12 shows that if postsynaptic cells are depolarized by current, the release of the inhibitory transmitter onto the membrane causes a hyperpolarization; in contrast, if the membrane potentials are artificially increased, inhibition causes a depolarization. In all three cell types in Figure 12 there is one particular value, the reversal potential, at which the inhibitory transmitter produces no potential change (or current flow). This usually lies close to the resting potential.

Ionic channels activated during inhibition

Which ions are involved during inhibition? At inhibitory synapses in crustaceans and in the motoneurons there is strong evidence that the permeability is increased primarily to chloride (see below). The circuit of the membrane can be completed by inserting a chloride conductance (Δg_{Cl}) in series with a chloride battery (E_{Cl}) and a switch that is closed by the inhibitory transmitter (Figure 12*D*). If the membrane potential is the same as the equilibrium potential for chloride, closing the switch produces no net flow of current (although the permeability is greatly increased). In the hyperpolarized state, the inhibitory transmitter action gives rise to movement of chloride out of the cell; conversely, in the depolarized cell it gives rise to inward movement of chloride. In the experiments that show the involvement of chloride, either the ion is injected into the cell through a micropipette or the chloride concentration of the bath solution is reduced. In each case the voltage of the chloride battery is decreased and the reversal potential is altered, as expected from the Nernst equation:

$$E_{Cl} = -58 \log_{10} \frac{Cl_o}{Cl_i}$$

Whether a particular transmitter substance causes excitation or inhibition depends principally on the ionic channels that are opened in the postsynaptic membrane. For example, ACh excites at the motor end plate by opening permeability channels for sodium and potassium. At other synapses, such as those on heart muscle, ACh can be inhibitory. Here the postsynaptic parasympathetic nerve releases ACh, which causes an increase in potassium conductance alone.[44] It thereby hyperpolarizes heart muscle fibers and slows or stops the heartbeat. In the sea hare, *Aplysia,* Kandel and his colleagues[45] have shown that one transmitter can produce two different postsynaptic actions—one excitatory and the

[41]Otsuka, M., Iversen, L. L., Hall, Z. W., and Kravitz, E. A. 1966. *Proc. Natl. Acad. Sci. U.S.A. 56*:1110–1115.
[42]Takeuchi, A., and Takeuchi, N. 1967. *J. Physiol. 191*:575–590.
[43]Coombs, J. S., Eccles, J. C., and Fatt, P. 1955. *J. Physiol. 130*:326–373.
[44]Harris, E. J., and Hutter, O. F. 1956. *J. Physiol. 133*:58*P*–59*P*.
[45]Wachtel, H., and Kandel, E. 1971. *J. Neurophysiol. 34*:56–68.

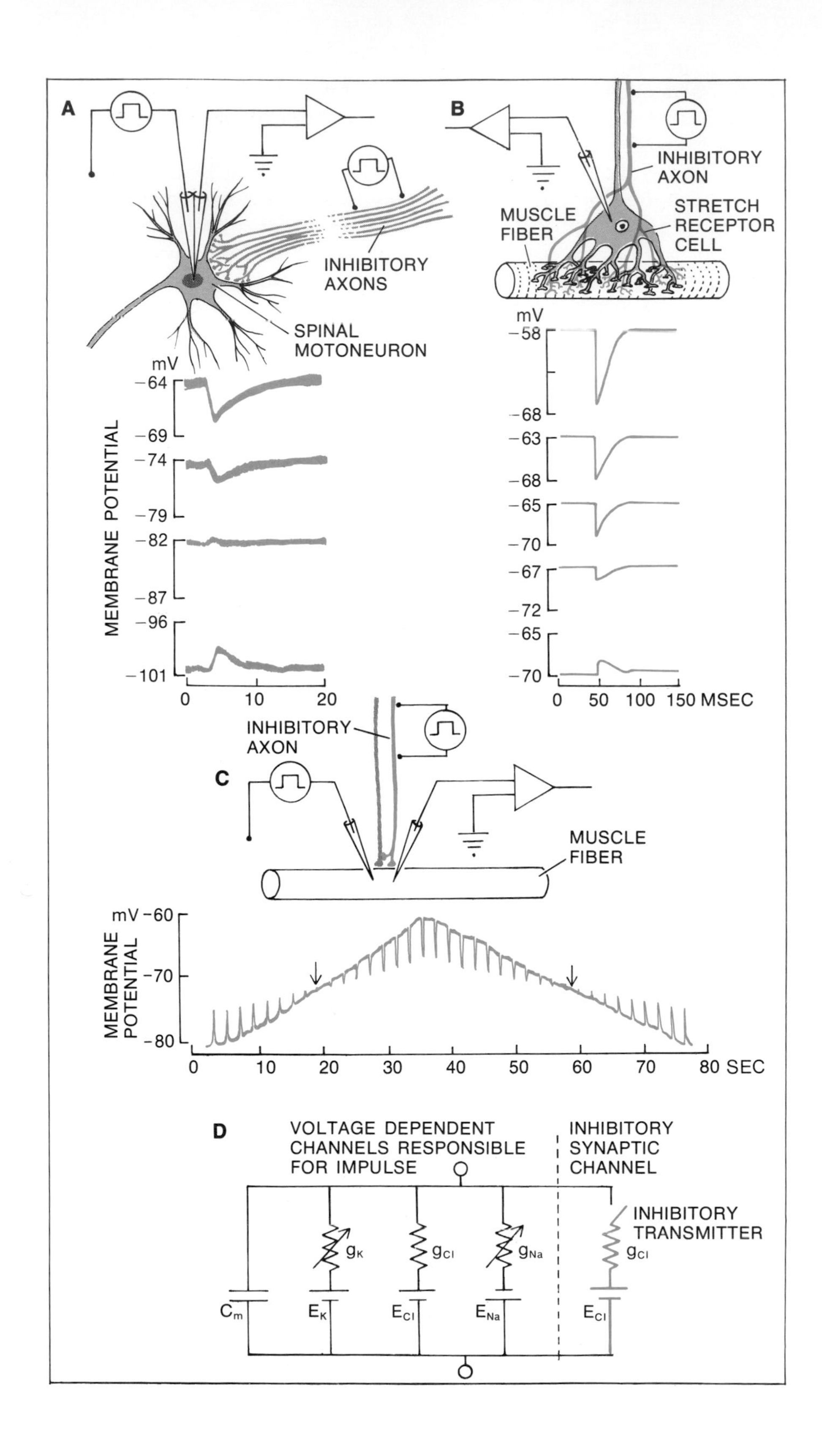
A
INHIBITORY
AXONS
SPINAL
MOTONEURON
mV
MEMBRANE POTENTIAL
-64
-69
-74
-79
-82
-87
-96
-101
0
10
20
B
INHIBITORY
AXON
MUSCLE
FIBER
STRETCH
RECEPTOR
CELL
mV
-58
-68
-63
-68
-65
-70
-67
-72
-65
-70
0
50
100
150 MSEC
C
INHIBITORY
AXON
MUSCLE
FIBER
mV -60
MEMBRANE
POTENTIAL
-70
-80
0
10
20
30
40
50
60
70
80 SEC
D
VOLTAGE DEPENDENT
CHANNELS RESPONSIBLE
FOR IMPULSE
INHIBITORY
SYNAPTIC
CHANNEL
INHIBITORY
TRANSMITTER
g_K
g_{Cl}
g_{Na}
g_{Cl}
C_m
E_K
E_{Cl}
E_{Na}
E_{Cl}

other inhibitory. During inhibition the transmitter reacts with different receptors that give rise to increases in either chloride or potassium conductances or in both.[45,46]

The postsynaptic action produced in a neuron is therefore determined by (1) the particular transmitter released by the presynaptic cell or (2) the special properties of the receptors in the postsynaptic membrane. The transmitter interacts with the receptor molecules to open permeability channels for certain ions whose movement leads either to excitation or to inhibition. Since a transmitter may have different actions, it is misleading to refer to substances as either "excitatory" or "inhibitory" without reference to the particular receptors on which they act.

PRESYNAPTIC INHIBITION

A means for selectively excluding excitatory pathways

In the examples of synaptic transmission described in this chapter the postsynaptic effect of the transmitters is stressed. Over the years, however, a number of experiments indicated that it is difficult to account for all the inhibition solely in terms of the postsynaptic permeability changes.[47,48] Eventually, in 1961, an additional inhibitory mechanism was described in mammals by Eccles and his colleagues[49] and in crustaceans by Dudel and Kuffler.[50] Figure 13 shows that the action of the inhibitory nerve at a crustacean neuromuscular junction can also be exerted on excitatory terminals, reducing their output of transmitter. In Chapter 9 we shall see that transmitter is released in multimolecular packets. The presynaptic inhibitory transmitter acts by reducing the number rather than the size of packets of transmitter that are released. The presynaptic effect of an inhibitory impulse is relatively brief, its

[46]Kehoe, J. 1972. *J. Physiol.* *225*:85–114, 115–146, 147–172.
[47]Fatt, P., and Katz, B. 1953. *J. Physiol.* *121*:374–389.
[48]Frank, K., and Fuortes, M. G. F. 1957. *Fed. Proc.* *16*:39–40.
[49]Eccles, J. C., Eccles, R. M., and Magni, F. 1961. *J. Physiol.* *159*:147–166.
[50]Dudel, J., and Kuffler, S. W. 1961. *J. Physiol.* *155*:543–562.

12

EFFECT OF INHIBITORY TRANSMITTER on a cat spinal motoneuron (A), a crayfish stretch receptor cell (B), and a muscle fiber in the crayfish (C). In the motoneuron and muscle fiber the membrane potential is set to different levels by passing current through a micropipette; in the stretch receptor neuron it is altered by adjusting the amount of stretch on the dendrites. Each cell has a reversal potential at which inhibition causes no potential change. In A it is between —72 and —82mV; in B it is between —67 and —70 mV. In C brief trains of inhibitory stimuli every 2 secs depolarize at the resting potential, but reverse polarity at —72 mV (arrows) as the membrane potential is shifted from —80 mV to —61 mV and back again. D shows an equivalent circuit including the inhibitory conductance channel. (A after Coombs, Eccles, and Fatt, 1955; B after Kuffler and Eyzaguirre, 1955; C after Dudel and Kuffler, 1961)

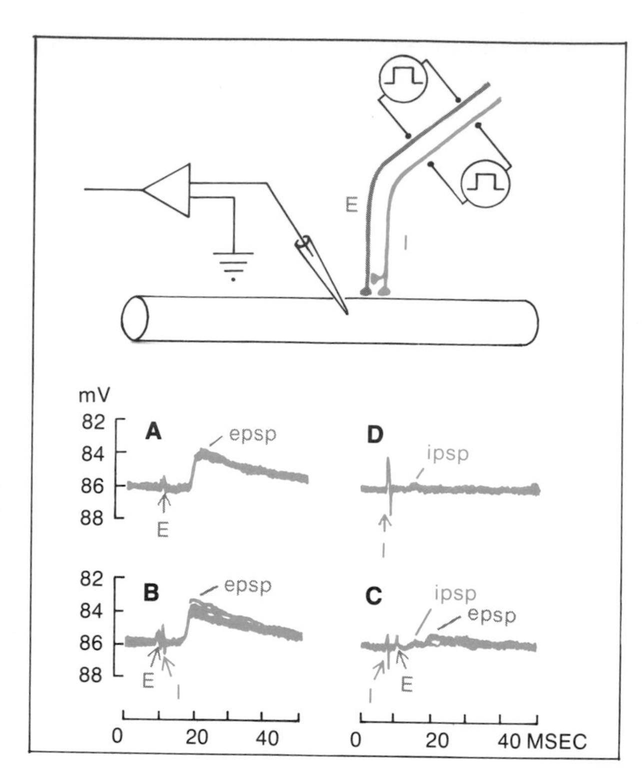

13

PRESYNAPTIC INHIBITION in a crustacean muscle fiber innervated by one excitatory and one inhibitory axon. At the resting potential, —86 mV, the excitatory impulse (E) causes a 2 mV postsynaptic potential (A), while the inhibitory impulse (I) in D depolarizes by about 0.2 mV. If I arrives at the synapse just after E (as in B), it is too late to exert an inhibitory effect. If however, I precedes E by several milliseconds (as in C), it greatly reduces the excitatory potential. (After Dudel and Kuffler, 1961)

peak effect lasting a few milliseconds and its total duration being about 6 to 7 msec. To be most potent, inhibitory impulses must arrive at the nerve terminal several milliseconds before the excitatory impulse. The importance of accurate timing is illustrated in Figure 13, where *A* and *D* show the excitatory and inhibitory potentials following stimulation of each of the two nerves on its own. In *B* the inhibitory impulse follows the excitatory one by 1.5 msec and arrives too late to act; in *C*, however, it precedes the excitatory impulse, in time to markedly reduce the size of the excitatory postsynaptic potential. There is evidence that γ-aminobutyric acid, the transmitter that acts postsynaptically in crustacean synapses, also mediates presynaptic inhibition at these axo-axonic synapses.[50]

Presynaptic inhibition is common in mammals, in which it was discovered by Eccles and his colleagues.[51] It is associated with a depolarization of the presynaptic nerve fibers and a considerable reduction of the excitatory postsynaptic potential, indicating that the output of transmitter has been reduced. In mammalian neurons the duration of the presynaptic inhibitory action after several stimuli may be 100 to 200 msec—much longer than in crustaceans. The mechanism of this prolonged effect, best investigated in the spinal motor neurons, is not known. It may be due to a continued action of the transmitter or to a repetitive firing of inhibitory interneurons, which then act on the ter-

---

[51]Eccles, J. C. 1964. *The Physiology of Synapses.* Springer Verlag, Berlin.

minals that release the excitatory transmitter. In any event, presynaptic inhibition in the mammalian spinal cord, as at crustacean inhibitory synapses, appears to result from a reduction in the number of packets of transmitter.

The relative importance of pre- and postsynaptic inhibition in the nervous system has not yet been established. However, one can readily see how presynaptic inhibition could be useful for information processing in a neuron innervated by many pathways converging from different regions. For such a cell it may be important under certain conditions to reduce or eliminate selectively some of the inputs without affecting others. This cannot be done with the use of the postsynaptic conductance mechanism that affects the entire cell. Presynaptic inhibition goes a long way toward fulfilling these requirements.

Structural basis of presynaptic inhibition

The experiments demonstrating presynaptic inhibition naturally presupposed the existence of axo-axonic synapses between the inhibitory and excitatory terminals. These have since been demonstrated directly by electron microscopy at the crustacean neuromuscular junction,[52] and appropriate axo-axonic synapses have been described on nerve terminals in numerous locations in the mammalian central nervous system.[53] However, a rigorous correlation of these structures and presynaptic inhibition remains difficult. A possibility also exists that inhibitory neurons themselves can be inhibited presynaptically, since synapses have been demonstrated electron microscopically on inhibitory terminals that end on stretch receptors.[54] Such structural evidence, on its own, might also mean that the synapses serve presynaptic facilitation.

Electrical inhibition

In a series of rigorous experiments Furukawa and Furshpan[55] have described a form of electrical inhibition that can be explained in terms of current flow generated by presynaptic nerve fibers. These observations were made on the Mauthner cell in the medulla of the goldfish (Chapter 16). In the case of electrical inhibition, the fluid space just OUTSIDE the Mauthner cell is made more POSITIVE by current spread from presynaptic impulses. This occurs in the critical region of the axon cap, where impulses are normally initiated. The effect is similar to that of hyperpolarizing the membrane through an extrinsic current produced by connecting the positive pole of a battery to the outside of the membrane at the axon cap and the negative pole to another region; no special conductance change in the postsynaptic cell is required to drive the potential away from threshold. It is not known how widespread this type of inhibition is. It apparently requires specialized anatomical features, and once recognized, these may provide criteria for locating other sites for physiological study.

Unusual mechanisms of synaptic transmission

So far the common features of each mode of transmission have been emphasized. At electrical synapses excitation occurs because depolariz-

[52]Atwood, H. L., and Morin, W. A. 1970. *J. Ultrastruct. Res.* *32*:351–369.

[53]Schmidt, R. F. 1971. *Ergeb. Physiol.* *63*:20–101.

[54]Nakajima, Y., Tisdale, A. D., and Henkart, M. P. 1973. *Proc. Natl. Acad. Sci. U.S.A.* *70*:2462–2466.

[55]Furukawa, T., and Furshpan, E. J. 1963. *J. Neurophysiol.* *26*:140–176.

ing currents pass directly from the presynaptic to the postsynaptic cell. At chemical synapses the conductance for one or more ions is increased. There is, however, no reason to suppose that these are the only mechanisms operating at synapses. Various unusual types of transmissions have been under investigation for some years. "Unusual" here does not necessarily mean uncommon, but merely indicates that current knowledge is limited and that generalizations cannot yet be proposed. Excitation could be attained if, for example, the transmitter DECREASED the resting POTASSIUM PERMEABILITY of the postsynaptic neuron. According to such a scheme, the potential would then move toward E_{Na}, since the resting sodium conductance, even if small, would predominate.[56] On the other hand, inactivation or REDUCTION of the SODIUM PERMEABILITY could lead to a hyperpolarization and inhibition, because the membrane potential would then move toward E_K. An illustration of the variety and range of mechanisms that can operate comes from experiments on neurons in ganglia of snails. In these cells 5-hydroxytryptamine, a likely transmitter, can cause excitation or inhibition, acting through both an increase and a decrease of ionic conductances.[57] Similarly, in *Aplysia* synaptically evoked potentials of surprisingly long duration, persisting for minutes or even hours, have been found. The ions involved appear to be potassium and chloride.[58] A further example of unconventional synaptic transmission is mentioned in the description of the vertebrate retina (Chapter 2).

Additional unusual forms of synaptic transmission are being studied in sympathetic ganglia, in which the stimulation of presynaptic fibers causes a sequence of rapid and slow potentials. The permeability changes responsible for the slow events in these structures are not yet well understood, largely because in both amphibians and mammals sympathetic ganglia are not simple relay stations. The ganglia contain not only the well-known postsynaptic cells (principal cells) but also interneurons whose connections have not yet been sufficiently worked out. Synaptic potentials recorded from principal neurons may persist for many seconds or minutes as a result of a few presynaptic impulses. Evidence is available that attributes the slow excitatory and inhibitory potentials in ganglia to decreases in conductance[59] and to electrogenic pumps.[60]

A novel mechanism, involving cyclic AMP, for producing long-lasting signals in synaptic ganglia has also been proposed in recent years. It is known, largely from the work of Sutherland, that certain hormones, including epinephrine, exert their effects on cells by stimulating an enzyme, adenylcyclase. This is responsible for producing adenosine-3′,5′-phosphate (cyclic AMP), which activates a protein kinase, leading to the phosphorylation of a protein within the postsynaptic

---

[56]Dudel, J., and Kuffler, S. W. 1960. *Nature 187*:246–247.
[57]Gerschenfeld, H. M., and Paupardin-Tritsch, D. 1974. *J. Physiol. 243*:427–456.
[58]Parnas, I., and Strumwasser, F. 1974. *J. Neurophysiol. 37*:609–620.
[59]Weight, F. F., and Padjen, A. 1973. *Brain Res. 55*:225–228.
[60]Libet, B. 1970. *Fed. Proc. 29*:1945–1956.

membrane. This in turn leads to ionic permeability changes and alterations in membrane potentials. Support for this or a similar hypothesis is accumulating—for example, in the work of Greengard[61] and his colleagues in sympathetic ganglia. Their results show that following presynaptic stimulation cyclic AMP can increase in the postsynaptic cells and possibly give rise to slow potential changes in the membrane.

This entire area of unusual modes of transmission is in a developmental stage, and rigorous evidence for mechanisms is not available. Nevertheless, this state of the art almost assures one that in the course of clarification new and interesting developments will emerge.

SUGGESTED READING

Papers marked with an asterisk (*) are reprinted in Cooke, I., and Lipkin, M. 1972. *Cellular Neurophysiology.* Holt, New York.

General reviews

Eccles, J. C. 1964. *The Physiology of Synapses.* Springer Verlag, Berlin.

Gerschenfeld, H. M. 1973. Chemical transmission in invertebrate central nervous systems and neuromuscular junctions. *Physiol. Rev. 53*:1–119.

Hall, Z. W., Hildebrand, J. G., and Kravitz, E. A. 1974. *Chemistry of Synaptic Transmission.* Chiron Press, Newton. (Contains many original papers quoted in this chapter.)

Katz, B. 1966. *Nerve, Muscle, and Synapse.* McGraw-Hill, New York.

Krnjević, K. 1974. Chemical nature of synaptic transmission in vertebrates. *Physiol. Rev. 54*:418–540.

The Synapse. 1976. *Cold Spring Harbor Symp. Quant. Biol. 40.* (A collection of about 60 articles.)

Electrical transmission

GENERAL REVIEW

Bennett, M. V. L. 1974. Flexibility and Rigidity in Electronically Coupled Systems. In M. V. L. Bennett (ed.). *Synaptic Transmission and Neuronal Interaction.* Raven Press, New York, pp. 153–178.

ORIGINAL PAPER

*Furshpan, E. J., and Potter, D. D. 1959. Transmission at the giant motor synapses of the crayfish. *J. Physiol. 145*:289–325.

Excitatory mechanisms

*Fatt, P., and Katz, B. 1951. An analysis of the end-plate potential recorded with an intra-cellular electrode. *J. Physiol. 115*:320–370.

Magleby, K. L., and Stevens, C. F. 1972. The effect of voltage on the time course of end-plate currents. *J. Physiol. 223*:151–171.

*Takeuchi, A., and Takeuchi, N. 1960. On the permeability of the end-plate membrane during the action of transmitter. *J. Physiol. 154*:52–67.

---

[61]Beam, K. G., and Greengard, P. 1976. In *The Synapse. Cold Spring Harbor Symp. Quant. Biol. 40*:157–168.

Inhibitory mechanisms

*Coombs, J. S., Eccles, J. C., and Fatt, P. 1955. The specific ion conductances and ionic movements across the motoneuronal membrane that produce the inhibitory post-synaptic potential. *J. Physiol.* *130*:326–373.

*Dudel, J., and Kuffler, S. W. 1961. Presynaptic inhibition at the crayfish neuromuscular junction. *J. Physiol.* *155*:543–562.

Eccles, J. C., Eccles, R. M., and Magni, F. 1961. Central inhibitory action attributable to presynaptic depolarization produced by muscle afferent volleys. *J. Physiol.* *159*:147–166.

Furukawa, T. Y., and Furshpan, E. J. 1963. Two inhibitory mechanisms in the Mauthner neurons of goldfish. *J. Neurophysiol.* *26*:140–176.

Unusual synapses

Bennett, M. V. L. (ed.). 1974. *Synaptic Transmission and Neuronal Interaction.* Raven Press, New York.

CHAPTER NINE

RELEASE OF CHEMICAL TRANSMITTERS

The mechanism by which the impulse at a presynaptic nerve terminal causes transmitter release has been studied in great detail at the frog neuromuscular junction. The stimulus for secretion of the transmitter is depolarization of the presynaptic ending produced either by an impulse or by artificially applied current. Invariably, a delay (about 0.5 msec at 22° C) intervenes between the voltage change and secretion. For release to occur, calcium ions must be present in the bathing fluid at the time of the depolarization, and there is good evidence that calcium enters the terminal to trigger release.

Changes in the postsynaptic membrane potential provide a convenient assay for estimating the timing and the quantity of transmitter release. A key finding is that transmitter is secreted in multimolecular packets (thousands of molecules), or quanta, of fixed size. These constitute the fundamental physiological units of release.

Variations in release occur through more or fewer packets being secreted by the terminal. The synaptic potential at the skeletal neuromuscular junction is normally made up of about 300 packets of acetylcholine (ACh) that are almost synchronously released. In a low-calcium medium, this release can be reduced in a stepwise manner or completely prevented. At rest, the nerve spontaneously releases packets of the transmitter, which give rise in the postsynaptic cell to miniature synaptic potentials. In most neurons and muscles these are in the millivolt range.

In adrenergic nerves the transmitter norepinephrine is packaged in vesicles that can be seen in electron micrographs clustered in the terminals. The same holds for ACh. Quantal release appears to be the general mechanism for secretion of transmitter at chemical synapses in the central as well as the peripheral nervous systems of vertebrates and invertebrates.

The mechanism of transmitter release has been explored in detail by Katz and his colleagues. It is discussed here at some length because the experiments have given new insights that seem to apply to all chemical synapses. Further, the study of release provides another example of how the cellular approach can supply much detailed knowledge about a fundamental physiological process.

A number of questions arise concerning the way in which the presynaptic neuron releases transmitter. Experimental analysis requires a highly sensitive, quantitative, and reliable measure of the transmitter released by the nerve terminals, with a time resolution in the millisecond scale. In the experiments described below the membrane potential of the postsynaptic cell is used as an index or bioassay for the transmitter liberated by the presynaptic neuron. Once again, the neuromuscular junction of the frog, where acetylcholine (ACh) is known to be the transmitter, offers many advantages. However, it is often necessary to record intracellularly from presynaptic endings, as well as from the postsynaptic cell—for example, to establish the precise relation between membrane potential and transmitter release. The frog presynaptic terminals are too small to be impaled by microelectrodes, but a synapse where this can be done is the stellate ganglion of the squid.[1] This synapse operates by a chemical mechanism, but the transmitter has not yet been identified.[2] As at the neuromuscular junction, an action potential in the presynaptic fiber normally gives rise, after a delay, to a large depolarizing potential in the postsynaptic membrane, which usually reaches threshold and produces an action potential.

In the stellate ganglion of the squid, Katz and Miledi worked out the precise relation between the membrane potential of the presynaptic terminal and the amount of transmitter released.[3] In one set of experiments they applied the puffer fish poison tetrodotoxin (TTX) to the synapse and thereby reduced the amplitude of the presynaptic impulse

1

PRESYNAPTIC IMPULSE AND POSTSYNAPTIC RESPONSE relationship. A. Sketch of the stellate ganglion in the squid emphasizing two large axons that form a chemical synapse. The cells can be impaled with microelectrodes. B. Simultaneous recordings from the presynaptic (lower records) and postsynaptic nerves during the development of conduction block after application of TTX. C. Relation between the size of the presynaptic impulses and postsynaptic responses (filled circles) plotted on a linear and a semilogarithmic scale (inset). The open and half-filled circles show the similarity in relation after complete TTX block. The presynaptic fiber potential was artificially changed to various levels through the intracellular current electrode. (A after Bullock and Hagiwara, 1957; C after Katz and Miledi, 1965)

---

[1]Bullock, T. H., and Hagiwara, S. 1957. *J. Gen. Physiol.* *40*:565–577.
[2]Hagiwara, S., and Tasaki, I. 1958. *J. Physiol.* *143*:114–137.
[3]Katz, B., and Miledi, R. 1967. *J. Physiol.* *192*:407–436.

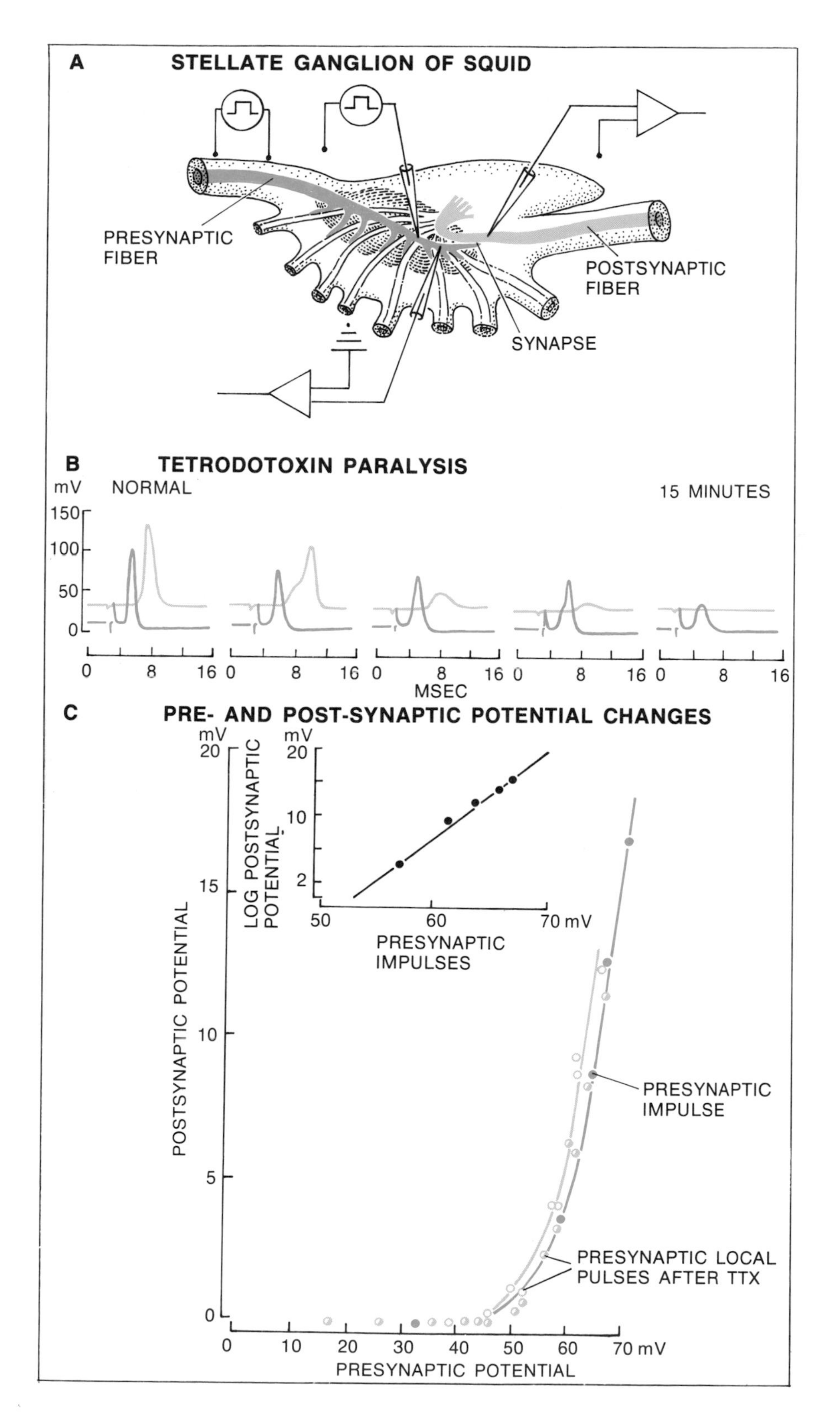
A
STELLATE GANGLION OF SQUID
PRESYNAPTIC
FIBER
POSTSYNAPTIC
FIBER
SYNAPSE
B
TETRODOTOXIN PARALYSIS
mV
NORMAL
15 MINUTES
150
100
50
0
0
8
16
MSEC
C
PRE- AND POST-SYNAPTIC POTENTIAL CHANGES
mV
20
15
10
5
0
POSTSYNAPTIC POTENTIAL
LOG POSTSYNAPTIC
POTENTIAL
2
50
60
70 mV
PRESYNAPTIC
IMPULSES
PRESYNAPTIC
IMPULSE
PRESYNAPTIC LOCAL
PULSES AFTER TTX
10
20
30
40
PRESYNAPTIC POTENTIAL

until it failed altogether. The relation between the size of the progressively failing presynaptic action potential and the size of the postsynaptic potential is shown in Figure 1. Smaller action potentials cause smaller postsynaptic potentials. This indicates that less transmitter has been released, since TTX does not reduce the sensitivity of the postsynaptic receptors.

Once the impulse has been totally abolished by TTX, currents can be applied to the presynaptic fiber to change the membrane potential to any desired level and thereby obtain a synaptic input-output curve. In an experiment of this sort, depolarizations of 1- to 2-msec duration mimic the effect of the naturally occurring impulse (Figure 1). Characteristically, a delay still intervenes between the time of application of the voltage and the release of transmitter. The artificial depolarizations are not accompanied by a specific regenerative increase in sodium permeability that is characteristic of the normal nerve impulse. The size of the postsynaptic potential (which indicates the amount of transmitter released) varies logarithmically with the size of the presynaptic depolarization.

Synaptic delay

A clue that some intermediary event occurs between the depolarization and the liberation of transmitter is provided by the synaptic delay, which is markedly influenced by temperature. For example, at the frog neuromuscular junction at room temperature at least 0.5 msec elapses between the depolarization of presynaptic endings and the first sign of a postsynaptic depolarization of the motor end plate. This time is too long to be accounted for by diffusion across the synaptic cleft, several tens of nanometers wide, which requires at most 50μsec. In fact, when ACh is applied iontophoretically from a micropipette to the postsynaptic membrane, the delay may be as brief as 150μsec even though the pipette is more distant than the nerve endings from the postsynaptic membrane. Furthermore, if the preparation is cooled to 2° C, the synaptic delay is increased to as long as 7 msec, while the delay associated with iontophoretically applied ACh is not prolonged.[4] Two hypotheses, singly or in combination, can explain the sensitivity of the delay to temperature: (1) since diffusion is relatively unaffected by such temperature changes, a metabolic process may intervene between depolarization and release; (2) some substance may enter the nerve before release occurs.

Evidence that calcium entry is essential for transmitter release

Calcium has long been known as an essential link in the process of transmission. When its concentration in the extracellular fluid is decreased, release of ACh at the neuromuscular junction is reduced and eventually abolished. The importance of calcium for release has been established at all chemical synapses tested, irrespective of the nature of the transmitter. Its role is further generalized to other secretory processes—for example, to the liberation of hormones by cells in the posterior pituitary gland, the adrenal medulla, and salivary glands.[5]

---

[4]Katz, B., and Miledi, R. 1965. *Proc. R. Soc. Lond.* B *161*:483–495.
[5]Douglas, W. W. 1968. *Br. J. Pharmacol.* *34*:451–474.

Magnesium ions antagonize the action of calcium and reduce the secretion of transmitters and hormones.

Katz and Miledi made decisive experiments showing not only that calcium is essential, but also that it must be present at the time when the depolarization of the presynaptic fiber occurs (Figure 2). The principle of the experiment was to block the release of ACh at the neuromuscular junction of the frog by reducing the calcium concentration in the bathing fluid. Under these conditions, presynaptic action potentials still propagated into the motor nerve terminals, but failed to give rise to postsynaptic potentials because no transmitter was released. Next, the ability of the terminal to liberate transmitter was restored by positioning a micropipette containing calcium close to the nerve ending. Positively charged calcium ions were ejected from the micropipette onto a restricted region of the terminal by passing a pulse of current. It was found that calcium, to be effective, must be applied to the nerve terminal just before the depolarization produced by the action potential. If calcium is applied during the period of the synaptic delay (after the end of depolarization of a terminal, but before the time when the transmitter would have been liberated from the ending), it is without effect. Similar results are obtained in the presence of TTX if the terminals are depolarized by current. Again, calcium is required just before the presynaptic depolarization.

Subsequently, evidence was obtained to suggest that calcium itself enters the nerve terminal to facilitate the release. Other experiments on nerve fibers (Chapter 6) have shown that membrane permeability to calcium (g_{Ca}) is increased with depolarization of the membrane and that some calcium enters with each action potential. In certain invertebrate neurons and in the presynaptic terminals of motor nerve fibers, calcium current has been shown to contribute to the impulse current. The idea that calcium enters was strengthened by experiments of Katz and Miledi[6] on the squid synapse. They reasoned that if the presynaptic terminal was depolarized to E_{Ca} or beyond, no calcium would enter during the pulse and no release should occur (just as internal positivity during the action potential tends to prevent sodium from running in). The inside of the presynaptic terminal was made strongly positive by using large currents and adding tetraethylammonium (TEA) to the bath. This reduces g_K and makes it easier to achieve large changes in membrane potential. As expected, release was blocked during these large voltage pulses to the presynaptic ending. Only at their conclusion, when the membrane was repolarizing, was transmitter liberated from the presynaptic ending. This provides indirect evidence for the idea that a positively charged ion, calcium, must enter for transmitter to be released.

A more direct test has been made by Miledi,[7] who injected calcium iontophoretically into the presynaptic nerve terminal of the squid giant axon synapse and thereby caused release, as shown by the appearance

[6]Katz, B., and Miledi, R. 1971, *J. Physiol.* *216*:503–512.
[7]Miledi, R. 1973. *Proc. R. Soc. Lond.* B *183*:421–425.

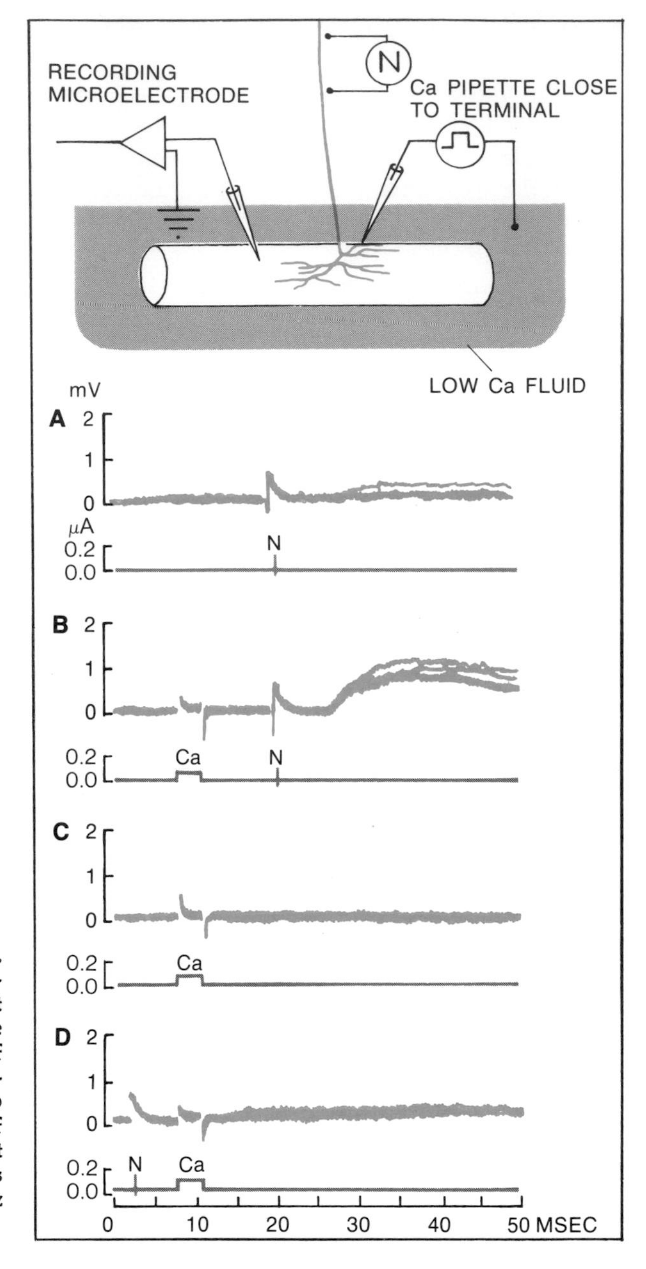

2

EFFECT OF CALCIUM on transmitter release. A. In low-calcium medium a motor nerve impulse causes little or no transmitter release at the skeletal neuromuscular synapse. B. If the nerve impulse is preceded by application of calcium from a pipette to the terminal, transmitter is released. C. Calcium alone has no effect. D. If calcium is applied after arrival of the presynaptic impulse in the terminal, but before the expected synaptic potential, it has no influence on transmitter release. (After Katz and Miledi, 1969)

of postsynaptic depolarizations (see below). Under these conditions, as long as calcium was available WITHIN the terminal it was not required in the outside medium.

So far the general scheme can be depicted as depolarization of presynaptic terminal → calcium entry → release of transmitter.

Quantal release

Once the general framework for release has been established, it remains to be shown how the transmitter is secreted from the terminals. It will be shown later that the effect of a depolarization is to release the transmitter in MULTIMOLECULAR PACKETS. The term QUANTUM can be used interchangeably with "packet"; it simply describes the smallest unit in which transmitter is normally secreted. To take some imaginary figures, it is as if 0, 1000, 2000, or 3000 molecules of ACh could be released, but not smaller amounts than 1000 or intermediate amounts like 1254 molecules. In general, the number of packets liberated by a depolarization may vary, but the packet or unit size is fixed. Translated into the terms used so far, the normal synaptic potential is made up of discrete smaller units, and therefore the size of the synaptic potential is a measure of the amount of transmitter secreted.

Release in the form of quanta was surprising, because one could have imagined other mechanisms for secreting transmitter at nerve endings. For example, in theory, depolarization might simply increase the permeability of the presynaptic nerve ending to ACh, allowing the transmitter to diffuse out. A process of this sort would give rise to a continuously graded release, whereas quantal release results in synaptic potentials that are always integral multiples of the size produced by a single quantum.

The far-reaching conclusions about the release mechanism are based on two main findings: (1) at rest, small spontaneous fluctuations of membrane potential, the MINIATURE SYNAPTIC POTENTIALS, were recorded by Fatt and Katz[8] at the motor end plate; (2) statistical variations of the same order of magnitude as the miniature synaptic potentials occurred when the nerve impulse liberated only a small amount of transmitter. Incidentally, the original accidental observations are also instructive in that they would seem trivial to many investigators, who perhaps would attach Greek letters to these small irregular signals before proceeding to other problems.

Spontaneous miniature synaptic potentials

In a fiber of the frog sartorius muscle, the miniature potentials are less than 1 mV in size and occur with a mean frequency of about 1/sec. In their time course, they resemble little synaptic potentials (Figure 3). Their importance derives from the observations that they are the result of spontaneous release of packets of ACh from the nerve endings and that these packets containing many molecules constitute the units out of which the normal synaptic potential is synthesized.

The evidence that the spontaneous potentials are caused by ACh release was provided by showing that they (1) are decreased in size by curare, which blocks the receptors in the motor end plate, and prolonged by inhibitors of acetylcholinesterase (AChE), an enzyme that normally hydrolyzes ACh and thereby shortens its action (Figure 2, Chapter 8); (2) are abolished by botulinum toxin, which prevents ACh release; and (3) can no longer be recorded after the nerves are cut and have degenerated.[9]

One of the clearest additional signs that the miniature potentials arise as the result of the transmitter liberated by the nerve endings, and are not produced in the muscle itself, is that depolarization of the

[8]Fatt, P., and Katz, B. 1952. *J. Physiol.* *117*:109–128.
[9]Birks, R., Katz, B., and Miledi, R. 1960. *J. Physiol,* *150*:145–168.

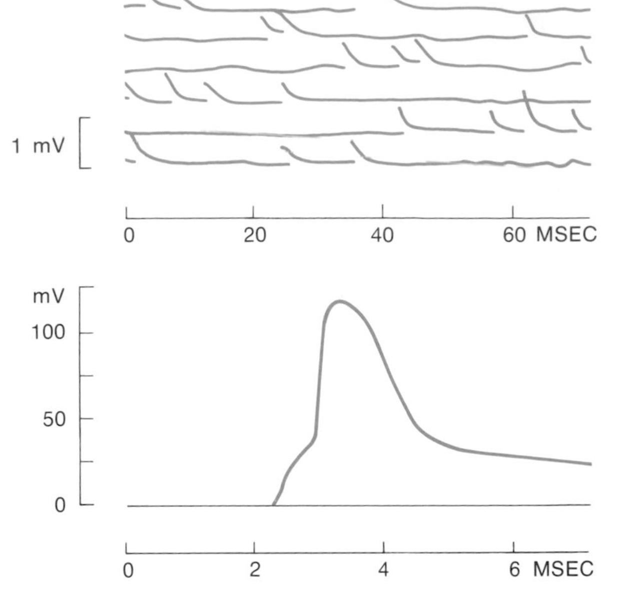

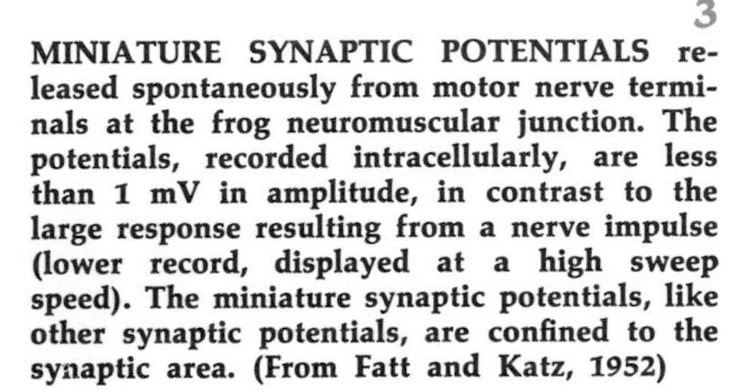
3

MINIATURE SYNAPTIC POTENTIALS released spontaneously from motor nerve terminals at the frog neuromuscular junction. The potentials, recorded intracellularly, are less than 1 mV in amplitude, in contrast to the large response resulting from a nerve impulse (lower record, displayed at a high sweep speed). The miniature synaptic potentials, like other synaptic potentials, are confined to the synaptic area. (From Fatt and Katz, 1952)

presynaptic terminals causes an increase in the frequency (but not in the size) of the miniature potentials. Depolarization of the postsynaptic membrane, on the other hand, has no effect on the frequency of miniature potentials.[10]

Two difficulties have militated against the evidence that miniature potentials result from the release of transmitter: (1) external calcium is not required for their appearance, and (2) the potentials have been seen in denervated muscles. Miniature potentials arise after chronic denervation of muscles from an unexpected source—the Schwann cells that have phagocytotic properties and engulf segments of the degenerating nerve terminal (Chapter 13). Similarly, the finding that the spontaneous potentials continue to occur in calcium-free solutions does not contradict the notion that calcium is involved in the coupling between membrane potential and ACh release. This is supported by tests at frog neuromuscular junctions in which various metabolic inhibitors, such as ruthenium red, cause mitochondria to release their stored calcium into the cytoplasm of the terminals. The result is an increased release of miniature potentials.[11] Spontaneous release, like the amount of nerve-evoked release, therefore, can be interpreted as a reflection of the calcium concentration within terminals, which is normally kept at a low and constant value.

That spontaneous miniature potentials arise from the release of multimolecular packets rather than from the effect of individual molecules of ACh is clear from a number of experiments. Adding ACh to the bathing fluid does not give rise to higher frequencies of miniature

[10]del Castillo, J., and Katz, B. 1954. *J. Physiol.* *124*:586–604.
[11]Alnaes, E., and Rahamimoff, R. 1975. *J. Physiol.* *248*:285–306.

potentials, as would be expected if each miniature potential were the result of a collision between an individual ACh molecule and a receptor in the end plate. Furthermore, a graded depolarization that is considerably smaller than a miniature synaptic potential can be produced by applying varying concentrations of ACh to the receptors iontophoretically or by adding ACh to the bathing fluid. Finally, curare does not decrease the number of miniature potentials, but reduces their size.

Obviously, each time a molecule of ACh interacts with a receptor, a tiny "subquantal" depolarization of the postsynaptic membrane must be produced, but the resolving power of conventional recording techniques is inadequate to show such small effects. Recently, however, statistical analysis of the small voltage fluctuations that occur in the muscle fiber during continued application of ACh has shown that each molecule contributes about $0.3 \mu V$ to the membrane depolarization, far smaller than the miniature potentials. These results have cast light on the molecular interactions of ACh with receptors and are more fully discussed in Chapter 10.

It can be concluded that the spontaneous miniature potentials result from the synchronous release of many molecules of ACh from the nerve endings. It remains to be shown that these miniature potentials represent the basic unit of the release mechanism and that the synaptic potential evoked by nerve stimulation results from the synchronous release of many packets of ACh.

Statistical fluctuations of the synaptic potential

The normal synaptic potentials at the skeletal neuromuscular junction depolarize the postsynaptic membrane by 50 to 70 mV. To produce a potential of this size, about 200 to 300 miniature potentials are required. Although the average amplitude of miniature potentials is 0.5 mV, the potentials that result from increases in conductance do not sum linearly with progressively larger depolarizations because, as discussed in Chapter 8, the driving force producing the potentials decreases as the equilibrium potential is approached with increasing depolarization. It is therefore difficult to detect the small variations corresponding to the contribution of individual packets in a full-sized synaptic potential. On the other hand, if the output of ACh is reduced to a very low level so that only a few packets are released, producing a potential of a few millivolts, then the separate contribution of each quantum can be resolved.

Release of ACh can be reduced by decreasing the external calcium concentration, by increasing the magnesium in the bathing fluid, or by a combination of both. Under these conditions successive impulses produce small synaptic potentials of the same order as a spontaneously occurring miniature potential.[12] Another way of releasing small amounts of transmitter is by depolarizing the presynaptic fiber with graded amounts of current in the presence of TTX so as to cause potentials of about the same size as spontaneously occurring miniature potentials.[13]

---

[12]del Castillo, J., and Katz, B. 1954. *J. Physiol.* *124*:560–573.

[13]Katz, B., and Miledi, R. 1967. *Proc. R. Soc. Lond.* B *167*:23–38.

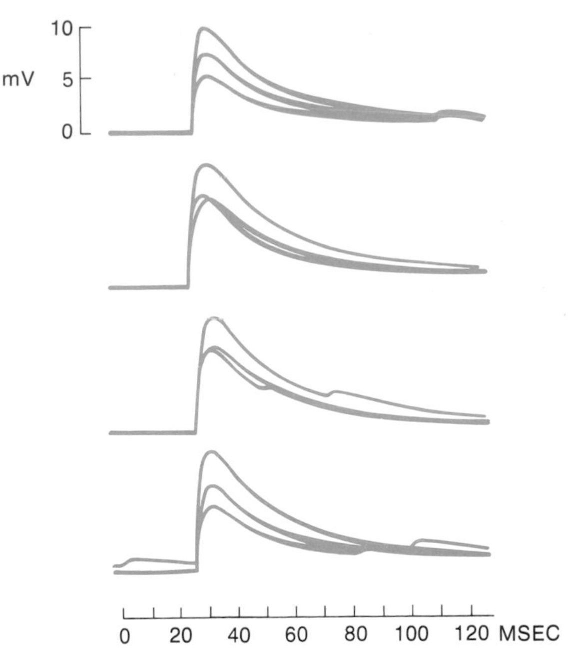

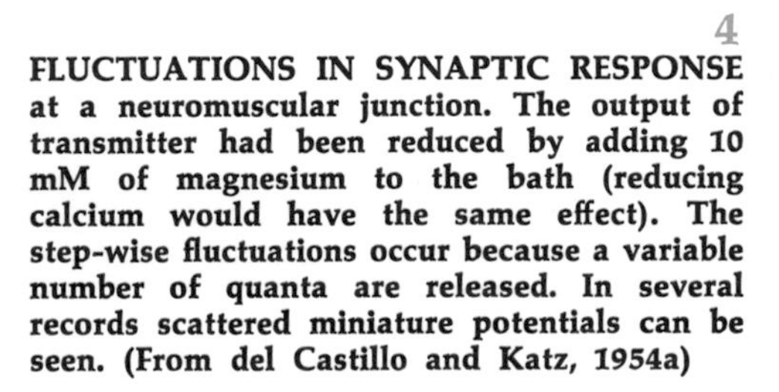
4
FLUCTUATIONS IN SYNAPTIC RESPONSE at a neuromuscular junction. The output of transmitter had been reduced by adding 10 mM of magnesium to the bath (reducing calcium would have the same effect). The step-wise fluctuations occur because a variable number of quanta are released. In several records scattered miniature potentials can be seen. (From del Castillo and Katz, 1954a)

The small synaptic potentials produced by these methods show obvious fluctuations in amplitude. In the example shown in Figure 4, the amplitude of the potential clearly varies in discrete steps. Often there are failures; that is, the motor nerve impulse releases no transmitter at all and no synaptic response is seen. Such fluctuations are expected if transmitter is released only in packets and if only a small number of packets is released each time. (Once again, in these experiments, the postsynaptic membrane is being used to assay the amount of transmitter liberated by the presynaptic nerve fiber.)

The results with nerve-produced small synaptic potentials are in marked contrast to what is observed if ACh is released from a pipette onto the postsynaptic membrane by fine increments of current. The size of the resulting potential can be smoothly graded, giving smaller or larger responses than those caused by neural release (Chapter 10).

Figure 4 makes it appear as though the synaptic potential under these conditions really does consist of multiple units about the size of a spontaneous miniature potential. In fact, the situation is not so simple: there is a considerable variation in the size of the fluctuations that must be treated statistically. A statistical test is needed to ascertain how many synaptic potentials are to be expected at each amplitude. For example, with a fixed number of stimuli, how many failures would occur and how many potentials correspond in size to single or multiple (2, 3, 4 etc.) miniature potentials? An appropriate test for such variations in synaptic potentials is the POISSON DISTRIBUTION. This holds for a situation where (1) packets are released from a large pool of identical units and the chance that a given packet is released in a given trial is very small, and (2) packets are released independently so that release of one does not influence the probability of release of the next.

One of the best known original applications of the Poisson distribution is an analysis of the number of Prussian cavalry officers hurt each year by horse kicks. In some years there were failures—none was kicked; in others one or two were kicked, but always an integral number. Another convenient analogy is a slot machine that accepts dimes, but not pennies, nickels, or quarters. The unit size is fixed at 10 cents, and the probability of release, in real life, is low. Any one trial may release nothing, 10 cents, 20 cents, or 30 cents, but never 3 cents, 15 cents, or 22 cents.

Application of the test requires some way of combining results on miniature potentials and the fluctuations. One could then predict the fluctuations in the amplitude of the synaptic potential from knowledge of the miniature potential size.

From the Poisson equation, the probability of observing a particular end plate potential containing x packets (P_x) is:

$$(P_x) = \left(\frac{m^x}{x!}\right)(e^{-m}) = \frac{n_x}{N}$$

where m equals the mean number of packets liberated by one impulse (the mean number of units in N trials). The probability that any particular synaptic potential contains x packets (P_x) is the number of times such an event occurs (n_x) in N trials of nerve stimulation. This equation can show how well the values predicted for n_x agree with those found in practice. What one needs is m (the mean number of quanta per impulse). To return to the analogy of the slot machine, the mean number of dimes released per trial must be known; in the experimental situation, however, prior knowledge about the type of coin the machine releases is lacking.

There are two independent ways of measuring m. The first estimate of m depends on the assumption that the miniature synaptic potential is equal to one unit of the total synaptic potential. By definition, therefore:

$$m = \frac{\text{mean amplitude of synaptic potential}}{\text{mean amplitude of miniature potential}}$$

By analogy again, this is equivalent to dividing the mean number of cents released per trial by 10 cents. The second estimate involves counting the number of failures, when $x = 0$ (when the slot machine fails to deliver a coin). In this case, the Poisson distribution becomes simplified to:

$$\frac{\text{number of failures } (= n_0)}{\text{number of trials } (= N)} = e^{-m}$$

that is, $m = \log_e(N/n_0)$.

The agreement between m estimated in these two entirely different ways is excellent. If the hypothesis is correct, one should now be able

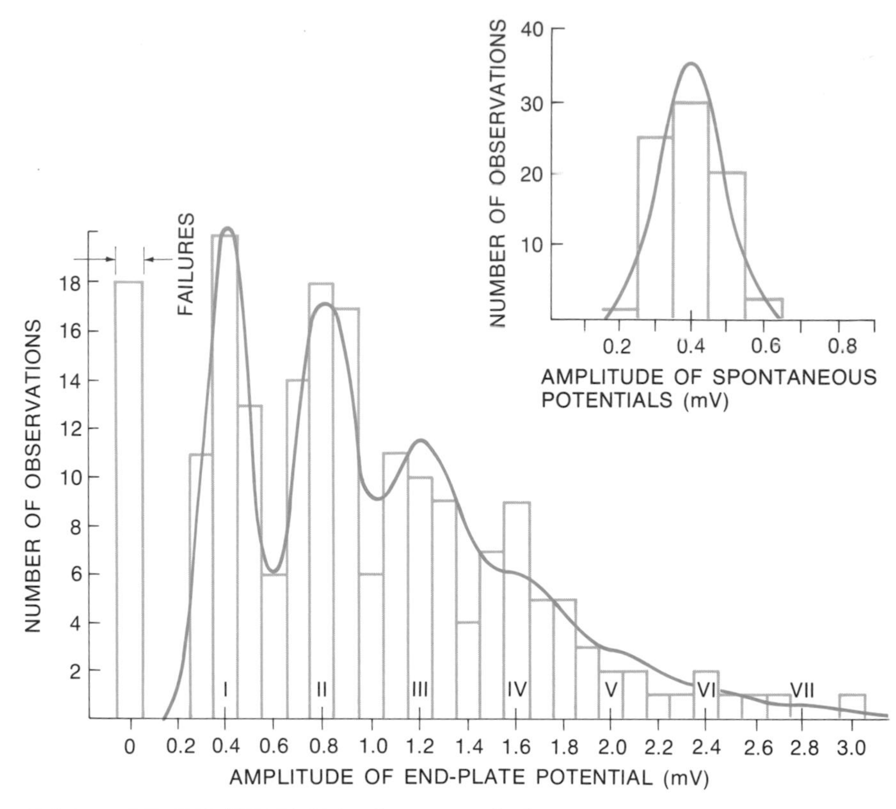

5 **AMPLITUDE DISTRIBUTION of nerve-evoked synaptic potentials at a mammalian skeletal neuromuscular junction in which the release of ACh had been reduced. The histogram shows the numbers of synaptic potentials observed at each amplitude. The peaks in the histogram occur at one, two, three, and four times the mean amplitude of the spontaneous miniature potentials plotted in the inset. The solid line represents the theoretical distribution of synaptic potential amplitudes calculated according to the Poisson equation. Arrows indicate the predicted numbers of failures (zero amplitude) following nerve stimulation. (From Boyd and Martin, 1956)**

to predict the number of synaptic potentials of each particular amplitude in a total of N trials, using the value of m calculated from the miniature potentials. Figure 5 shows the good agreement between the observed amplitude of the synaptic potentials and those predicted from the Poisson equation. The spread of the peaks comes about because the miniature potentials are not all exactly the same size but are distributed normally.

The Poisson distribution described above constituted the first direct demonstration of quantal release. The Poisson statistics do not apply, however, if a large fraction of the available packets is released per trial. Other more complex statistical distributions are needed for experiments in which the probability of release for a given packet is high. Under these conditions a general way of describing release characteristics statistically is that $m = np$; where m, as before, is the mean number of quanta released per trial; n is the number of quanta available for release; and p is the average probability that any one quantum will be released in one trial. The Poisson distribution is then a good description when p is small, as in a low-calcium medium. With higher values of p the binomial distribution is a more accurate description of the fluctuations.[14–16] These statistical treatments are interesting because they confirm that release occurs in quantal units and provide in principle a method for studying the mechanisms responsible for changes observed in synaptic transmission in terms of probability and available transmitter. For example, a decrease in the size of a synaptic potential after prolonged stimulation could be due to a lowered probability of release (p) or to a decrease in available quanta (n). Whichever way one applies the statistics, the mechanisms remain the same.

General significance of quantal release

One of the most important applications of the concept of quantal release is for analysis of synaptic transmission in subcellular terms. At the same time, the quantal phenomena bear directly on mechanisms that operate during integration in the nervous system, because quanta provide the units that make up synaptic signals. For example, at the neuromuscular junction of crustaceans and of frogs, a second impulse delivered shortly after the first gives rise to a bigger synaptic potential (Figure 6). Similarly, during a train of nerve impulses successive synaptic potentials may increase in size. The effect, called FACILITATION, has been shown to be caused by an increase in the amount of transmitter liberated by the presynaptic terminal.[17] During such facilitation each succeeding presynaptic impulse secretes a greater number of packets, while the size of individual packets remains unchanged.[18] In the course of prolonged stimulation one may also observe that less transmitter than usual is liberated. Such a DEPRESSION results from a reduction in the number of packets secreted and not in the amount of transmitter in each individual packet.

The depression, or reduction, of release observed during presynaptic inhibition in crustaceans operates in the same manner, and at many other synapses facilitation and depression of release can be shown to be general mechanisms in inhibitory as well as excitatory terminals. They provide a good example of the flexible nature of chemical transmission in which the number of units released is markedly influenced by previous activity.

Under certain abnormal conditions, the packet size can be reduced, There is evidence in skeletal muscle that regenerating nerve terminals release ACh in smaller quantal packages than normal.[19] A second example is the effect on nerve terminals of hemicholinium, a drug that

[14]Christensen, B. N., and Martin, A. R. 1970. *J. Physiol.* *210*:933–945.
[15]Wernig, A. 1972. *J. Physiol.* *226*:751–759.
[16]Zucker, R. S. 1973. *J. Physiol.* *229*:787–810.
[17]del Castillo, J., and Katz, B. 1954. *J. Physiol.* *124*:574–585.
[18]Dudel, J., and Kuffler, S. W. 1961. *J. Physiol.* *155*:543–562.
[19]Dennis, M. J., and Miledi, R. 1974. *J. Physiol.* *239*:571–594.

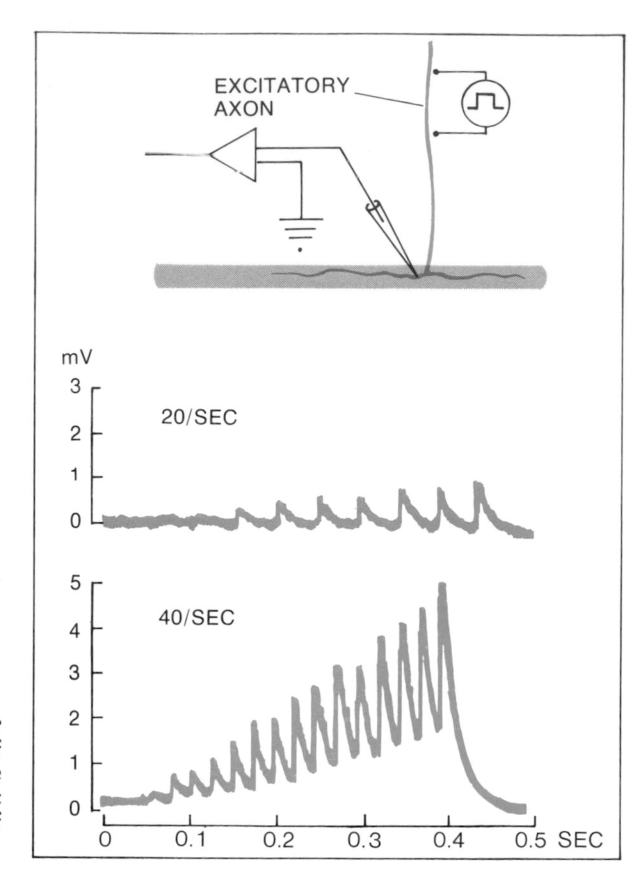

6

FACILITATION at crustacean neuromuscular junction brought about by the liberation of increased numbers of quanta with successive nerve stimuli at 20/sec and 40/sec. The extent of facilitation depends on the frequency of stimulation. (From Dudel and Kuffler, 1961)

interferes with ACh synthesis by reducing or preventing the uptake of choline that is needed for the replenishment of transmitter in the terminals.[20,21] Finally, in some of the myoneural diseases that afflict man, such as myasthenia gravis, it is important to determine whether individual packets contain fewer ACh molecules than normal or whether the postsynaptic receptor system is impaired.[22,23]

SUGGESTED READING

Papers marked with an asterisk (*) are reprinted in Cooke, I., and Lipkin, M. 1972. *Cellular Neurophysiology.* Holt, New York.

General reviews

Hubbard, J. I. 1973. Microphysiology of vertebrate neuromuscular transmission. *Physiol. Rev. 53*:674–723.

[20]Birks, R., and MacIntosh, F. C. 1961. *Can. J. Biochem. Physiol. 39*:787–827.

[21]Ceccarelli, B., and Hurlbut, W. P. 1975. *J. Physiol. 247*:163–188.

[22]Fambrough, D. M., Drachman, D. B., and Satyamurti, S. 1973. *Science 182:* 293–295.

[23]Fields, W. S. (ed.). 1971. Myasthenia gravis. *Ann. N.Y. Acad. Sci. 183*:1–386. (A collection of papers related to the disease.)

Katz, B. 1969. *The Release of Neural Transmitter Substances.* Liverpool University Press. Liverpool.

Krinjevic, K. 1974. Chemical nature of synaptic transmission in vertebrates. *Physiol. Rev. 54*:418–540. (Contains more than 1200 references.)

Kuno, M. 1971. Quantum aspects of central and ganglionic synaptic transmission in vertebrates. *Physiol Rev. 51*:647–678.

Kuno, M. 1974. Factors in Efficacy of Central Synapses. In M. V. L. Bennett (ed.). *Synaptic Transmission and Neuronal Interaction.* Raven Press, New York, pp. 79–86.

Calcium and transmitter release

Dodge, F. A., and Rahamimoff, R. 1967. Cooperative action of calcium ions in transmitter release at neuromuscular junction. *J. Physiol. 193*:419–432.

Douglas, W. W. 1968. Stimulus-secretion coupling: the concept and clues from chromaffin and other cells. *Br. J. Pharmacol. 34*:451–474. (A review of the role of calcium in secretory processes.)

*Katz, B., and Miledi, R. 1967. The timing of calcium action during neuromuscular transmission. *J. Physiol. 189*:535–544.

*Katz, B., and Miledi, R. 1967. A study of synaptic transmission in the absence of nerve impulses. *J. Physiol. 192*:407–436.

Miledi, R. 1973. Transmitter release induced by injection of calcium ions into nerve terminals. *Proc. R. Soc. Lond.* B *183*:421–425.

Quantal release

Boyd, I. A., and Martin, A. R. 1956. The end-plate potential in mammalian muscle. *J. Physiol. 132*:74–91.

*del Castillo, J., and Katz, B. 1954. Quantal components of the end-plate potential. *J. Physiol. 124*:560–573.

*Katz, B., and Miledi, R. 1967. The release of acetylcholine from nerve endings by graded electric pulses. *Proc. R. Soc. Lond.* B *167*:23–38.

Presynaptic modification of quantal release

del Castillo, J., and Katz, B. 1954. Statistical factors involved in neuromuscular facilitation and depression. *J. Physiol. 124*:574–585.

*Dudel, J., and Kuffler, S. W. 1961. Presynaptic inhibition at the crayfish neuromuscular junction. *J. Physiol. 155*:543–562.

Zucker, R. 1973. Changes in the statistics of transmitter release during facilitation. *J. Physiol. 229*:787–810.

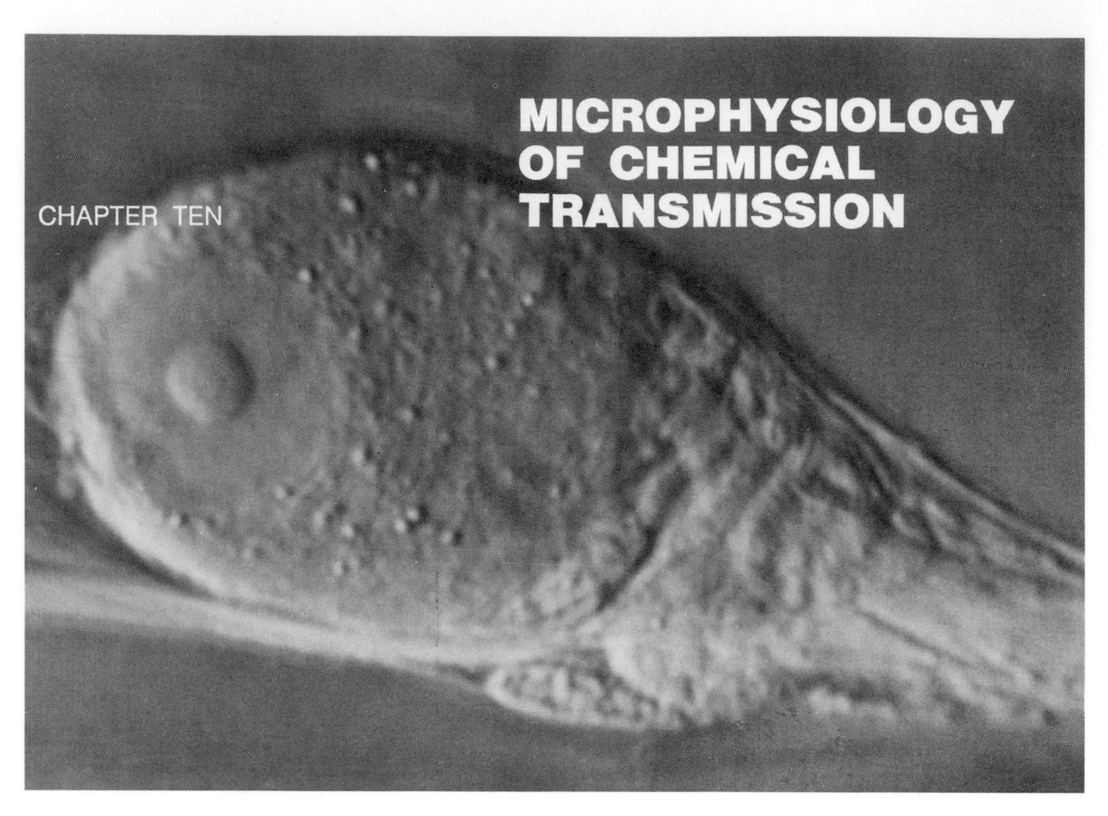

1 **LIVING NEURON in the heart of the frog. (From McMahan and Kuffler, 1971)**

Several steps in the release of the chemical transmitter and its postsynaptic action have been correlated with subcellular structures and with molecular constituents of synapses at neuromuscular junctions. Transmitters are concentrated within nerve terminals in membrane-bound vesicles which can be depleted by sustained neural activity. Electron microscopic evidence supports the hypothesis that vesicles fuse with the cell membrane to release their contents into the synaptic cleft.

Once acetylcholine (ACh) molecules have been released by motor terminals, they diffuse across the synaptic cleft and combine with receptor molecules in the subsynaptic membrane. This is the area where the ACh receptors are concentrated. In snake and frog muscles the sensitivity to ACh decreases by a factor of about 100 within a distance of a few microns from the edge of the nerve terminal. ACh receptors have been extracted from various membranes and purified; they are proteins whose properties are being characterized.

Electrical recording techniques have provided insight into the molecular mechanisms involved in the interactions of ACh with chemoreceptors and the subsequent opening of ionic gates. The events associated with the opening and closing of single ionic channels can be correlated with fluctuations in membrane potential produced during a steady application of ACh. Such analyses of

"ACh noise" have revealed that the unitary event consists of the opening of a channel which allows about 5×10^4 ions to flow.

Quanta, the smallest physiological units of nerve-released transmitter, consist of fewer than 10,000 molecules of ACh. The ionic flow initiated by a quantum results from the activation of approximately 1000 ionic channels that remain open for about 1 msec. The synaptic current produced by a nerve impulse consists of the linear summation of responses to synchronously released quanta. The enzyme acetylcholinesterase (AChE) hydrolyzes ACh molecules and thereby restricts the spread of the transmitter within the synaptic clefts. When AChE is inhibited, however, quanta are free to diffuse within the synaptic cleft so that they cover overlapping areas and potentiate each other's postsynaptic effects.

Measurements of membrane noise and quantal release have advanced our understanding of the mechanism of action of transmitters and drugs that influence synaptic transmission. Much evidence suggests that principles similar to those operating at neuromuscular synapses apply to neurons within the central nervous system.

Chapters 8 and 9 treat the synapse as a simple functional unit made up of two components: the presynaptic fiber and the postsynaptic membrane. This concept provides an adequate framework for discussing the basic principles of chemical synaptic transmission, focusing chiefly on the secretion of transmitters and the permeability changes they initiate in the postsynaptic membrane. At the subcellular level, however, there is now much detailed structural information about the sites where transmitter is released and about the location of postsynaptic receptors. New aspects of the process of transmission have been revealed by correlating the function of synapses with their fine structure and the molecular components of cell membranes.

This chapter emphasizes studies on neuromuscular synapses. Much progress has been made recently, particularly as a result of Katz and Miledi's analysis of acetylcholine (ACh) produced voltage fluctuations within the microvolt range that appear in the form of electrical "noise." This has led to a recognition of electrical events associated with the opening by ACh of individual ionic channels in the postsynaptic membrane. Further, the use of improved optical methods by McMahan and his colleagues has made it possible to determine with considerable accuracy the distribution of chemoreceptors on the surface of nerve and muscle membranes. Other physiological experiments have explored such questions as: What is the ACh concentration in the synaptic cleft when a quantum is released? How do receptors react to varying concentrations of the transmitter? How many molecules make up a quantum?

Structural elements of synapses

A consistent feature of chemical synapses is the cleft, a few tens of nanometers wide, that separates the presynaptic nerve terminals from the postsynaptic membrane. The internal architecture of nerve terminals at many diverse synapses shows common elements such as mitochondria, neurofilaments, microtubules, an agranular endoplasmic reticulum, and synaptic vesicles. This discussion concentrates chiefly on the SYNAPTIC VESICLES, which were tentatively linked with chemical transmission shortly after their discovery. This correlation of structure and function has since been well supported, and the presence of vesicles alone is one of the most useful criteria for the recognition of chemical synapses.[1,2] The vesicles range from about 40 to 200 nm in diameter, and their characteristic morphology is often associated with a particular chemical transmitter. For example, terminals that release catecholamines contain large vesicles, each with an electron-dense granule at its core,[3] while cholinergic terminals contain clear (agranular) vesicles. One cannot, however, conclude from the presence of agranular vesicles that ACh is necessarily the transmitter at an unknown synapse.

Nerve terminals possess a certain degree of autonomy and self-regulation for maintaining adequate levels of transmitter.[4] Thus, transmitters can be synthesized in the terminals at a rate depending on the amount of release. The continued secretory activity of terminals also depends on the maintenance of an axonal transport system from the cell body, because the terminals contain no ribosomes and receive their proteins by axonal transport from the cell body (Chapter 11).

Sites of transmitter release at neuromuscular synapses

Electron microscopic observations of the morphology of synapses provide a basis for studying the life cycle of synaptic vesicles and their role in the release of transmitter, and also serve as a guide for understanding the postsynaptic events.

At a typical skeletal neuromuscular synapse in the frog, the myelinated axon gives off branches that run in shallow grooves on the surface of the muscle. In electron micrographs of the terminals some of the vesicles, instead of being randomly distributed, are lined up in a double

---

[1]Palay, S. L., and Palade, G. E. 1955. *J. Biophys. Biochem. Cytol.* *1*:69–88.
[2]DeRobertis, E. 1967. *Science* *156*:907–914.
[3]Bloom, F. E. 1972. *Handbuch Exp. Pharmacol.* *33*:46–78.
[4]Collier, B., and MacIntosh, F. C. 1969. *Can. J. Physiol. Pharmacol.* *47*:127–135.

2

NEUROMUSCULAR JUNCTION of the frog. A. Low-power view of several muscle fibers and their innervation. Below is a three-dimensional sketch of part of a synaptic contact area. Synaptic vesicles are clustered in the nerve terminal in special regions opposite the opening of postsynaptic folds. These active zones are the sites where transmitter is released into the synaptic cleft. Processes of Schwann cells (S) usually run between the nerve terminal and the postsynaptic membrane, separating active zones. B. Electron micrograph of a portion of a motor nerve terminal. It shows many of the features seen in the sketch, including active zones (arrows) and Schwann cell processes (S). Within the synaptic cleft and the junctional folds the basement membrane is visible. (Courtesy of U. J. McMahan)

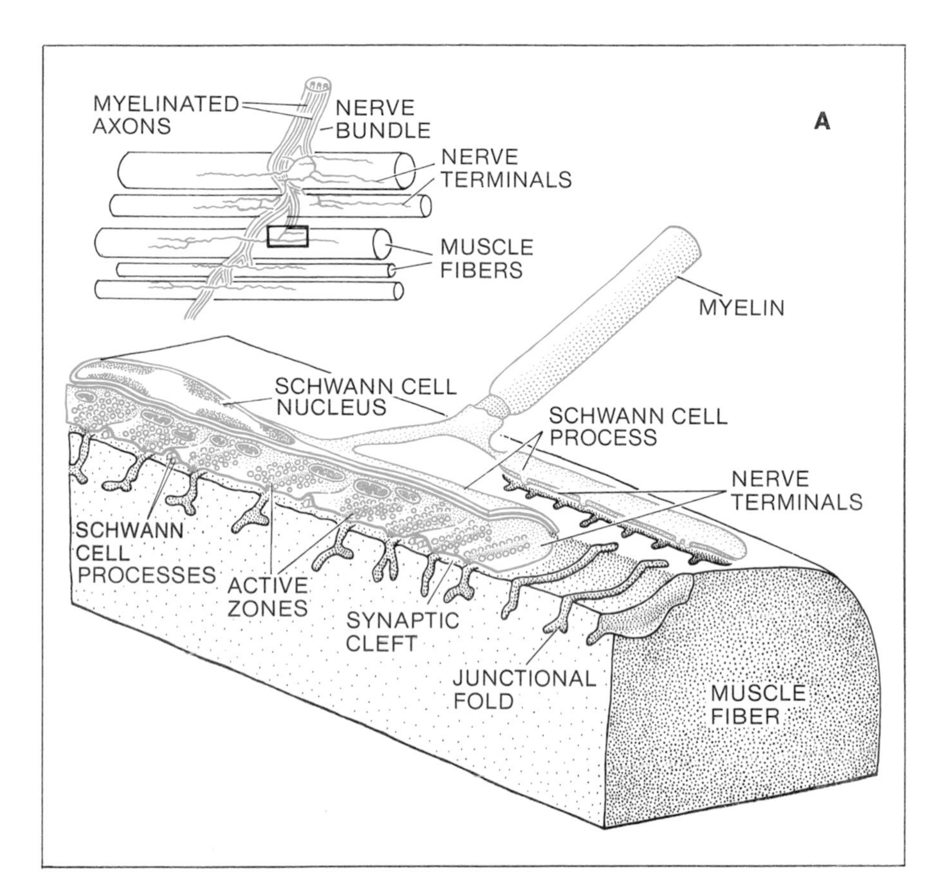
A
MYELINATED AXONS
NERVE BUNDLE
NERVE TERMINALS
MUSCLE FIBERS
MYELIN
SCHWANN CELL NUCLEUS
SCHWANN CELL PROCESS
NERVE TERMINALS
SCHWANN CELL PROCESSES
ACTIVE ZONES
SYNAPTIC CLEFT
JUNCTIONAL FOLD
MUSCLE FIBER

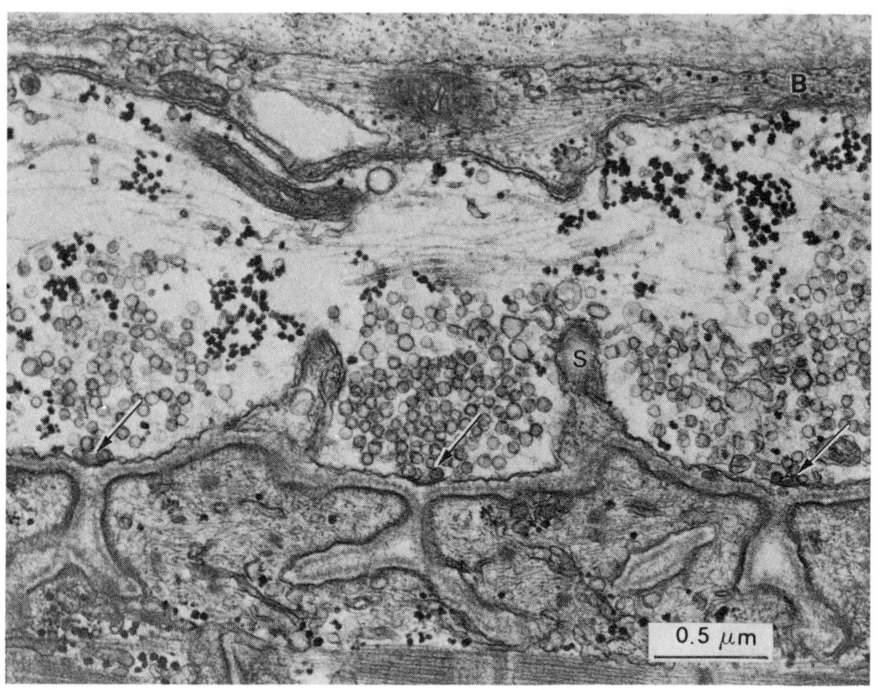
B
S
0.5 μm

row along narrow transverse bars of electron-dense material attached to the presynaptic membrane. This presynaptic region is usually called the ACTIVE ZONE; much evidence indicates that it is the site where transmitter is released.[5–7] On the opposite side of the cleft occur other characteristic specializations, the POSTSYNAPTIC FOLDS, which radiate out from the cleft at regular intervals. Figure 2 shows these features, including several active zones and corresponding postsynaptic areas. In addition, within the cleft can be seen the basement membrane, which follows the contours of the muscle surface. Schwann cell lamellae cover the terminal, sending fingerlike processes around it, thereby creating regularly spaced subdivisions. The principal subcellular elements shown in Figure 2 also occur in nerve-nerve synapses (Figure 5*C*).

The relatively recent technique of freeze-fracturing has revealed many additional features of cell membranes. In this procedure portions of frozen tissue are broken apart and prepared for microscopy. The line of cleavage does not occur along the intercellular clefts between the surfaces of two adjacent cells; instead, freeze-fracture lines split along a plane between the two bilayers of a plasma membrane to produce two artificial surfaces. The resulting pictures show the two fracture surfaces, one belonging to the cytoplasmic half and the other to the external half of the membrane leaflet. Thus, one can view broad areas of the interior of membranes containing particles of varying sizes.

Figure 3 gives a three-dimensional presentation of the results of the freeze-fracture method. Figure 3*A* shows some of the main features of several active zones in a terminal and the corresponding postsynaptic structures, as seen in an electron microscopic section. If the pre- and postsynaptic membrane bilayers are split open (at the arrows), one obtains the situation depicted in Figure 3*B*. The upper portion of Figure 3*B* shows the opened up cytoplasmic half of the presynaptic membrane with a lineup of synaptic vesicles on one side and particles protruding on the other. Pits, corresponding to the particles, appear in the fracture face of the outer leaflet that faces the synaptic cleft. Similarly for the postsynaptic membrane that has been split in the area of a synaptic fold the protruding particles within the membrane are on the cytoplasmic leaflet's fracture face; their counterparts show up as pits in the outer leaflet that borders the synaptic cleft.

In Figure 4*A* a thin section has been made in the region of the active zone within the presynaptic terminal. An orderly row of synaptic vesicles can be seen lined up along a band of dense material. The area of the presynaptic membrane, just under and parallel to the structures shown in Figure 4*A*, has been exposed by the freeze-fracture procedure in Figure 4*B*. There the fracture surface of the cytoplasmic portion of the presynaptic membrane is seen to be lined by a row of particles

---

[5]Couteaux, R., and Pécot-Dechavassine, M. 1970. *C. R. Acad. Sci. (Paris) 271:* 2346–2349.

[6]Heuser, J. E., Reese, T. S., and Landis, D. M. D. 1974. *J. Neurocytol. 3*:109–131.

[7]Peper, K., Dreyer, F., Sandri, C., Akert, K., and Moor, H. 1974. *Cell Tissue Res. 149*:437–455.

(10 to 12nm in diameter); on both sides are indentations, presumably caused by fusion of several of the synaptic vesicles with the membrane. In Figure 4C several pre- and postsynaptic components are apparent as the fracture line passes through the outer leaflet of the nerve terminal (*T*), across the synaptic cleft (*C*), and exposes the cytoplasmic half of the postsynaptic muscle membrane. A population of particles can be resolved at the edges or lip of synaptic folds (*F*), just across from the active zone in the area where one expects ACh to react with the postsynaptic membrane; these particles are assumed to correspond with ACh receptors that are concentrated in this area (see later).[6–8]

Vesicles as the sites of transmitter storage and release

When brain tissue is homogenized and fractionated, some of the fractions contain large numbers of membrane-enclosed structures called SYNAPTOSOMES. These are pinched-off nerve terminals, rich in vesicles and transmitters. As a rule, part of the subsynaptic membrane remains attached during fractionation. The vesicles within synaptosomes are often similar to those seen in motor nerves of the frog neuromuscular junction (Figure 2). Particularly good sources of ACh-containing vesicles are the electric organs of the electric eel and *Torpedo*,[9] where transmission is cholinergic.

The correlation between vesicles and transmitter storage and release is further supported by electron microscopic studies which show that nerve stimulation reduces the number of vesicles in nerve terminals.[10,11] In control experiments nerves were stimulated in fluid containing low calcium and high magnesium concentrations to prevent transmitter release. Under these conditions the vesicles were not depleted. In other experiments spider venom was applied and caused a massive increase in the frequency of miniature potentials (that is, transmitter release), and also reduced the number of vesicles.[12]

If the release of transmitter results from the fusion of vesicles with the presynaptic membrane, as indicated in Figure 4B, one would expect the process to result in an expansion of the surface area of the terminal. The hypothesis postulates that to keep the number of vesicles and the terminal's membrane area constant, either the vesicle membrane is taken back into the interior of the presynaptic terminal or else new vesicles are pinched off. Experimental evidence supports the latter idea. Thus, when terminals are incubated in a solution containing horseradish peroxidase (which provides a marker that can be seen in electron micrographs), vesicles containing the enzyme appear within the terminal. A reasonable explanation for this is that as vesicles are pinched off they contain peroxidase, which is present in the external fluid. Subsequently these vesicles appear to coalesce to form membrane-bound cisternae which also contain peroxidase. If the nerve is then stimulated

---

[8]Porter, C. W., and Barnard, E. A. 1975. *J. Membrane Biol.* *20*:31–49.

[9]Whittaker, V. P. 1970. In P. Anderson and J. K. S. Jansen (eds.). *Excitatory Synaptic Mechanisms.* Universitetsforslaget, Oslo.

[10]Birks, R. I. 1974. *J. Neurocytol.* *3*:133–160.

[11]Heuser, J. E., and Reese, T. S. 1973. *J. Cell Biol.* *57*:315–344.

[12]Ceccarelli, B., Hurlbut, W. P., and Mauro, A. 1973. *J. Cell Biol.* *57*:499–524.

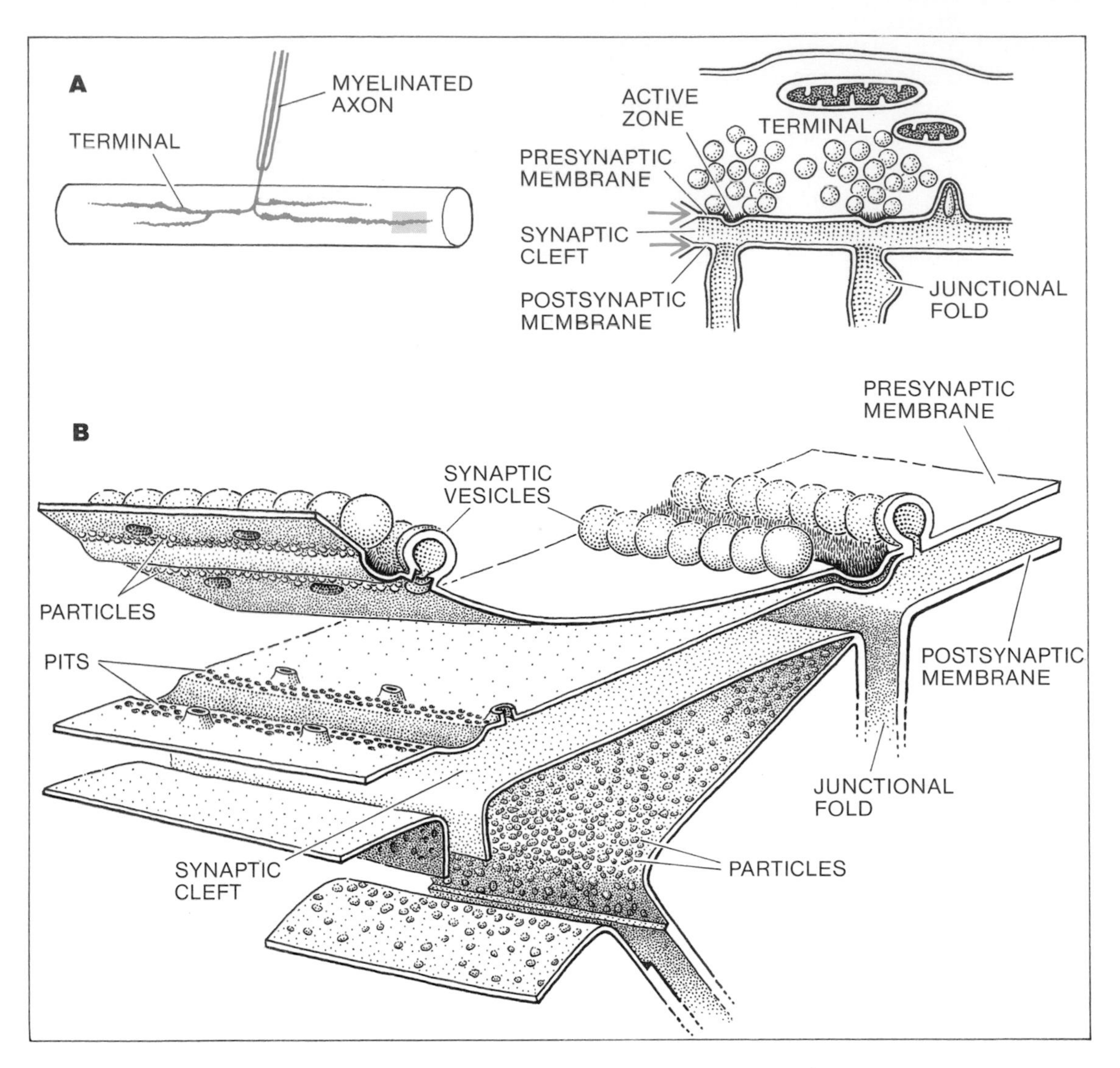

3 **SYNAPTIC MEMBRANE STRUCTURE. A. Entire frog neuromuscular junction (left) and longitudinal section through portion of nerve terminal (colored square right). B. Three-dimensional view of pre- and postsynaptic membranes with active zones and immediately adjacent rows of synaptic vesicles. The plasma membranes are split (at colored arrows in A) to illustrate structures observed upon freeze fracturing. The cytoplasmic half of the presynaptic membrane at the active zone shows on its fracture face protruding particles whose counterparts are seen as pits on the fracture face of the outer membrane leaflet. Vesicles which fuse with the presynaptic membrane give rise to characteristic protrusions and pores in the fracture faces. The fractured postsynaptic membrane in the region of the folds shows a high concentration of particles on the cytoplasmic leaflet. The particles are probably ACh receptors. (Courtesy of U. J. McMahan)**

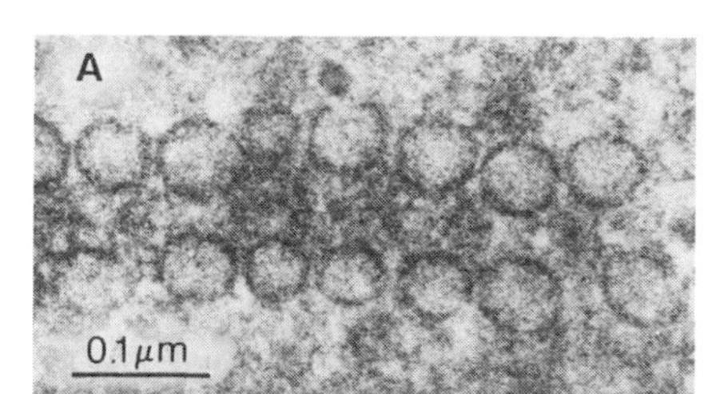

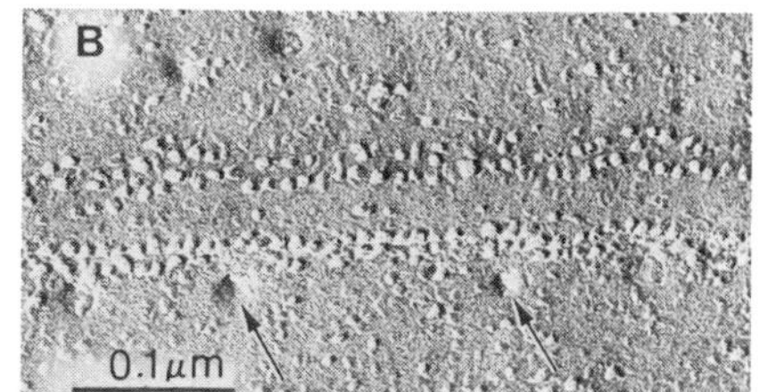

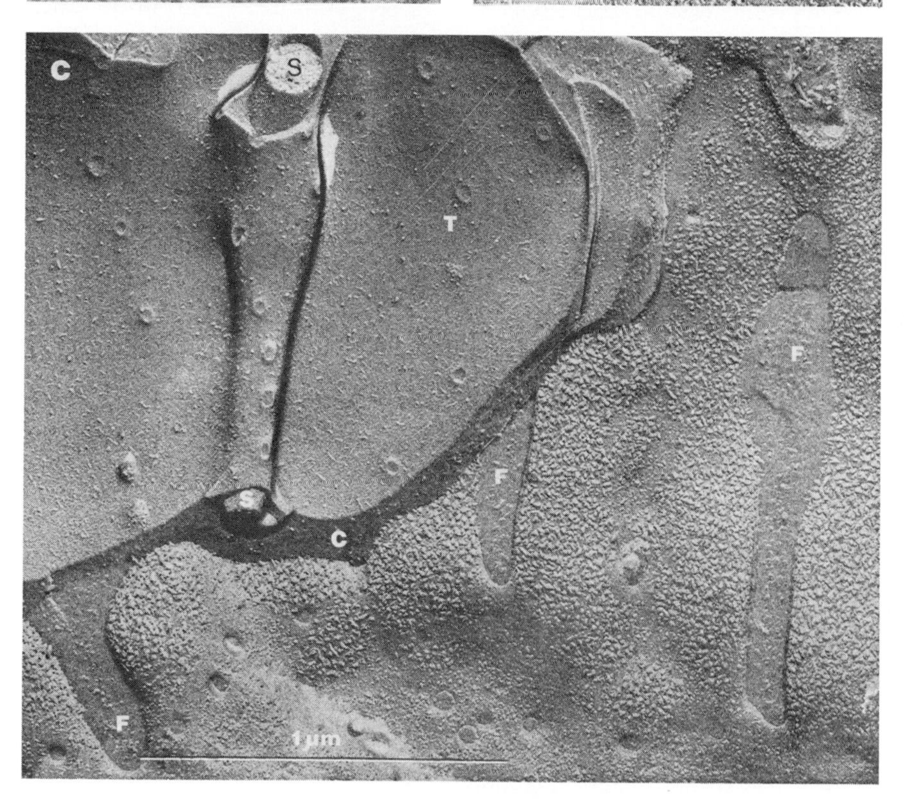

4

REGION OF TRANSMITTER RELEASE and postsynaptic action at the neuromuscular synapse of the frog. A. Electron micrograph of a section through the nerve terminal parallel to an active zone portion of the presynaptic membrane. B. Fracture face of the cytoplasmic half of the presynaptic membrane in the active zone, revealed by freeze-fracture. The region of the active zone is lined by membrane particles (about 10 nm in diameter); further to the side are deformations caused by the fusion of synaptic vesicles with the membrane (arrows). C. Fracture passes first through the presynaptic membrane of the terminal (T) whose outer fracture face is viewed; then it crosses the synaptic cleft (C) and enters the postsynaptic membrane. On the fracture face between synaptic folds (F) one sees, on the cytoplasmic leaflet, particles that are characteristic of the subsynaptic membrane where ACh receptors are concentrated. A Schwann cell process (S) passes between terminal and muscle. (A from Couteaux and Pécot-Dechavassinne, 1970; B, C from Heuser, Reese, and Landis, 1974)

again in normal fluid, vesicles are depleted of peroxidase as they release their contents into the synaptic cleft.[11]

Many of the details of such a recycling scheme remain to be resolved. Nevertheless, at this stage a reasonable picture is emerging of the sequence of formation of vesicles and secretion of transmitter. The evidence that vesicles contain transmitter is satisfactory and fits well with most of the physiological observations. It is not yet clear, however, whether one vesicle corresponds to one quantum[13] and how much transmitter can be stored in the cytoplasm outside vesicles.

CHEMORECEPTORS AND THEIR DISTRIBUTION

Chapter 8 describes how excitatory and inhibitory synaptic potentials result from the opening of specific permeability channels by the action of the transmitter. The effect of the transmitter is produced by its interaction with specific chemoreceptor molecules in the cell membrane. This situation raises some general questions about the manner in which the cell surface is constructed. Are the chemoreceptors and the ionic channels they activate sharply confined to structurally distinct synaptic areas? If this is so, receptors should be concentrated on cell surfaces in restricted areas according to the distribution of synapses. For example, in neurons whose cell bodies and dendrites are densely covered by synaptic boutons which release either excitatory or inhibitory transmitters (Figure 9, Chapter 16) patches of receptors with distinct properties should be found. Each type of transmitter should control specific permeability channels. These questions can be studied most readily in neurons and in muscle fibers in which ACh mediates excitation.

The existence of special properties of skeletal muscle fibers in the region of their innervation has been known since the beginning of this century. For example, Langley assumed the presence of a "receptive substance" around motor nerve terminals where a localized sensitivity exists for various chemical agents, such as nicotine.[14] Gradually, the term RECEPTOR has become accepted to indicate a site where a transmitter, hormone, or drug exerts its effects. Only recently have receptor molecules for ACh been identified. Their isolation has been helped greatly by the discovery of snake toxins that bind specifically and with high affinity to ACh receptors.[15] Once occupied by toxin, the receptors cannot react with ACh, and this accounts for the paralyzing action that results from the bite of certain snakes. One of the most useful toxins has been α-bungarotoxin, obtained from *Bungarus* snakes. The toxin can be radioactively labeled without losing its activity. Such labeled toxin is then bound to receptors, and the complex can be isolated and chemically characterized. Many of the studies have been made on receptors of the receptor-rich electric organs of the fish *Electrophorus*

---

[13]Kriebel, M. E., and Gross, C. E. 1974. *J. Gen. Physiol.* *64*:85–103.
[14]Langley, J. N. 1907. *J. Physiol.* *36*:347–384.
[15]Lee, C. Y. 1972. *Annu. Rev. Pharmacol.* *12*:265–286.

and *Torpedo.* ACh receptors are glycoproteins whose molecular weight is probably about 250,000 to 300,000, made up of five to six subunits.[16]

Chemical specialization of synaptic areas

Chemosensitivity of the surface membrane of the postsynaptic cell can be determined by releasing a substance iontophoretically from the tip of a pipette which is apposed to the cell surface (Chapter 8). Sensitivity can be expressed numerically by measuring the postsynaptic depolarization caused by the passage of a measured amount of charge from the pipette. For example, if 1 pC of charge is applied to a pipette filled with ACh, a depolarization of 5 mV may result; the sensitivity can thus be designated as 5 mV/pC or 5000 mV/nC.

The iontophoretic method has recently been further refined as a result of the availability of several new preparations. The common feature of these is that they are thin and transparent, so that detail of living functioning synapses can be resolved with transmitted illumination.[17] This allows the tip of a micropipette to be closely apposed to the synaptic membranes under visual guidance, and the areas of chemosensitivity to be delineated.

Preparations in which the surface of neurons with their synapses can be viewed come from amphibians, in particular from the INTERATRIAL SEPTUM in the heart of the frog. The septum is a thin transparent sheet in which are embedded nerve cells that receive their innervation from the vagus nerve. Usually one or more vagal axons distribute 12 to 15 synaptic boutons on a cell body (Figure 5*A*). The boutons are the sites where ACh is released by the vagus fibers onto the nerve cell body, causing excitatory synaptic potentials that lead to the initiation of impulses along the axons of the neurons. A synaptic bouton on a living nerve cell is readily recognized in Figure 5*B* (arrow). Additional boutons on the same neuron can be seen at different focal depth. In Figure 5*C* is depicted a cross section through such a bouton showing the typical constituents of a chemical synapse, particularly the synaptic vesicles and pre- and postsynaptic membrane thickenings.

The right hand portion of Figure 5 also shows a picture of synapses, this time from a muscle fiber in the body wall of a snake. The end plates in snake muscles resemble in their compactness those in mammals; they are about 50μm in diameter with 50 to 70 terminal swellings that are analogous to synaptic boutons. The boutons that release ACh rest in craters sunk into the surface of the muscle fiber (Figure 5*E*). A cross section through one of these boutons (Figure 5*F*) shows once more the characteristic features of a chemical synapse. To the right in Figure 5*F* is an electron micrograph of a typical micropipette used to apply transmitter in iontophoretic experiments. The outer diameter of the tip is about 100 nm, while the opening is approximately 50 nm, about the size of a synaptic vesicle.

---

[16]Papers dealing with receptors can be found in a recent symposium, *The Synapse.* 1976. *Cold Spring Harbor Symp. Quant. Biol. 40.*

[17]McMahan, U. J., Spitzer, N. C., and Peper, K. 1972. *Proc. R. Soc. Lond.* B *181*:421–430.

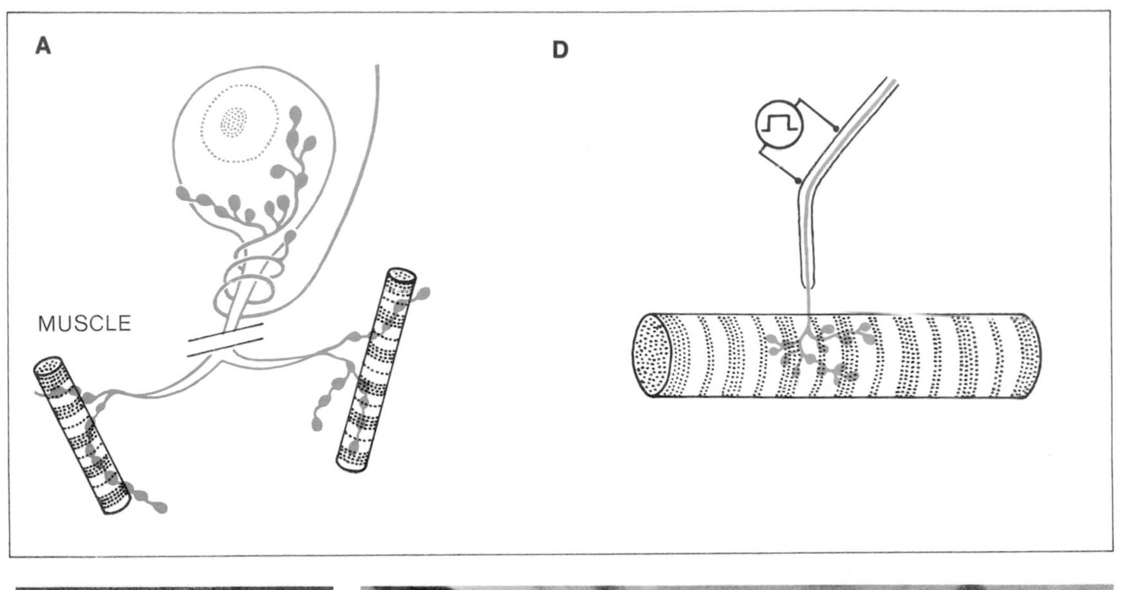
A
D
MUSCLE

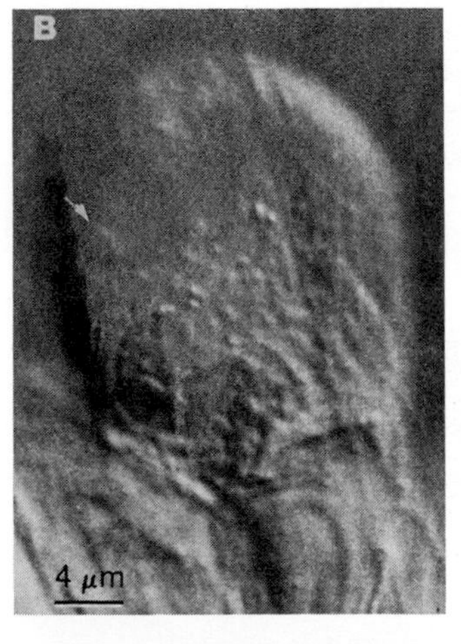
B
4 μm

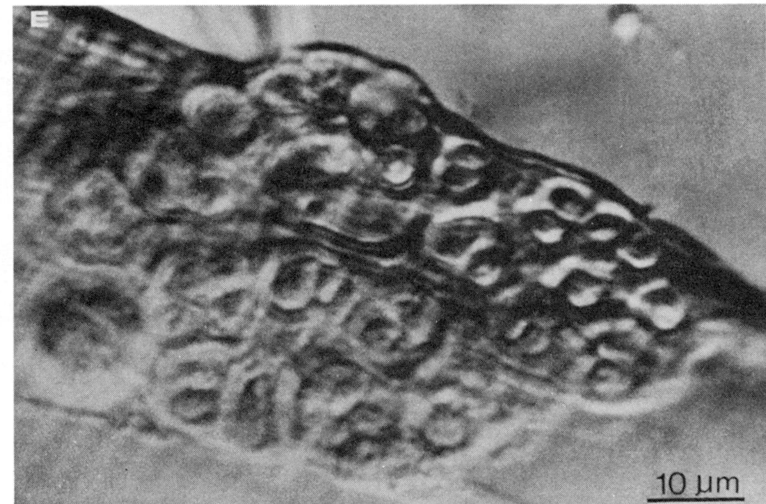
E
10 μm

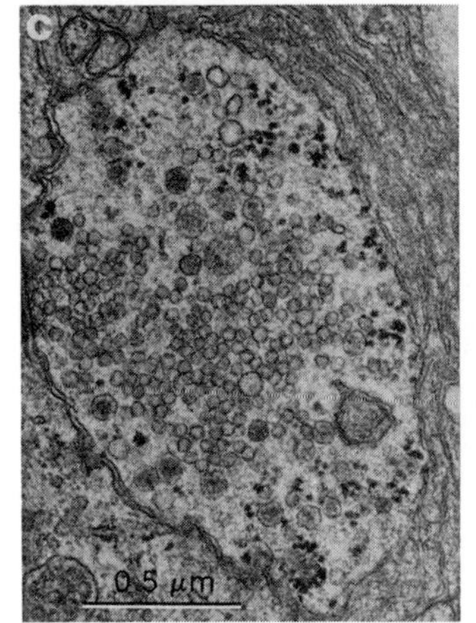
C
0.5 μm

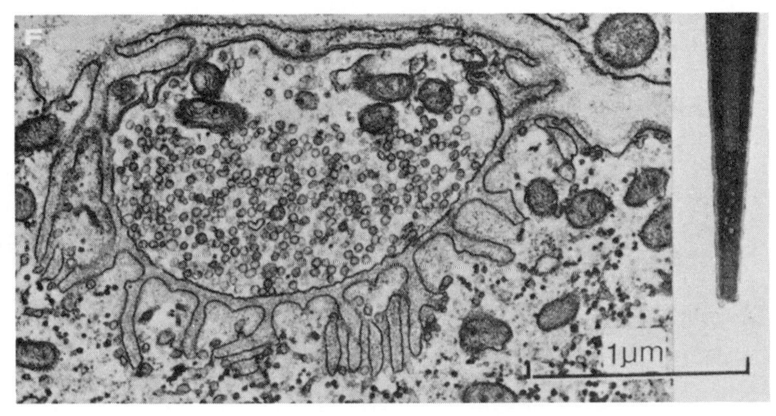
F
1μm

As expected, the synaptic membrane of neurons is highly sensitive to application of ACh where it is released by boutons of the vagus nerve. This is in contrast to the relative insensitivity of the extrasynaptic area, even several microns from the edge of a bouton.[18] For technical reasons muscle fibers are more suitable for a demonstration of the sharp delineation of synaptic areas. In muscles the nerve terminals with their synaptic boutons can be removed by applying to the synaptic area the enzyme collagenase, which frees the attachments that hold the nerve terminals without injuring the muscle fibers. The process of lifting off the terminals is shown in Figure 6*A*. Each bouton leaves behind a circumscribed crater lined with the exposed subsynaptic membrane; this is shown in greater detail in Figure 6*B* in which an ACh-filled micropipette points at an empty crater. The tracings in Figure 6*C* give further detail, emphasizing a distinct rimlike border area of about 1.5μm between the subsynaptic and extrasynaptic surfaces of the muscle. This border is probably an optical effect of the synaptic folds lining the walls of the craters. If one places the tip of an ACh-filled pipette onto the exposed subsynaptic membrane, one finds that 1 pC of charge applied to the pipette causes on the average a 5.0 mV change (5000 mV/nC) in the muscle membrane. On the other hand, at a distance of about 2μm, in the extrasynaptic region just beyond the rim, the same amount of ACh produces a response that is 50 to 100 times smaller.[19] In the rim itself sensitivity flucturates over a wide range (Figure 6*D*). We shall see below that the number of molecules of ACh carried in 1 pC in such experiments can be determined.

Experiments on neuronal and muscle membranes allow the conclusion that there exists a gradient of chemosensitivity that falls off within microns from the edge of a synaptic bouton. The chemosensitivity measured by the iontophoretic method provides a good index for the relative densities of chemoreceptors, and the findings are fully supported by an independent chemical method. When radioactive α-bungarotoxin

[18] Harris, A. J., Kuffler, S. W., and Dennis, M. J. 1971. *Proc. R. Soc. Lond.* B *177*:541–553.

[19] Kuffler, S. W., and Yoshikami, D. 1975 *J. Physiol.* *244*:703–730.

5

NEURONAL AND NEUROMUSCULAR SYNAPSES. A. A ganglion cell in the interatrial septum sends its axon to innervate the heart muscle and receives synapses from the vagus nerve. B. Live ganglion cell viewed with Nomarski optics. A conspicuous synaptic bouton is marked by the arrow. Other boutons can be seen at different focal depths. C. Cross section through a synaptic bouton that contains all the elements typical of a chemical synapse. D. Sketch of an end plate on a skeletal muscle of the snake. E. Living end plate. Individual synaptic boutons rest in craters sunk into the muscle surface. F. Cross section through a bouton. To the right is an electron micrograph of a micropipette used for iontophoresis of ACh; it has an outer diameter of 100 nm and an opening of about 50 nm. (B, C, from McMahan and Kuffler, 1971; F from Kuffler and Yoshikami, 1975a)

is used for labeling, the sites of receptors can be detected by autoradiography. From counts in photographs such as that in Figure 6*E* (an end plate of the boa constrictor), the density of receptors is found to be about 10^4/sq μm. It is likely that the receptor density is actually greatest around the lip or crest of the postsynaptic folds (Figure 4*C*). The

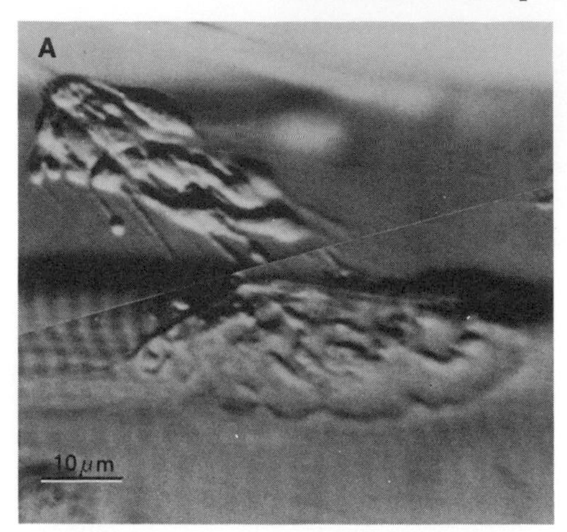

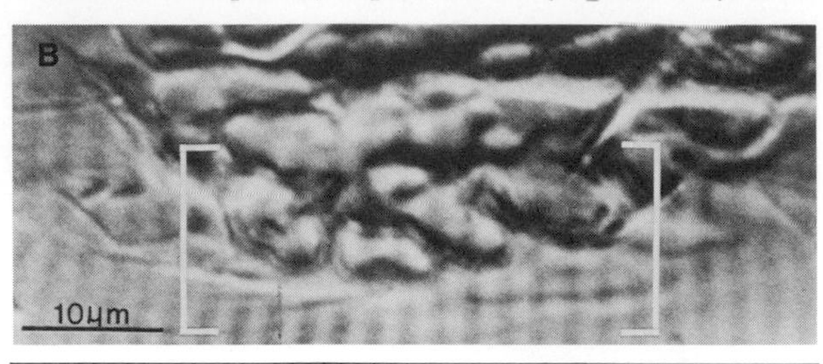

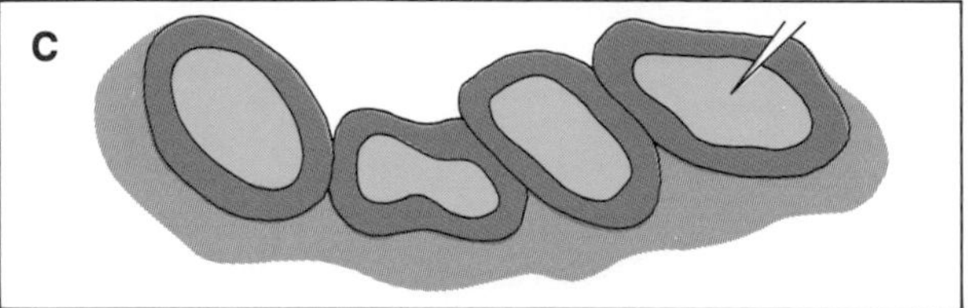

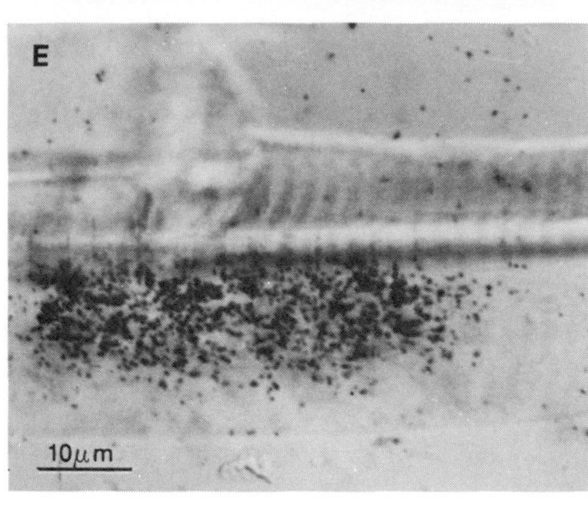

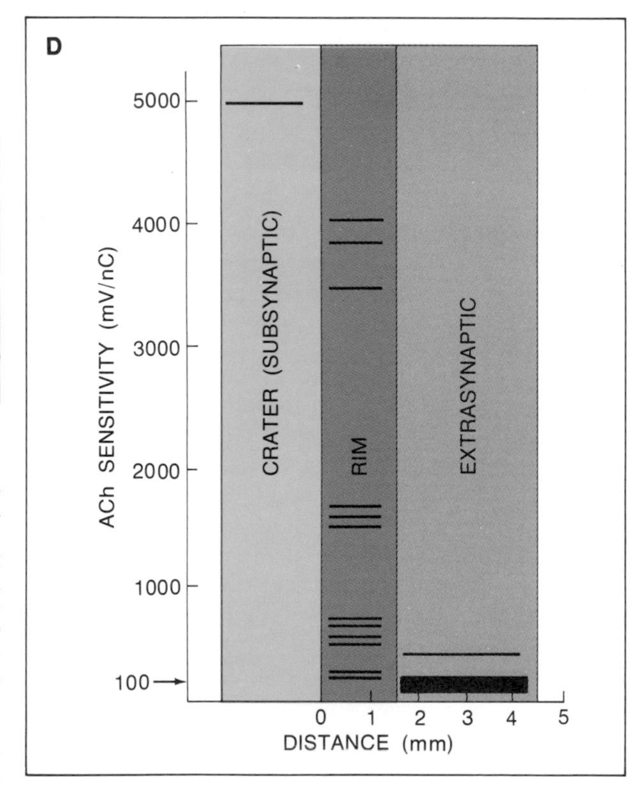

6
ACETYLCHOLINE RECEPTOR DISTRIBUTION at neuromuscular synapses in skeletal muscles of the snake. A. Process of removal of nerve terminal with boutons from a muscle fiber treated with collagenase. B. Remaining empty synaptic craters showing the exposed subsynaptic membrane. An ACh-filled pipette (coming in from the right) points to the floor of a crater. C. Enlarged drawing of the bracketed area in B. D. Determination of sensitivity to ACh. In the subsynaptic region, sensitivity (indicated by horizontal lines) is uniformly high (5000 mV/nC); in the extrasynaptic region beyond the rim of the craters, it is uniformly low (about 100 mV/nC). In the rim area the sensitivity fluctuates. E. An end plate in a boa constrictor muscle. Receptors are labeled with radioactive α-bungarotoxin, which is almost exclusively confined to synapses. (D from Kuffler and Yoshikami, 1975a; E from Burden, Hartzell, and Yoshikami, 1975)

extrasynaptic area, on the other hand, has a far lower density, estimated as around 5/sq μm using the same methods.[20-22]

Comparably detailed studies on chemosensitivities are not yet available for inhibitory synapses in vertebrates. One expects that their receptors are dense in the synaptic membrane since, as mentioned in Chapter 16, in crustacean muscle fibers the chemosensitivities for both the excitatory and the inhibitory transmitters are sharply confined to postsynaptic regions near nerve terminals. Measurements are also available in ganglia of a variety of snail neurons for ACh as well as for catecholamines and amino acids. In those preparations, however, the synapses are out of sight in the neuropil, which consists of a maze of fine fibers, and one cannot establish differences in sensitivities between synaptic and extrasynaptic regions.

The sharp boundaries between special areas of cell membranes are not immutable. If the input from the nerve is removed by cutting nerves or by blocking impulses in them, a relatively high density of new receptors appears in the extrasynaptic membranes of the postsynaptic cells (Chapter 18).

COMPONENTS OF THE SYNAPTIC RESPONSE

Having discussed the chemical specialization of the postsynaptic membrane, we can now reexamine more closely the signals that are generated there. At synapses where ACh is released one generally assumes a sequence of reversible steps—the first of these consists of the binding of ACh to the receptor protein. This leads sequentially to a conformational change, to the opening of ionic channels, and to a depolarization of the postsynaptic membrane. Two events combine to terminate the depolarization: as ACh comes off the receptor, it diffuses away and its further action is reduced or prevented through hydrolysis by acetylcholinesterase (AChE).

The following discussion starts with the smallest components of synaptic potentials, the elementary molecular responses, which supply much information about the ACh-receptor interaction and its consequences, and then continues with the quantal and full synaptic responses.

Elementary response of the synaptic membrane

The smallest physiological unit of transmitter release is a quantum, which in muscles usually produces a potential change of less than 1 mV (Chapter 9). A quantum is made up of more than 1000 molecules (discussed later), and it is of obvious physiological interest to measure directly the effects produced by individual ACh molecules. Such knowledge at the molecular level would provide new insights into the mechan-

---

[20]Fambrough, D. M. 1974. *J. Gen. Physiol.* *64*:468–572.

[21]Fertuck, H. C., and Salpeter, M. M. 1974. *Proc. Natl. Acad. Sci. U.S.A.* *71*: 1376–1378.

[22]Barnard, E. A., Dolly, J. O., Porter, C. W., and Albuqurque, E. X. 1975. *Exp. Neurol.* *48*:1–28.

ism of action of transmitters and drugs that affect synaptic transmission.

Because the signals are so small, conventional recording techniques cannot resolve the events associated with the interactions of ACh molecules with individual receptors. Katz and Miledi overcame this difficulty by analyzing the ELECTRICAL NOISE produced by ACh.[23] This analysis was an original and unusual approach, since electrical noise is not as a rule regarded as a welcome phenomenon in one's laboratory. Their basic procedure was to record the membrane potential at the endplate by means of an intracellular electrode. Recordings were made first from a resting muscle fiber (Figure 7*A*) and then again after a steady dose of ACh had been applied to the end plate (Figure 7*B*). During the ACh-induced depolarization, the electrical noise in the membrane clearly showed multiple fluctuations that were much larger than those seen in the absence of ACh. These fluctuations were not due simply to the change in the membrane potential, since they did not appear when a similar depolarization was produced by an electric current injected into the fiber by a second intracellular microelectrode.

To analyze the elementary events that make up the noise, Katz and Miledi used a method based on a theory developed more than 30 years ago for telephone communication.[24] The underlying assumption is that the steady synaptic depolarization results from the opening and closing of individual ionic channels, or gates, as the ACh molecules collide with receptors. Since collision with receptors is a random process, the number of channels open at any one time is not constant. Therefore, the conductance varies with the concentration of ACh and the number of open channels. This, in turn, creates fluctuations in the synaptic current (and potential). A steady ACh-induced depolarization means that many channels are open at any one time. From the fluctuations around the mean depolarization, as shown in Figure 7, one can determine certain properties of the elementary components (also called SHOT EVENTS).

Three main parameters can be extracted by noise analysis. These are the magnitude (a), the time course (τ), and the number of channels open at any given time (n). For the solution of the underlying equations, the shape of the elementary shot event must be known. For simplicity, Katz and Miledi assumed that the channel opening reaches its peak (a) instantaneously and declines exponentially with a time constant of τ. The subsequent estimates would differ by a factor of no more than two if the channel remains open and then closes abruptly, that is, if the shape of the shot event is a square pulse.

To obtain the amplitude of the voltage change that occurs as a result of a single channel opening, the mean depolarization (V) and the variance of the membrane potential ($\overline{E^2}$) must be measured. Then:

$$a = 2\overline{E^2}/V$$

Experimentally, in frog muscle, the mean value of a was about 0.3μV.

---

[23]Katz, B., and Miledi, R. 1972. *J. Physiol.* 224:665–699.
[24]Rice, S. O. 1944. *Bell Syst. Tech. J.* 23:282–332.

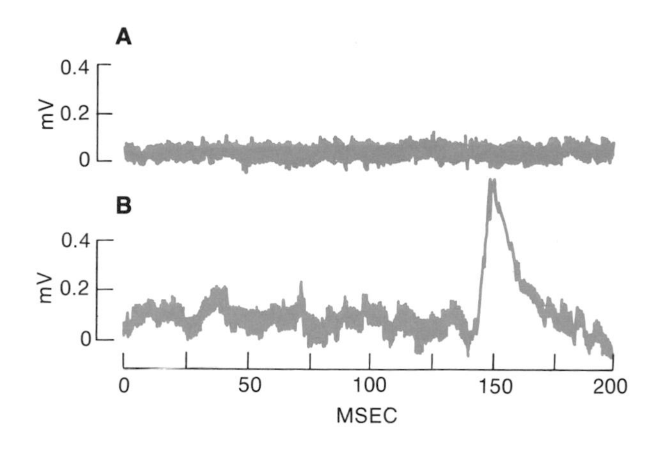

7
ACh NOISE. A. Intracellular recording from a muscle fiber at an end plate at rest. B. ACh diffuses onto the end plate causing a steady depolarization onto which are superimposed fluctuations resulting from the opening and closing of ionic channels. The large deflection is a spontaneous miniature synaptic potential. (After Katz and Miledi, 1972)

As expected, the size of the elementary voltage response (a) decreases with depolarization of the membrane potential, owing to a reduced driving force, in the same way as the excitatory synaptic potentials whose reversal potential is near −15 mV (discussed in Chapter 8). Other factors that affect a are temperature, the size of the postsynaptic cell, and denervation, all of which are accompanied by a change in input resistance.

In order to find τ, the duration of the elementary event, the noise is treated as if it consists of the sum of a series of sine waves of various frequencies. From the relative contribution of each frequency, the POWER SPECTRUM is derived. The frequency (f) at which the power is reduced to one half is related to τ by the following equation:

$$\tau = \frac{1}{2\pi f}$$

The value of τ, obtained by intracellular recordings, is mainly determined by the time constant (resistance × capacitance) of the postsynaptic membrane and is about 10 msec (8.4 to 12.5 msec, depending on temperature). To obtain the lifespan of the open channel, one measures synaptic currents whose time course does not depend on the time constant of the membrane. Such currents are most readily recorded with extracellular microelectrodes and indicate that the duration of the elementary current flow is about 1 msec.

Having determined a and τ, one can calculate n, the number of events in any given time:

$$n = \frac{V}{a\tau}$$

For a depolarization of 10 mV, the frequency of opening and closing of gates ranges from 2.0×10^6/sec to 5.5×10^6/sec.

These data provide information about a number of physiologically significant questions, such as the conductance of a single channel and the charge transfer across it. It was found that about 5×10^4 univalent ions flow through an open channel. The basic findings of Katz and

Miledi were confirmed and expanded by the rigorous studies of Stevens and his colleagues.[25] Interestingly, the value for the conductance change in a synaptic channel is not much different from that calculated by Hille for the sodium channel in the activated axon membrane (Chapter 6). A later section of this chapter discusses how many ACh molecules are involved in the opening of synaptic channels.

Noise analysis can also be applied to investigation of the drug-receptor interaction of various pharmacological substances at a molecular level. For example, curare and α-bungarotoxin act as blocking agents at the postsynaptic membrane. They could, in theory, act by decreasing a, n, or τ. Noise analysis reveals that the main action of these blocking agents is on n (the number of events in a given time), and that they interfere in the first step of the reaction by occupying sites at which ACh binds to the receptor. They prevent activation of channels but do not change significantly the opening and closing once the process has been initiated. The blockage of gates, therefore, occurs in an all-or-none fashion. In contrast, atropine, another blocking agent, acts by shortening the duration of the molecular gating action by about 85 percent and by reducing the elementary voltage change (a).[26]

Effects of drugs on elementary responses

Of particular physiological interest is the effect of inhibitors of AChE, such as prostigmine, which greatly prolong the effects of nerve-released ACh by preventing its hydrolysis; as a result its activity terminates only after it has diffused out of the synaptic cleft. The question arises, however, whether part of the prolonged action of ACh is due to an increase in the duration of the opening of individual ionic gates. Noise analysis has shown that inhibition of AChE does not significantly alter either the magnitude or the duration of the elementary gating event. In the absence of hydrolysis, ACh molecules remain in the cleft for relatively long periods and therefore are repeatedly bound by receptors; in this way one ACh molecule can sequentially open many channels. This repeated binding of ACh to receptor delays diffusion and prolongs the action of ACh.[27]

Quantal response

From the smallest unit response of the synaptic membrane we now proceed to the response of the smallest physiological unit, the effect of a quantum of transmitter released by a presynaptic terminal.

The significance of quantal release is underscored by its occurrence at all the vertebrate and invertebrate chemical synapses at which the necessary tests have been made. The effectiveness of synaptic transmission is regulated by the number of quanta secreted at both excitatory and inhibitory synapses. Studying the responses to individual quanta and the effect of many simultaneously released quanta supplies an opportunity for a functional dissection of the full synaptic potential. In particular, does the full synaptic response represent the simple sum of individual quantal components or can two or more quanta of ACh interact and influence each other's effect at postsynaptic sites?

---

[25] Anderson, C. R., and Stevens, C. F. 1973. *J. Physiol.* *235*:655–691.
[26] Katz, B., and Miledi, R. 1973. *Proc. R. Soc. Lond.* B *184*:221–226.
[27] Katz, B., and Miledi, R. 1973. *J. Physiol.* *231*:549–574.

To test precisely whether quanta interact, experiments are made on voltage-clamped muscle fibers (Chapter 6); with this technique no complications are introduced by the time constant of the membrane or by changes in the driving force during the synaptic response. To obtain synaptic currents initiated by the release of one or only a few quanta, it is convenient first to reduce the calcium and increase the magnesium content in the bathing solution until the nerve impulse completely fails to secrete transmitter (Chapter 9). Then a solution containing a normal concentration of calcium is slowly reintroduced, so that progressively more quanta are liberated with each impulse.

In Figure 8*A* an example is illustrated in which successive synaptic current responses resulting from the release of 5 to 12 quanta are superimposed. Each quantum in Figure 8*A* adds a peak current of about 4 nA. Eventually, when the bathing solution attains its normal concentration

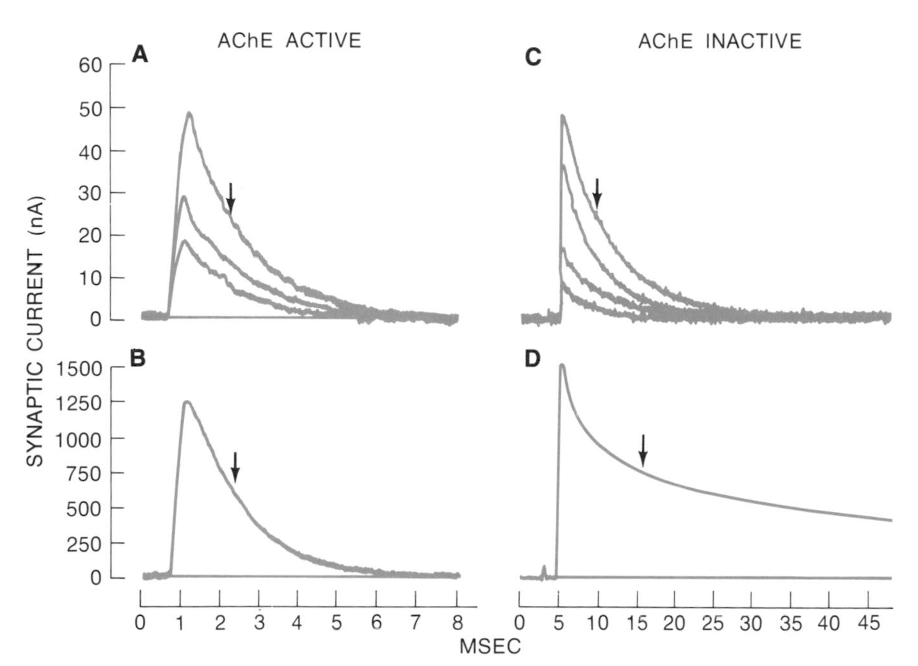

8

INDEPENDENT ACTION AND INTERACTION among quanta. The number of quanta released is regulated by adjusting the calcium and magnesium content of the Ringer's solution. A. Nerve-evoked synaptic currents in an end plate of the snake caused by the release of 5 to 12 quanta of ACh. Records are superimposed. B. Synchronous secretion of about 300 quanta. The half-times in A and B are all close to 1.2 msec (arrows). C. Release of 2 to 12 quanta when the hydrolyzing action of AChE is inhibited. The half-time (arrow) of synaptic currents are longer (3 to 4 msec) compared with those in A and B, but similar in all records (note different time scales). In A, B, and C the quanta act independently of one another. D. Release of 300 quanta prolongs the half-time of the current to 9.5 msec (arrow), indicating that postsynaptic interaction had occurred. (After Hartzell, Kuffler, and Yoshikami, 1975)

of calcium, the response reaches a peak current of about 1250 nA (Figure 8*B*), representing the release of about 300 quanta distributed over the entire end plate. A comparison of the time course of all the responses in Figure 8*A* and *B* shows that the half-times of decay (1.2 msec at arrows) do not change, irrespective of how many quanta are released. The only difference is in the magnitude of responses.[25]

This experiment indicates that QUANTA ACT INDEPENDENTLY of each other; the peak of the full postsynaptic current simply represents the linear summation of all quantal components. Further, the total synaptic current is normally no longer than the duration calculated for the conductance in individual ionic channels (about 1 msec half-time). The normal synaptic response, then, is made up of the sum of components that are all similar in time course. At the bottom of the scale are the elementary molecular events, followed by the larger quantal response, and finally the full multiquantal response. The brevity of synaptic currents indicates that a single ACh molecule has, on the average, only one opportunity to open an ionic gate. Apparently the AChE normally present at the end plate prevents repetitive opening of channels by limiting the lifetime of ACh, so that after a molecule of ACh has initiated a conductance change it is hydrolyzed.

In contrast to the functional isolation of quanta that occurs normally, one can observe clear interaction among them when AChE is inhibited, by prostigmine, for example. Under these circumstances ACh survives longer in the cleft and molecules are free to diffuse over a larger area and to act repeatedly upon receptors. As a result, the synaptic potentials and synaptic currents are significantly prolonged. This is demonstrated in Figure 8*C* in which 2 to 13 quanta are superimposed, as in Figure 8*A*. The half-times of quantal currents in Figure 8*C* are prolonged to 3 to 4 msec (note the different time base in *A* and *C*). Yet they all have a rather similar time course whether released singly or in unison; that is, they do not significantly influence each other. The result differs, however, when large numbers of quanta are released. In Figure 8*D* the half-time of the full synaptic current, caused by more than 300 quanta, is more than doubled. The explanation of this result is discussed below.

Postsynaptic potentiation

Chapter 9 mentions that as a rule changes in the magnitude of synaptic responses are the result of increased or decreased amounts of transmitter released from terminals. This does not apply to the example shown in Figure 8*D*. The reason for the prolonged or potentiated response is not presynaptic; it occurs only when release sites are close to each other, so that when 300 quanta are liberated, each quantum acts upon a postsynaptic area that partially overlaps with the areas covered by neighboring quanta. Such a situation is brought about if a nerve terminal is made to release a small number of quanta very close together; for example, from one or two synaptic boutons. First, transmitter secretion is abolished altogether by bathing the preparation in a solution low in calcium and high in magnesium. Release is then restored by applying graded amounts of calcium from a pipette according to the method of Katz and Miledi (Chapter 9).[28] In this way varied numbers of quanta can be released by the bouton close to the tip of the calcium-filled pipette. Figure 9 displays superimposed records when 2 to 12 quanta are released in such a manner. The peak sizes of individual quantal currents (about 4 nA) are similar to those shown in Figure 8*C*, but the duration of multiquantal currents is progressively prolonged (up to 16.5 msec, arrows) each time a quantum is added. The only difference between the experimental situations in Figures 8*C* and 9 is that in the first case the quantal release is scattered over the relatively large total end plate area, while in the second case a similar number of quanta are released, but act close together on a restricted region. This leads to the conclusion that during focal release the site of potentiation occurs in the postsynaptic membrane.

More detailed ideas about the basis for postsynaptic potentiation can be derived from experiments of Katz and Thesleff in which ACh is applied artificially onto overlapping areas of the postsynaptic membrane.[29] Two separate, but identical, low concentrations of ACh are released iontophoretically from two micropipettes covering the entire motor end plate area (Figure 10*A*). Each, on its own, produces the same slow synaptic currents that are superimposed in record *a,b* of Figure 10*B*. With simultaneous release from both pipettes (record $a + b$) the response is not simply the linear sum of *a* and *b* (dotted line) but is almost four times as great as either response alone. This potentiation between two doses of ACh results from the nonlinear behavior of the synaptic receptors, which give a sigmoid dose-response curve. Low concentrations sum nonlinearly, while at higher concentrations the shape becomes linear and eventually saturates.[30]

Several mechanisms may be responsible for the nonlinear po-

[28] Katz, B., and Miledi, R. 1965. *Proc. R. Soc. Lond.* B *161*:483–495.

[29] Katz, B., and Thesleff, S. 1957. *J. Physiol.* *138*:63–80.

[30] Hartzell, H. C., Kuffler, S. W., and Yoshikami, D. 1975. *J. Physiol.* *251*:427–463.

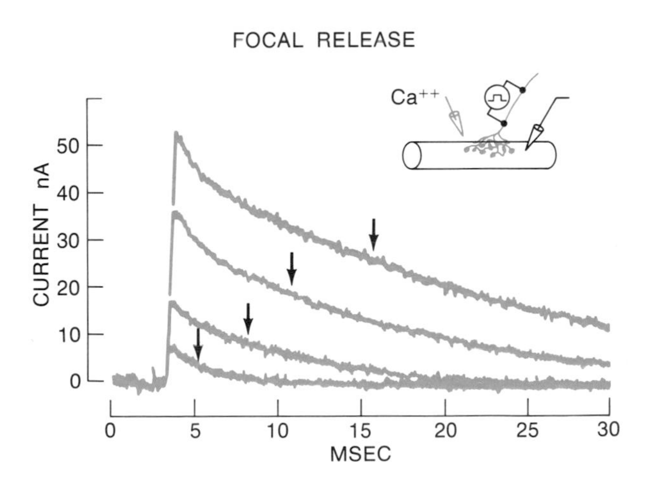

9

POSTSYNAPTIC POTENTIATING INTERACTION between quanta. Release of transmitter is blocked by a solution low in calcium and high in magnesium, except for a small area near the tip of a calcium-filled pipette poised at the edge of a synaptic bouton. The activity of acetylcholinesterase in inhibited. Records of synaptic currents caused by 2 to 12 quanta are superimposed. The duration of synaptic currents becomes progressively prolonged (arrows) as quanta are added. This facilitation results from the interaction of quanta of ACh that are free to diffuse within the synaptic clefts and cover overlapping areas of the subsynaptic membrane. (From Hartzell, Kuffler, and Yoshikami, 1975)

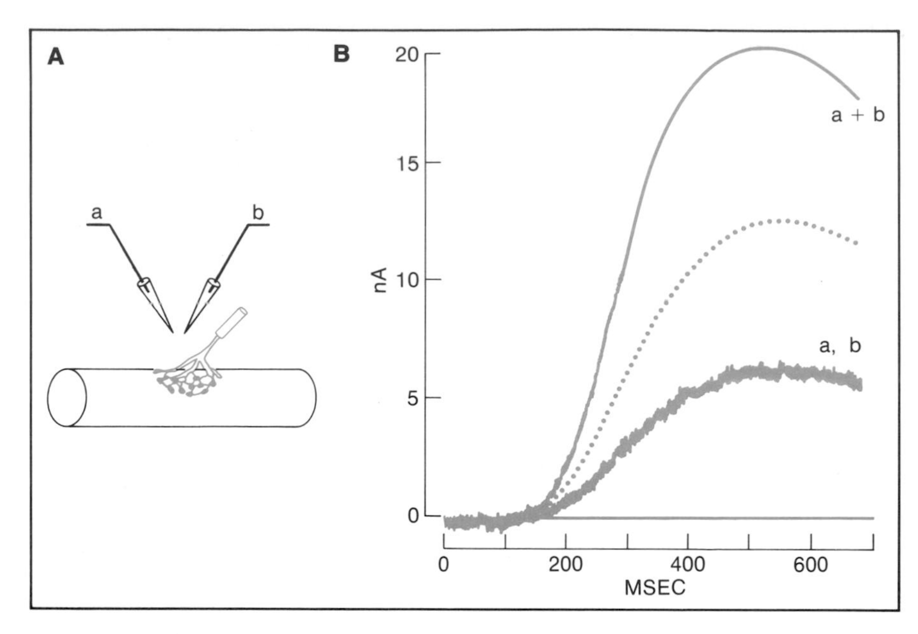

10 **POSTSYNAPTIC POTENTIATING INTERACTION when ACh is applied to overlapping areas of the synaptic membrane. A. Two pipettes (a and b) release identical small quantities of ACh that cover the entire end plate area. B. Identical synaptic currents are caused by two consecutive superimposed traces (a,b). Doubling the dose of ACh by releasing a and b simultaneously results in a response (a+b) more than threefold larger than the response to a single dose. The dotted line indicates the response expected if the postsynaptic currents had summed in a linear manner. Acetylcholinesterase was inactivated. (From Hartzell, Kuffler, and Yoshikami, 1975)**

tentiating behavior of the synaptic membrane. One likely possibility is that the binding of one ACh molecule to a receptor increases the probability of binding another one.[31–33] Alternatively, more than one ACh molecule may be required to activate a receptor. Cooperative mechanisms may also be involved at the level of activation of ionic channels.

In the case of quantal release, when AChE is inhibited, one can conclude that as ACh is liberated at neighboring sites it spreads laterally, so that the low concentration fringes of quanta overlap and potentiate each other's effect.

AChE and quantal action

The findings on postsynaptic potentiation provide a tool for detecting spatial overlap of the action of ACh, pipette-applied or nerve-released, on the postsynaptic membrane. When AChE is fully active, no potentiating interaction occurs between simultaneously released quanta. From this independence of quantal action it follows, perhaps surprisingly, that regardless of the number of quanta released from a

[31]Magleby, K. L., and Terrar, D. A. 1975. *J. Physiol.* *244*:467–495.
[32]Cohen, J. B., and Changeux, J.-P. 1975. *Annu. Rev. Pharmacol.* *15*:83–103.
[33]Karlin, A. 1967. *J. Theor. Biol.* *16*:306–320.

terminal, the maximal concentration of ACh in the cleft at any one point cannot exceed that produced by a single quantum.

The results mentioned earlier also provide an approximation of the area activated by one quantum. From the total synaptic area (neglecting the folds), and the number of quanta liberated, one finds that the mean distance between release sites of quanta is less than 2μm; and since neighboring quanta do not act upon overlapping areas, a quantum normally affects less than 4 sq μm.

Figure 11 represents a hypothetical scheme based on the preceding discussion of the action of quantal responses. It shows the relative density of activated ionic channels and the area of the postsynaptic membrane that is involved at different times after release of transmitter. Three quanta, out of a normal total of 300, serve as samples. The x and y plane gives the dimensions of the subsynaptic membrane area under the presynaptic release sites of these quanta, and the z axis represents the density of open ionic channels when AChE is fully active (*A*) and when it is inhibited (*B*). Above each sketch is the time course of the synaptic current. In *A*, each quantum acts independently in a punctate manner on the subsynaptic membrane. About 0.2 to 0.3 msec after release at T_1, when the ACh action and synaptic currents are at their peak, most of the ACh molecules are likely to be bound to receptors. The remainder of the current flow (T_2 and T_3) represents the closing time of the ionic channels. Therefore, about 1.0 and 1.5 msec after the maximal conductance change, relatively few ionic gates are still open and no free ACh remains in the synaptic cleft.

When AChE is inhibited (Figure 11*B*), the initial situation at T_1 is not much different from normal immediately after release of a packet of ACh, except that in addition to receptor-bound ACh, free ACh is also present in the cleft; but 3 to 5 msec later (T_2 and T_3) much of the ACh has spread within the cleft, and during that time diffusing molecules can bind repeatedly to receptors before escaping into open space. In this way quanta can act on overlapping areas, as indicated in Figure 10*B*, and bring about potentiation.

In conclusion, functional dissection of synaptic responses emphasizes two aspects of cholinergic transmission: (1) AChE normally ensures that quanta act rapidly and independently of each other by confining the synaptic area over which each quantum can spread (in the process AChE also serves the essential task of providing choline for reuptake by the terminals); and (2) in the absence of AChE activity, a postsynaptic potentiating mechanism becomes apparent. The functional significance of this mechanism is not yet clear. However, the nonlinear properties of the membrane on which potentiation is based have been seen at diverse synapses in vertebrates and invertebrates and may be a widespread property of synaptic membranes.[30]

Artificial synaptic response and the number of molecules in a quantum

The chemical transmitter converts the electrical presynaptic nerve impulse into a postsynaptic signal, and it is natural to inquire how effective the SYNAPSE is as an ELECTROCHEMICAL TRANSDUCER. How many charges are transferred across the postsynaptic membrane in response to the action of one or more ACh molecules? Other questions concern the buildup of concentration of the transmitter on release into the synaptic cleft and the numerical relation between postsynaptic receptors and ACh molecules.

Three approaches have been used to estimate the number of mole-

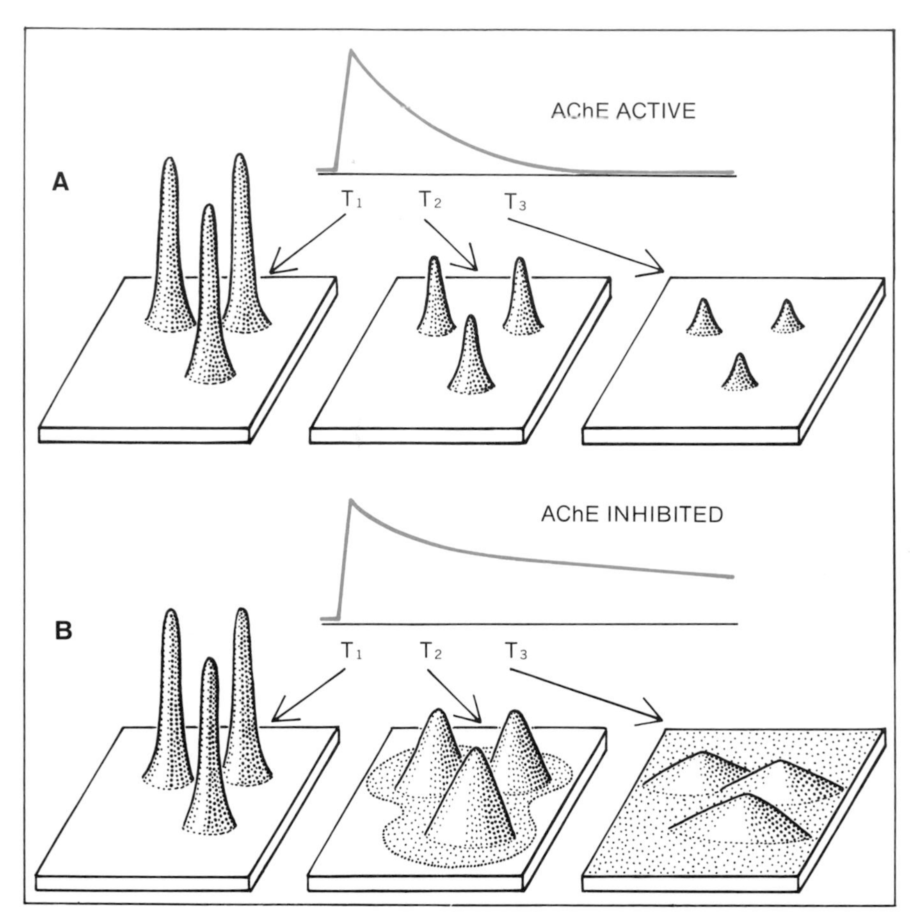

11 **HYPOTHETICAL DISTRIBUTION OF ACTIVATED ACh RECEPTORS after release of three quanta of ACh onto neighboring sites on the subsynaptic membrane. Density of activated receptors, or open ionic channels, is shown on the z axis; the x,y plane represents the postsynaptic membrane. The progressive change in the number of open channels is drawn at three different times: T_1 at the peak of the synaptic current response, and T_2 and T_3 during the falling phase. When AChE is fully active, there is little opportunity for lateral diffusion from the point of release, and within less than 0.3 msec the ACh is bound to receptors. The declining phase of the synaptic current represents the closing time of the ionic channels. When AChE is inhibited, ACh is able to diffuse along the synaptic membrane so that at T_2 and T_3 it covers overlapping areas, leading to potentiation (see Figure 9). (From Hartzell, Kuffler, and Yoshikami, 1975)**

cules in a quantum. First, from the work on elementary responses it is known that a quantum of ACh opens about 1000 ionic conductance channels. On the simplest assumption of one molecule combining with one receptor and opening one conductance channel, no fewer than 1000 molecules react with the postsynaptic membrane. Since not all released molecules reach synaptic receptors, the minimum number of molecules in a packet to activate a channel must be larger than 1000. Second, estimates can also be derived from experiments in which nerves are stimulated and the perfusate collected and assayed for the transmitter. Such calculations for skeletal muscles have ranged from 10,000 to several hundred thousand molecules per quantum.[34] This large range stems in part from the diversity of assumptions on which the calculations are based, such as the uncertainty of the number of quanta released per impulse during prolonged stimulation. Third, recent improvements in the iontophoretic method have put a maximal value on the number of ACh molecules in a quantum. These experiments employ an ACh-filled pipette with an outer diameter of about 100 nm and an opening that is not much larger than a synaptic vesicle (inset, Figure 5*F*). When such a pipette is accurately apposed to the synaptic membrane, a pulse of 0.5 to 1.0 msec duration ejects ACh that reproduces the effect of a quantum of transmitter.[19] This method of mimicking with a pipette the release of ACh from the synaptic bouton creates in effect an artificial synapse in which release of transmitter and the synaptic response are controlled at will. Figure 12 shows a comparison of postsynaptic responses caused by spontaneous quantal release and by a pulse of ACh. In *A* the tip of the ACh pipette releases transmitter onto a muscle fiber at the edge of a motor nerve terminal where its synaptic cleft opens; in *B*, ACh is applied in the same way to a synaptic bouton on a nerve cell in a frog's interatrial septum. In both cases the artificial synaptic potential is a close imitation of the nerve-released one (miniature synaptic potential). Therefore, establishing the amount of ACh released by the pipette under the conditions illustrated in Figure 12 provides an estimate of the quantity secreted by nerve terminals.

A method for counting relatively small numbers of ACh molecules has recently been devised. It utilizes small sample volumes (droplets of about 0.5 nl) and the end plate of the snake as an ACh concentration detector. Pulsing ACh from a pipette into the sample droplet allows assay of the droplet's content by applying it to an end plate. About 10,000 molecules reproduce the effect of a quantum at the skeletal neuromuscular synapse (Figure 12). This is an upper limit for the number of ACh molecules contained in a quantum because the tips of micropipettes are actually farther away from the synaptic membrane than are the nerve terminals.[35]

Knowing the approximate number of molecules in a quantum is helpful in drawing tentative conclusions about the role of vesicles and

[34]Krnjević, K. 1974. *Physiol. Rev.* 54:418–540.
[35]Kuffler, S. W., and Yoshikami, D. 1975. *J. Physiol.* 251:465–482.

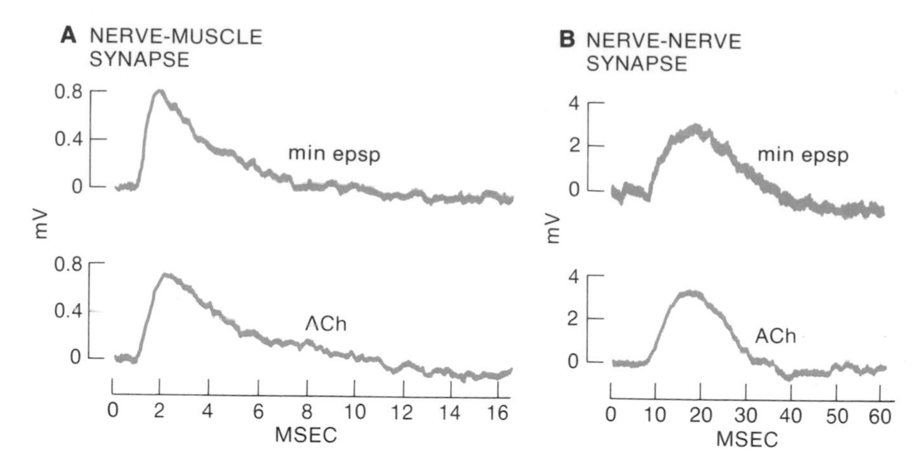

12 **SYNAPTIC RESPONSES to artificial and nerve-released ACh. A. Synaptic potential in a frog muscle caused by the spontaneous release of a quantum of ACh from the nerve terminal (miniature excitatory postsynaptic potential) and a synaptic response resulting from a 0.5 msec pulse of ACh onto the synaptic membrane. B. A quantal response (miniature excitatory postsynaptic potential) recorded from a nerve cell in the septum of a frog's heart and below it a response to ACh released from a pipette onto the postsynaptic area at a synaptic bouton. The rise times of the artificial responses are usually a fraction of a millisecond slower because the pipette is not perfectly apposed to the synaptic membrane. (A from Kuffler and Yoshikami, 1975a; B from Dennis, Harris and Kuffler, 1971)**

about events in the synaptic clefts and the postsynaptic membrane. The evidence that ACh is contained in vesicles is good, and the hypothesis that the vesicles release their contents directly into the synaptic cleft has been greatly strengthened. If the contents of a vesicle with an inner diameter of about 50 nm are isoosmotic with snake Ringer solution, which is about 320 mOsm, then the ACh concentration in the vesicle is 160 mM and the vesicle can accommodate about 6000 molecules. This is in the range of experimental findings. As an approximation one can calculate that if 10,000 molecules are evenly distributed within 0.5 msec after release in the volume of the synaptic cleft over an area of 1 sq μm, 50 nm in depth, the ACh concentration will be close to 3×10^{-4} M, an unexpectedly high concentration. These molecules face in the subsynaptic area receptors that have been estimated from autoradiography to be distributed with a density of about 10^4/sq μm. Not surprisingly, therefore, it was found that receptors are not saturated when a quantum of ACh is released.[80]

Concerning the efficiency, or amplification factor, of the electrochemical process at chemical synapses, the charge transfer across the synaptic membrane in muscles of the snake caused by the release of a quantum of ACh is about 5 pC, equivalent to 30 million univalent ions. On the basis of 10,000 molecules per quantum the conductance change for every molecule of ACh released results in a net flow of 3000 univalent ions. The number of ions that move is actually greater, since the number of ACh molecules released per quantum is expected to be smaller and, further, not all released molecules combine with receptors. The upper limit is set by the transfer of charge, equivalent to about 50,000 univalent ions through a single channel, possibly brought about by no fewer than two ACh molecules.

Application to neuronal synapses

This discussion focuses on the neuromuscular synapse, and to a lesser extent on neuronal junctions, because recently much microstructural and physiological work has been done on these relatively simple synapses. The expectation is that the results are directly relevant to synapses in the central nervous system. Mechanisms of release and quantities of transmitters and their actions on postsynaptic chemoreceptors, as understood in peripheral synapses, are likely to have their parallels in neurons of the higher centers. At the same time many questions about differences arise. In a typical central neuron the cell body and its dendrites are densely invested with synaptic boutons from a great variety and number of presynaptic fibers (Chapter 16), and more than one transmitter acts on neighboring patches of the cell membrane.

The conditions of inactivation or removal of some of the transmitters in nerve-nerve synapses may be less efficient than at neuromuscular synapses, and interactions between quanta are perhaps a normal feature. Further, synaptic elementary events at neuronal synapses have not yet been analyzed, and detailed studies of quantal responses have been somewhat restricted. In these and other respects of synaptic microphysiology we are only at the threshold of developments in the higher centers.

SUGGESTED READING

General reviews

The Synapse. 1976. *Cold Spring Harbor Symp. Quant. Biol. 40.* (A series of articles on chemoreceptor isolation and synaptic vesicles.)

Cohen, J. B., and Changeux, J.-P. 1975. The cholinergic receptor protein in its membrane environment. *Annu. Rev. Pharmacol. 15:*83–103.

Couteaux, R. 1974. Remarks on the Organization of Axon Terminals in Relation to Secretory Processes at Synapses. In B. Ceccarelli, F. Clementi and J. Meldolesi (eds.). *Advances in Psychopharmacology,* Vol. 2. Raven Press, New York, pp. 369–379.

Stevens, C. F. 1976. Membrane noise fluctuation analysis. In *The Synapse. Cold Spring Harbor Symp. Quant. Biol. 40:*169–174.

Verveen, A. A., and DeFelice, L. J. 1974. Membrane noise. *Prog. Biophys. Mol. Biol. 28:*189–265.

Original papers

Anderson, C. R., and Stevens, C. F. 1973. Voltage clamp analysis of acetylcholine produced end-plate current fluctuations at frog neuromuscular junction. *J. Physiol. 235:*655–691.

Dennis, M. J., Harris, A. J., and Kuffler, S. W. 1971. Synaptic transmission and its duplication by focally applied acetylcholine in parasympathetic neurons in the heart of the frog. *Proc. R. Soc. Lond.* B *177*:509–539.

Hartzell, H. C., Kuffler, S. W., and Yoshikami, D. 1975. Postsynaptic potentiation: Interaction between quanta of acetylcholine at the skeletal neuromuscular synapse. *J. Physiol.* *251*:427–463.

Heuser, J. E., and Reese, T. S. 1973. Evidence for recycling of synaptic vesicle membrane during transmitter release at the frog neuromuscular junction. *J. Cell Biol.* *57*:315–344.

Katz, B., and Miledi, R. 1972. The statistical nature of the acetylcholine potential and its molecular components. *J. Physiol.* *244*:665 699.

Katz, B., and Miledi, R. 1973. The characteristics of 'end-plate noise' produced by different depolarizing drugs. *J. Physiol.* *230*:707–717.

Katz, B., and Miledi, R. 1973. The binding of acetylcholine to receptors and its removal from the synaptic cleft. *J. Physiol.* *231*:549–574.

Magleby, K. L., and Stevens, C. F. 1972. A quantitative description of end-plate currents. *J. Physiol.* *223*:173–197.

CHAPTER ELEVEN

THE SEARCH FOR CHEMICAL TRANSMITTERS

Several substances have been shown unequivocally to act as transmitters at chemical synapses. They include acetylcholine, norepinephrine, epinephrine, and γ-aminobutyric acid (GABA). A number of other substances are present in high concentrations in neurons and are likely to be released at synapses. The steps required to establish the role of GABA as an inhibitory transmitter are described to illustrate how microchemical and physiological studies of individual neurons can be used to identify a transmitter.

Action potentials in inhibitory nerves supplying crustacean muscle release GABA and increase the chloride permeability of the postsynaptic membrane. Application of GABA mimics the action of the neural transmitter, including its presynaptic inhibitory effects.

Analysis of individual neurons has shown that GABA is highly concentrated in inhibitory axons, which contain about 100 mM, while 1 mM or less is found in excitatory or sensory neurons. The enzymes that synthesize and degrade GABA have been extracted from inhibitory and excitatory axons. Both types of cells contain similar amounts of GABA transaminase, which converts GABA to succinate; glutamic decarboxylase, however, which synthesizes GABA from glutamate, is far more concentrated in inhibitory nerves. This apparently accounts for the higher level of GABA in inhibitory cells.

GABA is also found in relatively high concentrations in various parts of the mammalian brain. In the cerebellum up to 8 mM are contained in large isolated Purkinje cells which are known to be inhibitory. Their synaptic actions on identified postsynaptic cells are mimicked by iontophoretically applied GABA. Other identified neurons in the brain that are not inhibitory contain little or no GABA.

The amino acid glycine also occurs in relatively high concentrations in parts of the brain and may function as a second inhibitory transmitter.

The preceding account of synaptic transmission has emphasized the release of transmitter, its interaction with receptors, and the conductance changes that follow. Transmitter physiology and chemistry remain developing and expanding areas; conspicuous gaps still exist in knowledge about the identity of transmitters and their actions. This chapter discusses the general problem of identifying chemical transmitters and the chemistry of their synthesis and degradation in the presynaptic neuron. We have chosen to restrict the discussion to one particular example, γ-aminobutyric acid (GABA), instead of reviewing comprehensively neurochemical studies on a variety of transmitters.

There are two main reasons why chemical studies on GABA are well suited to the cellular approach we stress in this book. First, in crustaceans the cell bodies and axons of many neurons are large enough to be identified in living preparations. These neurons can then be isolated and used to work out with micromethods the enzymatic steps in GABA metabolism. Second, the same microchemical procedures can also be used successfully to study GABA metabolism in individual neurons within the mammalian brain. Moreover, on physiological and chemical grounds, GABA is one of the relatively few substances known to be a transmitter at specific synapses and the first to be identified by a systematic chemical survey of a nervous system. An account of the search illustrates well the general problems encountered in establishing the identity of a transmitter at a synapse and determining the characteristics of the enzymes in the metabolic pathway.

Similar neurochemical approaches at the cellular level have been used for various transmitter compounds in other preparations. Schwartz, Kandel, and their colleagues have carried out detailed and elegant studies on the synthesis, transport, and release of acetylcholine (ACh) in single neurons of the sea hare, *Aplysia*.[1,2]

Establishing the identity of a chemical transmitter

Little direct information exists about the identity of the transmitters at the majority of synapses within the vertebrate brain. There are obvious technical reasons for this gap in knowledge. Nevertheless, it is possible to determine whether transmission across a particular synapse operates by a chemical mechanism even in higher centers. The criteria are morphological and physiological. For example, electron microscopy reveals clusters of vesicles and electron-dense material in the presynaptic terminal, and across the synaptic cleft postsynaptic thickenings are generally visible (Chapter 10). Physiological testing can identify a reversal potential and specific changes in ionic conductance which, together with pharmacological tests, provide reliable evidence for a chemical synapse. This is fortunate, because even before a transmitter is identified, the knowledge that certain cells have chemical rather than electrical synapses fashions the experimental approach (Chapter 8).

To obtain firm evidence for the identity of the substance liberated

---

[1]Koike, H., Kandel, E. R., and Schwartz, J. H. 1974. *J. Neurophysiol.* *37*:815–827.

[2]Eisenstadt, M. L., and Schwartz, J. H. 1975. *J. Gen. Physiol.* *65*:293–313.

by the presynaptic terminals, a number of different experimental approaches are required. It is not enough to know that a particular chemical mimics the action of the presynaptic nerve. To conclude that it is the transmitter, one must demonstrate in addition that the terminals (1) normally contain it, (2) release it at the right time in response to stimulation, and (3) release sufficient quantities to produce the appropriate effects in the postsynaptic cell. These procedures require the microanalysis of very small amounts (often as small as 10^{-14} moles) of a number of candidate compounds, together with physiological recordings from individual cells. The methodological difficulties are so great that it has not yet been determined with any assurance which chemical is responsible for transmission even at synapses where much detailed physiological information is available, such as the squid giant synapse and at endings on the mammalian spinal motoneuron.

In contrast to many peripheral structures, from which most information is derived, the central nervous system is composed of an inhomogeneous population of nerve cell bodies and terminals that are closely apposed and intermingled. The transmitter that is liberated cannot easily be collected from the release sites, but often has to diffuse over large distances before it appears at the surface of the brain or in the cerebrospinal fluid. It is often difficult to stimulate selectively a particular homogeneous group of fibers known to be excitatory or inhibitory; and further difficulties are associated with attempts to record intracellularly from the postsynaptic cells and to mimic the action of nerves by applying chemicals through a micropipette. Nevertheless, as a working hypothesis, it often seems plausible to accept the mere presence and appropriate postsynaptic effects of substances in neurons as adequate evidence for their presumed transmitter actions. To this group belong such substances as 5-hydroxytryptamine (5-HT or serotonin), dopamine, octopamine, substance P (see below), adenosine triphosphate (ATP), and the amino acids glutamate and glycine.

Another approach, already mentioned, uses the specialized metabolic machinery of different neurons to make and degrade particular substances. During these activities neurons use special enzymes whose end products can be marked by various tracers. For example, cells that use GABA as a transmitter are known to have high levels of the enzyme glutamic decarboxylase. Recently the enzyme was purified and an antibody made. When appropriately tagged, the antibody can be utilized to localize the enzyme in tissue sections; thus, GABA-containing neurons can be visually identified.[3] In this way chemical maps can be made of the distribution of neurons with special chemical properties.

Identified transmitters

Transmitters that have been unequivocally identified by criteria already mentioned include ACETYLCHOLINE, EPINEPHRINE, NOREPINEPHRINE, and GABA (Figure 1). The direct demonstration of their role as transmitters, even for these substances, has been obtained only at relatively

[3]Matsuda, T., Wu, J.-Y., and Roberts, E. 1973. *J. Neurochem.* *21*:159–166, 167–172.

few synapses. The following paragraphs present first a thumbnail sketch of the evidence for a transmitter role of acetylcholine, epinephrine, and norepinephrine; and later a detailed description of GABA.

The case for cholinergic transmission (synapses where acetylcholine is the transmitter) in the central nervous system is only indirect. The evidence is best for synapses within the spinal cord made by motoneurons. These cells give off branches that end on interneurons (named Renshaw cells after their discoverer). The Renshaw cells in turn feed back onto the motor nerve cells where they form inhibitory synapses (Chapter 16). Chapter 8 discusses in detail the evidence that motoneurons release acetylcholine from their terminals, where they make synapses with skeletal muscles, and now we can add that branches of the motoneurons ending within the spinal cord appear to release the same transmitter.[4,5] In many other central neurons application of acetylcholine produces prompt excitation that is enhanced by inhibitors of acetylcholinesterase. This, on its own, is suggestive, but by no means conclusive, evidence.

As a second example, we shall consider norepinephrine (also called noradrenaline). Of all transmitters its chemistry has been most thor-

---

[4]Eccles, J. C., Fatt, P., and Koketsu, K. 1954. *J. Physiol.* 126:524–562.
[5]Kuno, M., and Rudomin, P. 1966. *J. Physiol.* 187:177–193.

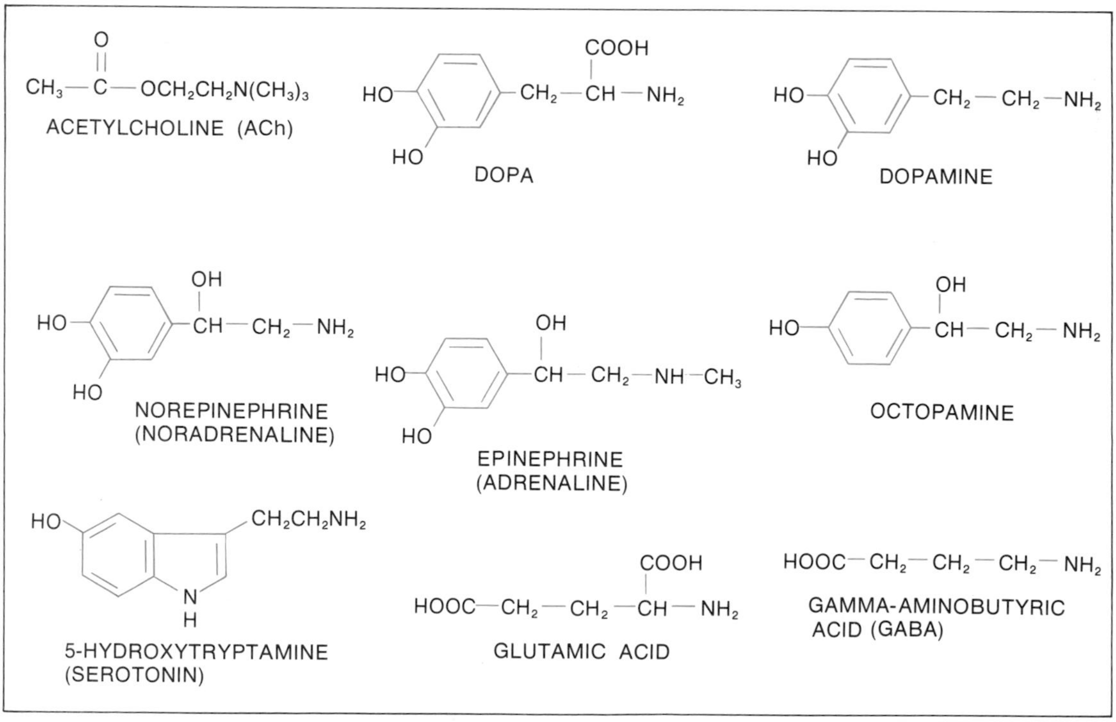

$CH_3-C(=O)-OCH_2CH_2N(CH_3)_3$
ACETYLCHOLINE (ACh)

DOPA

DOPAMINE

NOREPINEPHRINE
(NORADRENALINE)

EPINEPHRINE
(ADRENALINE)

OCTOPAMINE

5-HYDROXYTRYPTAMINE
(SEROTONIN)

$HOOC-CH_2-CH_2-CH(COOH)-NH_2$
GLUTAMIC ACID

$HOOC-CH_2-CH_2-CH_2-NH_2$
GAMMA-AMINOBUTYRIC ACID (GABA)

1 **CHEMICAL STRUCTURES OF TRANSMITTER SUBSTANCES. ACh, norepinephrine epinephrine, and GABA are established transmitters.**

oughly studied. It is released by certain nerves in all internal organs, including the gut, spleen, and heart. The processes involved in the synthesis, storage, and release of this substance have been worked out in great detail by Von Euler, Axelrod, Udenfriend, and their colleagues.[6] A particularly significant advance has been the development of a fluorescence method by Falck and Hillarp,[7] which enables the recognition under the light microscope of axons or terminals containing epinephrine, norepinephrine, dopamine, and serotonin. The term CATECHOLAMINE is often used to designate collectively the substances dopa, dopamine, epinephrine, and norepinephrine, which contain a catechol nucleus (a benzene ring with two adjacent hydroxyl groups). In the terminals of such nerves electron micrographs show characteristic vesicles containing granules about 40 to 140 nm in diameter that can be readily identified (Chapter 10). These fluorescence methods have paved the way for striking advances because they have opened up for easier exploration the entire central nervous system. The techniques enable the tracing in the brain of many pathways and groups of adrenergic neurons whose terminals are likely to liberate norepinephrine or related amines.

Serotonin is another substance for whose role as a transmitter there is much suggestive evidence at many sites in the brain. Well-documented examples at identified synapses were obtained in the sea hare, *Aplysia,*[8] and in snails.[9] In these preparations the presynaptic cell contains serotonin, whose application onto the postsynaptic cell mimics the effect of presynaptic stimulation. Both the drug and the synaptic action are blocked by a specific antagonist, bufotenine. As already mentioned (Chapter 8), in neurons of *Aplysia* serotonin can exert excitatory and inhibitory actions, depending on the type of chemoreceptors in the postsynaptic membrane.

GABA: AN INHIBITORY TRANSMITTER

GABA was first found in extracts of the mammalian brain in 1950 almost simultaneously in the laboratories of Awapara, Udenfriend, and Roberts.[10] At that time it was of interest mainly as a possible metabolite in the glutamic acid cycle. The role of GABA as a transmitter was first suggested by Florey and his colleagues, who used a sensory nerve cell in the peripheral nervous system of a crustacean to assay inhibitory substances. This cell responds to stretch with a steady discharge of impulses (Chapter 15). It was shown that the firing is inhibited by GABA and also by an extract of the mammalian brain which Florey called factor I. Extracts containing factor I were subsequently analyzed by Bazemore, Elliott, and Florey, who found that GABA was the prin-

---

[6]Molinoff, P., and Axelrod, J. 1971. *Annu. Rev. Biochem. 40*:465–500.

[7]Falck, B., Hillarp N.-Å., Thieme, G., and Thorp, A. 1962. *J. Histochem. Cytochem. 10*:348–354.

[8]Gerschenfeld, H. M., and Paupardin-Tritsch, D. 1974. *J. Physiol. 243*:457–481.

[9]Cottrell, G. A., and Macon, J. B. 1974. *J. Physiol. 236*:435–464.

[10]Florey, E. 1961 *Annu. Rev. Physiol. 23*:501–528.

cipal blocking agent. Florey and coworkers also found inhibitory activity in extracts of crustacean nerves containing inhibitory axons, but the active compound could not be identified.[10]

Florey's physiological finding of the inhibitory effect of GABA on the crustacean stretch receptor was confirmed in 1958 by Kuffler and Edwards,[11] who showed in addition that GABA mimics closely the action of the neurally released transmitter on the membrane. On its own this was not adequate to establish GABA as a transmitter. Extracts of the entire nervous system of lobsters were therefore systematically fractionated and the extracts tested for inhibitory activity, using crustacean nerve-muscle junctions as well as stretch receptor neurons for the bioassay.

The initial chemical work was carried out by Kravitz and his colleagues[12,13] on a large sample of the pooled central and peripheral nervous systems of about 500 lobsters. After a long series of fractionations, 10 amino acids that produce inhibition were found. Of these the following eight were identified (in order of potency): GABA, taurine, betaine, alanine, β-alanine, aspartate, glutamine, and homarine. Taurine was present in the highest concentration, accounting in some extracts for half the inhibitory activity. Yet, in its specific inhibitory action, GABA (mole for mole) was about 10 times more potent.

At this stage it became essential to track the various inhibitory substances down to their cells of origin, because it was expected that the neural transmitter would be preferentially concentrated in inhibitory neurons. Fortunately, the locations of excitatory and inhibitory axons innervating crustacean leg muscles were known, chiefly through the studies of Wiersma.[14] As a result, one could isolate either an individual excitatory axon that on electrical stimulation makes the muscles contract or an inhibitory axon that inhibits contraction (Chapter 8). A pair of such fibers, each about 50μm in diameter, is shown in Figure 2.

Large numbers of individual inhibitory and excitatory axons were identified, dissected, extracted, and fractionated. The results showed that the concentrations of inhibitory amino acids, with the notable exception of GABA, were roughly equal in both excitatory and inhibitory axons. In contrast, GABA was present in the surprisingly high concentration of approximately 100 mM in the inhibitory axon; this corresponds to about 0.5 percent of the wet weight. But in excitatory axons less than 1 mM of GABA could be detected.

The great asymmetry in distribution of transmitter in the axons of excitatory and inhibitory cells suggests that the inhibitory nerve cell bodies should contain much larger amounts of GABA than the excitatory cells. This was shown to be so for single cells dissected from ganglia

---

[11]Kuffler, S. W., and Edwards, C. 1958. *J. Neurophysiol.* *21*:589–610.

[12]Kravitz, E. A., Kuffler, S. W., Potter, D. D., and van Gelder, N. M. 1963. *J. Neurophysiol.* *26*:729–738.

[13]Kravitz, E. A., Kuffler, S. W., and Potter, D. D. 1963. *J. Neurophysiol.* *26*:739–751.

[14]Wiersma, C. A. G., and Ripley, S. H. 1952. *Physiol. Comp. Oecol.* *2*:391–405.

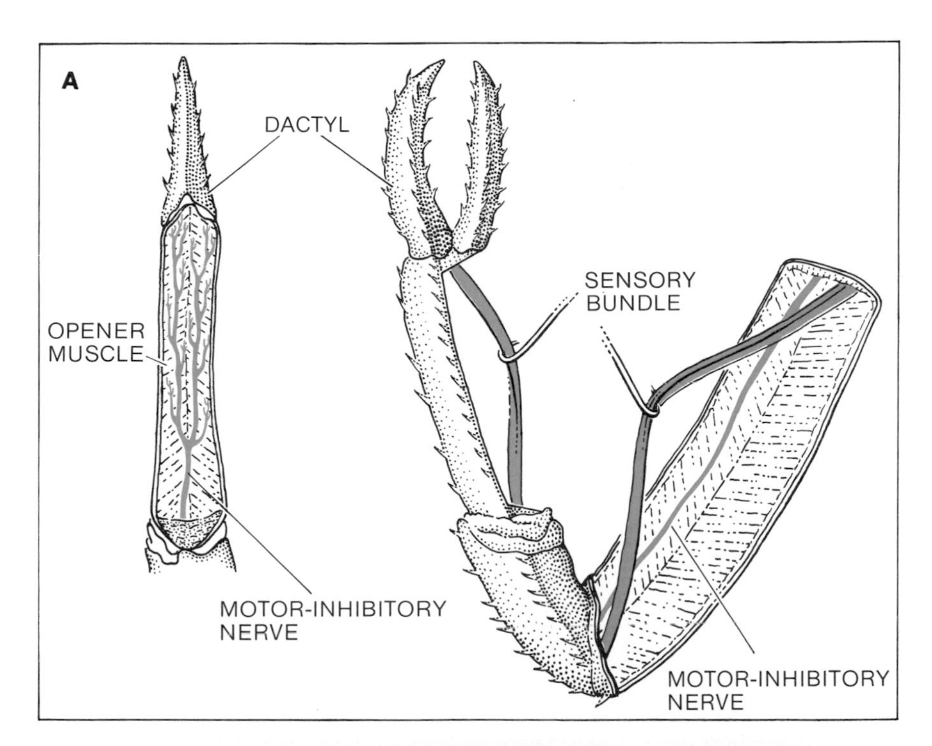

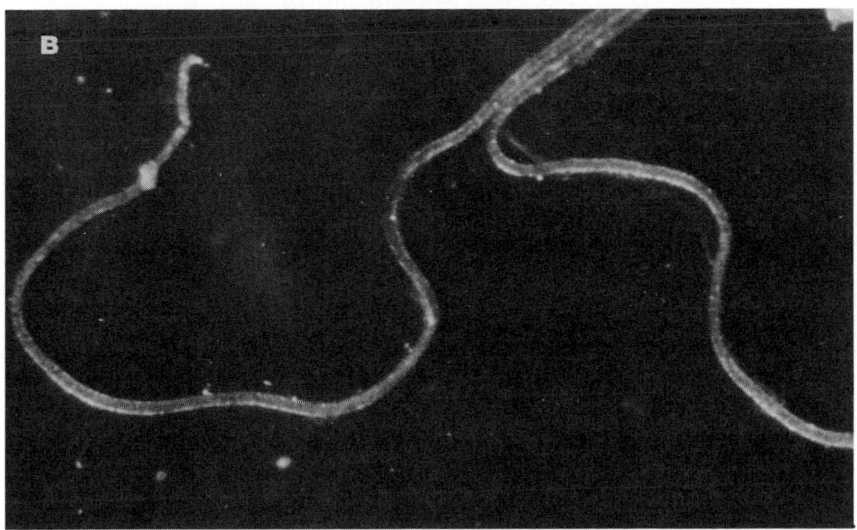

WALKING LEG OF LOBSTER and its various nerve bundles (A). Some muscles are cut away. The opener muscle of the dactyl is supplied by one excitatory and one inhibitory fiber. Long segments of these can be dissected out during their course before they reach their target and are available for stimulation or chemical analysis (B). The excitatory axon is marked by a knot. The axons are about 50 μm in diameter. (A from Hall, Bownds and Kravitz, 1970; B from Kravitz, Kuffler, Potter and Van Gelder, 1963)

2

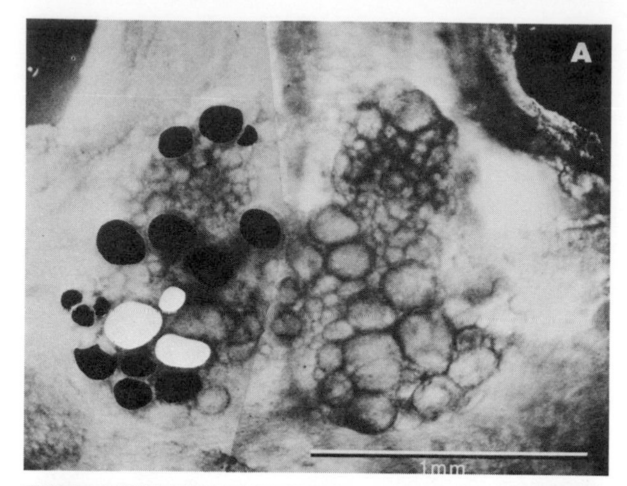

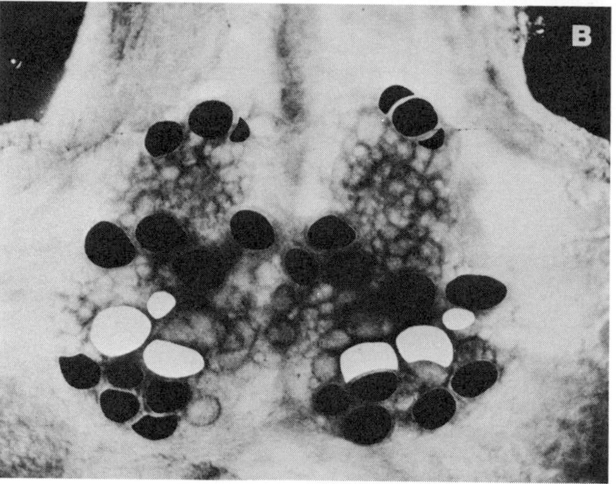

3
CHEMICAL AND PHYSIOLOGICAL ARCHITECTURE OF NEURONS in a ganglion of the lobster. In A inhibitory neurons are marked in white; black markings indicate nerve cells that excite muscles. B. Chemical map of homologous neurons in both halves of the same ganglion. Inhibitory neurons (white) contain more than 2 × 10^{-11} moles of GABA. Excitatory nerves containing no detectable GABA are indicated in black. (After Otsuka, Kravitz, and Potter, 1967)

of lobsters, in which cells can be reliably identified as excitatory or inhibitory by physiological criteria.[15] With some practice the cells can also be recognized by their shapes and positions in different ganglia (Figure 3*A*). The inhibitory cell bodies contain 13 to 15 mM of GABA; excitatory motoneurons contain less than 1 mM. The procedure for isolating cells is shown in Figure 4; a single inhibitory cell body is shown first outside the ganglion and then in the microtube used for chemical analysis. From such a direct microchemical analysis, a partial chemical map of excitatory and inhibitory neurons of a lobster ganglion can be constructed (Figure 3*B*).

It seems surprising that the inhibitory axon should contain about six times more GABA than the cell body. One possible reason for the relatively low GABA estimate in cell bodies may be technical. The calculations were made from the total volume of the cell. However, organelles occupy a large fraction of the cell body but not of the axon;

---

[15]Otsuka, M., Kravitz, E. A., and Potter, D. D. 1967. *J. Neurophysiol. 30:* 725–752.

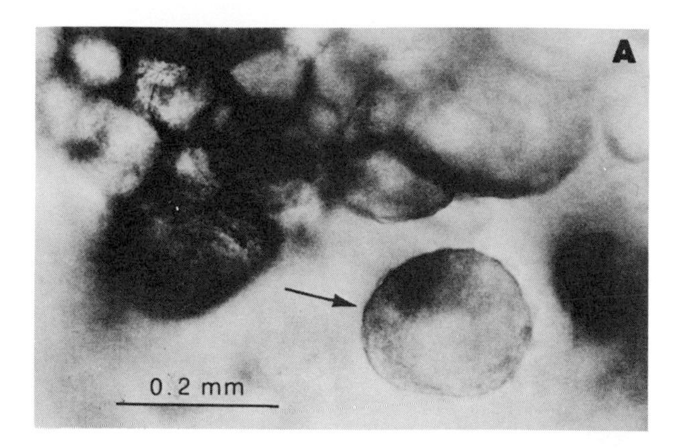

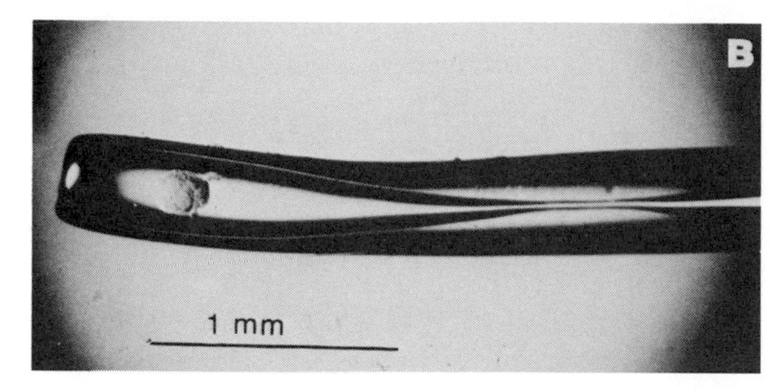

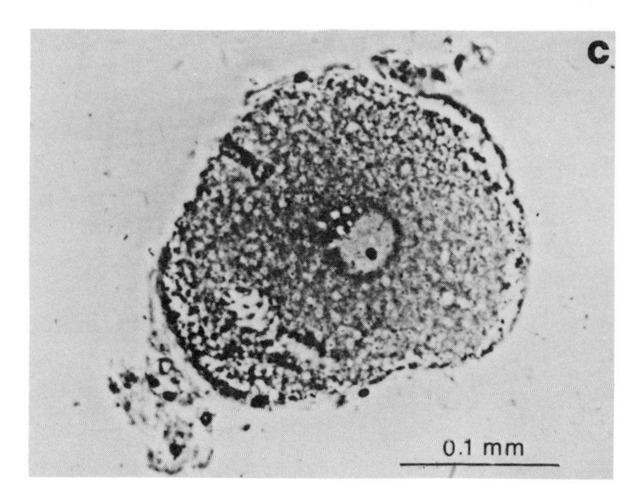

4
SINGLE NEURON ANALYSIS. A. Single cell body (below) isolated from a lobster ganglion is transferred (B) for chemical analysis or histological examination (C). (From Otsuka, Kravitz and Potter, 1967)

the difference in concentration between the cytoplasm of axon and cell body may therefore be smaller than the figures indicate.

In addition to demonstrating the presence of GABA in inhibitory neurons, Kravitz and his colleagues made tests to see what other transmitters, if any, these cells could synthesize.[16] The principle was first to

[16]Hildebrand, J. G., Barker, D. L., Herbert, E., and Kravitz, E. A. 1971. *J. Neurobiol.* 2:231–246.

bathe a lobster ganglion in a solution containing radioactive precursors of various transmitters and then to dissect out individual cells and see what they had synthesized. Since the precursors are common metabolites, they are taken up by all cells and metabolized. Only in cells using a particular transmitter is the precursor converted to the transmitter product. For example, cholinergic cells incorporate choline and make acetylcholine, while noradrenergic cells take up tyrosine for eventual production of norepinephrine, and other cells convert glutamate to GABA. In the ganglia of the lobster, as expected, inhibitory cells make radioactive GABA but not acetylcholine or catecholamines.

The virtual restriction of GABA to inhibitory neurons greatly increased the likelihood of its being a transmitter. There still remained the problem of showing that GABA is in fact released from nerve terminals. An experiment similar to that originally performed by Loewi on the heart was therefore performed on lobster muscles.[17]

The inhibitory nerve to the opener muscle of the claw was stimulated while the muscle was perfused. The perfusate was collected and analyzed for GABA using an enzymatic assay. (Testing the activity on a stretch receptor or on a nerve-muscle preparation would have been less specific and too insensitive.) Only small amounts of GABA were released (about $1-4 \times 10^{-14}$ moles per nerve impulse) from all the inhibitory terminals in the opener muscle of the claw. In solutions containing low calcium concentrations nerve conduction was unimpaired, but transmission was blocked and extra GABA no longer appeared in the perfusate following stimulation. This result supports the assumption that in normal fluid nerve stimulation causes the release of GABA from the inhibitory terminals.

At present the actual amount of GABA released cannot be estimated because in lobster preparations much of the GABA that is liberated from the terminals is taken up into tissues before it can be collected. It is intriguing that most of the uptake is by Schwann cells and connective tissue around the inhibitory nerve fibers, rather than by the nerve terminals themselves, as might be expected by analogy with the norepinephrine system (Chapter 13).[18]

Similarity between the actions of GABA and the nerve-released transmitter

Several experiments are described below which show that GABA acts on the postsynaptic membrane in a manner similar to that of the neurally released transmitter.

Experiments by several groups have demonstrated that applied GABA, like nerve stimulation, produces its effects by increasing permeability to chloride, so that both procedures shift the postsynaptic membrane potential toward E_{Cl}, the equilibrium potential for chloride. At E_{Cl} the synaptic current and potential reverse (Chapter 8). Potassium ions, on the other hand, do not contribute significantly to the inhibitory actions of applied or nerve-released GABA, since changing the potas-

---

[17]Otsuka, M., Iversen, L. L., Hall, Z. W., and Kravitz, E. A. 1966. *Proc. Natl. Acad. Sci. U.S.A.* *56*:1110–1115.

[18]Orkand, P. M., and Kravitz, E. A. 1971. *J. Cell Biol.* *49*:75–89.

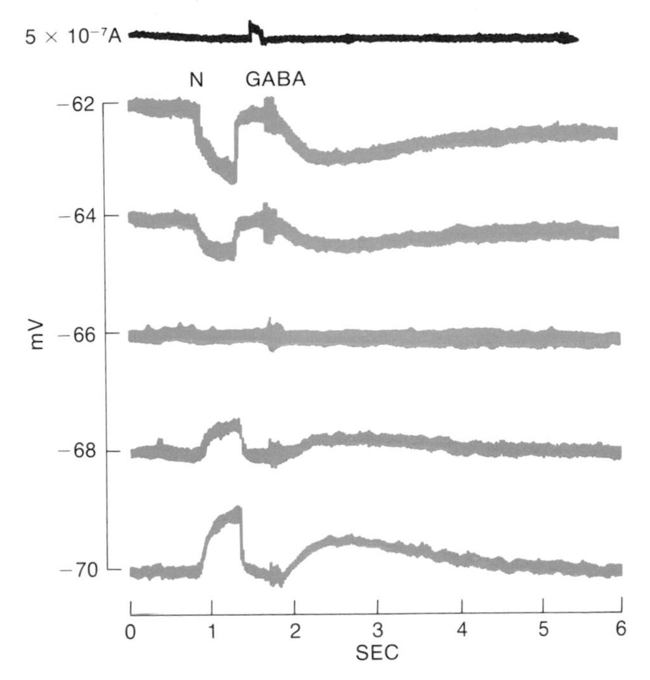

5

SIMILARITY OF CONDUCTANCE CHANGES resulting from the action of nerve-released transmitter and from applied GABA in a muscle fiber of the crayfish. Brief trains of inhibitory nerve stimuli (N) and iontophoretically applied GABA appear on each record. As the resting membrane potential (—70 mV) is depolarized, the synaptic and the GABA potentials reverse at the same level, near —66 mV. Trace at top indicates current through GABA-filled pipette. (From Takeuchi and Takeuchi, 1965)

sium concentration in the bathing fluid does not alter the reversal potential.[19–21] As expected, however, removal of chloride ions abolishes inhibitory synaptic potentials without disturbing excitatory transmission.

Figure 5 shows an example in which the membrane potential of a muscle fiber in the crayfish claw was depolarized from its resting level of −70 mV by passing current through a microelectrode. At various potentials the inhibitory axon was stimulated by a short train of pulses at 30/sec and its effect compared with that of GABA applied iontophoretically. There was a unique value of membrane potential, in this case −66 mV, at which no potential change was caused by the neural transmitter or by GABA. This is the reversal potential.

It has also been shown that when GABA is released from the micropipette, its action is confined to the region of the postsynaptic membrane where the inhibitory terminals release their transmitter and where the inhibitory potentials are set up.[21] Apparently, nerve-released and artificially applied GABA act on the same synaptic receptors and cause identical conductance changes. The similarity has been further demonstrated in tests using a pharmacological agent, picrotoxin, which blocks inhibitory synapses in crustacean muscle and also abolishes the response to GABA.[22] In conclusion, at crustacean neuromuscular synapses GABA

---

[19]Boistel, J., and Fatt, P. 1958. *J. Physiol.* *144*:176–191.
[20]Dudel, J., and Kuffler, S. W. 1961. *J. Physiol.* *155*:543–562.
[21]Takeuchi, A., and Takeuchi, N. 1967. *J. Physiol.* *191*:575–590.
[22]Takeuchi, A., and Takeuchi, N. 1969. *J. Physiol.* *205*:377–391.

fulfills many of the same criteria as acetylcholine at the vertebrate neuromuscular synapse.

Inhibitory neuromuscular junctions in the earthworm and several insects also seem to be operated by GABA, although the studies have not been so extensive as on the lobster and crayfish.[23] There are cells, however, in which GABA has an excitatory depolarizing effect. These include dorsal root ganglion cells[24] and cells from the superior cervical ganglion in culture.[25]

GABA action on excitatory presynaptic terminals

The inhibitory action of crustacean peripheral nerves is not confined exclusively to the postsynaptic muscle membrane. Inhibitory nerve impulses can also reduce the amount of transmitter released by excitatory nerve terminals. Identical presynaptic inhibitory effects are observed when GABA is applied to the bathing fluid in low concentrations that do not significantly affect the postsynaptic membrane (Chapter 8).[20,26]

GABA action in vertebrate brains

GABA was originally discovered in the mammalian brain, where its distribution is inhomogeneous, suggesting that some cells contain relatively high concentrations. In addition, Curtis,[27] Krnjević,[28] and their colleagues have shown that many nerve cells in various regions of the brain are inhibited from discharging when GABA is applied diffusely or by iontophoresis. These observations, together with the finding that GABA can be collected from fluid on the surface of the cortex or from ventricles after electrical stimulation,[29] have encouraged studies at a more detailed cellular level, mainly in cats.

Tests made by Otsuka, Ito, Obata, and their colleagues are of particular interest because they focus on specific identified cells. These investigators examined the inhibitory action of the Purkinje cells of the cat cerebellum. These are large, conspicuous neurons with a wide arborization onto which converges most of the circuity of the cerebellum (Figure 2, Chapter 1). The entire output of the cerebellum is by way of the Purkinje axons, which form inhibitory synapses in several distinct nuclei. One of these includes a collection of large cells called the lateral vestibular nucleus of Deiters. This direct pathway provides an exceptional opportunity to study an inhibitory neuron that can be identified, together with its target cells, as discussed below.

To estimate the GABA content of Purkinje cells, single cells were dissected out and analyzed by an enzymatic assay that measured down to 2×10^{-14} moles. The value found in Purkinje cells was 5 to 8 mM. In control experiments motoneurons were found to contain less GABA, about 1 mM.[30]

---

[23]Otsuka, M. 1972. In G. H. Bourne (ed.). *Structure and Function of Nervous Tissue*, Vol. IV. Academic Press, New York, pp. 266–268.

[24]De Groat, W. C. 1972. *Brain Res. 38*:429–432.

[25]Obata, K. 1974. *Brain Res. 73*:71–88.

[26]Takeuchi, A., and Takeuchi, N. 1966. *J. Physiol. 183*:433–449.

[27]Curtis, D. R., and Johnston, G. A. R. 1974. *Ergeb. Physiol. 69*:97–188.

[28]Krnjević, K. 1974. *Physiol. Rev. 54*:418–540.

[29]Jasper, H., and Koyama, I. 1969. *Can. J. Physiol. Pharmacol. 47*:889–905.

[30]Obata, K. 1969. *Experientia 25*:1283.

Two considerations arise in relation to these analyses on isolated cell bodies: (1) as in crustaceans, the cell body may have a lower concentration of transmitter than the axon or the terminals where it is released; (2) some GABA is present in cells that are not inhibitory (for example, in spinal motoneurons), and in these the GABA may be contributed by presynaptic nerve terminals. Pursuing such considerations, Otsuka and coworkers analyzed neurons in the ventral and dorsal portions of the lateral vestibular nucleus of Deiters. Most Purkinje cells make synapses within the dorsal portion of the nucleus. Neurons dissected from this region contained as much GABA per unit volume as did Purkinje cells (about 6 mM). Neurons from the ventral nucleus, which receive no innervation from the Purkinje cells, had about half as much GABA. When Purkinje axons were cut and degenerated, the GABA levels in the dorsal nucleus were greatly reduced, while those in the ventral nucleus were not affected. This suggests that the GABA in Deiters's nucleus is not derived from the cell bodies but from the synaptic endings upon them. The results reinforce the view that GABA (and other transmitters) may be highly concentrated in terminals.[31]

If the cerebellar output is stimulated electrically and recordings are made with an intracellular electrode from a Deiters neuron, inhibitory hyperpolarizing synaptic potentials appear. Electrophoretic application of GABA to the postsynaptic Deiters's cells mimics these synaptic potentials in the same way as at the crustacean neuromuscular junction.[32] Thus, hyperpolarizing potentials, set up neurally and electrophoretically, decrease and reverse at the same membrane potential when the postsynaptic membrane is depolarized by current. Furthermore, with cerebellar stimulation, GABA appears in the fluid that diffuses out of the medulla into a nearby ventricle (Chapter 14). As expected, localization of glutamic decarboxylase to the Purkinje cells by the antibody technique (see above) provides further confirmation for a role of GABA as the inhibitory transmitter.

Metabolic basis of GABA accumulation

In parallel series of experiments the groups of Roberts[33] and of Kravitz[34] have shown that the pathways of GABA metabolism are similar in crustaceans and in the mammalian brain. In crustaceans, both excitatory and inhibitory axons possess large stores of glutamate (the precursor of GABA), but only the inhibitory axons are able to accumulate GABA. Figure 6 illustrates the components of the metabolic pathways in individual excitatory and inhibitory neurons. All the metabolites and enzymes have been isolated and purified. The most conspicuous difference between excitatory and inhibitory neurons is a high glutamic acid decarboxylase activity in inhibitory axons. An interesting property of the decarboxylase is that the product of the reaction, GABA, inhibits the activity of the enzyme. These features have enabled

[31]Otsuka, M., Obata, K., Miyata, Y., and Tanaka, Y. 1971. *J. Neurochem.* *18*:287–295.

[32]Obata, K., Takeda, K., and Shinozaki, H. 1970. *Exp. Brain Res.* *11*:327–342.

[33]Susz, J. P., Haber, B., and Roberts, E. 1966. *Biochemistry* *5*:2870–2877.

[34]Hall, Z. W., Bownds, M. D., and Kravitz, E. A. 1970. *J. Cell Biol.* *46*:290–299.

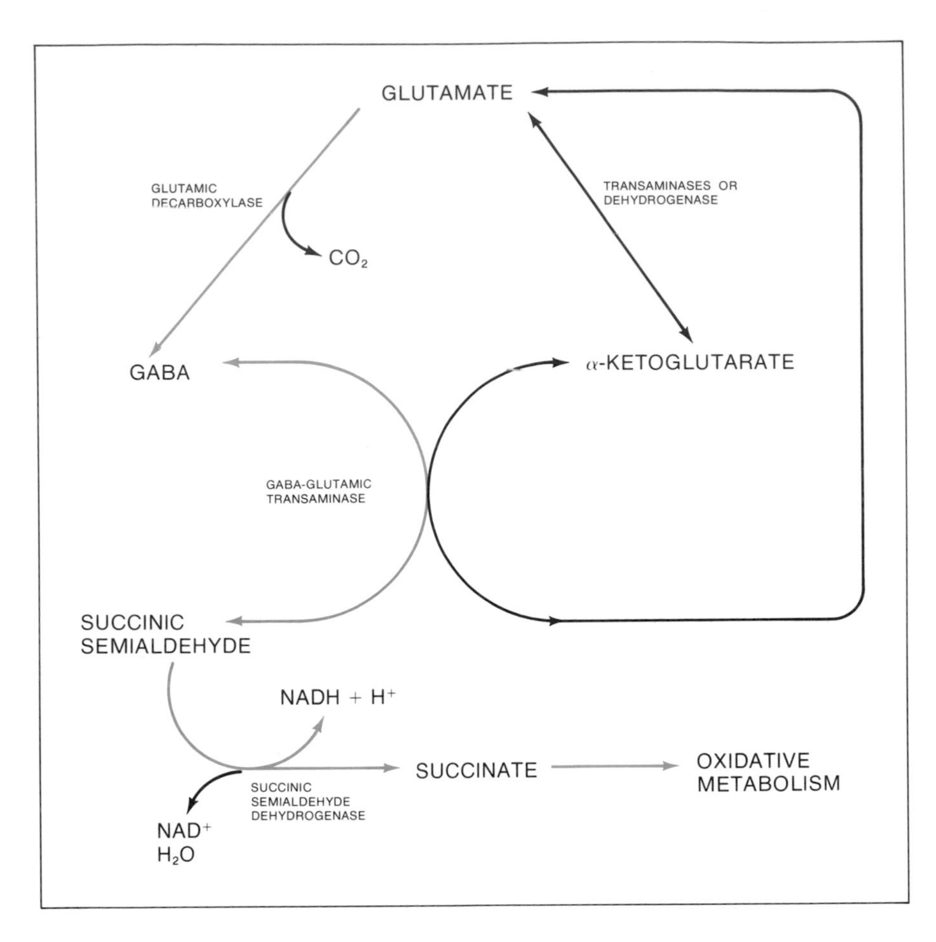

6 **METABOLISM OF GABA. Main pathways are indicated in color. (After Kravitz, Kuffler, and Potter, 1963)**

Kravitz and his colleagues to propose an explanation for the basis of GABA accumulation: according to their scheme the inhibitory axon synthesizes GABA until the concentration reaches about 100 mM, a stage at which GABA inhibition of the decarboxylase prevents further accumulation.

The scheme, based on the difference in the amount of decarboxylase, does not explain the presence of a small, but apparently real amount of GABA in some excitatory axons or cell bodies in which little or no decarboxylase is detected. The explanation may lie in a failure to assay small amounts of the enzyme. Alternatively, the small amounts of GABA may be contained in cells that surround neurons in the lobster.[18] Schwann cells and connective tissue in peripheral nerves have been shown to possess an active GABA transport system, and some neurons may take up GABA in relatively small amounts. Similarly, glial cells in the mammalian brain also have an active GABA-uptake mechanism in areas where many GABA-containing neurons are found (Chapter 13).

Glycine, an amino acid with two carbons in the chain instead of four, as in GABA, was proposed by Werman, Davidoff, and Aprison[35] as an inhibitory transmitter in the mammalian spinal cord. Glycine, like GABA, is unevenly distributed in the spinal cord; if pipetted onto several types of neurons, it mimics potentials set up by inhibitory stimulation. Although the evidence is not as good for GABA, glycine is a candidate that satisfies several of the requirements for being a transmitter. Much experimental work, therefore, has been concerned with distinguishing between glycine and GABA on the assumption that both amino acids may be used in the nervous system at different sites.

Is glycine a second transmitter mediating inhibition?

Three useful agents that block inhibition pharmacologically are strychnine, picrotoxin, and bicucculine.[27,28] At present the main questions concern the specificity of each of these drugs and whether their actions provide diagnostic criteria for recognizing the transmitter by selectively and reliably blocking either GABA or glycine. For example, in spinal motoneurons the inhibitory actions of iontophoretically applied glycine and of the neural transmitter are both blocked by strychnine but not by bicucculine; in contrast, GABA action is not significantly affected by strychnine but is blocked by bicucculine. Such experiments can be interpreted in favor of glycine as the transmitter at these synapses, but other questions remain. First of all the drug concentrations with electrophoretic application are not known at the critical sites, and comparative effectiveness may be difficult to interpret; the possibility also exists that some blocking drugs may act preferentially on the presynaptic terminals to prevent transmitter release.[36] Thus, the postsynaptic membrane may still be available for the action of electrophoretically applied inhibitory substances and thereby weaken much of the present pharmacological evidence. Recently, the case for glycine has been questioned on the basis that in rat spinal motor neurons a high background concentration (close to 3 mM) of glycine does not significantly affect the inhibitory synaptic potentials produced by the nerve-released transmitter.[37] Such a background may be expected to desensitize and thereby reduce the effectiveness of neural transmission; or, if desensitization is absent, adding glycine should enhance neural inhibition. Therefore, the pharmacological criteria, although suggestive, are not yet sufficiently decisive.

The search for other transmitters

Considerable momentum has been gained from improved techniques for isolating identified cells and analyzing their contents. The aim of such studies at known synapses is usually twofold: (1) to analyze physiologically the processes of synaptic transmission, a prerequisite for a rational approach to testing or designing pharmacologically active compounds that can interfere at specific sites in a known step of a chain of reactions, and (2) to construct chemical maps of the nervous system

---

[35]Werman, R., Davidoff, R. A., and Aprison, M. H. 1968. *J. Neurophysiol.* *31*:81–95.

[36]Diamond, J., Roper, S., and Yasargil, G. M. 1973. *J. Physiol.* *232*:87–111.

[37]Otsuka, M., and Konishi, S. 1976. *The Synapse. Cold Spring Harbor Symp. Quant. Biol.* *40*:135–144.

as an additional tool for unraveling the organization or design of connections. The methods developed by Falck and Hillarp for visualizing cells and axons containing catecholamines, and the experiments mentioned here, provide some examples. The micromethods of Lowry for accurate measurements of enzymes and other subcellular components of individual neurons are further valuable techniques, as are those devised by Edström and Hydén, who have analyzed individual nerve cells dissected from Dieters's nucleus. A new, promising immunochemical approach, mentioned earlier, is to make antibodies to purified enzymes that participate in the synthesis of transmitters, such as glutamic decarboxylase in the case of GABA,[3] and thereby visualize the distribution of transmitters and the pathways in which they are active.

In conclusion, GABA plays an important role as the mediator of inhibition at many crustacean synapses. The evidence for its role as a transmitter in mammals is also good, but less complete. In addition, the finding that GABA can depolarize and excite certain neurons, such as dorsal root ganglion cells, raises the possibility that it may also act as an excitatory transmitter.[24,25]

It is likely that more transmitter substances remain to be discovered: Burnstock[38] has reviewed the evidence for ATP as a transmitter, and recent experiments by Leeman, Otsuka, and their colleagues[39,40] suggest that mammalian sensory fibers contain and probably release a peptide (substance P) that acts as the excitatory transmitter to motoneurons in the spinal cord of the frog, the cat, and the rat.[37] There also is considerable evidence for the role of dopamine as a transmitter in invertebrates,[41] in sympathetic ganglia, and in the nigrostriatal pathway in the mammalian brain.[42]

Axoplasmic flow and transport of transmitters

Nerve terminals continually secrete transmitter, and the question arises whether it is shipped to them ready-made, whether they assemble it from parts provided by the cell body, or whether they themselves synthesize some or many of the essential components. It is known that transmitter can be assembled in terminals provided the precursors are available. The terminals cannot, however, manufacture the needed enzymes either for transmitter synthesis or for the making of new parts of the cell membrane; this seems to be so, because no ribosomes have been found in either axons or terminals and it is unlikely that mitochrondia provide the needed proteins. These are apparently transported from the cell body to the rest of the neuron. In many instances distances are so great that if simple diffusion accounted for transport, it might take years for a substance to move, for example, from the spinal cord to the toe of a large animal. A relatively rapid transport mechanism must therefore be postulated.

---

[38]Burnstock, G. 1972. *Pharmacol. Rev. 24*:509–581.

[39]Takahashi, T., and Otsuka, M. 1975. *Brain Res. 87*:1–11.

[40]Leeman, S. E., and Mroz, E. A. 1974. *Life Sci. 15*:2033–2044.

[41]Wallace, B. G., Talamo, B. R., Evans, P. D., and Kravitz, E. A. 1974. *Brain Res. 74*:349–355.

[42]Hornykiewicz, O. 1973. *Br. Med. Bull. 29*:172–178.

There is now abundant evidence for a continuous movement of substances along axons in both directions. For forward transport of materials there are slow and fast components, from 1 to about 400 mm/day.[43] Examples of this are mentioned earlier in connection with the visual system, where movement of labeled amino acids along neurons from the eye to the lateral geniculate nucleus and the cortex is demonstrated. This occurs far more rapidly than simple diffusion. Similarly, the reverse movement of a protein, horseradish peroxidase, from terminal to cell is now used to mark specific pathways.[44]

A variety of approaches have been used to demonstrate movement of transmitters along the axon. For example, Schwartz and his colleagues[1,2] injected radioactive choline into a large cholinergic neuron in *Aplysia*. Choline was converted to acetylcholine, whose movement could be followed down the nerve to the terminals where it was released with nerve stimulation. Similarly, Dahlström,[45] Geffen,[46] and their colleagues have observed the movement of transmitters and enzymes along sympathetic neurons.

What at present is not clear is the mechanism for movements faster than diffusion. Simple bulk flow is difficult to understand in view of the two-way traffic in axons. The idea that microtubules may be involved is attractive, but incompletely defined and not yet proved. Further, what happens to products once they reach the nerve endings or the cell bodies? Materials do not continue to accumulate. Presumably a mechanism exists that regulates the turnover of the amount of various substances and that is sensitive to the metabolic needs of neurons in various physiological states.[47]

SUGGESTED READING

Papers marked with an asterisk (*) are reprinted in Cooke, I., and Lipkin, M. 1972. *Cellular Neurophysiology*. Holt, New York.

General reviews

Cooper, J. R., Bloom, F. E., and Roth, R. E. 1974. *The Biochemical Basis of Neuropharmacology*. Oxford University Press, London.

Hall, Z. W., Hildebrand, J. G., and Kravitz, E. A. 1974. *Chemistry of Synaptic Transmission*. Chiron Press, Newton, Mass. (A collection of reprinted papers with introductory essays.)

Krnjević, K. 1974. Chemical nature of synaptic transmission in vertebrates. *Physiol. Rev. 54*:418–540.

GABA

Hall, Z. W., Bownds, M. D., and Kravitz, E. A. 1970. The metabolism of gamma aminobutyric acid in the lobster nervous system. *J. Cell Biol. 46*:290–299.

---

[43] Grafstein, B., and Laurens, R. 1973. *Exp. Neurol. 39*:44–57.

[44] La Vail, J. H., and La Vail, M. M. 1974. *J. Comp. Neurol. 157*:303–358.

[45] Dahlström, A., 1971. *Phil. Trans. R. Soc. Lond. (Biol. Sci.) 261*:325–358.

[46] Geffen, L. B., and Livett, B. G. 1971. *Physiol. Rev. 51*:98–157.

[47] Collier, B., and MacIntosh, F. C. 1969. *Can. J. Physiol. Pharmacol. 47*:127–135.

*Kravitz, E. A., Kuffler, S. W., and Potter, D. D. 1963. Gamma-aminobutyric acid and other blocking compounds in crustacea. *J. Neurophysiol.* *26:*739–751.

Otsuka, M. 1972. γ-Aminobutyric Acid in the Nervous System. In G. H. Bourne (ed.). *The Structure and Function of Nervous Tissue,* Vol. IV. Academic Press, New York. Review.

Otsuka, M., Kravitz, E. A., and Potter, D. D. 1967. Physiological and chemical architecture of a lobster ganglion with particular reference to gamma-aminobutyrate and glutamate. *J. Neurophysiol.* *30:*725–752.

Otsuka, M., Obata, K., Miyata, Y., and Tanaka, Y. 1971. Measurement of γ-aminobutyric acid in isolated nerve cells of cat central nervous system. *J. Neurochem.* *18:*287–295.

Catecholamines and axoplasmic flow

Dahlström, A. 1973. Aminergic transmission. Introduction and short review. *Brain Res.* *62:*441–460.

Geffen, L. B., and Livett, B. G. 1971. Synaptic vesicles in sympathetic neurons. *Physiol. Rev.* *51:*98–157.

Grafstein, B. 1971. Transneuronal transfer of radioactivity in the central nervous system. *Science* *172:*177–179.

Iversen, L. L. 1967. *The Uptake and Storage of Noradrenaline in Sympathetic Nerves.* Cambridge University Press, London.

Iversen, L. L. (ed.). 1973. Catecholamines. *Br. Med. Bull.* *29:*91–178. (A collection of papers.)

CHAPTER TWELVE

ACTIVE TRANSPORT OF IONS

In the resting state sodium tends to move into neurons and potassium tends to leave the interior. This process is greatly accelerated when nerves conduct impulses for long periods. To sustain their concentration-powered batteries, that is, the appropriate ionic distribution, energy must be expended by cells to pump sodium out and potassium in. A continually acting restorative pump is particularly important for small cells because their available store of ions is relatively limited.

Experiments on a variety of invertebrate nerve cells and unmyelinated mammalian axons have shown that the transport mechanism gives rise to a greater sodium movement outward than potassium movement inward. Activity of the pump, therefore, produces a net outward movement of positive charges. When sodium is extruded following trains of impulses, this current can hyperpolarize the membrane potential and thereby change the signaling properties of neurons.

The energy source for the sodium-potassium pump is derived from the hydrolysis of adenosine triphosphate (ATP), and there is now much evidence to suggest that the pump itself is an enzyme. Characteristically, the pump and the enzyme activity depend on the presence of sodium and potassium; both are blocked by the cardiac glycoside ouabain. Analogous operations of metabolic pumps have been seen in red blood cells. In squid axons a separate transport mechanism extrudes the calcium that enters during impulses.

All the electrical signals of the nervous system described in earlier chapters can be explained in terms of permeability mechanisms that allow ions to move passively along their electrochemical gradients, without the nerve cell having to perform metabolic work. During action potentials, generator potentials, and excitatory potentials, sodium and potassium ions run rapidly into and out of the cell; at the resting potential similar movements proceed continuously, but much more slowly. Inevitably, the concentration gradients across the membrane of a nerve cell will run down progressively unless restitution is made.

For many years, until more information became available about ion transport in other types of cells, it was not obvious how nerve cells could adjust the internal concentrations of ions. As early as 1902, Overton[1] recognized that though the problem applied to all types of cells in the body, it was especially serious in nerve and muscle fibers that conduct millions of impulses throughout the life of an animal, but still maintain high internal potassium and low sodium concentrations. The mechanism for achieving this must be quite different from passive ionic movements governed by permeability. Thus, to move out of the cell, sodium must be transported UPHILL, against the electrochemical gradient.

In early electrical measurements the only obvious sign of an active recovery process was provided by certain slow changes in membrane potential that followed trains of impulses in small, unmyelinated nerve fibers. It was known that these AFTERPOTENTIALS were markedly reduced by metabolic poisons. Their possible functions and the underlying mechanisms were not, however, understood, and they were largely ignored.

A change developed as more became known about the processes used for actively transporting ions against their electrochemical gradients. Such active transport processes are often referred to as PUMPS, and their mode of operation is of particular interest for a number of reasons: (1) pumps are essential for maintaining the long-term functions of nerve cells; (2) their effects can lead to large changes in membrane potentials that persist for seconds or even minutes, and these potentials are, therefore, far slower than those (in the millisecond time scale) usually associated with signaling; (3) the pump-evoked potentials inevitably influence signaling; for example, the threshold of a cell, the conduction of an action potential along its branches, and the way in which the cell integrates are altered in a significant manner; and (4) an enzymatic basis for sodium pump activity has been demonstrated, and this offers an opportunity to link ionic movements with a specific enzyme within the membrane.[2]

Before the experimental evidence for metabolically driven active transport is described, it is convenient to discuss the changes in ionic concentration that occur inside and outside cells following impulses.

IONIC CONCENTRATION CHANGES RESULTING FROM NERVE IMPULSES

Chapter 5 explains that during each action potential about 3×10^{-12} moles of sodium and potassium move into and out of the squid axon

---

[1]Overton, E. 1902. *Pflügers Arch.* *92*:346–386.
[2]Skou, J. C. 1957. *Biochim. Biophys. Acta* *23*:394–401.

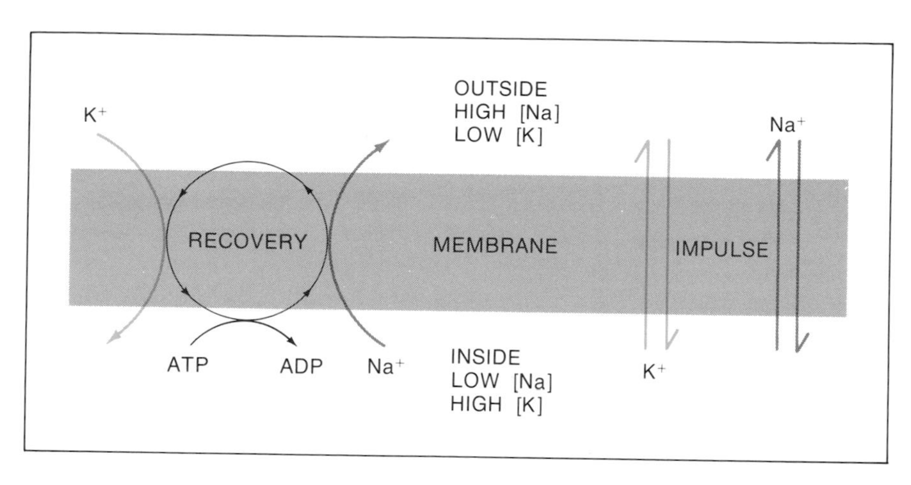

1 **MOVEMENT OF IONS through the nerve membrane. The downhill movements which occur during the impulse are shown on the right by the heavy arrows; uphill movements during recovery are shown on the left. (After Hodgkin and Keynes, 1955)**

through each square centimeter of membrane. These changes are insignificant over short periods in a large cell like the squid axon, especially since the natural frequency of firing is low and occurs in short bursts of only about 8 to 10 impulses per second. In contrast, high frequencies of firing in small nerve fibers would be expected to lead to larger changes in ion concentration, owing to their smaller reservoir of axoplasm in relation to the membrane surface.

As sodium accumulates inside, potassium is lost to the outside. Almost all neurons within the central nervous system are surrounded by narrow clefts a few tens of nanometers wide, where potassium can accumulate. The increase of potassium concentration in this extracellular space can build up to such an extent during repetitive firing that E_K can change significantly (Chapter 13). For the original state to be restored, the sodium gained by the cell during impulse activity must be extruded by some process involving energy expenditure (Figure 1).

A special pump is not necessarily required to maintain the low internal chloride concentration in neurons. Provided the sodium and potassium concentrations are prevented from changing, chloride passively distributes itself according to the Nernst equation, so that E_{Cl} is near the resting potential of the neuron. There is evidence, however, for a chloride pump in epithelial cells in the stomach, the toad bladder, and in certain neurons.[3]

Experimental evidence for active transport of sodium and potassium in axons

In nerves, the evidence for ion transport driven by metabolic energy was provided in a series of experiments by Hodgkin, Keynes, Shaw,

[3]Russell, J. M., and Brown, A. M. 1972. *J. Gen. Physiol.* *60*:499–518.

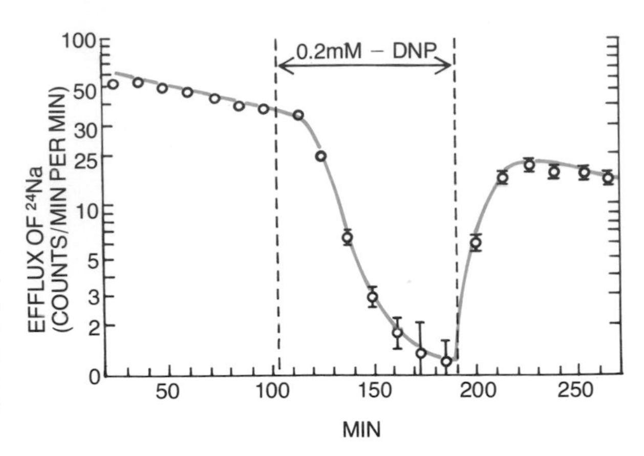

2

SODIUM EFFLUX from a cuttlefish (Sepia) axon during treatment with a metabolic inhibitor, dinitrophenol (DNP). The axon was loaded with ^{24}Na before the experiment and was maintained in artificial seawater. The ordinate gives the rate at which ^{24}Na leaves the axon; vertical lines are ± 2 x SE of the mean. (From Hodgkin and Keynes, 1955)

Caldwell, Brinley, Mullins, and others. The principle of the experiments was to load a giant axon with radioactive sodium by stimulating it in seawater containing the isotope or by intracellular injection and then to measure the efflux into nonradioactive seawater.[4,5] The results (Figure 2) indicate that there is at rest a continuous uphill movement of sodium out of the axon, against both the concentration gradient and the electrical potential. This outward movement of sodium has the following properties:

1. It is rapidly and reversibly reduced by a number of poisons that interfere with the oxidative metabolism of the cell, notably dinitrophenol (DNP), cyanide, and azide.
2. It is prevented (in squid axons) by the cardiac glycoside, ouabain, which is chemically related to digitalis, a drug used extensively for treating heart failure in man. Ouabain specifically blocks ion pumps that transport sodium and potassium, but does not interfere with the oxidative metabolism of the cell. Interestingly, this agent acts only on the outside of the nerve fiber and is ineffective when introduced into the axoplasm.
3. It is blocked by removing potassium from the bathing fluid (see below).
4. Lithium ions, which can substitute for sodium in producing action potentials, are not extruded from the cell.
5. Agents that block active transport cause little immediate change in the resting or action potentials, which continue, as before, for long periods. The squid axon, as we have mentioned, is so large that over short periods the alterations in the intracellular contents are small and the potential across the membrane is reduced by only a few millivolts after blockage of the sodium pump.[4]

[4]Hodgkin, A. L., and Keynes, R. D. 1955. *J. Physiol. 128*:28–60.

[5]Baker, P. F., Blaustein, M. P., Keynes, R. D., Manil, J., Shaw, T. I., and Steinhardt, R. A. 1969. *J. Physiol. 200*:459–496.

In addition to extruding sodium at rest, the axon takes in potassium from the outside medium, against its electrochemical gradient. These two processes, extrusion and uptake, are linked and depend upon the breakdown of adenosine triphosphate (ATP) by the cell (see below). Both are reduced by metabolic inhibitors and by ouabain. The interdependence of the processes used by the membrane for transporting one type of ion in and the other out is further illustrated by the finding (mentioned above) that some potassium must be present in the outside fluid for the neuron to extrude sodium. When the outside potassium concentration is reduced to zero, the outward movement of sodium is greatly decreased, though it is not completely abolished (Figure 5). Measurements made on squid axons have shown that the mechanism does not involve a simple one-to-one exchange, since the ratio of sodium extruded to potassium taken up is of the order of 3:2. Transport mechanisms that involve the exchange of one ion species with another have been found in many types of cells and are commonly referred to as COUPLED SODIUM-POTASSIUM PUMPS.

Energy sources for active transport of sodium and potassium

How are metabolic processes of the nerve cell used to drive sodium and potassium against their electrochemical gradients? In recent years it has been shown that the immediate energy source is ATP. In experiments on squid axons metabolism was first blocked by cyanide or DNP to prevent the active transport of sodium and potassium. Next, a variety of energy-rich compounds were injected into the cytoplasm to determine which could serve as substrate for the pump and restore its activity. In early experiments it appeared that a number of substances were able to reactivate coupled sodium-potassium transport.[6] Later, however, Mullins and Brinley[7] showed (Figure 3) that in axons depleted of small organic ions by internal dialysis only ATP could serve as a direct source of energy. This has recently been confirmed in perfused axons.[8]

The finding that ATP is consumed while ions are being transported suggests that the pump itself may be an enzyme, hydrolyzing ATP. This idea is reinforced by Skou's demonstration in crab nerves that an ATPase fulfilled many of the requisite criteria. The enzyme hydrolyzed ATP, but only in the presence of sodium and potassium. As with the pump, the enzyme activity was inhibited by ouabain, and lithium ions could not substitute for sodium.[9]

The correlation of an ATPase with sodium and potassium transport has been more directly demonstrated by three types of evidence obtained in red blood cells: (1) the sodium and potassium concentrations required for pumping are similar to those needed for optimum activity

---

[6]Caldwell, P. C., Hodgkin, A. L., Keynes, R. D., and Shaw, T. I. 1960. *J. Physiol.* *152*:561–590.

[7]Mullins, L. J., and Brinley, F. J. 1967. *J. Gen. Physiol.* *50*:2333–2355.

[8]Baker, P. F., Foster, R. F., Gilbert, D. S., and Shaw, T. I. 1971. *J. Physiol.* *219*:487–506.

[9]Skou, J. C. 1964. *Prog. Biophys. Mol. Biol.* *14*:133–166.

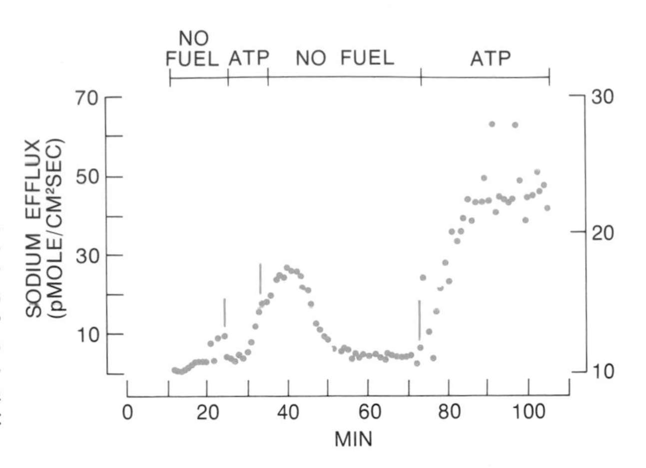

3

EFFECT OF ATP ON SODIUM EFFLUX from an internally dialyzed squid axon. The composition of the axoplasm is regulated by perfusion through an intracellular dialysis capillary that removes substrates used for metabolism. While "fuel"-free fluid flows through the tubing, the rate of sodium extrusion is low. When ATP in a concentration of 5 mM is applied internally, sodium efflux increases. (After Mullins and Brinley, 1967)

of the enzyme; (2) the specificity for activation of the pump and the enzyme in red blood cells is similar to that in nerve, with respect to other ions such as lithium; and (3) ouabain blocks both the transport of ions and the enzyme.

In addition to these similarities, the enzyme in red cells has the same directional sensitivity as the pump, with respect to the concentrations inside and outside the cell. In the red cells both the transport mechanism and its ATPase are activated by a high concentration of sodium inside and of potassium outside, but not by the opposite concentration ratios. To show this, Glynn[10] and his colleagues have made preparations consisting of red cell membranes, called ghosts. These are red cells from which the hemoglobin has been extruded by decreasing the osmotic pressure of the medium. A valuable feature of the technique is that the ghosts can be reconstituted by immersing them in an appropriate solution. This enables one to fill the cell interior with solutions of different sodium and potassium compositions. In the presence of high internal sodium and high external potassium, the ATPase activity of reconstituted ghosts was stimulated; this indicates that the enzyme, like the transport mechanism, can recognize concentration differences between the inside and the outside of the cell.

A further link between pump and ATPase is provided by experiments in which the enzymatic reactions are driven in the opposite direction from usual, resulting in the synthesis of ATP from ADP and phosphate.[11] With an unusually high concentration of sodium outside red blood cell ghosts and a high potassium concentration inside them, the gradients can be increased to such an extent that the ionic movements drive the pump backward, reversing its activity and thus leading to a net synthesis of ATP (Figure 1).

The close correspondences between the coupled sodium-potassium

---

[10]Glynn, I. M. 1968. *Br. Med. Bull.* *24*:165–169.

[11]Garrahan, P. J., and Glynn, I. M. 1967. *J. Physiol.* *192*:237–256.

active transport and the enzyme provide the basis for a new and promising approach, the solubilization and subsequent purification of the ATPase molecule to determine more precisely how it acts to transport ions.[12,13]

Production of current flow by ionic pumps: general considerations

In theory, a scheme can be constructed in which no current need flow for a restoration of the original concentrations inside and outside a neuron. If exactly equal amounts of sodium and potassium were carried out of and into the cell, the net transfer of charge would be zero, and the concentrations would be restored without net charge transfer through the membrane. Turning the pump on or off would, therefore, produce no immediate change in membrane potential. A pump of this type would be electrically neutral. With the pump off, however, the potential would slowly decline, owing to leakage of sodium and potassium, thereby leading to changes in the values of E_K and E_{Na}.

In contrast to the mechanism described above, a pump could give rise to current flow if it transported one ion alone through the cell membrane. If sodium were transported outward without an accompanying anion and without inward transport of potassium, there would be a net outward flow of positive charges—an outward sodium current. Activity of the pump would therefore make the inside of the cell more negative and lead to hyperpolarization. If an ongoing hyperpolarizing pump of this sort were suddenly stopped by an agent that blocked its activity, an immediate depolarization would follow (through removal of hyperpolarizing current). The term ELECTROGENIC has been coined to describe a pump which by its activity transfers charges across the membrane and thereby causes a change in membrane potential.

In practice, the changes in membrane potential produced by an electrogenic pump depend primarily on its rate of activity and the COUPLING RATIO, a term denoting the excess of transport of one ion over another across the cell membrane. Another important factor is the permeability of the cell membrane. Unidirectional transport of ions across the membrane is equivalent to current. The circuit has to be completed by current flowing in the opposite direction through ionic permeability channels. An equivalent circuit is shown in Figure 4. Outward sodium current gives rise to only a small hyperpolarization across a membrane that has a low resistance; for example, in a cell with a high resting chloride conductance, chloride may leak out as fast as sodium is pumped out and no voltage change develops. The same pump current in a cell that is only slightly permeable to chloride can give rise to a large change in voltage.

The transport mechanisms observed in nerves differ from these simplified theoretical schemes by being neither neutral nor completely uncoupled. As already pointed out, approximately three sodium ions are carried out for every two potassium ions that move in, and this results

[12]Glynn, I. M., and Karlish, S. J. D. 1975. *Annu. Rev. Physiol.* 37:13–55.
[13]Dahl, J. L., and Hokin, L. E. 1974. *Annu. Rev. Biochem.* 43:327–356.

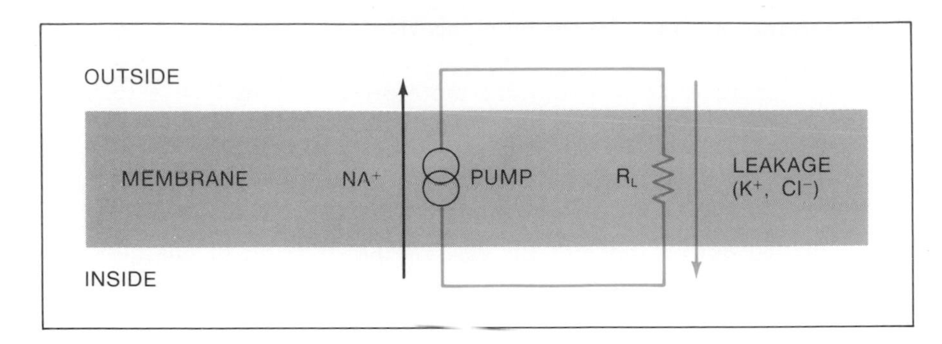

4 **PATHWAYS FOR CURRENT FLOW produced by an electrogenic Na pump. The pump behaves as a constant current source driving positive charges outward. For the circuit to be complete, inward current must flow back into the cell through leakage pathways (R_L) provided by chloride and potassium channels. The voltage generated by the pump current is proportional to R_L. The extent of the hyperpolarization therefore depends on the resistance of a cell as well as on the amount of charges being transported out.**

in a NET outward movement of positive charge. The voltage produced across the cell membrane varies in different cells, depending on the rate of activity and the membrane properties (see below).

Hyperpolarization by electrogenic sodium transport

In squid axons the net charge transfer, by pumping out more sodium than bringing in potassium, has only a small effect on the membrane potential. This is because the fiber is large, the membrane resistance is relatively low, and the internal sodium concentration is changed little in the short run either from leakage or by a few impulses. It has been estimated that the current produced by the pump to maintain steady concentrations in the face of leakage would keep the membrane hyperpolarized by only about 1.8 mV.[4] In other types of cells with different membrane properties, a pump similar to that in the squid axon can contribute to the resting potential and generate large changes in membrane potential.[14]

A clear example is provided by neurons in snail ganglia; their cells become markedly hyperpolarized when the internal sodium concentration is increased as a result either of impulses or of direct injection of sodium through a microelectrode. In detailed studies Thomas has correlated changes in membrane potential with the pump current and the internal sodium concentration.[15,16]

For the demonstration of an electrogenic sodium pump, Thomas used two intracellular electrodes to deposit ions into the cell—one filled with sodium acetate and the other with potassium acetate; a third intracellular electrode recorded the membrane potential. With this arrangement (Figure 5*A*) current flowed out through the sodium electrode and

---

[14]Gorman, A. F. L., and Marmor, M. F. 1970. *J. Physiol.* *210*:897–917.
[15]Thomas, R. C. 1969. *J. Physiol.* *201*:495–514.
[16]Thomas, R. C. 1972. *J. Physiol.* *220*:55–71.

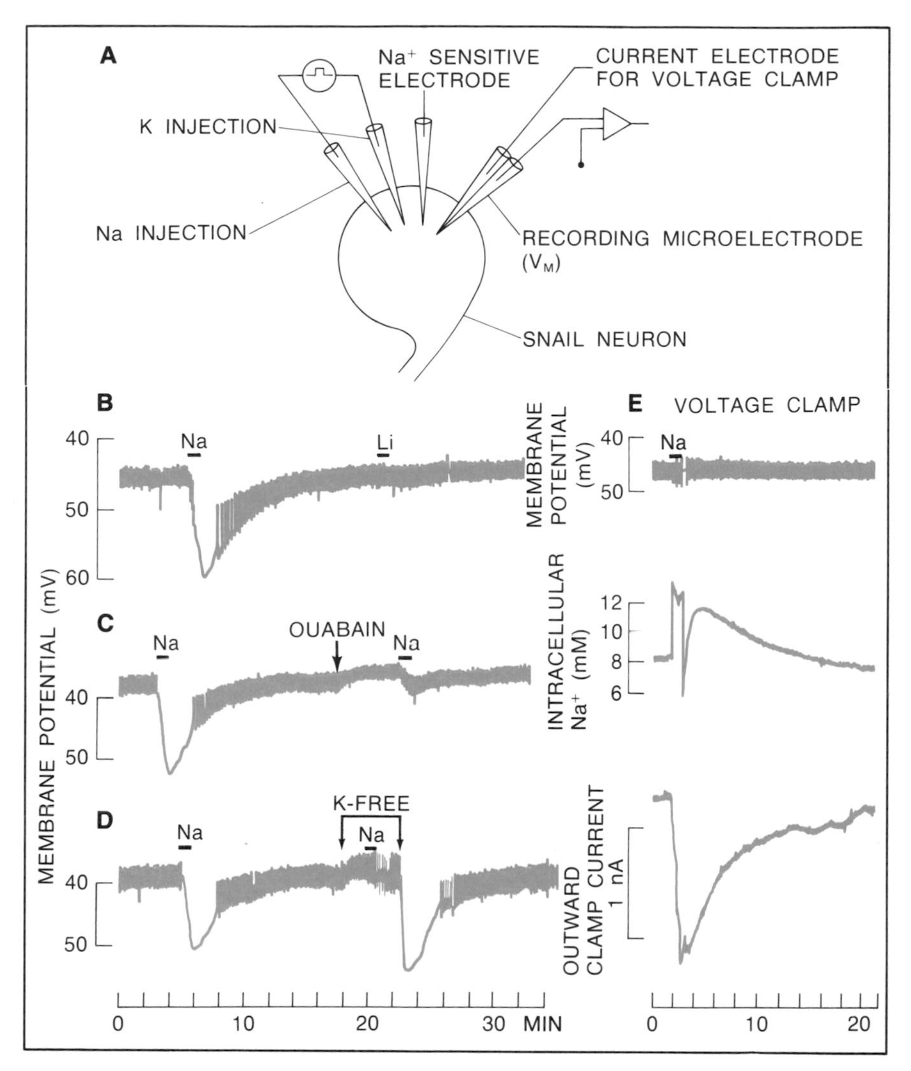

5

EFFECT OF SODIUM INJECTION. Changes in intracellular sodium concentration, membrane potential, and current following injection of sodium into snail neurons. A. Sodium is injected by passing current between two electrodes filled with sodium acetate and potassium acetate (see text). A Na^+ -sensitive electrode measures Na^+_i. Two other electrodes register the membrane potential and voltage clamp the cell. B. Hyperpolarization of membrane following intracellular injection of sodium. This record was made without voltage clamp. (The small rapid deflections on the records are impulses, reduced in size by the pen recorder.) Injection of Li^+ does not produce a hyperpolarization. C. Ouabain, 2×10^{-5} g/ml, which blocks the sodium pump, reduces the hyperpolarization by the sodium injection. D. Removal of potassium from the bathing fluid blocks the sodium pump so that injection of sodium produces no hyperpolarization. The hyperpolarization appears only when the potassium concentration in the bathing fluid returns to normal. E. Correlation or Na^+_i with outward pump current in a voltage-clamped snail neuron. (After Thomas, 1969)

the circuit was completed by negatively charged acetate ions leaving the potassium electrode. Thus, sodium was injected into the cell without current being passed through its membrane. The result is shown in Figure 5*B*. After a brief injection of sodium the cell became hyperpolarized by about 20 mV and gradually recovered over the next few minutes. The injection of lithium (Figure 5*B*), acetate, or potassium did not produce a hyperpolarization.

Several lines of evidence show that the potential change after sodium injection was due to the action of a pump and not to permeability changes of the membrane. For example, the electrical resistance of the membrane did not decrease as would be expected if the hyperpolarization were the result of an increase in the permeability to potassium or chloride. The hyperpolarization could be greatly reduced or prevented, however, by bathing the ganglion in ouabain, which acts specifically on the sodium pump (Figure 5*C*). Similarly, removal of potassium from the bathing fluid markedly reduced the effect of sodium injection, as would be expected for a partially coupled pump that requires the presence of potassium outside the cell (Figure 5*D*). Only when potassium was reintroduced into the bathing fluid did the hyperpolarization appear.

In snail neurons, which are very large, it was possible to voltage-clamp the potential and so measure directly the ionic current generated by sodium transport across the membrane. A fourth microelectrode (Figure 5*A*) (connected to a feedback amplifier of the type described in Chapter 6) was used to hold the membrane potential of the cell constant by passing an inward current that was equal and opposite to the outward pump current after sodium had been injected. At the same time, the sodium concentration inside the cell was continuously monitored by a fifth electrode made of special sodium sensitive glass. The potential developed at the tip of a sodium electrode acts as an index of sodium concentration in the fluid with which it is in contact.

The results in voltage-clamped neurons showed that sodium injection gave rise to an outward surge of current whose amplitude and duration mirrored the internal sodium concentration. From the current the number of sodium ions carried out of the cell electrogenically could be calculated and compared with the total number of sodium ions injected from the pipette. The net charge carried out of the cell was only about one third of that injected in the form of sodium ions. Therefore, this evidence suggests that two out of every three sodium ions are transported out in conjunction with potassium transport inward. In this respect, the pump in snail neurons appears to resemble the coupled sodium-potassium pump in squid axons and red blood cells. In snail neurons, however, the current density per square centimeter of membrane and the membrane resistance are high enough for the potential to change when the pump is active.

Significance of electrogenic pumps for signaling

Long-lasting hyperpolarizing potentials that can be attributed to the activity of sodium pumps have now been found in a number of vertebrate and invertebrate nerve cells. These include unmyelinated axons in the vagus nerve of the rabbit, neurons in the cat spinal cord, sympathetic ganglion cells, crustacean stretch receptors, and neurons in the central nervous systems of the leech and a variety of mollusks.[17] In such cells the hyperpolarization following trains of impulses can be as large as 35 mV and persist for many minutes. For technical reasons, the ion specificity of these transport systems has not been studied in the same detail as in squid or snail neurons. It is possible that different variants of coupled sodium-potassium pumps exist in different cells.

Changes produced by electrogenic pumps can have important consequences. Most significant is that hyperpolarization moves the membrane potential away from threshold and thereby causes a decrease in the sensitivity of neurons to excitatory synaptic stimulation or, in the case of sensory nerve endings, to external stimuli.[18] In leech neurons even impulse conduction can become blocked at branch points (Chapter 17). In the context of signaling that is emphasized in these discussions, the most significant general consideration is the demonstration that the previous activity of a neuron influences its subsequent performance over relatively long periods.

How many pump sites are there per unit area of membrane?

Chapter 6 demonstrates that it is possible to estimate the number of sodium permeability channels by electrical techniques and by the binding of tetrodotoxin. A question of obvious interest is the number of sites used for pumping sodium out of the cell. The results of such an analysis reinforce the concept of an inhomogeneous membrane with a mosaic arrangement of different elementary properties.

The principle of the experiment is to bathe a nerve in radioactive ouabain in a concentration sufficient to block the sodium pump, and then to measure how much is bound per unit weight of tissue. From this the numbers of ouabain molecules bound can be measured and pump sites per unit area of membrane estimated. For example, in the vagus nerve of the rabbit in which the electrogenic pumping of sodium has been extensively studied the estimate of the number of pump sites was about 750 sq μm of membrane,[19] or considerably more than the calculated number of sodium channels (Chapter 6).

CALCIUM EXTRUSION FROM SQUID AXONS AT THE EXPENSE OF SODIUM ENTRY

During the action potential the permeability to calcium is increased, and like sodium, calcium moves into the cell, down its concentration and electrical gradients. One can calculate that if calcium were distributed passively, its internal concentration is squid axons would be very high. The actual value of free calcium measured by aequorin is approximately 0.3μM. For this low internal concentration to be maintained, the cell must transport calcium out against a very steep gradient.[20,21]

[17]Thomas, R. C. 1972. *Physiol. Rev.* 52:563–594.
[18]Nakajima, S., and Onodera, K. 1969. *J. Physiol.* 200:161–185.
[19]Landowne, D., and Ritchie, J. M. 1970. *J. Physiol.* 207:529–537.
[20]Baker, P. F., Hodgkin, A. L., and Ridgway, E. B. 1971. *J. Physiol.* 218:709–755.
[21]Baker, P. F. 1972. *Prog. Biophys. Mol. Biol.* 24:177–223.

The mechanism for removing the calcium that leaks into nerve fibers is different in principle from the sodium-potassium transport described above. The experimental evidence suggests that instead of being driven by energy derived directly from the hydrolysis of ATP, the outward movement of calcium is coupled to an inward flux of sodium.

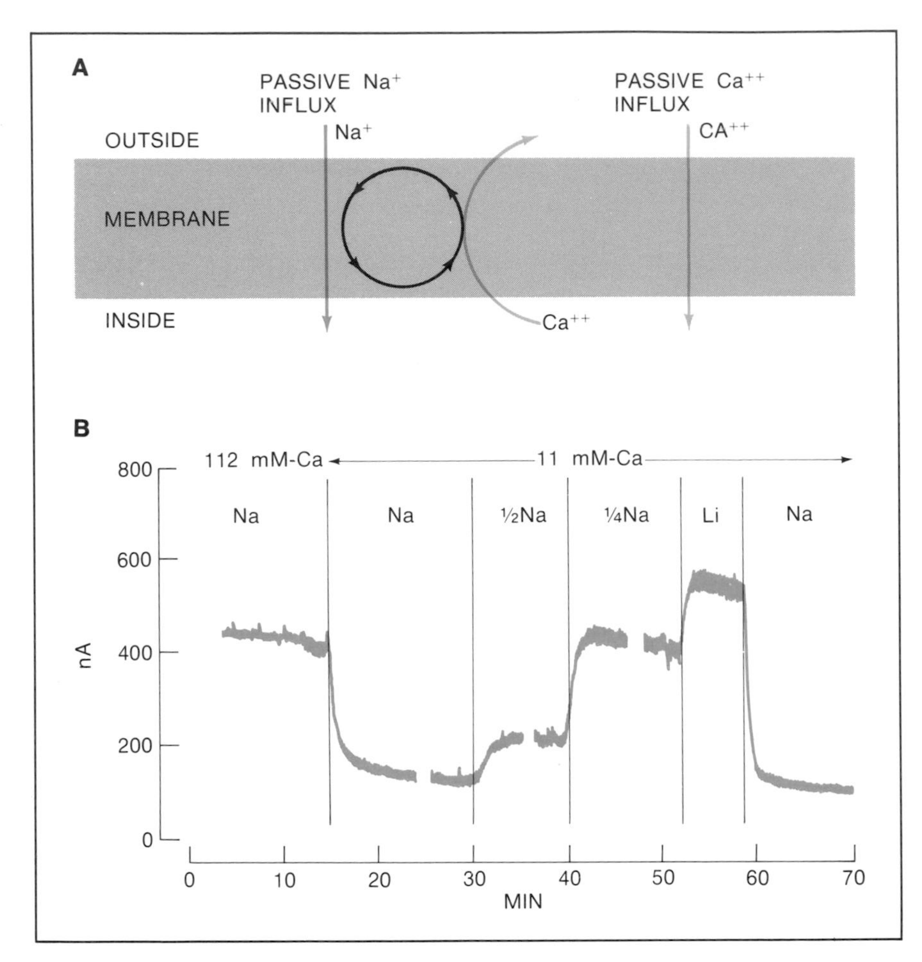

6 **TRANSPORT OF CALCIUM IONS from nerve fibers. A. Scheme for coupling sodium entry to calcium efflux. The influx of sodium, which occurs passively downhill along the electrochemical gradient, is coupled to calcium extrusion against the electrochemical gradient. B. Effect of changes in external sodium and calcium on intracellular calcium activity in a squid axon. To determine changes in calcium activity, the light produced by injected aequorin is measured by a photomultiplier (nA). Increased light indicates an increase in intracellular Ca^{++}. Reducing external calcium from 112 mM to 11 mM reduces light emission. Reducing external sodium to one half or one fourth of the value in normal seawater diminishes calcium efflux and thereby increases intracellular calcium and light emission. Lithium ions cannot substitute for sodium. (From Baker, Hodgkin, and Ridgway, 1971)**

Figure 6 shows the scheme whereby one sodium ion, diffusing downhill along its electrochemical gradient, is used to finance the uphill movement of a calcium ion. The cell gains some sodium in the process, but the amount is small and can be taken care of by the major outward-directed sodium pump described earlier. This transport mechanism for calcium, therefore, depends on metabolic energy only indirectly, insofar as it is a subsidiary part of the regulation of sodium concentration.

The above scheme for a transport mechanism is based on the following observations.[21,22] In the squid axon at rest there is an efflux of calcium which is approximately equal to the rate at which it leaks in. This uphill, outward movement is increased when the level of ionized calcium inside the cell rises. The coupling of sodium entry to calcium efflux has been demonstrated in experiments in which the concentrations were changed; for example, when the sodium concentration in the external fluid is reduced to one half or one fourth of the normal value, the transport of calcium out of the cell is also reduced. As a consequence, the intracellular calcium activity rises. This is illustrated in Figure 6*B*, where the light emitted by aequorin inside the axon (Chapter 6) is used to measure the level of ionized calcium. A decrease in the external sodium concentration causes the intracellular calcium activity to rise. With an unusually high sodium concentration inside the axon and a high calcium concentration outside, the transport of calcium may change direction, so that calcium INFLUX is enhanced. These effects of changed sodium concentration are specific; other ions such as lithium (Figure 6*B*) are not as effective as sodium in promoting the outward or inward movement of calcium. The system behaves as though there is a carrier within the membrane, with an affinity for sodium and calcium, that transports the two ions in opposite directions. At reduced temperatures, outward transport of calcium is depressed. In contrast to the effects on the sodium-potassium pump, ouabain and a low external potassium concentration do not immediately reduce calcium transport.

The actions of metabolic inhibitors, such as cyanide, are at first glance surprising.[20,23] In the presence of cyanide the outward movement of calcium against its electrochemical gradient is dramatically INCREASED. Paradoxically, a process that requires energy seems to function better when the ATP level of a cell has fallen too low to maintain other pumping activities of the cell. The probable explanation is that the depression of metabolism by cyanide allows the level of ionized calcium in the cell to rise. This is because mitochondrial enzymes are inhibited, allowing calcium stored within mitochondria to escape. When calcium is liberated from this store, the internal concentration rises, tending to enhance the efflux (see above).

The system for calcium-sodium exchange can be represented in

---

[22]Baker, P. F., Blaustein, M. P., Hodgkin, A. L., and Steinhardt, R. A. 1969. *J. Physiol.* *200*:431–458.

[23]Blaustein, M. F., and Hodgkin, A. L. 1969. *J. Physiol.* *200*:497–528.

terms of a simple analogy. The main coupled sodium-potassium transport system, which depends on energy production by a cell, can be compared to a mechanical pump that efficiently removes water seeping into a leaky boat. At the same time, part of the inward leakage is utilized for the purpose of driving a second pump that empties a small bucket of water out of the boat (analogous to an outward movement of calcium): this constitutes only a small amount compared to the total water entering the boat and can be accommodated by the main pump.

In red blood cells and the sarcoplasmic reticulum of muscle there is evidence that calcium is extruded by a metabolic pump dependent on ATP. This raises the possibility that similar processes may yet be found in nerve.[21]

Entry of calcium ions plays a number of fundamental roles in various aspects of neural signaling: (1) calcium entry into nerve terminals is a link in the coupling process between changes in membrane potential and the secretion of transmitter; (2) calcium entry is an intermediary step in initiating muscular contraction; and (3) in certain nerve cells, calcium entry dominates the inward current during the action potential, and in motor nerve terminals contributes to it. In addition, calcium movements may be crucial in generating electrical responses of photoreceptors and in processes that involve the activation of cyclic AMP. We have also stressed the general role of calcium in secretory activity.

SUGGESTED READING

Papers marked with an asterisk (*) are reprinted in Cooke, I., and Lipkin, M. 1972. *Cellular Neurophysiology*. Holt, New York.

General reviews

Baker, P. F. 1972. Transport and metabolism of calcium ions in nerve. *Prog. Biophys. Mol. Biol. 24*:177–223.

Dahl, J. L., and Hokin, L. E. 1974. The sodium-potassium adenosinetriphosphatase. *Annu. Rev. Biochem. 43*:327–356. (Deals with recent work on the properties and enzymatic aspects of the sodium-potassium pump in non-neural cells.)

Glynn, I. M. and Karlish, S. J. D. 1975. The sodium pump. *Annu. Rev. Physiol. 37*:13–55. (Discusses recent work on the properties and enzymatic aspects of the sodium-potassium pump in non-neural cells.)

Ritchie, J. M. 1973. Energetic aspects of nerve conduction: The relationships between heat production, electrical activity and metabolism. *Prog. Biophys. Mol. Biol. 26*:149–187.

Thomas, R. C. 1972. Electrogenic sodium pump in nerve and muscle cells. *Physiol. Rev. 52*:563–594.

Original papers

Baker, P. F., Blaustein, M. P., Hodgkin, A. L., and Steinhardt, R. A. 1969. The influence of calcium on sodium efflux in squid axons. *J. Physiol. 200*: 431–458.

Baker, P. F., Blaustein, M. P., Keynes, R. D., Manil, J., Shaw, T. I., and Steinhardt, R. A. 1969. The ouabain-sensitive fluxes of sodium and potassium in squid giant axons. *J. Physiol. 200*:459–496.

*Hodgkin, A. L., and Keynes, R. D. 1955. Active transport of cations in giant axons from *Sepia* and *Loligo*. *J. Physiol. 128*:28–60.

Mullins, L. J., and Brinley, F. J. 1967. Some factors influencing sodium extrusion by internally dialyzed squid axons. *J. Gen. Physiol. 50*:2333–2355.

Rang, H. P., and Ritchie, J. M. 1968. On the electrogenic sodium pump in mammalian non-myelinated nerve fibres and its activation by various external cations. *J. Physiol. 196*:183–221.

Thomas, R. C. 1969. Membrane current and intracellular sodium changes in a snail neurone during extrusion of injected sodium. *J. Physiol. 201*:495–514.

Thomas, R. C. 1972. Intracellular sodium activity and the sodium pump in snail neurons. *J. Physiol. 220*:55–71.

PART THREE

SPECIAL ENVIRONMENT OF NERVE CELLS IN THE BRAIN FOR SIGNALING

CHAPTER THIRTEEN

PHYSIOLOGY OF NEUROGLIAL CELLS

Most nerve cells in the central and the peripheral nervous systems are surrounded by satellite cells. These satellites are divided according to anatomical criteria into (1) neuroglial cells in the brain, which are further subdivided into oligodendrocytes and astrocytes, and (2) Schwann cells in the periphery. Taken together, the neuroglial cells make up almost half the volume of the brain and greatly outnumber neurons. To some glial cells can be assigned a definite functional role; for example, the oligodendrocytes and Schwann cells form myelin around the larger axons and speed up conduction of nerve impulses. Definite roles for the other glial cells remain to be clarified.

The basic membrane properties of glial cells differ in several essential aspects from those of neurons. Glial cells behave passively in response to electric current and, unlike neurons, their membranes do not generate conducted impulses. The glial membrane potential is higher than that of neurons and depends primarily on the distribution of potassium, the principal intracellular cation. Further, glial cells are linked by low-resistance connections which permit the direct passage of ions and perhaps small molecules between the cells. Neurons and glial cells, on the other hand, are separated from each other by narrow, fluid-filled extracellular spaces about 20 nm wide which prevent currents generated by nerve impulses from spreading into neighboring glial cells.

Neurons transmit signals to glial cells by releasing potassium into the intercellular spaces during the conduction of impulses and thereby depolarizing the glial membrane potential. There is no difference in the potassium-mediated signaling that occurs between excitatory or inhibitory neurons and glial cells. Such signaling is nonsynaptic.

Potassium-mediated glial depolarization creates potential changes that can be recorded from the surface of tissue; as a result glial cells contribute to the electroencephalogram and the electroretinogram.

Some of the intriguing questions that await answers include

how glial cells respond metabolically to potassium-mediated depolarization and whether they play a role in supplying materials to neurons.

Nerve cells in the brain are intimately surrounded by satellite cells called neuroglial cells. From counts of cell nuclei it has been estimated that they outnumber neurons by at least 10:1 and make up about one half of the bulk of the nervous system. Studies of glial cells are in a peculiar state. The importance of these cells is stressed frequently; yet, except for their role in speeding conduction, no significant function has been firmly established for them. It is remarkable that one should have to discuss the performance of the brain in terms of neurons only, as if glial cells did not exist. Even if in the end this neglect should turn out to be correct, which we doubt, one ought to be able to justify the reasons for ignoring the most numerous cell types in the nervous system.

Glial cells were first described in 1846 by Rudolf Virchow, who later gave them their name. He clearly recognized that they differed fundamentally from neurons and from interstitial tissue elsewhere in the body. Several excerpts from a paper by Virchow give the flavor of his approach and thinking.[1] He pointed out many aspects of glial tissue that later became important for formulating various hypotheses.

> Hitherto, considering the nervous system, I have only spoken of the really nervous parts of it. But . . . it is important to have a knowledge of that substance also which lies *between the proper nervous parts,* holds them together and gives the whole its form. (our italics)

Speaking of the ependyma (see below), he continued:

> This peculiarity of the membrane, namely, that it becomes continuous with interstitial matter, the real cement, which binds the nervous elements together, and that in all its properties it constitutes a tissue different from the other forms of connective tissue, has induced me to give it a new name, that of *neuro-glia.* (nerve glue; our italics)

Later on he stated:

> Now it is certainly of considerable importance to know that in all nervous parts, in addition to the real nervous elements, a second tissue exists, which is allied to the large group of formations, which pervade the whole body, and with which we have become acquainted under the name of connective tissues. In considering the pathological or physiological conditions of the brain or spinal marrow, the first point is always to determine how far the tissue which is affected, attacked or irritated, is nervous in its nature, or merely an interstitial substance. . . . Experience shows us that this very interstitial tissue of the brain and spinal marrow is one of the most frequent seats of morbid change, as for example, of

---

[1]Virchow, R. 1859. *Cellularpathologie.* Trans. F. Chance. Hirschwald, Berlin. Excerpts are from pp. 310, 315, and 317.

> fatty degeneration. . . . Within the neuroglia run the vessels, which are therefore nearly everywhere *separated from the nervous substance* by a slender intervening layer, and are not in immediate contact with it. (our italics)

In the subsequent 100 years, neuroglial cells were intensively studied chiefly by neuroanatomists and also by pathologists, who knew them to be the most common source of tumors in the brain. This is perhaps not surprising, because normally, unlike neurons, glial cells can still divide in the mature animal.

This chapter considers glial cells from a cellular standpoint. The morphology and the various functions attributed to glial cells are reviewed, but the main emphasis is on their physiological properties, about which a good deal is known. Without background knowledge about their membrane potentials and how they are influenced by neuronal signaling, it appears difficult to deal with the wider issues that inevitably arise.

The experiments on glial cells illustrate once again the basic unity of the principles by which the nervous system works in higher and lower animals. Thus, for technical reasons the membrane properties of glial cells were most easily determined in especially large cells of the leech brain. Once this had provided a clue to what to look for, it became simpler to investigate amphibian and then mammalian glial cells, which share many key properties with those in the more modest nervous system of the leech. Far from being a roundabout approach, this turned out to be a shortcut.

Appearance and classification of glial cells

One of the most distinct structural features of neuroglial cells as compared with neurons is the absence of axons, but many other differences have been demonstrated by light and electron microscopy. A representative picture of mammalian neuroglial cells is shown in Figure 1. The cytoplasmic contents suggest that glial cells are metabolically active structures containing the usual organelles, including mitochondria, endoplasmic reticulum, ribosomes, lysosomes, and often deposits of glycogen and fat.

Glial cells are classified on histological grounds, and in the vertebrate central nervous system they are usually subdivided into two main groups (astrocytes and oligodendrocytes) and several subgroups.

ASTROCYTES can be classified into two subgroups: (1) fibrous astrocytes, which contain filaments and are more prevalent among bundles of myelinated nerve fibers, the white matter of the brain; and (2) protoplasmic astrocytes, which contain less fibrous material and are more abundant in the gray matter around nerve cell bodies, dendrites, and synapses. Both types of astrocytes make contacts with capillaries and neurons (see below).

Recently, a clear distinction has been made between fibrous and protoplasmic astrocytes by an immunological method. Fibrous astrocytes contain a protein against which specific antibodies have been prepared. When this antibody is labeled with a fluorescent marker, the fibrous astrocytes, to which it binds selectively, can be clearly distin-

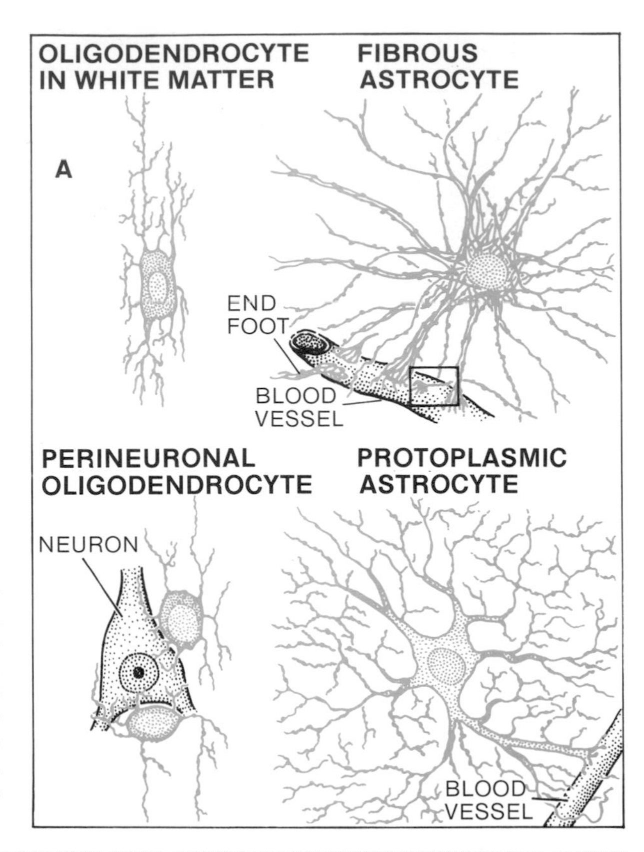

1

NEUROGLIAL CELLS in the mammalian brain. A. Neuroglial cells stained with silver impregnation. Oligodendrocytes and astrocytes represent the principal neuroglial cell groups in the vertebrate brain. They are closely associated with neurons and form end-feet on blood vessels. B. Electron micrograph of glial cells in the optic nerve of the rat. In the lower portion is the lumen of a capillary (CAP) lined with endothelial cells (E). The capillary is surrounded by end-feet formed by processes of fibrous astrocytes (As). Between the end-feet and the endothelial cells is a space filled with collagen fibers (Col). In the upper portion is part of a nucleus of an oligodendrocyte (Ol) and to the right are axons surrounded by myelin wrapping (see Figure 3). (A after Penfield, 1932; B from Peters, Palay, and Webster, 1976)

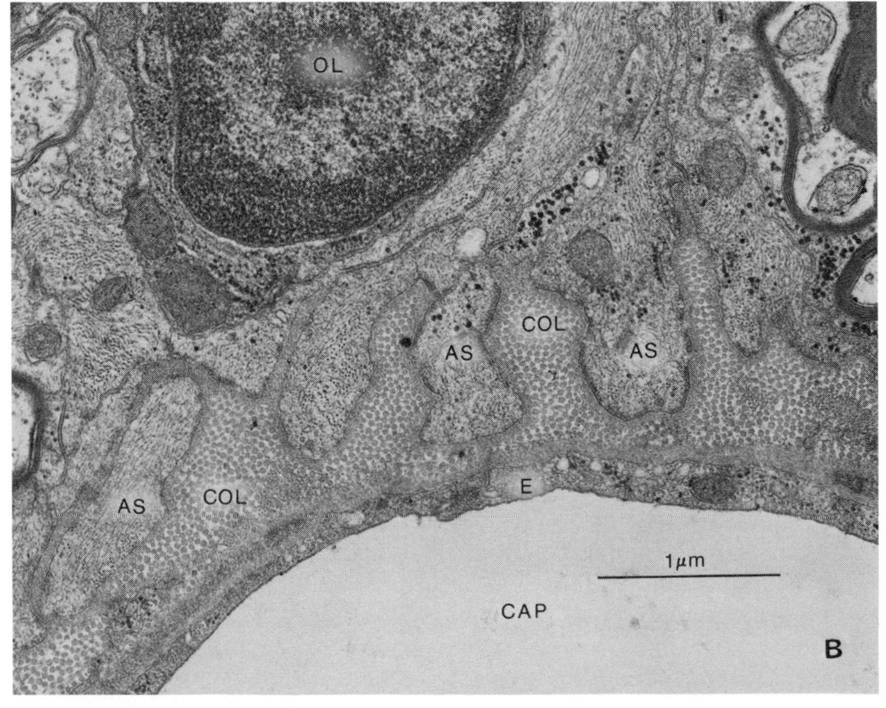

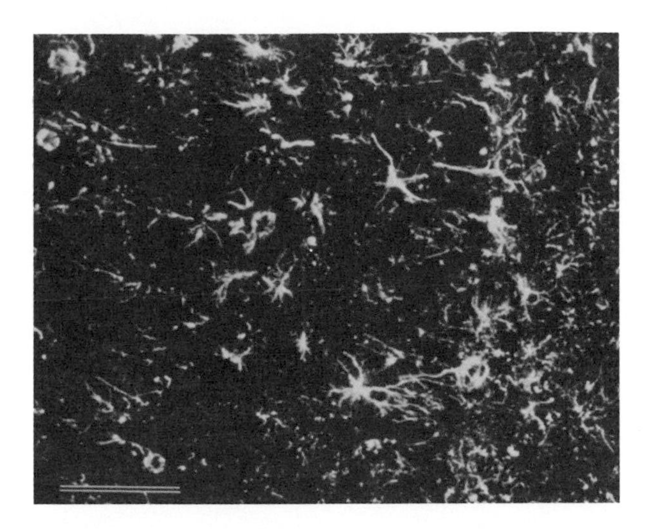

2
FIBROUS ASTROCYTES specifically labeled in the brain of a rat. These cells contain a protein against which an antibody is formed. The antibody is then made to fluoresce. Scale, 0.1 mm. (After Bignami and Dahl, 1974)

guished in micrographs. The protein is present in fibrous astrocytes in all vertebrates that have been examined.[2] An example is shown in Figure 2. The role of this fibrillary protein in the function of the astrocytes is not yet known.

OLIGODENDROCYTES are predominant in the white matter, where they form myelin around the larger axons. This is a wrapping of glial cell processes with practically all the cytoplasm squeezed out in between, so that the membranes are tightly apposed as they spiral around the axon (Figure 3*A*). The large numbers of smaller diameter axons (1μm or less) that are unmyelinated are also surrounded by glial cells, singly or in bundles.[3]

In peripheral nerves the SCHWANN CELLS are analogous to oligodendrocytes in that they form MYELIN around the larger fast-conducting axons (up to 20μm in diameter). Smaller nonmedullated axons (usually below 1μm in diameter), as in the brain, have a Schwann cell envelope without myelin.

The glial cells of the brain and of peripheral nerves have different embryological origins: glial cells in the central nervous system are derived from precursor cells that line the inner surface of the brain; Schwann cells arise from the neural crest. An interesting immunological distinction between oligodendrocytes and myelin-forming Schwann cells in the periphery has also been seen. Injection into animals of extracts of the brain produces antibodies which lead to demyelination within the central nervous system.[4] In contrast, antibodies made to peripheral nerves selectively destroy the myelin-forming Schwann cells.

---

[2]Bignami, A., and Dahl, D. 1974. *J. Comp. Neurol.* *153*:27–38.
[3]Bunge, R. P. 1968. *Physiol. Rev.* *48*:197–251.
[4]Brostoff, S. W., Sacks, H., and DiPaola, C. 1975. *J. Neurochem.* *24*:289–294.

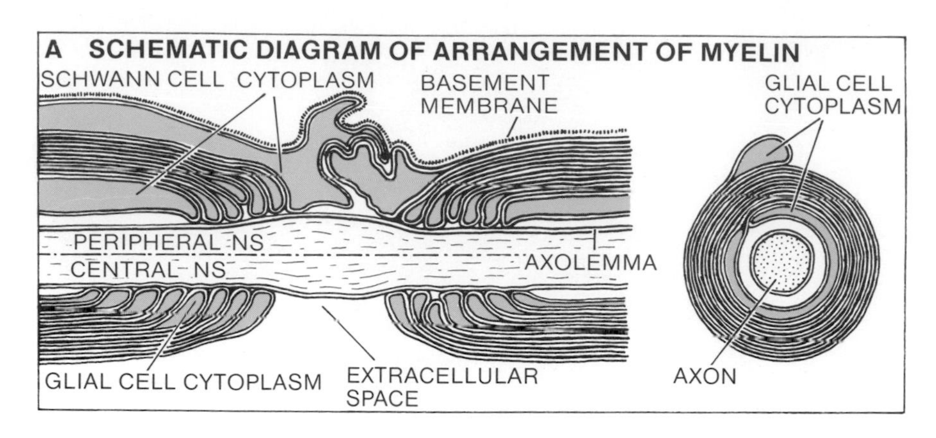

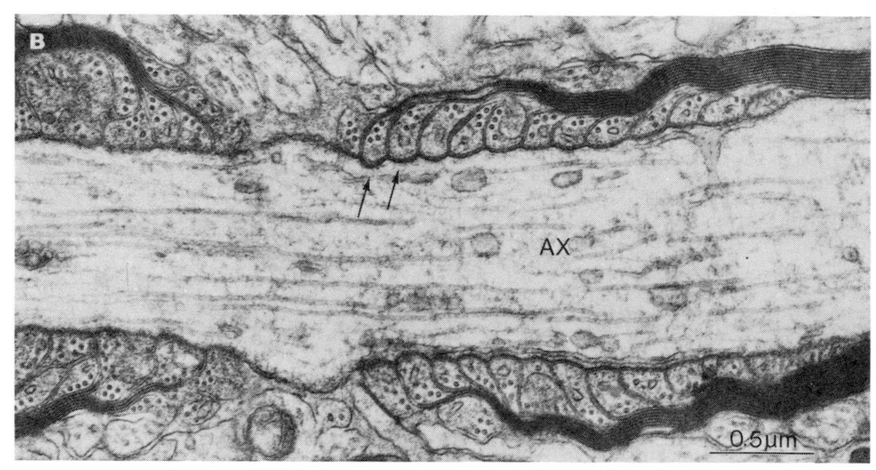

3 **MYELIN AND NODES OF RANVIER. Oligodendroglial cells and Schwann cells form wrappings of myelin around segments of axons. A. At periodic intervals, at the nodes of Ranvier, the myelin covering is interrupted and the axon is exposed. The upper half of the nodal region, with a loose covering of processes, is typical of the arrangement in peripheral nerves. To the right is a transverse section through a myelin-covered axon portion. B. An electron micrograph of a nodal region in the optic nerve of the rat. Compare with the lower portion of the drawing in A. At the edge of the node, where the Schwann cell lamellae terminate, there occurs a specialized close contact area between the membrane of the axon (Ax) and the membrane of the myelin wrapping (arrows). (A after Bunge, 1968; B from Peters, Palay, and Webster, 1976)**

EPENDYMAL CELLS that line the inner surfaces of the brain, in the ventricles, are also usually classed as glial cells. No physiological role has been assigned to them.

In invertebrates, the classification of satellite cells into distinct groups is not well established. However, there is no question about the functional analogy of the various glial structures.

Structural relation between neurons and glia

A glance at almost any electron micrograph of brain tissue brings home the difficulty of making physiological and chemical studies of

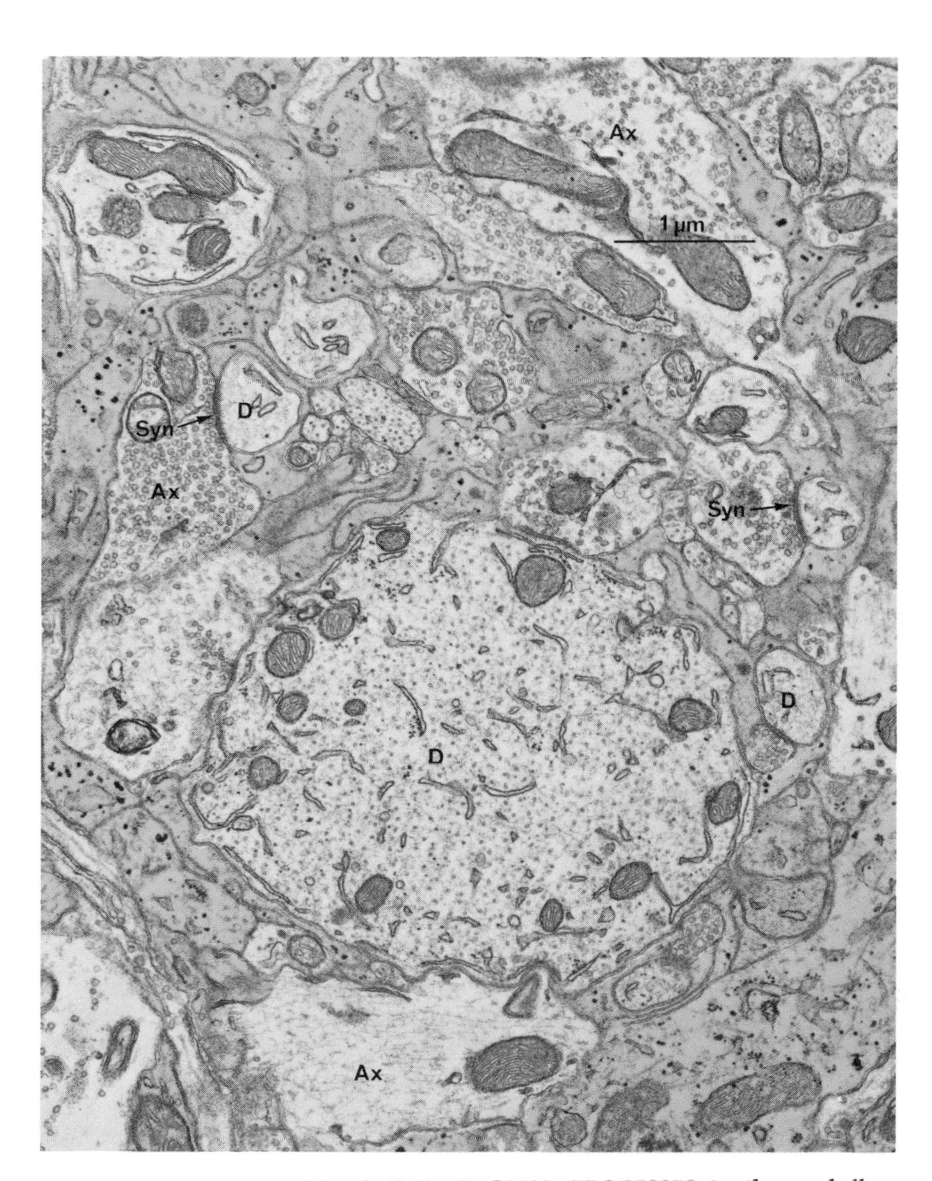

4

NEURONS AND GLIAL PROCESSES in the cerebellum of the rat. The glial contribution is lightly colored. The neurons and glial cells are always separated by clefts of about 20 nanometers in width. The neural elements are dendrites (D) and axons (Ax). Two synapses (Syn) are marked by arrows. (After Peters, Palay, and Webster, 1976)

neuroglial cells. Figure 4 shows an example from the cerebellum of a rat. The section is filled with neurons and glial cells, which can be distinguished from each other only after considerable experience. To make the task simpler, the glial contribution is highlighted. The extracellular space is restricted to narrow clefts, about 20 nm wide, that separate all cell boundaries. Astrocytic processes generally surround neurons except where synaptic contacts are made. Many of the axons

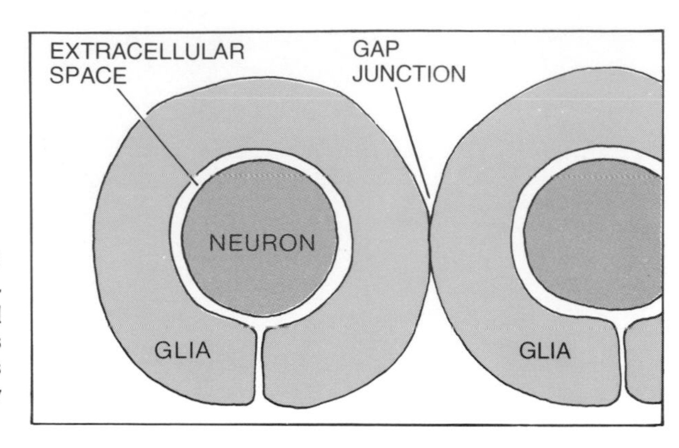

5

NEURONS AND GLIA. Diagramatic presentation of the relationship between neurons and glia and between glial cells. While neurons are always separated from glia by a continuous cleft, the interiors of glial cells are linked by gap junctions.

are typically grouped together and instead of each having an individual covering, entire bundles of axons are surrounded by a common glial envelope. This arrangement is usual in the central nervous system.

From a comparison of neurons and glial cells, one sees that in some regions the cross-sectional area is about equally divided between neurons and astrocytes, while in others, as in Figure 4, the glial contribution is smaller. Glial processes tend to be thin, at times less than 1μm thick. Only around the glial nuclei are there larger volumes of nonlamellated glial cytoplasm.

Electron microscopy has clarified the neuron-glia relation by demonstrating the intimate apposition of their membranes. Special connections, such as gap junctions, are not seen between the two cell types. The clefts (several tens of nanometers wide) seen in Figure 4 always intervene between the surface membranes of neurons and glial cells. This visible separation agrees with results of physiological tests that fail to reveal direct low-resistance pathways between the neurons and glial cells. Such pathways do, however, link glial cells. The special connections are made by gap junctions (Chapter 8). The relation of glial cells, neurons, and extracellular space is shown diagrammatically in Figure 5. An interesting exception, mentioned earlier (Figure 3), is the close apposition of myelin to axons at the edge of the node where specialized contacts are seen between the two structures in the freeze-fracture preparation.[5]

Basic to the difficulty of studying glial function is the lack of a method for separating neurons from glia. The tissues are so interwoven (Figure 4) that it is difficult to make separations into pure glial and neuronal fractions. Some promising methods, however, are becoming available that enable distinction of various glial and neuronal components, such as nuclei or membrane fractions and their chemical constituents. One example is the protein associated with fibrous astrocytes.

Myelin and the role of neuroglial cells in axonal conduction

One well-established role for oligodendrocytes and Schwann cells is the production of a myelin sheath around axons—a high-resistance

[5]Livingston, R. B., Pfenninger, K., Moor, H., and Akert, K. 1973. *Brain Res.* *58*:1–24.

covering akin to insulating material around wires. The myelin is interrupted at the nodes of Ranvier (Figure 3), which occur at regular intervals of about 1 mm in most nerve fibers. Since the ionic currents associated with the conducted nerve impulse cannot flow across the myelin, the ions move in and out at the nodes between the insulation (Chapter 7). This leads to an increased conduction velocity and seems an ingenious solution for acquiring speed. The alternative is to make nerve fibers larger, but this is less effective. For example, in the squid the largest axons are between 0.5 and 1.0 mm in diameter and conduct at rates no faster than about 20 m/sec; a myelinated axon 20μm in diameter conducts at 120 m/sec.

The association of satellite cells with axons to form myelin raises a number of interesting problems. For example, what are the genetic factors that enable glial cells to select the appropriate axons to surround at the right time and to maintain myelin sheaths around them? What are the characteristics of some of the neurological disorders caused by genetic abnormalities? Sidman and his colleagues have studied the latter question in strains of mutant mice, called "jimpy," in which myelin fails to develop normally.[6]

At the fine structural level (Figure 3), the elaborate membrane specializations at the nodes of Ranvier can be cited. There the satellite cell is closely apposed to axons and limits longitudinal spread of current.[5]

The fact that myelin is not a static substance was shown in studies that followed the uptake and movement of radioactive cholesterol into Schwann cells and then along the lamellae of myelin.[7] The idea of a dynamic role of the satellites is reinforced by observations in diseases such as allergic encephalomyelitis. Following demyelination, which need not be accompanied by axonal disintegration, conduction is greatly slowed.[8] With recovery, the normal conduction velocity is restored.

In the past dozen years glial cells have been burdened by almost every nervous system task for which no other obvious explanation has been found. For example, they have been implicated in the processes of learning and memory and in specifically regulating the chemical environment of nerve cells. The principal difficulty until relatively recently has been the lack of physiological and chemical methods adequate for testing the various proposals. Understandably, the majority of the long-established hypotheses relied primarily on histological observations, but these can hardly on their own reveal the physiological interaction between glial cells and neurons that involve biochemical changes or signaling. In the following paragraphs we summarize and comment briefly on possible functions of glial cells.

[6]Sidman, R. L., Dickie, M. M., and Appel, S. H. 1964. *Science 144*:309–311.
[7]Rawlins, F. 1973. *J. Cell Biol. 58*:42–53.
[8]Rasminsky, M., and Sears, T. A. 1972. *J. Physiol. 227*:323–350.

HYPOTHESES FOR FUNCTIONAL ROLES OF NEUROGLIAL CELLS

At the beginning of this century several hypotheses assigning various roles for glial cells were formulated, and some of them have gained wide acceptance. The proposals are still current in most treatises on the subject.

Structural support

Virchow, around 1850, recognized that neuroglial cells are interspersed among neurons and therefore form part of the "structure" of the brain. In one sense, physical support is provided, but neuroglial cells may not necessarily "bind" the brain structure together.

Isolation and insulation of neurons

Ramón y Cajal very early proposed that glial cells prevent "cross talk" by current spread during conduction of nerve impulses. This plausible suggestion has now been tested in several instances, but it lacks force. The fluid-filled clefts between cells seem adequate to prevent effective current flow between cells. However, a role as a spatial barrier for the spread of various substances, such as potassium and transmitters, has found some recent experimental support (see below).

Uptake of chemical transmitters by glia

There is now evidence from autoradiographic methods that glial cells can take up γ-aminobutyric acid (GABA) in mammalian dorsal root ganglia, in the spinal cord, in autonomic ganglia, and at crustacean

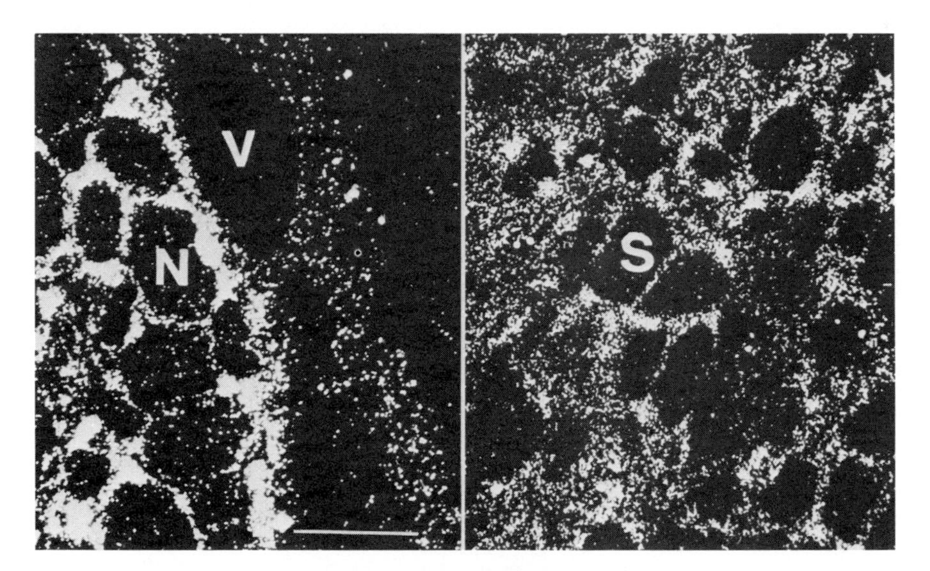

6 **GLIAL CELLS take up labeled gamma aminobutyric acid (GABA). Silver grains appear white in this autoradiograph observed with dark field illumination. Left: Schwann cells surrounding neurons (N) in a dorsal root ganglion of a rat have preferentially accumulated GABA labeled with ^{3}H. Only sparse labeling occurs around the vessels (V). Right: Similar situation in a sympathetic ganglion in which the neurons (S, dark roundish structures) are practically devoid of radioactivity but the satellite cells are heavily labeled. White bar is 25 μm long. (Courtesy of F. Schon and J. S. Kelly)**

neuromuscular junctions.[9,10] An example is shown in Figure 6. The preparation was bathed in radioactive GABA, which then is visualized in the form of silver grains over the cytoplasm of the satellite cells. Intriguing also is the suggestion that GABA may be accumulated preferentially in the brain by glial cells in areas in which the neurons are rich in GABA and presumably release it.[11] This raises the possibility that transmitters may be taken up and stored in glial cells, to be handed back again as the need arises.

Secretory function

Such a role was clearly suggested by Nageotte[12] in 1910. He wrote that "la nevroglie est une glande interstitielle annexée au système nerveux." One example has now been well documented for the secretion of neurotransmitter by Schwann cells. In chronically denervated skeletal muscles Schwann cells come to occupy the site of the degenerated nerve terminals. At rest they secrete quanta of acetylcholine (ACh) that produce miniature synaptic potentials in the muscle. Release of acetylcholine can also be provoked by passing large currents through Schwann cells.[13] Unfortunately, it is not known whether Schwann cells can normally secrete transmitter or acquire the ability only after the nerve has degenerated (see below and Chapter 18).

A second example of transmitter release from glial cells is provided by the experiments of Brown, Iversen, Kelly, and their colleagues.[14–17] They have shown that radioactive GABA is released from satellite cells in sympathetic ganglia and dorsal root ganglia when the potassium concentration of the bathing fluid is increased. Glial cells are depolarized by potassium concentration changes of the same order as those following nerve impulses (see later). The functional role of GABA release by glial cells is not yet clear, but is likely to be significant in view of the physiological occurrence of potassium buildup in the central nervous system (see below).

Role in repair and regeneration

Unlike neurons, glial cells retain the ability to divide throughout the life of an animal. When neurons disappear during aging or due to injury, glial cells divide and occupy the vacant spaces. In addition, they take part in the formation of scar tissue and have phagocytotic properties. At synapses in autonomic ganglia and in the spinal cord, glial cells undergo characteristic changes after section of the postsynaptic axons.[18,19] In electron micrographs they can be seen to invade the

---

[9]Orkand, P. M., and Kravitz, E. A. 1971. *J. Cell Biol. 49*:75–89.
[10]Schon, F., and Kelly, J. S. 1974. *Brain Res. 66*:275–288.
[11]Ljungdahl, A., and Hokfelt, T. 1973. *Brain Res. 62*:587–595.
[12]Nageotte, J. 1910. *C. R. Soc. Biol. (Paris) 68*:1068–1069.
[13]Dennis, M. J., and Miledi, R. 1974. *J. Physiol. 237*:431–452.
[14]Bowery, N. G., and Brown, D. A. 1972. *Nature New Biol. 238*:89–91.
[15]Young, J. A. C., Brown, D. A., Kelly, J. S., and Schon, F. 1973. *Brain Res. 63*:479–486.
[16]Minchin, M. C. W., and Iversen, L. L. 1974. *J. Neurochem. 23*:533–540.
[17]Iversen, L. L., and Kelly, J. S. 1975. *Biochem. Pharmacol. 24*:933–938.
[18]Hunt, C. C., and Nelson, P. 1965. *J. Physiol. 177*:1–20.
[19]Purves, D. 1975. *J. Physiol. 252*:429–463.

synaptic cleft; at the same time electrical recordings indicate that transmission is impaired. These results demonstrate that glial cells can be influenced by changes in the properties of the neurons they surround. In the course of regeneration, peripheral axons can grow back to their destinations along a route marked by residual Schwann cells. This can be interpreted as a sign of chemical affinity that guides the regenerating axon to its proper site. In this context it is interesting that glial tumor cells have been shown to contain nerve growth factor—a protein that causes growth and sprouting in sympathetic and dorsal root ganglion cells[20] (Chapter 18). Whether they can secrete it is not yet known.

Neuroglial cells and the development of the nervous system

A role for glial cells in the growth of neurons and the formation of connections has frequently been suggested. In a comprehensive series of experiments, Rakic and Sidman[21,22] have studied the development of the cerebral cortex and the cerebellum in monkeys and in man. The formation of the various cell types and their migration to their final destinations have been followed by light and electron microscopy. Their evidence shows that the neurons move along the glial processes during development. The close association of the two cell types suggests that glial cells provide an initial framework around which subsequent neuronal organization takes place. A generalization along such lines, however, is not at present warranted.

The role of glial cells in synapse formation is even less certain. In the embryonic spinal cord of the monkey formation of chemical synapses is in progress before glial cells appear in the area of synapses at motoneurons.[23] It is also known that functioning synapses can be formed in tissue culture without satellite cells.[24]

Nutritive role and transfer of substances to neurons

This has been the most popular hypothesis up to the present. It was proposed by Golgi in about 1883. He wrote:[25]

> I find it convenient to mention that I have used the term connective tissue with regard to neuroglia. I would say that "neuroglia" is a better term, serving to indicate a tissue which, although connective because it connects different elements and for its own part *serves to distribute nutrient substances,* is nevertheless different from ordinary connective tissue by virtue of its morphological and chemical characteristics and its different embryological origin. (our italics)

Coupled with Golgi's histological staining methods, these ideas seemed so reasonable and had such force that they were hardly questioned through the years.

While the proposal for glial transfer of substances is attractive, it still lacks direct experimental support. Figure 7 shows graphically the

---

[20]Longo, A. M., and Penhoet, E. E. 1974. *Proc. Natl. Acad. Sci. U.S.A. 71:* 2347–2349.

[21]Rakic, P. 1971. *J. Comp. Neurol. 141*:283–312.

[22]Sidman, R. L., and Rakič, P. 1973. *Brain Res. 62*:1–35.

[23]Bodian, D. 1966. *Bull. Johns Hopkins Hosp. 119*:129–149.

[24]Fischbach, G. D., and Dichter, M. A. 1974. *Dev. Biol. 37*:100–116.

[25]Golgi, C. 1903. *Opera Omnia,* Vols. I, II. U. Hoepli, Milan.

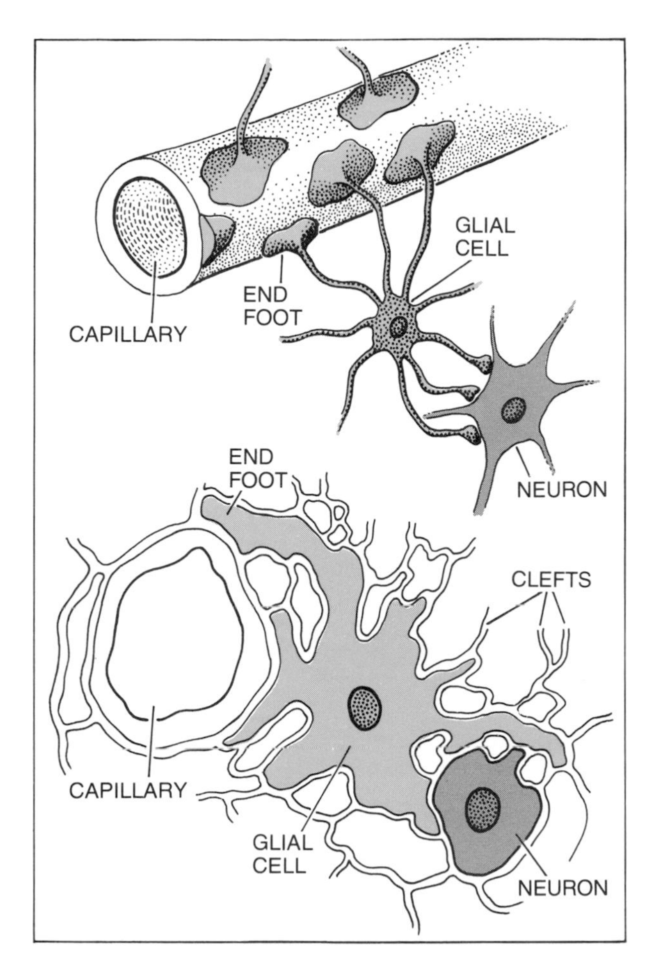

7
RELATIONS OF CAPILLARY, GLIA, AND NEURONS as seen in the light and electron microscopes. The most direct pathway from the capillary to the neuron is through the aqueous intercellular clefts which are open for diffusion. Cell dimensions are not in proportion. (After Kuffler and Nicholls, 1966)

considerations that relate to the Golgi hypothesis, which implies the following events. A molecule that leaves a capillary is either preferentially taken up by a glial cell or diffuses into the cell's interior. Once inside the glial cell, the molecule must be conveyed across the glial cytoplasm by some directional internal transport mechanism to the region where it meets the neuron; the molecule must then leave the glial cell for uptake by the neuron. The nutritive hypothesis also assumes implicitly that no adequate direct pathways for free diffusion of substances exist which would provide easier and preferential access to the neuron. Rather than crossing glial cell boundaries which intervene between capillary and neuron, perhaps a molecule could diffuse straight to the target. In fact, the glial membrane has a relatively high electrical resistance and the cell has its own particular internal environment (see below). Materials cannot therefore diffuse readily into the interior of glial cells. Further, free channels that are open for diffusion exist all around the neurons and glial cells, so that the most direct access to nerve cells is through the aqueous cleft system (Chapter 14).

One recent experimental test for transfer of materials from glia to neurons was to label precursors of protein whose fate could then be followed.[26] In such experiments the time sequence of events must be accurately resolved. For example, if a label appears "simultaneously" in nerve and glia, one assumes that it was made or taken up independently by the two tissues. If, however, a distinct time sequence occurs, such as the label's turning up first in glia and then in neurons, it may suggest that a protein was handed on from the first to the second cell type. The conclusions to be drawn from such experiments do not yet seem compelling.

The hypothesis that neurons need glial cells in their vicinity to synthesize transmitters is supported by experiments in cultured dissociated nerve cells from sympathetic and dorsal root ganglia.[27] In the absence of satellite cells, neurons reversibly lose their ability to synthesize acetylcholine. While these results suggest once more a role for glia in the transfer of materials, the effect is not specific. Other cells, such as fibroblasts, produce similar results on transmitter levels.

The conclusions about a specific transfer of various molecules between neurons and glial cells, therefore, remain tentative and are supported only by indirect evidence. The difficulties appear to be largely methodological, and for the time being such attractive propositions still await direct experimental support.

PHYSIOLOGICAL PROPERTIES OF NEUROGLIAL CELL MEMBRANES

The study of the physiological properties of neuroglia has been greatly aided by two preparations in which the glial cells are large and accessible and in which they can be conveniently investigated in their normal relation to nerve cells. One is the CENTRAL NERVOUS SYSTEM OF THE LEECH, the other is the OPTIC NERVE OF THE MUD PUPPY, *Necturus*. In both preparations glial cells can be impaled with microelectrodes and their membrane properties studied. Such studies have allowed physiological criteria to be defined by which glial cells can be recognized within the mammalian brain.

Simple preparations used for intracellular recording from glia

The leech nervous system, described more fully in Chapter 17, consists of a chain of ganglia joined by connectives and lies within a vascular sinus; it is less than 1 mm in diameter, contains no blood vessels, and is quite transparent. Ganglia can readily be removed from the animal and survive well in Ringer's solution. A photograph of a leech ganglion is shown in Figure 8. The relative sizes of neurons and glia in this nervous system are the reverse of what is seen in vertebrates, for in the leech the glial cells are larger than the nerve cell

[26]Lasek, R., Gainer, H., and Przybylski, R. J. 1974. *Proc. Natl. Acad. Sci. U.S.A.* *71*:1188–1192.

[27]Patterson, P. H., and Chun, L. L. Y. 1974. *Proc. Natl. Acad. Sci. U.S.A*: *71*:3607–3610.

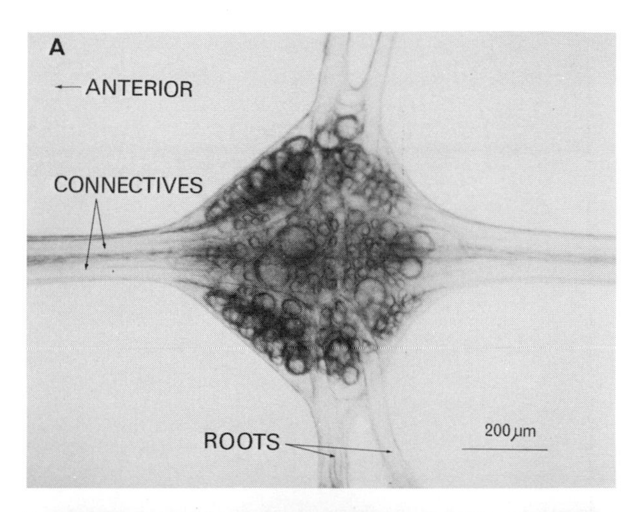

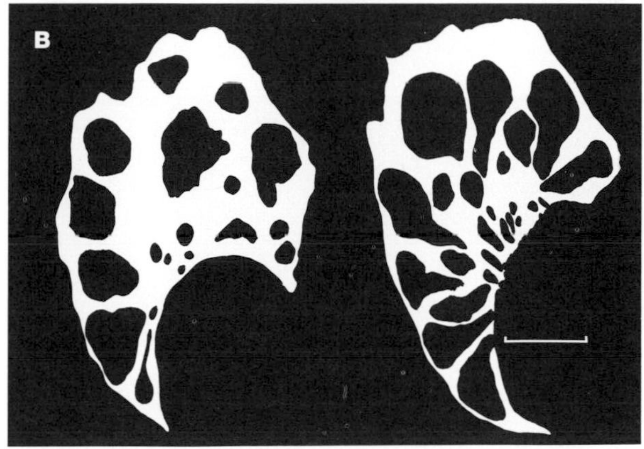

8
LEECH GANGLION. A. Living isolated ganglion in the central nervous system of the leech under transmitted illumination. The shapes and sizes of nerve cells can be seen; the glial cytoplasm between the neurons is not visible because it is transparent. Connectives and roots run to neighboring ganglia and body wall respectively. B. The cytoplasm of one glial cell is traced in two histological sections and appears white. It fills all the spaces not occupied by neurons (the enclosed black areas). Scale, 50 μm. (A from Nicholls and Baylor, 1968; B from Kuffler and Potter, 1964)

bodies and are few in number.[28] Ten large glial cells are present in each ganglion and its connectives. Six of these are associated with the nerve cell bodies in each ganglion, two with the synaptic regions, and two with several thousand axons running in the connectives. The glial cells in the ganglion are transparent and therefore appear under the dissecting microscope as spaces between nerve cells. They can be impaled consistently by a microelectrode inserted next to a neuron.

Another simple central nervous system preparation that has been studied extensively is the optic nerve of *Necturus*.[29] It is about 0.15 mm in diameter and is covered by a layer of connective tissue containing blood vessels that run parallel to the surface; no blood vessels occur within the tissue. The glial cells are large and intimately surround the nerve fibers. As in the leech, the neurons and glia receive their nutrients principally from the surface. Figure 9*A* presents a cross section of the *Necturus* optic nerve stained with toluidine blue. The glial nuclei

---

[28]Coggeshall, R. E., and Fawcett, D. W. 1964. *J. Neurophysiol.* 27:229–289.

[29]Kuffler, S. W., Nicholls, J. G., and Orkand, R. K. 1966. *J. Neurophysiol.* 29:768–787.

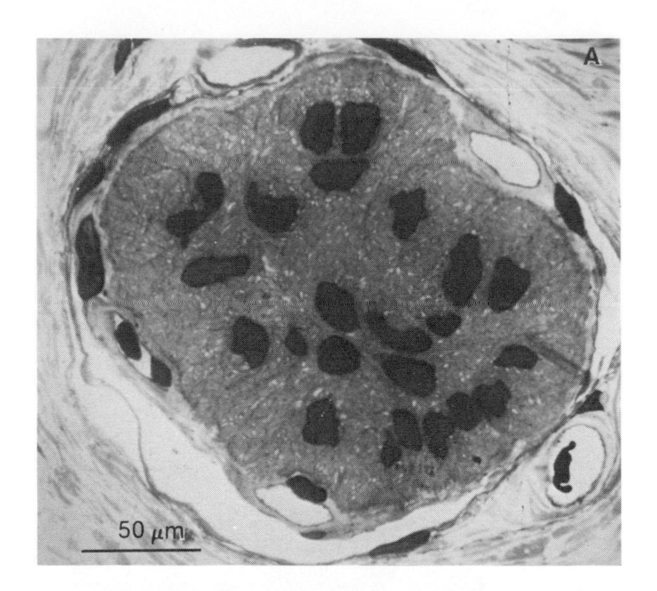

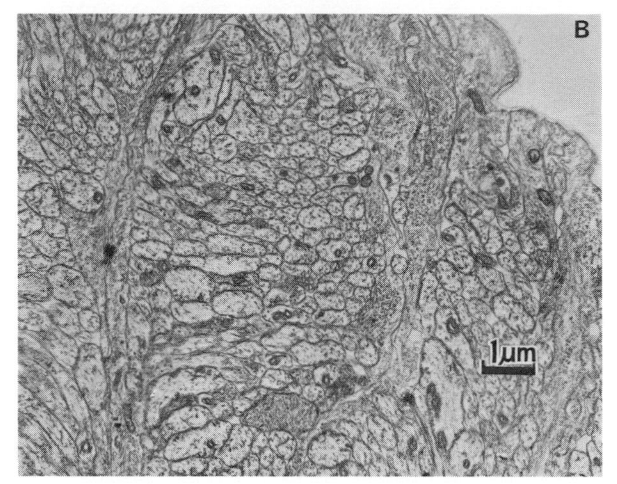

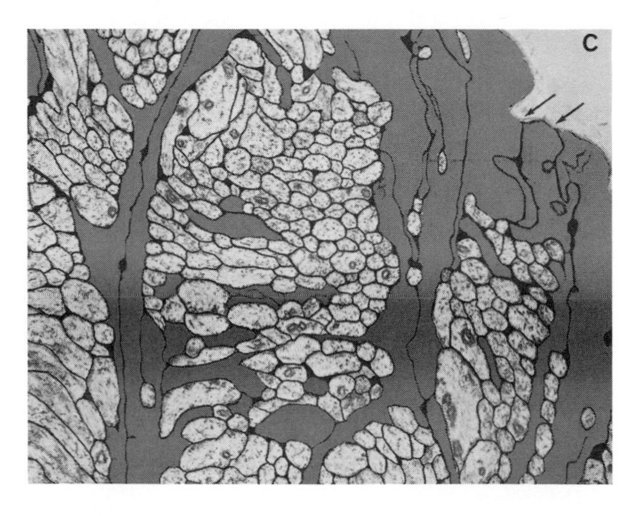

9

OPTIC NERVE of the mud puppy (Necturus) is shown in cross-section (A). Nuclei of glial cells are stained black. The outlines of the glial cytoplasm and bundles of nonmedullated axons can not be resolved. The nerve is surrounded by connective tissue containing capillaries. B and C: Two identical electron micrographs of part of the optic nerve. In one the glial processes are lightly shaded and the clefts that separate the cells are traced in black. Two cleft openings reaching the surface are marked by arrows. Axons run in closely packed bundles, as in the mammalian system (Figure 4). (From Kuffler, Nicholls, and Orkand, 1966)

are dark and prominent, while the glial cytoplasm remains unstained. The outlines of the bundles of nonmedullated axons (fiber diameters 0.1 to 1.0μm) are too small to be seen in the light micrograph. The blood vessels run close to the surface of the nerve within the loose connective tissue. The neuron-glia relation is illustrated in Figure 9*B* and *C*; spaces about 20 nm wide exist between the cells, as in the mammalian (Figure 4) and the leech nervous systems. Also visible is part of a cluster of tightly packed axons where glia surround the bundle, but not the individual axons.

The volume occupied by glial cells in both the central nervous system of the leech and the optic nerve of *Necturus* constitutes 35 to 55 percent of the cross-sectional area. For emphasis the glial cells with their processes between the nervous elements are shaded in Figure 9*C*, while the tortuous intercellular clefts are traced in black. Note that in two places (arrows) the clefts open to the outside. The fine structure of the leech nervous system appears strikingly similar in many respects, with narrow spaces intervening between neuronal and glial membranes (Chapter 17).

Glial membrane potentials

Recordings of the membrane potentials in the central nervous system of the leech show directly that glial cells have higher resting potentials than the neurons they surround; about −75 mV compared with −50 mV.[30] In vertebrates, including the frog, mudpuppy, cat, and rat, the highest membrane potentials recorded for neurons are about 70 to 75 mV, while the values for glial cells consistently approach 90 mV. Their membrane resistance (R_m, Chapter 7) is about 1000$\Omega\cdot$sq cm or more, which is comparable to the value measured in neurons.

Dependence of membrane potential on potassium

To study the origin of the high resting potentials, the membrane potentials of glial cells have been measured in different potassium concentrations (Figure 10). In isolated optic nerves of *Necturus* the potassium concentrations in the bathing fluid can be varied over a wide range. The results are unexpected, because the glial membrane behaves like a perfect potassium electrode and accurately follows the Nernst equation (Chapter 5):

$$E = \frac{RT}{F} \log_e \frac{K_o}{K_i} = 58 \log_{10} \frac{K_o}{K_i}$$

Changes in sodium and chloride concentrations do not produce significant changes in potential. One can conclude that ions other than potassium make a negligible contribution to the membrane potential, which is determined by the ratio K_o/K_i. Figure 10*C* shows a series of membrane potential measurements plotted against K_o on a logarithmic scale. The solid line is the theoretical slope of 59 mV predicted by the Nernst equation (at 24° C), and it agrees excellently with the experimental points. A particular feature of the relation is the good fit at low concentrations of K_o, down to 1.5 mM. In this respect glial cells differ

[30]Kuffler, S. W., and Potter, D. D. 1964. *J. Neurophysiol.* 27:290–320.

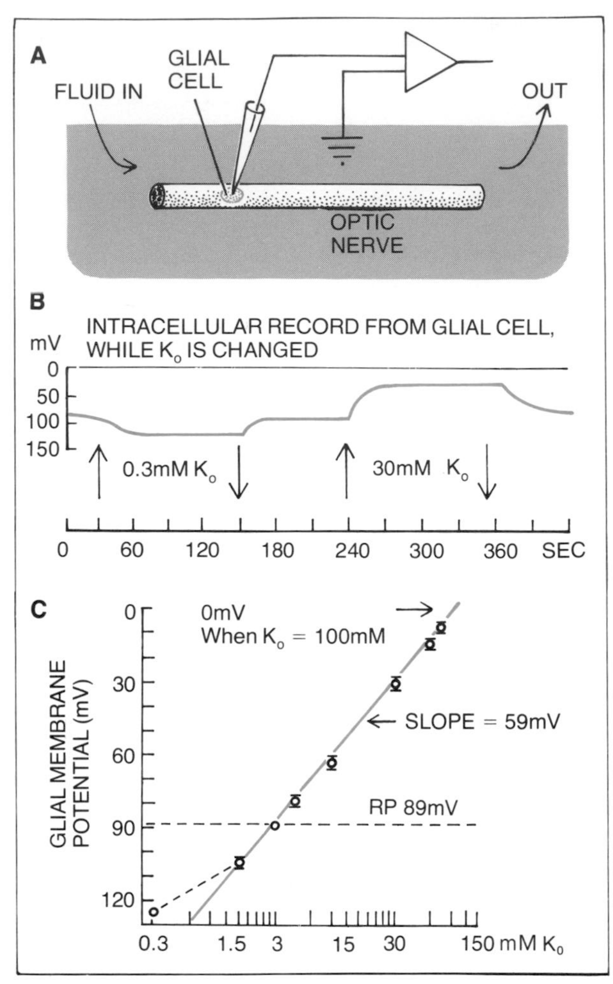

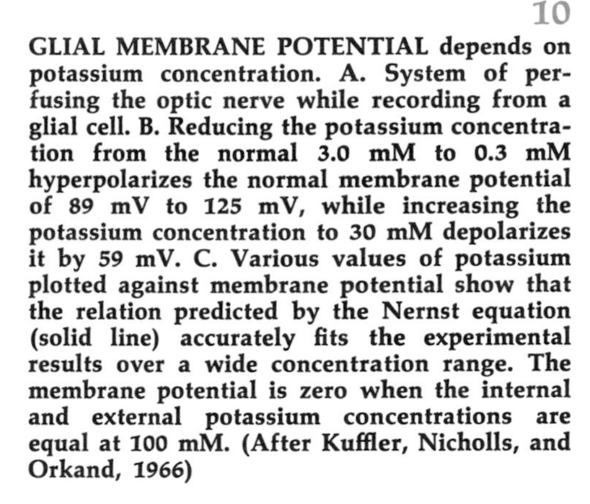

10

GLIAL MEMBRANE POTENTIAL depends on potassium concentration. A. System of perfusing the optic nerve while recording from a glial cell. B. Reducing the potassium concentration from the normal 3.0 mM to 0.3 mM hyperpolarizes the normal membrane potential of 89 mV to 125 mV, while increasing the potassium concentration to 30 mM depolarizes it by 59 mV. C. Various values of potassium plotted against membrane potential show that the relation predicted by the Nernst equation (solid line) accurately fits the experimental results over a wide concentration range. The membrane potential is zero when the internal and external potassium concentrations are equal at 100 mM. (After Kuffler, Nicholls, and Orkand, 1966)

significantly from most neurons, which deviate from the Nernst prediction in the physiological range of 2 to 4 mM K_o (Chapter 5).[29]

The experiments illustrated in Figure 10 provide a good estimate of the internal concentration of potassium (K_i). The Nernst equation indicates that when K_o is the same as K_i, the membrane potential is zero. This occurred when the outside potassium concentration was increased to 100 mM. Similar determinations of K_i (110 mM) were made in glial cells of the leech, but there the conclusions were further strengthened by direct flame photometric measurements.[31]

Glial cells therefore contain a high potassium concentration and have a negligible ionic permeability for ions other than potassium. On their own, these findings do not exclude the presence of other cations, such as sodium, in the glial cytoplasm. The concentration could be appreciable in cells containing part of the sodium in a bound form.

---

[31]Nicholls, J. G., and Kuffler, S. W. 1965. *J. Neurophysiol.* 28:519–525.

Absence of impulses in glial cells

A salient feature of glial cells is their widespread distribution throughout the nervous system and the absence of an axonlike process. This indicated to many of the earlier workers that glial cells were unlikely to conduct impulses in the manner of axons. However, the possibility remained that they might give some active electrical responses, possibly slow ones. It should be added that no synapses, formed by axons, have as yet been seen on glial cells in electron microscopic studies.

In the leech, frog, mudpuppy, and mammals, it has been shown that identified glial cells do not produce impulses. The results in the leech and in *Necturus* are particularly clear-cut, because the glial membrane behaves passively even when its potential is displaced over a range of about 200 mV. The cell membrane resistance remains constant throughout, like an ohmic resistance.[29,30] "Active" regenerative responses are therefore excluded. As expected, such properties make glial cells fundamentally different from the excitable neurons.

Schwann cells that surround squid axons have been impaled with electrodes, and their membranes also behave passively; that is, they are inexcitable. Their measured resting potentials (about 40 mV) are, however, lower than those found in glial cells. The low membrane potentials may be only apparent, the result of leakage current during electrode penetration of the cells which form a very thin layer, mostly several microns thick.[32]

Physiological identification of glial cells in the mammalian brain

When intracellular recordings are made from the mammalian central nervous system, certain cells appear to have properties quite different from those of conventional neurons. Their resting potentials are high, often near 90 mV, without the fluctuations typical of continued background synaptic bombardment, and they appear to be inexcitable by electrical currents.[33,34] These characteristics are basically the same as those of glial cells in the leech and *Necturus*. A further unambiguous physiological criterion for identifying glial cells is that they become depolarized through potassium accumulation when neurons in their vicinity discharge impulses (see below).

Such "silent" cells have been injected with various dyes, in particular with the fluorescent dye procion yellow, and have been shown to be glia by subsequent histological examination.[35]

Special connections between glial cells

In the vertebrate brain only relatively few neurons are linked by the low-resistance gap junctions that enable current to pass readily from one nerve cell to its neighbor (Chapter 8). Adjacent glial cells, however, including those of mammals, are linked to each other by gap junctions.[36] In this respect they resemble epithelial and gland cells and heart muscle fibers. In the leech, frog, and *Necturus* glial cells are al-

[32]Villegas, J. 1972. *J. Physiol.* *225*:275–296.

[33]Karahashi, Y., and Goldring, S. 1966. *Electroencephalogr. Clin. Neurophysiol.* *20*:600–607.

[34]Dennis, M. J., and Gerschenfeld, H. M. 1969. *J. Physiol.* *203*:211–222.

[35]Kelly, J. P., and Van Essen, D. C. 1974. *J. Physiol.* *238*:515–547.

[36]Brightman, M. W., and Reese, T. S. 1969. *J. Cell Biol.* *40*:668–677.

ways found to be coupled electrically. For technical reasons such tests have not yet been done in mammals.

The functional significance of the electrical coupling of glial cells is not known. Clearly, ions can exchange directly between cells without passing through extracellular space, and such interconnections may be useful for equalizing concentration gradients that may arise. The possibility is open that there exists some metabolic interaction between coupled cells, linked with demand that is induced by activity. It is shown later that the low-resistance coupling between the cells is essential for generation of current flow by glial cells that can be recorded from the surface of nervous tissues by extracellular electrodes.

Absence of electrical connections between neurons and glia

No special structural links between nerve and glia have been detected. This is of physiological interest because it is natural to wonder about the manner in which neurons and glial cells may interact. Direct tests have been made in the leech nervous system, where the potentials of neurons can be changed in a controlled manner by passing currents through them while recording from adjacent glial cells.[30] The reverse procedure has also been done—recording from nerve cells while changing the membrane potential of glial cells. Analogous tests in the optic nerve of *Necturus* show similar results: current flow around glial cells created by synchronized conducting nerve impulses has no effect on the neighboring glial membrane (above 1 mV), and therefore an electrical interaction between nerve and glia seems quite unlikely.

The simplest explanation of the failure of current to flow from one cell to another is their separation by intercellular channels (several tens of nanometers wide) filled with low-resistance fluid (about $100\Omega \cdot cm$) (Figures 4 and 9). These channels provide pathways for currents to flow out to the bathing fluid without passing through the relatively high-resistance cell membranes of neurons or glia.

A SIGNALING SYSTEM FROM NEURONS TO GLIAL CELLS

Most speculation about the role of glial cells entails some kind of interaction with neurons. Such a mutual influence might be expected between two cell types that are so intimately apposed to each other. The effect of nerve activity on glial cells can be most simply illustrated by experiments made in the brain of *Necturus;* similar results have been obtained in the leech and in mammals.

Glial depolarization by neuronal activity

The basic observation is illustrated in Figure 11. Recordings are made from a glial cell in the optic nerve of the mudpuppy, and a volley is set up in the nerve fibers so that impulses travel past the impaled glial cell. Each volley of nerve impulses is followed by a depolarization of the glial cell, rising to a peak in about 150 msec and declining slowly over several seconds. The size of the potential is graded, depending on the number of nerve fibers activated. With repeated stimulation the potentials in the glial cells sum, depending on the frequency of the stimulation (Figure 11*B*, *C*). If stimulation is maintained, surprisingly large glial membrane depolarizations of up to 48 mV can be seen. At

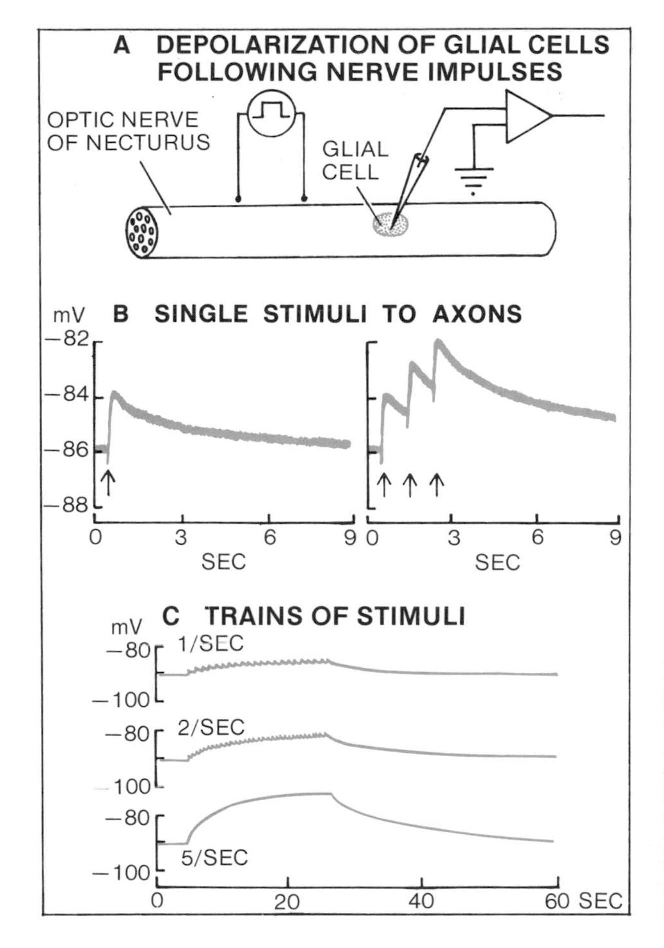

11

EFFECT OF NEURAL ACTIVITY on glial cells in the optic nerve of the mudpuppy. Synchronous impulses in nerve fibers cause glial cells to become depolarized. Each volley of impulses leads to a depolarization that takes seconds to decline. The amplitude of the potentials depends on the number of axons activated and the frequency of stimulation, as shown in B and C. (After Orkand, Nicholls, and Kuffler, 1966)

the end of a train of stimuli, a residue of such large potentials may persist for 30 sec or longer.[37]

To test whether potential changes recorded in glial cells are a physiological occurrence, experiments were done with natural stimulation of optic nerve fibers in anesthetized mudpuppies with intact circulation. A single brief flash of light caused a glial depolarization of about 4 mV (Figure 12). Repeated flashes added distinct but smaller potentials. The glial potential declined progressively during maintained illumination, but reappeared when the light was turned off. These results are in good agreement with the conclusion that discharges in the optic nerve are responsible for the glial potentials, since the discharge rate declines during maintained illumination but a renewed burst follows when the light is turned off.

In the cortex, glial cells become depolarized when neurons in their vicinity are activated by the stimulation of neural tracts, peripheral nerves, or the surface of the cortex. The magnitude of the potentials recorded from glia once more depends on the strength of stimulation as

[37]Kuffler, S. W., and Nicholls, J. G. 1966. *Ergeb. Physiol.* 57:1–90.

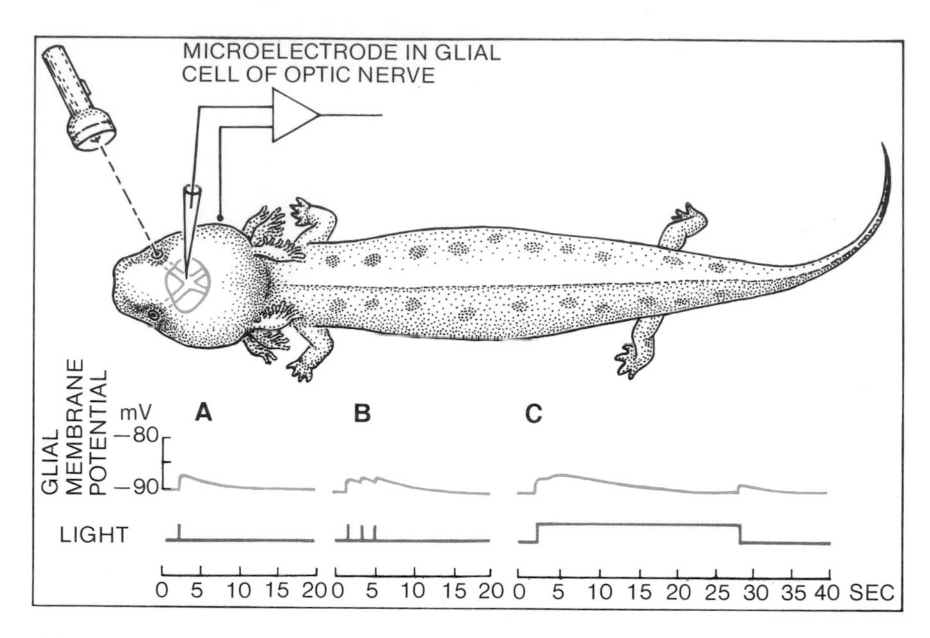

12 **EFFECT OF ILLUMINATION of the eye on the membrane potential of glial cells in the optic nerve of an anesthetized mudpuppy. With intact circulation. A. Single flash of light 0.1 sec long. B. Three flashes. C. Light stimulus maintained for 27 sec. During such prolonged illumination the initial glial depolarization declines as the nerve discharge adapts. At the end of illumination, a burst of "off" discharges initiates a renewed glial depolarization. Lower beams monitor light. (After Orkand, Nicholls, and Kuffler, 1966)**

additional axons conducting into the region are activated. An example from experiments by Ransom and Goldring[38] is shown in Figure 13. The slow time course of the glial potentials and their magnitude resemble those seen in *Necturus*. Similar results have now been obtained from mammalian glial cells by a number of investigators.[33,35,39]

These results suggest that glial cells within the visual cortex should become depolarized only when certain specific patterns of light are shone into the eye to activate groups of neighboring neurons. Kelly and Van Essen[35] found such a situation in the visual cortex of the cat. Glial cells identified by physiological and morphological criteria became significantly depolarized only when a bar of light with an appropriate orientation was shone into the eye (Figure 14). Illumination of both eyes was effective if corresponding areas were illuminated, but diffuse lights or inappropriate orientations produced no appreciable change in potential. These results are in good agreement with the assumption that glial cells that respond to selected stimuli only are situated in a column of neurons whose receptive fields are all oriented at one particular angle to the vertical. Still unknown is whether glial cells are specifically

[38]Ransom, B. R., and Goldring, S. 1973. *J. Neurophysiol.* *36*:869–878.
[39]Grossman, R. G., and Hampton, T. 1968. *Brain Res.* *11*:316–324.

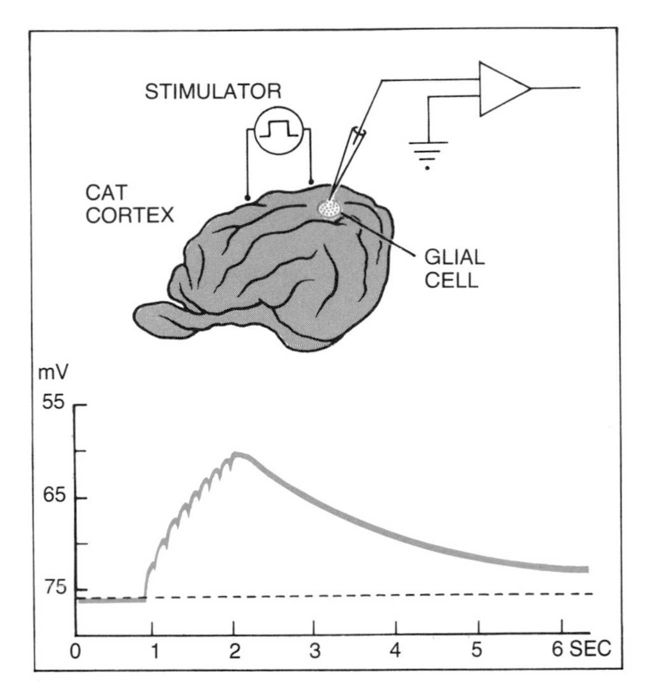

13
GLIAL DEPOLARIZATION resulting from stimulation of cortical neurons (at 8/sec). The amplitude of the glial depolarization is graded according to the strength and frequency of stimulation. (After Ransom and Goldring, 1973)

arranged in relation to neuronal columns (Chapter 3).

Potassium release as mediator of the effect of nerve signals on glial cells

The first possibility to be considered as the mechanism for the action of nerve on glial cells is the flow of current during nerve impulses. This is excluded because the time courses of the events are greatly mismatched. The peak current flow caused by impulses, such as a synchronous volley in the optic nerve, has already declined when the glial depolarization starts to rise. Further, passing current directly into nerve cells in the leech does not produce detectable responses in neighboring glial cells.

A more likely hypothesis is suggested by experiments on squid axons by Frankenhaeuser and Hodgkin,[40] who showed that following nerve impulses potassium accumulates in the spaces between axons and the surrounding Schwann cells. To check the hypothesis that glial cells are depolarized by potassium leakage from axons, use was made of the observation that the glial membrane potential provides a good quantitative assay for potassium in the environment of the cells (Figure 10). If potassium leaves axons and accumulates in the intercellular clefts, it changes the K_o/K_i ratio and alters the membrane potential of glial cells in a predictable way. The glial cells should depolarize as if potassium liberated by the axons into the clefts adds logarithmically to the K_o already present in the bathing fluid.

Figure 15 shows results that agree quantitatively with this prediction. The membrane potential was measured at three different external concentrations of potassium: normal Ringer's solution (K_o = 3 mM), 1.5 times normal (K_o = 4.5 mM), and half normal (K_o =

[40]Frankenhaeuser, B., and Hodgkin, A. L. 1956. *J. Physiol.* *131*:341–376.

1.5 mM). These are the three solid circles of Figure 15*C* that lie on the curve (solid line) predicted by the Nernst equation. If at the normal membrane potential (89 mV), in 3 mM potassium solution, a brief train of nerve stimuli was given, the membrane potential was reduced by 12.1 mV (plotted on the ordinate, middle open circle). This de-

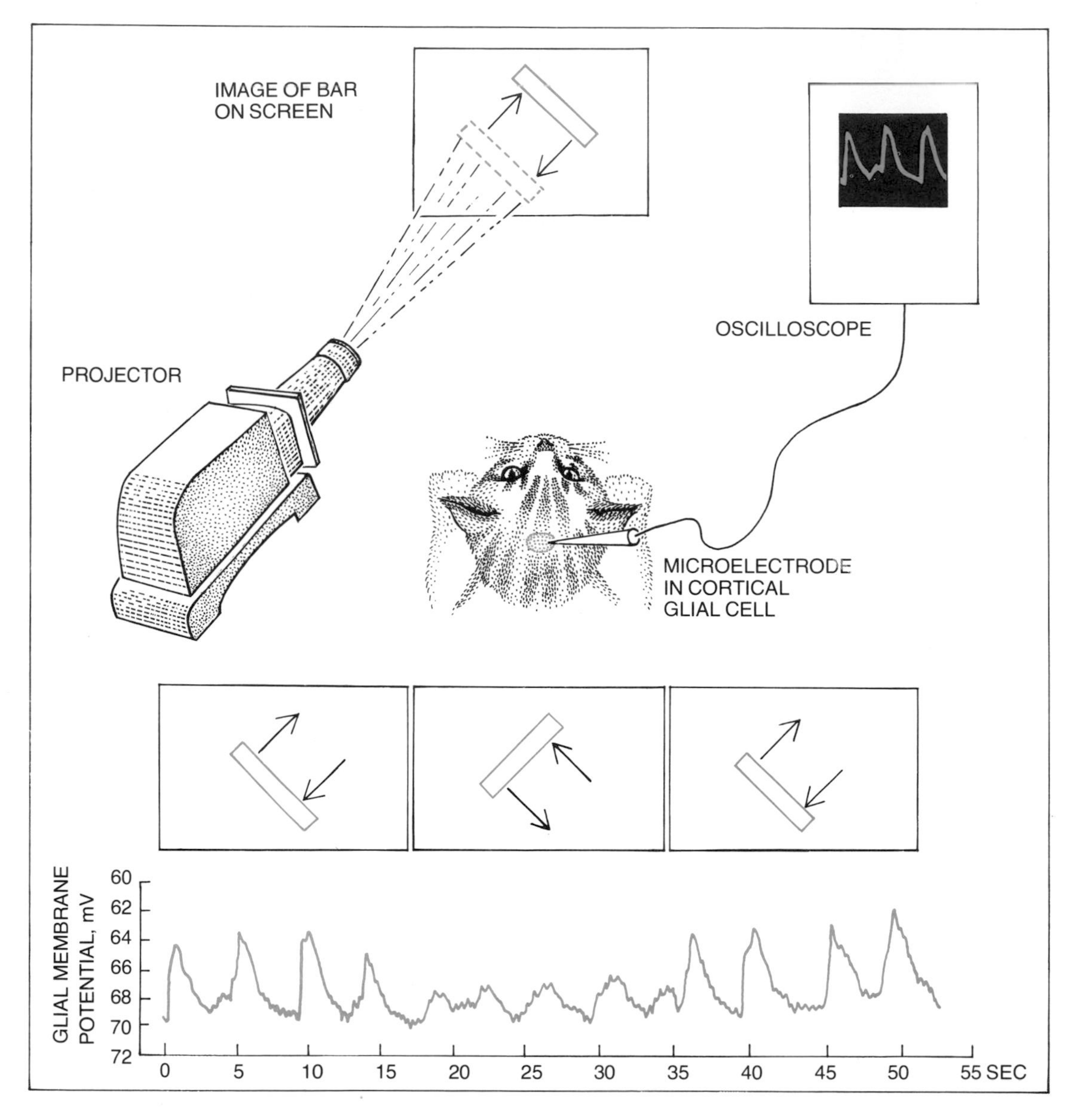

14 **DEPOLARIZATION OF GLIAL CELL in the cortex of a cat as a result of visual stimulation. The greatest depolarization was produced by moving slits of light having a certain orientation and confined to a part of the visual field. Neurons in the region near the glial cell had the same orientation preference. (After Kelly and Van Essen, 1974)**

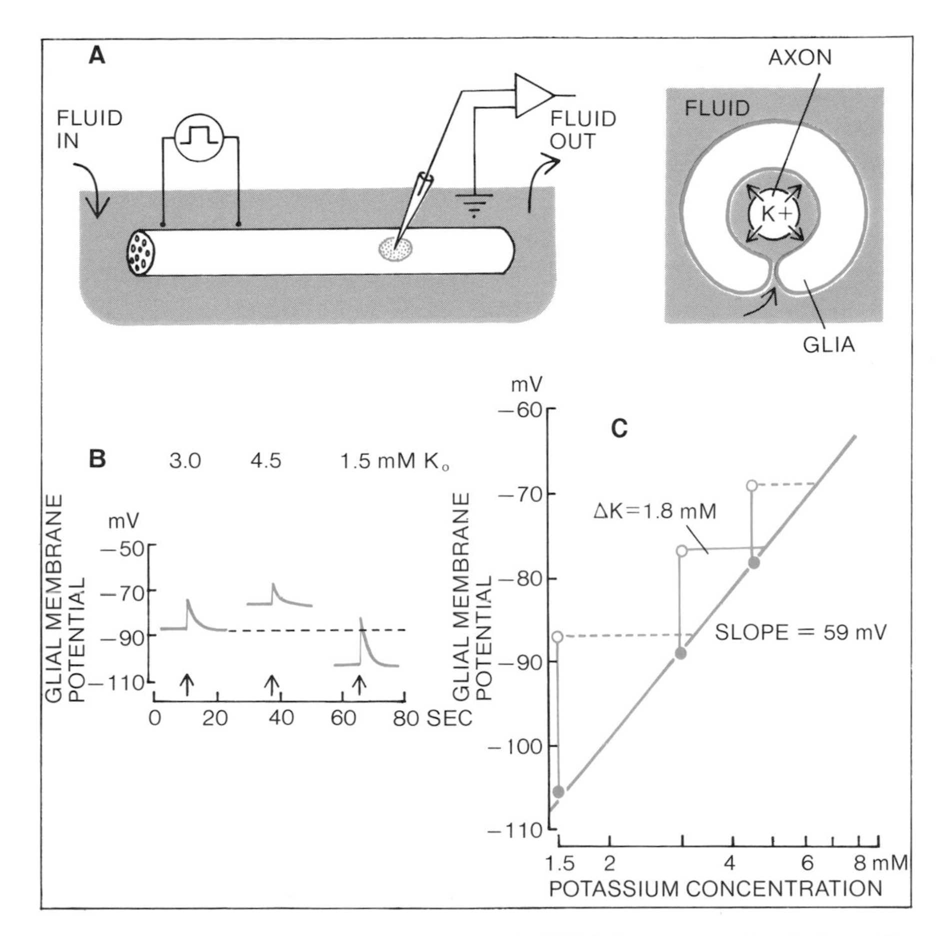

RELEASE OF POTASSIUM from axons depolarizes glia. A. System of perfusing the optic nerve of the mudpuppy and (right) a schematic representation of an optic nerve fiber surrounded by a glial cell. B. Three different concentrations of potassium in the bath shift the membrane potential to three different levels (Figure 10). When a standard train of nerve impulses is set up, it causes a smaller depolarization effect when K_o is higher (4.5 mM) and a larger effect when K_o is lower (1.5 mM). C. The observations in B plotted on a semilogarithmic scale. The solid circles indicate the membrane potential at K_o of 1.5, 3.0, and 4.5 mM, in quantitative agreement with prediction by the Nernst equation (solid line). The nerve-evoked peak depolarizations (open circles) are equivalent (dotted lines) to adding 1.8 mM of potassium to the bathing fluid. (After Orkand, Nicholls, and Kuffler, 1966)

15

polarization is equivalent to an increase of 1.8 mM of potassium in the bathing fluid. How much depolarization would 1.8 mM of potassium produce if it were added to a bathing solution that contained only 1.5 mM instead of 3 mM of potassium? This value is indicated by the lower horizontal broken line, which is within 0.4 mV of the value observed with a train of stimuli (lowest open circle). The upper horizontal dotted line shows the depolarization predicted if 1.8 mM of potassium is added to a solution that already contains 4.5 mM. Once more the depolarization (upper open circle) caused by the standard nerve stimulation is in excellent agreement with the prediction.[41]

These and other results can be explained by the simple assumption that in the physiological range of K_o nerve impulses release constant amounts of potassium into the intercellular clefts. As a result of the potassium that transiently accumulates, the membrane potential of the glial cell becomes depolarized, and its potential returns to normal as the potassium disappears by uptake and diffusion.

Recently, the introduction of potassium-sensitive glass electrodes has enabled several groups of investigators to measure directly the accumulation of potassium in the extracellular spaces of the brain during neuronal activity (Chapter 12). With repetitive stimulation of neurons, the potassium concentration increases, the value being comparable to that assumed from the glial cell depolarization.[42]

Effect on signaling of removal of glial cells

Do glial cells contribute to the signaling activity of those neurons they surround? A ready answer is obtained by cutting open ganglia in the leech and washing away the glial cytoplasm. This procedure leaves the neuronal surface exposed to the bathing fluid (Figure 16); yet, these naked neurons continued to give signals for many hours.[30] The experiment demonstrates that glia are not essential for the short-term survival of neurons or for impulse conduction. However, close scrutiny of the impulses in neurons deprived of glia shows that their manner of signaling is different when they are excited by a train of stimuli. The difference can be attributed to the absence of potassium accumulation. The analysis is based on the technique originally described by Frankenhaeuser and Hodgkin,[40] in which the undershoot of a nerve impulse is used to measure small changes of potassium concentrations in the bathing medium. During this phase of the action potential (g_K) is high and the cell is more sensitive to small changes in K_o than it is at rest (see below). As the potassium (K_o) in the environment increases, the undershoot becomes smaller. The undershoots of impulses can therefore be used as a bioassay for potassium concentration, just as the glial membrane potential was used in the tests described in Figure 15.

In the experiment illustrated in Figure 17 a train of nerve impulses was repeated with a neuron first in its normal environment with the glia and clefts intact, and then after the glial cytoplasm had been removed. In its normal environment the undershoots of the nerve impulses decrease as more and more potassium accumulates during a train of impulses, while the impulse peaks are but little altered. After removal of the glia and in the absence of potassium accumulation, each succeeding impulse has the same undershoot.[43]

These experiments bring out the significant fact that potassium

---

[41]Orkand, R. K., Nicholls, J. G., and Kuffler, S. W. 1966. *J. Neurophysiol.* *29*:788–806.

[42]Hotson, J. R., Sypert, G. W., and Ward, A. A. 1973. *Exp. Neurol.* *38*:20–26.

[43]Baylor, D. A., and Nicholls, J. G. 1969. *J. Physiol.* *203*:555–569.

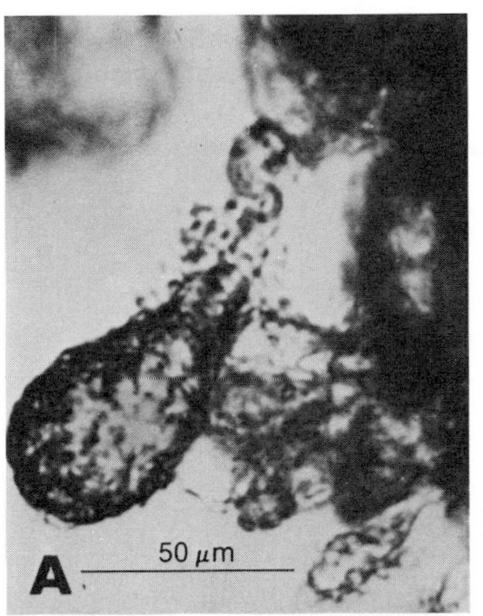

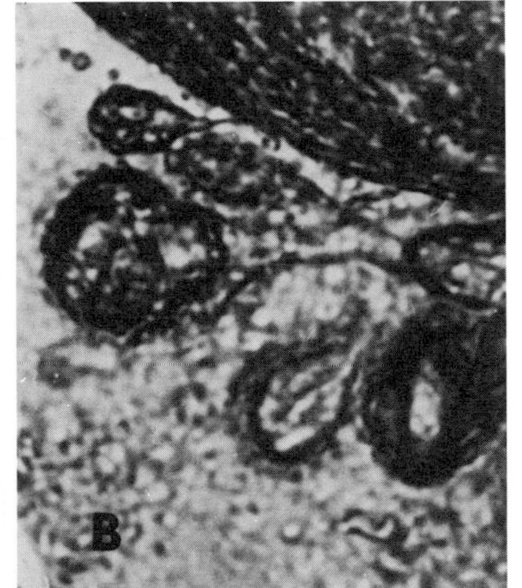

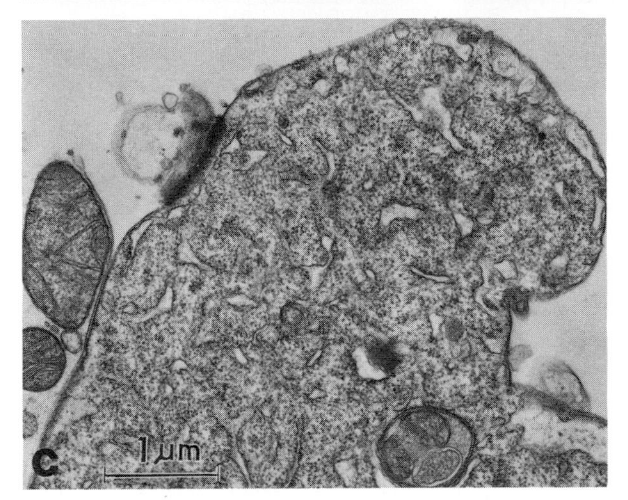

16

NEURONS DEPRIVED OF GLIA by making a cut in the connective tissue capsule surrounding a leech ganglion. A. A single nerve cell body and its initial axon segment after the glial cytoplasm was rinsed away. B. Same ganglion, just after the cut, before glial cytoplasm had been removed. C. Electron micrograph made by D. E. Wolfe, of part of a neuron that is devoid of its glial environment. Such "naked" neurons continue to conduct impulses. (From Kuffler and Potter, 1964)

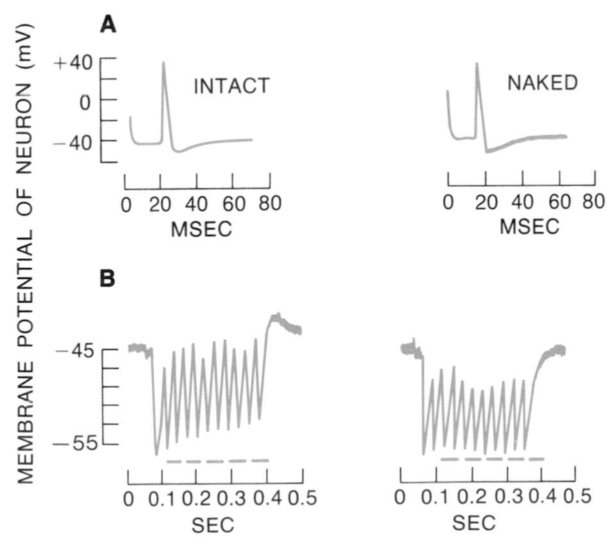

17

EFFECT OF GLIA ON NEURONAL SIGNALS in a leech ganglion. Records are from the same neuron before and after removal of the surrounding glia (Figure 16). In the intact ganglion the potassium accumulates in the clefts surrounding the neuron during a train of impulses, and the potassium-sensitive undershoot becomes progressively smaller (B, left record). In the naked neuron the undershoot does not change during a train of impulses (B, right record). (From Baylor and Nicholls, 1969)

accumulation during normal activity does not interfere with conduction, but does affect the slower afterpotentials and, therefore, alters signaling activity. Accumulation of potassium must, therefore, be considered in any analysis of the pattern of signaling. Secondarily, the results also mean that the mere presence of glial cells is significant, by delineating the clefts and at the same time providing a spatial buffer around neurons.

GLIAL CELLS AS A SOURCE OF POTENTIALS RECORDED WITH SURFACE ELECTRODES

General considerations

Recordings made with electrodes placed on the surface of a tissue measure currents flowing through extracellular fluids. Such currents are generated by a cell if various regions of its surface are at different potentials. Nerve cells, of course, use this principle as the mechanism for conduction. Thus, current flows from inactive regions of an axon into the part that is occupied by a nerve impulse. Positive charges are thereby drawn away from the uninvolved area in front of a nerve impulse so that the region ahead becomes depolarized and eventually "active." Glial cells do not extend over long distances like axons, but are linked to each other by low-resistance connections. The conducting properties of such contiguous coupled cells are therefore much the same as if the cells were in direct continuity. As a result, if several glial cells become depolarized by increased potassium concentrations in their environment, they draw current from the unaffected cells, thereby creating current flow. This, in turn, gives rise to a potential difference in the external fluid which can be recorded with external electrodes.

One would expect the patterns of current flow generated by glial cells in the brain to be complex because of their diffuse distribution in the tissue. Each cell has an elaborate arborization and makes contact at various points with other glial cells. We emphasize that the situation in glial cells would be quite different if they were electrically independent of each other and uniformly depolarized. If that were the case, no current would flow in the external circuits and the cells would be isolated from each other in the same way as two neighboring skeletal muscle fibers or nerve fibers that change their potentials independently without reference to each other.

Correlation between glial potential changes and current flow

A model situation suitable for demonstrating quantitatively the amount of current generated by glial cells compared with neurons occurs in the optic nerve of *Necturus*. The nerve's geometry is simple: a population of nerve fibers runs in parallel through the cylindrical nerve. The glial cells that are in electrical continuity, each coupled to its neighbors, also run in parallel through the nerve. For present purposes, therefore, the axons can be represented as a single continuous structure, and so can the glial cells (Figure 18).

The method of choice for measuring the relative contribution of neurons and glia to current flow is the sucrose gap technique. The chamber that holds the preparation is divided into three compartments

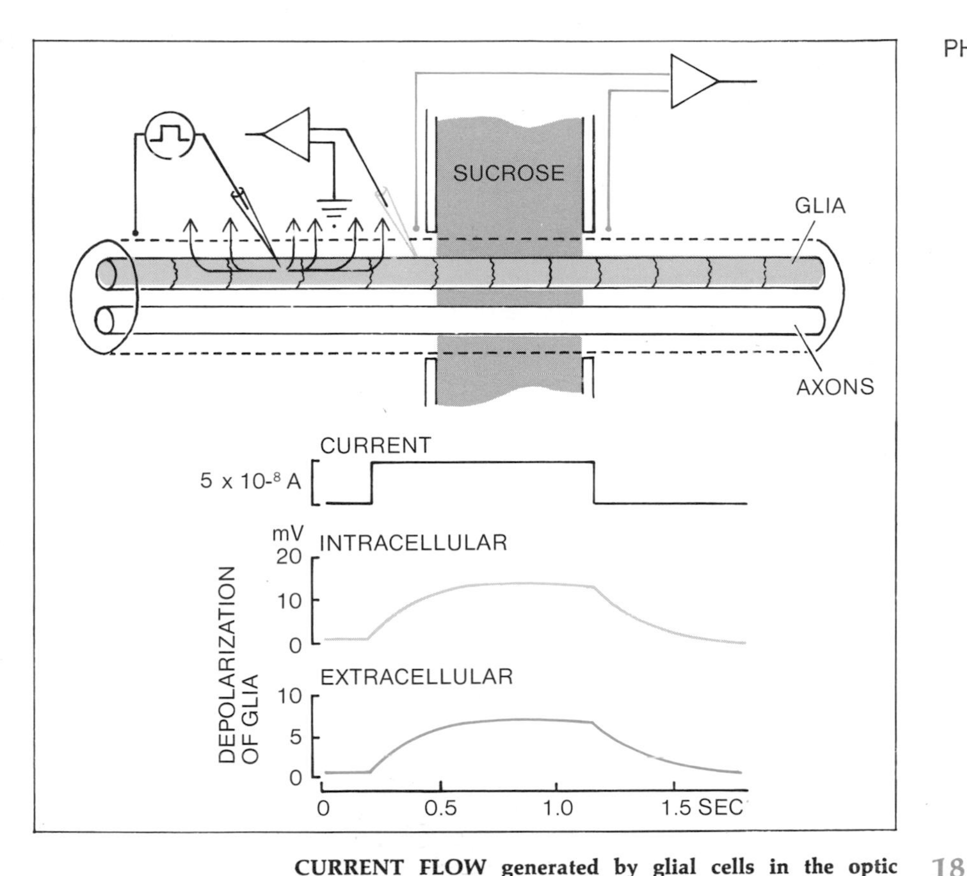

18

CURRENT FLOW generated by glial cells in the optic nerve of Necturus. In the sketch at top the axon bundles and glial cells are each represented as a continuous structure. Current was passed into the glial cells by an intracellular electrode, and the resulting potential change was recorded by another intracellular electrode as well as by electrodes across a sucrose gap. The extracellular potentials were about half the size of the intracellular ones and had a similar time course. This shows that glia and neurons make similar contributions to current flow. (After Cohen, 1970)

(Figure 18), so that a short segment of nerve is surrounded by isotonic sucrose. Since sucrose, which is not a conductor of current, fills all the extracellular spaces, the only electrical connection for current between the two Ringer's-solution-filled compartments is through the interior of the axons and glial cells. Under these conditions two surface electrodes across the sucrose gap register the voltage drop produced by current flow through the interior of all the cells. At the same time an electrode within a glial cell near the boundary between the Ringer's solution and the sucrose measures the membrane potential across the cell membrane.

In a series of experiments by Cohen,[44] the glial cells were depolarized by an intracellular current electrode while the glial membrane potential at the gap was registered by another electrode. External electrodes across the gap recorded a potential about half as great as that recorded by the intracellular microelectrode (Figure 18). This result is expected because in the optic nerve of *Necturus* the glial cells and axons are seen by electron microscopy to occupy about the same volume. The potentials recorded from glial cells with external electrodes are therefore reduced by 50 percent compared with intracellularly recorded potentials. In other nerves a variable contribution from glia is expected, depending on the volume ratio.

The conclusion that glial cells generate current flow is further strengthened by results of experiments in which the optic nerve was cut. After axons had degenerated, the glia/neuron ratio was changed so that a preparation of almost pure glial cells was left. In these nerves extracellular leads across the sucrose gap recorded almost the same glial potential (80 to 90 percent) as the intracellular electrode in glia at the boundary between the sucrose and the Ringer's solution.[44]

Such experiments show clearly that GLIAL CELLS DO CONTRIBUTE CURRENTS that can be measured by surface electrodes. By themselves, however, they do not enable prediction of the amount or time course of the resulting potential changes at other sites in the nervous system.

Glial contribution to the electroencephalogram

Electroencephalography is one of the few avenues available for obtaining objective information about activity in the human brain. Electroencephalograms (EEGs) are made routinely in fully conscious subjects or in animals under a variety of experimental conditions. Therefore, any method that separates the contribution of various elements, such as neurons and glia, is of potential interest. The same applies to recordings from the eye, electroretinograms (ERGs), in which one gross electrode is placed on the cornea and another indifferent electrode elsewhere on the body. In each case the potential changes represent the summed electrical activity of the underlying mass of neurons and glial cells.

How much is contributed by glial depolarizations to surface recordings from the brain? A quantitative answer about the contribution of glial cells to the EEG is difficult to obtain for a number of reasons: (1) both neurons and glial cells generate current flow, but depending on their distribution in the volume of the brain, the potentials recorded on the surface of the tissue may be either positive or negative; (2) the contributions of currents by neurons and glia sum algebraically; if they occur simultaneously, their potentials may be additive, or if they are in the opposite direction they may cancel each other; (3) slow potentials are by no means confined to glia but are common in neurons as well; thus, neurons may be hyperpolarized through the activity of an electrogenic pump at the same time as glia is depolarized by the buildup of potassium.[43]

One approach to assessing the contribution of glial cells to the EEG is provided by experiments in which slow glial membrane potentials are recorded with intracellular electrodes in the cortex of cats

[44]Cohen, M. W. 1970. *J. Physiol.* *210*:565–580.

following stimulation of neurons.[45] At the same time potentials are monitored with extracellular electrodes on the cortical surface and with other electrodes inserted at different depths within the cortex. Potential changes, fast as well as slow, that can be correlated with neuronal activity change their polarity as the probing extracellular electrode passes through different depths of the tissue. On the other hand, potentials attributed to the glial cells are more evenly distributed throughout the cortex and do not change their polarity.

Glial contribution to the electroretinogram

The ERG has been a useful tool in physiological studies of the eye and in clinical diagnosis. Literature going back over 100 years concerns slow potentials recorded from the eye. Intracellular and intraretinal recordings have helped greatly to determine the cellular origin of some of the surface potentials in the ERG. The main contributing elements are the photoreceptors, the neurons, and the glial cells. The glial cells, called Müller cells, extend through the whole depth of the retina. They have been impaled with microelectrodes and, as in the brain, have been shown to have higher membrane potentials (up to −85 mV) than the neurons. Illumination always depolarizes them, the response slowly declining if the light is kept on; when illumination is turned off, they give "off" responses rather like the records from a glial cell in the optic nerve of *Necturus* (Figure 12). Interestingly, the Müller cell potentials have a similar time course as one of the components (the b waves) in the ERG, obtained with extracellular recordings from the eye of the mudpuppy. They are likely to be responsible for most of that component of the standard ERG.[46]

The simplest interpretation of the Müller cell responses is that they are a secondary consequence of potassium leakage as neurons become active. Recall, however, that retinae of various fish and turtles contain nerve cells that give maintained potentials that will contribute to slow waves in the ERG; for example, the receptor cells and horizontal cells hyperpolarize with illumination and the bipolar cells can give maintained responses in either direction (Chapter 2).

MEANING OF SIGNALING FROM NEURON TO GLIA

A nonspecific system of communication

In general terms, glial cells indicate the level of impulse traffic in their environment. For example, the depolarization of a glial cell in the optic nerve is graded according to the number of axons and the frequency at which they carry nerve impulses. Impulse numbers are converted into potassium concentrations in the clefts, and these, in turn, into glial membrane potentials.

In many respects signaling between neurons and glia is radically different from specific synaptic activity. Synaptic action is confined to small specialized regions on the neuron cell body and the dendrites and

---

[45]Castellucci, V. F., and Goldring, S. 1970. *Electroencephalogr. Clin. Neurophysiol. 28*:109–118.

[46]Miller, R. F., and Dowling, J. E. 1970. *J. Neurophysiol. 33*:323–341.

may be excitatory or inhibitory. In contrast, signaling by the potassium mechanism is not confined to special structures, such as synapses, but occurs along the entire length of a neuron wherever potassium is released. Nor does it make any difference whether an impulse liberates an excitatory or an inhibitory transmitter. In this respect the NEURON-GLIA SIGNALING IS NONSYNAPTIC AND NONSPECIFIC, except that one particular glial cell is influenced preferentially by a discrete population of neurons that are in close proximity to it. One would therefore expect the physiological role of glial cells to be a generalized rather than a discriminating one. One might speculate that potassium-induced depolarization leads somehow to stimulation of enzymes in glial cells, causing them to produce a product the nerves need for their activity or for recovery afterward.

Effects of increased potassium on glial metabolism and transmitter release

The idea that external potassium affects the metabolic activity of glial cells has recently found support in experiments in which the overall metabolism was measured by studying changes in the fluorescence of a pyridine nucleotide (NADH).[47] The experiments were made on a pure glia preparation obtained from optic nerves of *Necturus* in which the axons had been previously cut and allowed to degenerate. Concentrations of cyanide, which decrease oxidative metabolism, produced marked reversible changes in NADH fluorescence. Interestingly, increasing the external potassium concentrations to 15 mM also increased the uptake and metabolism of ^{14}C-glucose by glial cells.[48] Clearly it is of interest to determine what significance this may have for neuron-glia interactions.

One specific possibility is that increased concentrations of potassium may lead to secretion of materials from satellite cells. Already mentioned is the fact that GABA, which is stored in glial cells, can be released by raised concentrations of potassium (20 mM or more). Glial cells may, therefore, play a role in regulating the concentrations of transmitter in the intercellular spaces—both by taking it up and by secreting it again in the presence of increased potassium produced by signaling.[9–11] Equally possible is the secretion of other, as yet undetermined, materials under the influence of altered potassium concentrations in the clefts.

It has also been suggested that the current flow produced by glial cells may affect the signaling of neighboring neurons. This does not seem a likely possibility because clefts have been shown to be effective barriers to current spread between cells. Another idea is a role for glia as a reservoir of electrolytes or as a device for rapid uptake of potassium from the clefts to regulate and preserve the constancy of the environment; again, critical experiments have not yet been done to test these notions.

---

[47]Orkand, P. M., Bracho, H., and Orkand, R. K. 1973. *Brain Res. 55*:467–471.

[48]Salem, R. D., Hammerschlag, R., Bracho, H., and Orkand, R. K. 1975. *Brain Res. 86*:499–503.

Consequences of potassium accumulation on synaptic and integrative activity of neurons

A single brief flash of light into the eye can depolarize glial cells by about 4 mV (Figure 12). This corresponds to an average increase of about 0.5 mM of potassium in the clefts around these cells. It seems reasonable to assume that similar fluctuations also occur around terminals near synapses. While such small changes have a negligible effect on nerve conduction, the consequences on synapses may be important. The accumulation of potassium becomes increasingly significant with synchronous activity in groups of axons. Thus, under rather artificial conditions when axons are driven electrically by stimuli at relatively high frequencies, the potassium concentration in the clefts may rise from the normal 2 to 3 mM to 20 mM. In fact, with prolonged synchronous activity, conduction can block itself through potassium accumulation.[41]

Fluctuations in potassium concentrations might produce effects at two sites. The first is at presynaptic nerve terminals where depolarization acts to modify the release of transmitter. In the squid stellate ganglion a depolarization of 2 mV depresses the release of transmitter and reduces the effect of a nerve impulse by about 15 percent.[49] At rat neuromuscular junctions, raising K_o by 1 mM increases the spontaneous release of quanta of transmitter by about 25 percent.[50,51] However, little is known about the effects of changes in K_o in the vertebrate central nervous system. At the second site of its action, on the postsynaptic neuron, potassium accumulation brings the cell closer to the firing level by depolarizing it.

In view of these considerations, one might think that increments of potassium concentration by even 2 to 4 mM, which is a physiological range, would alter signaling and integrative activity in neurons. It is, therefore, interesting that the membrane potential of nerve cells, in contrast to that of glial cells, is relatively insensitive to changes in the potassium concentration in its environment. For example, in neurons of the leech a fivefold increase in K_o (from 4 mM to 20 mM) results in a depolarization of only 5 mV, while the glial membrane potential changes by 25 mV.[52] The neuron's membrane potential, therefore, is at least in part protected from fluctuations, presumably because it has a significant resting permeability to other ions (chiefly sodium) besides potassium, in contrast to glial cells in which potassium practically determines the full membrane potential. In spite of this relative insensitivity of neurons, the effects of potassium increments on synaptic transmission may be considerable, especially since sensitivity to potassium may increase after prolonged activity.[53] Moreover, one would expect that in the central nervous system, with its continued inter-

[49]Katz, B., and Miledi, R. 1967. *J. Physiol.* *192*:407–436.
[50]Liley, A. W. 1956. *J. Physiol.* *134*:427–443.
[51]Cooke, J. D., and Quastel, D. M. J. 1973. *J. Physiol.* *228*:435–458.
[52]Nicholls, J. G., and Kuffler, S. W. 1964. *J. Neurophysiol.* *27*:645–671.
[53]Jansen, J. K. S., and Nicholls, J. G. 1973. *J. Physiol.* *229*:635–655.

actions between large numbers of neurons, even a very small change in synaptic efficiency might have profound consequences.

In the context of the release of potassium, an intriguing speculation put forward by clinical neurologists should be mentioned: local changes in potassium concentration may play a role in triggering synchronized abnormal neuronal activity, such as epileptic discharges.[54] A study of the distribution of glia in such situations seems of interest because the possibility exists that glial cells can act as buffers in shielding neurons from unduly influencing each other by their release of potassium.

SUGGESTED READING

General reviews

Bunge, R. P. 1968. Glial cells and the central myelin sheath. *Physiol. Rev. 48:* 197–251.

Iverson, L. L., and Kelly, J. S. 1975. Uptake and metabolism of γ-aminobutric acid by neurons and glial cells. *Biochem. Pharmacol. 24:*933–938.

Kuffler, S. W. 1967. Neuroglial cells: Physiological properties and a potassium mediated effect of neuronal activity on the glial membrane potential. *Proc. R. Soc. Lond.* B. *168:*1–21.

Kuffler, S. W., and Nicholls, J. G. 1966. The physiology of neuroglial cells. *Ergeb. Physiol. 57:*1–90.

Somjen, G. G. 1975. Electrophysiology of neuroglia. *Annu. Rev. Physiol. 37:* 163–190.

Original papers

Baylor, D. A., and Nicholls, J. G. 1969. Changes in extracellular potassium concentration produced by neuronal activity in the central nervous system of the leech. *J. Physiol. 203:*555–569.

Cohen, M. W. 1970. The contribution by glial cells to surface recordings from the optic nerve of an amphibian. *J. Physiol. 210:*565–580.

Kuffler, S. W., Nicholls, J. G., and Orkand, R. K. 1966. Physiological properties of glial cells in the central nervous system of amphibia. *J. Neurophysiol. 36:*855–868.

Kuffler, S. W., and Potter, D. D. 1964. Glia in the leech central nervous system: Physiological properties and neuron-glia relationship. *J. Neurophysiol. 27:*290–320.

Orkand, R. K., Nicholls, J. G., and Kuffler, S. W. 1966. Effect of nerve impulses on the membrane potential of glial cells in the central nervous system of amphibia. *J. Neurophysiol. 29:*788–806.

Ransom, B. R., and Goldring, S. 1973. Ionic determinants of membrane potential of cells presumed to be glia in cerebral cortex of cat. *J. Neurophysiol. 36:*879–892.

Schon, F., and Kelly, J. S. 1975. Selective uptake of [^{3}H] β-alanine by glia: Association with the glial uptake system for GABA. *Brain Res. 86:*243–257.

---

[54]Sypert, G. W., and Ward, A. A., Jr. 1971. *Exp. Neurol. 33:*239–255.

CHAPTER FOURTEEN

REGULATION OF THE COMPOSITION OF THE FLUID SPACES IN THE BRAIN

A homeostatic system controls the fluid environment of nerve and glial cells in the brain and keeps its chemical composition relatively constant, compared with that of the blood plasma. This is necessary for the proper integrative activity of the neural signaling system.

Within the brain are three different types of fluid: (1) the blood supplied to the brain through a dense network of capillaries; (2) the cerebrospinal fluid (CSF) that surrounds the bulk of the nervous system and is also contained in the internal cavities (the ventricles); and (3) the fluid in the intercellular clefts. The constitution of the blood plasma and of the CSF is known from direct chemical analysis. This chapter deals principally with the determination and regulation of the composition of the fluid in the intercellular spaces that are generally no more than 20 nm wide. The contents of these spaces provide the immediate environment of nerve and glial cells in the brain.

The intercellular clefts are open for diffusion of ions and certain large molecules. They constitute the main channels through which materials are distributed to reach neurons and glial cells. The membrane potential of neuroglial cells within the brain can be used as a quantitative bioassay for determining concentrations of potassium ions within the intercellular clefts and also for estimating the rates of diffusion of various ions and small molecules. By the use of electron-dense markers and electron microscopy, the distribution of larger molecules has been traced through the cleft system of the brain.

The homeostatic control that keeps the environment of neural elements relatively constant, and different from blood plasma, is provided by two cell systems: (1) the endothelial cells of capillaries and (2) the epithelial cells surrounding capillaries of the choroid plexus. These epithelial cells secrete cerebrospinal fluid and act as a barrier for ions and various molecules. In addition, their regulating action is brought about by an active transport system.

Once substances have found their way into the CSF, they are

free to diffuse into the tissues of the brain. Although the CSF dominates the composition of the intercellular spaces, there exist different regional chemical microclimates within the brain.

The problem

Nerve cells in the brain function in an environment that differs significantly from the fluid in which peripheral organs are bathed. It has long been known that the bulk of the brain and spinal cord is surrounded by a specially secreted, clear fluid called the CEREBROSPINAL FLUID (CSF) (Figure 1). This solution is almost devoid of protein and contains only about one-two hundredth of the amount present in blood plasma. Chemical substances such as metabolites move relatively freely from the alimentary canal into the bloodstream, but not into the CSF. As a result, the levels in the blood plasma of sugar, amino acids, or fatty acids fluctuate over a wide range, while their concentrations in the CSF remain relatively stable. The same is true for hormones, antibodies, certain electrolytes, and a variety of drugs. Injected directly into the bloodstream, they act rapidly on peripheral tissues such as the muscles, heart, or glands, but they have little or no effect on the central nervous system. When administered by way of the CSF, however, the same substances exert a prompt and strong action. The conclusion is that the substances do not reach the CSF and brain with sufficient rapidity and in effective concentrations.

It is hard to see how neurons could perform their essential signaling precisely unless the chemical environment were kept relatively constant. Constancy would seem particularly important in a system where the activity of many cells is integrated and where small variations may change the balance of delicately poised excitatory and inhibitory influences. In contrast, at myoneural junctions and in autonomic ganglia, where information is generally transmitted with a good safety margin, fluctuations in the environment are of relatively small consequence. Thus, homeostasis does not mean absolute constancy, but a setting of limits to fluctuations that are tolerable without disruption of function. It therefore seems reasonable that the brain should have a special and more closely adjusted homeostatic control than the rest of the body.

The way in which the brain keeps its environment constant is frequently discussed in terms of a BLOOD-BRAIN BARRIER. The reasoning is generally based on the following considerations. The sustenance of the brain is derived from its rich blood supply. The vertebrate brain must discriminate against the entry of some materials from the blood, and therefore it has developed certain barriers. This term "barrier," however, which represents structural impediments to the access of substances, should not obscure the fact that other mechanisms also regulate the fluid environment.

If materials enter, however slowly, without participation and control by an active cell transport system, the concentration of fluids in the brain will eventually equilibrate with that in the plasma. The fallacy of an exclusively mechanical barrier concept is readily illustrated by the proposal of a blood-urine barrier—which cannot explain the mechanism by which glucose is normally kept out of the bladder, by reuptake from the kidney tubules. It therefore follows that linked with structural barriers, there must exist active processes that regulate the environment of nerve cells in the brain.

Most of the general conclusions mentioned so far about access of materials to the brain have been derived from determining differences between the CSF and blood plasma, two fluids that are readily available in relatively large quantities for chemical analysis. Frequently the assumption is made implicitly that the CSF actually provides the "true" environment of nerve cells, because of the convincing evidence that there are fewer obstacles to an exchange between the CSF and neurons than between the blood and the CSF. Yet, the immediate environment of neurons is neither the blood plasma nor the CSF, which is contained in the large spaces, such as the ventricles from which the samples are taken. The crucial sites providing the fluid environment of neurons are the intercellular spaces between neurons and glial cells. These are, as discussed in Chapter 13, narrow clefts several tens of nanometers in width. They are too small to be accessible for direct chemical analysis, and studies of their fluid composition have depended on a prior determination of the total extracellular spaces in the brain, a task beset by difficulties of interpretation.

The main problem discussed in this chapter involves the ways in which one can determine in the brain the composition of the extracellular spaces that surround nerve cells, for comparison with the composition of the blood plasma and the CSF. To set the stage, it is necessary to discuss first the gross distribution of CSF and blood in the brain and then the various pathways from capillaries or CSF to neurons.

Distribution of CSF and blood

The brain and spinal cord are shock-mounted in a jacket of CSF. Between the mass of the brain and the skull are two layers of connective tissue, the pia-arachnoid and the dura mater, which is attached to the bone. The pia-arachnoid forms a trabecular meshwork providing fluid spaces on the surface of the brain (called subarachnoid spaces). A schematic illustration is shown in Figure 1. There is contact but no bulk fluid between the dura and the arachnoid. The CSF around the brain surface communicates with the CSF within the cavities of the brain, the ventricles, and the spinal canal. In man, the CSF is continuously replaced at an estimated rate of six to seven times a day. It drains into the sinus venosus through the arachnoid villi, which constitute a valve system. There is no lymphatic outflow from the brain.

The brain is supplied by many large blood vessels that run in the CSF-filled subarachnoid spaces before diving from the surface into the parenchyma, the bulk of the brain tissue, taking with them their surrounding connective tissue covering. Bulk CSF around the blood vessels thereby penetrates the brain at multiple places. As the blood vessels become progressively smaller, the space around them narrows until finally it is only a fraction of a micron around the capillaries (Figure 1*B*).

In considerations of the chemical environment of nerve cells one should bear in mind that the vascular bed in the brain is very dense and permeates the entire structure, so that each nerve cell is probably no farther than 40 to 50μm from a capillary.[1]

INTERCELLULAR CLEFTS AS CHANNELS FOR DIFFUSION IN THE BRAIN

Nerve cells and neuroglia are so closely packed everywhere in the brain that the spaces between them are reduced to about 20 nm. For some time, around the early 1960s, it was commonly thought that such narrow spaces were not functional pathways and could not play a part in the distribution of materials. It was even postulated that neuroglial cells actually constitute the extracellular spaces through which materials must pass to reach neurons, much as originally suggested by Golgi (Chapter 13).

[1]Scharrer, E. 1944. *Q. Rev. Biol.* *19*:308–318.

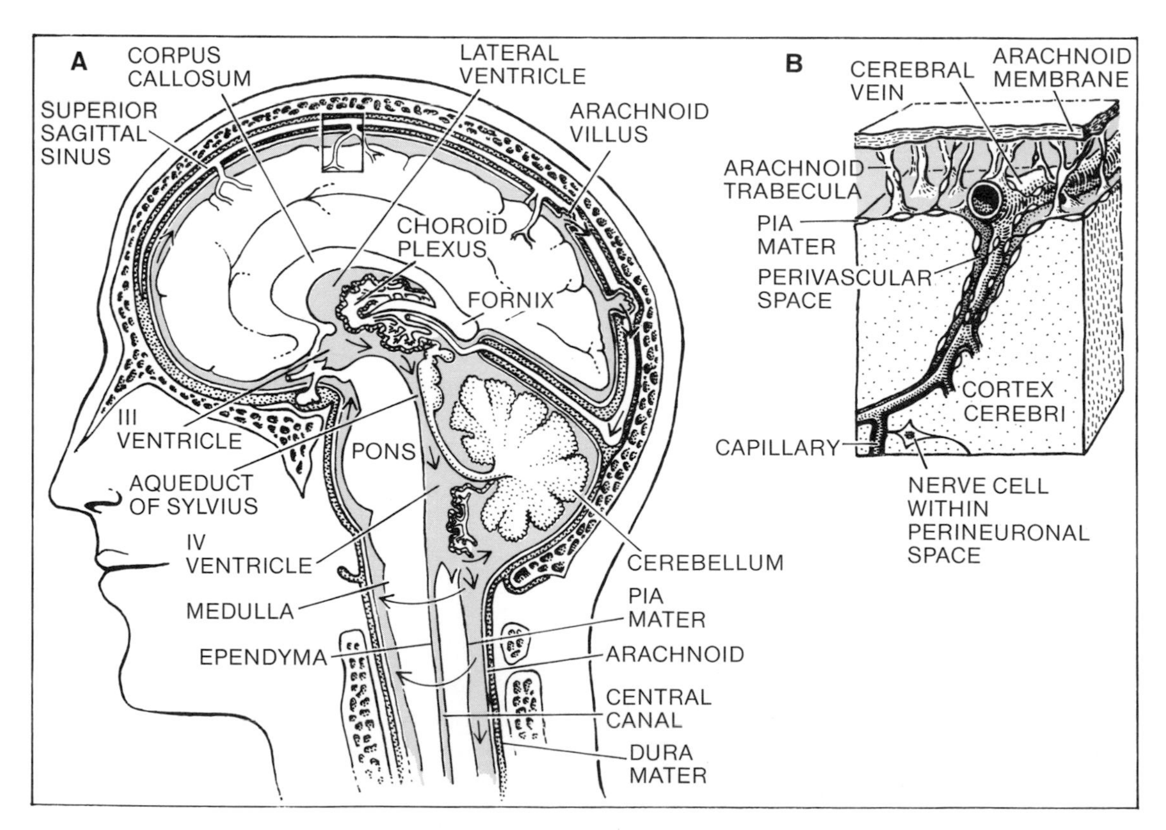

1 **DISTRIBUTION OF CEREBROSPINAL FLUID (CSF) and its relation to larger blood vessels and to structures surrounding the brain (A). All spaces containing CSF communicate with each other. CSF is drained into the venous system through the arachnoid villi (B).**

Several questions about intercellular clefts must be considered: (1) Do the intercellular spaces, if open, allow the diffusion of ions, like sodium, or of molecules of various dimensions, such as sucrose or proteins? (2) What is the rate of movement of materials through the nervous system? (3) What is the evidence that certain substances do not move through glial cells? If they do, what is the relative importance of the alternative pathways through the intercellular spaces and through the cytoplasm?

Theoretical considerations of diffusion through narrow channels

From the size of a molecule and the size of a fluid-filled channel, one can calculate rates of diffusion from simple diffusion equations. It is also possible to estimate the limiting size of the channels, if one assumes them to be tubes. The relative coefficient of diffusion, D'/D is given by:

$$D'/D = \frac{(1 - a/r)^2}{1 + 2.4\, a/r}$$

where D is the coefficient of diffusion, r is the radius of the tube, and a is the size of the particle.[2] As an example, if one takes a value of 0.44 nm for the radius of sucrose and 15 nm as the diameter of a tube, the relative coefficient of diffusion turns out to be only 18 percent less than that in free solution. The size becomes significantly limiting only for molecules having a radius of more than about 15 percent of the tube width. Hence, the actual dimensions of the clefts would not necessarily prevent movement of ions or of small molecules. These considerations would not apply if the intercellular clefts were (1) filled with material that slowed or prevented diffusion or (2) closed off at certain points (see below).

Diffusion of electron-dense molecules into the clefts

The distribution of a number of substances of different molecular weights through the brain tissues can be detected by electron microscopy. One tracer molecule frequently used in the past is the protein ferritin, which has a diameter of about 10 nm and a molecular weight of 900,000. Another is the enzyme horseradish peroxidase, with a diameter of about 4 nm and a molecular weight of 43,000. More recently, microperoxidase (molecular weight near 2,000) has become available. With such an enzyme as a marker, the product deposited as the result of the reaction of the enzyme can be detected. The substances are generally injected into the CSF of animals, and their distribution is determined after fixation.

The results clearly demonstrate that horseradish peroxidase and even ferritin can enter intercellular clefts from the CSF if given adequate time for diffusion.[3,4] The electron-dense molecules, deposited by the peroxidase reaction, are then lined up in the clefts and fill the extracellular spaces. Tracer materials may also be taken up into the interior of cells, usually enclosed in pinocytotic vesicles. Further, one

[2]Pappenheimer, J. R. 1953. *Physiol. Rev.* 33:387–423.
[3]Brightman, M. W. 1965. *Am. J. Anat.* 117:193–219.
[4]Brightman, M. W., and Reese, T. S. 1969. *J. Cell Biol.* 40:648–677.

can determine the location of sites where clefts are occluded or reduced in diameter because tracers stop spreading when they reach these regions. An example is provided in Figure 2*A*, which shows that when horseradish microperoxidase is injected into the ventricles, diffusion takes place throughout the clefts. The opposite situation is shown in Figure 2*B*, in which the enzyme was injected into the circulation. The capillary is filled with the enzyme, but no significant deposits appear in the intercellular spaces. As explained later (Figure 7), the junctions between the capillary endothelial cells provide the barrier.

Although the electron microscopic method provides clear evidence that intercellular spaces are open for diffusion, it is difficult for technical reasons to establish accurately the rates of movement of substances by means of electron microscopy.

Rates of movement through intercellular clefts

Physiological measurements have established the rates at which sucrose, sodium, and potassium move through the clefts in the central nervous system of the leech[5] and of the mudpuppy (*Necturus*).[6] These studies provide a better time resolution than do electron-dense tracers, and since the effects are reversible, controls can be made in live preparations.

The technique for measuring diffusion times depends on the use of neurons as indicators of the sodium and potassium in their environment. If the sodium around a neuron is reduced or replaced by sucrose or

[5]Nicholls, J. G., and Kuffler, S. W. 1964. *J. Neurophysiol.* *27*:645–671.

[6]Kuffler, S. W., Nicholls, J. G., and Orkand, R. K. 1966. *J. Neurophysiol.* *29*:768–787.

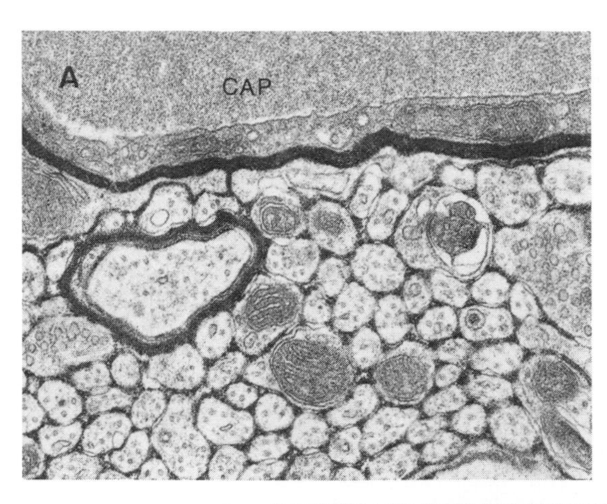

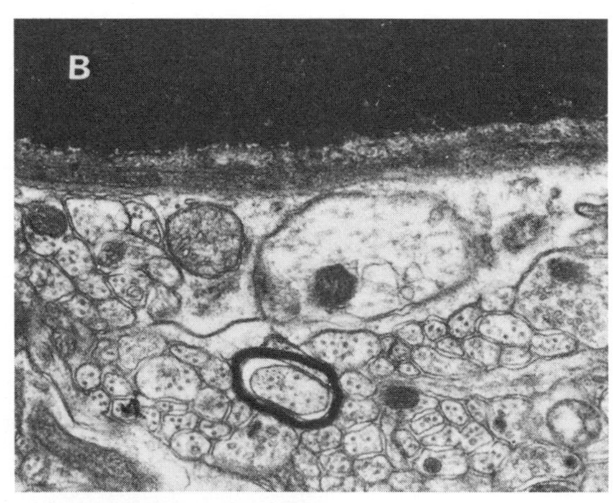

2 **PATHWAYS FOR DIFFUSION IN THE BRAIN. A. Demonstration in the mouse that the enzyme microperoxidase diffuses freely from the cerebrospinal fluid (CSF) into the intercellular spaces of the brain which are filled with the dark reaction product. No enzyme is seen in the capillary (Cap). B. The enzyme, injected into the circulation, fills the capillary but is prevented by the capillary endothelium from escaping into the intercellular spaces. (From Brightman, Reese, and Feder, 1970)**

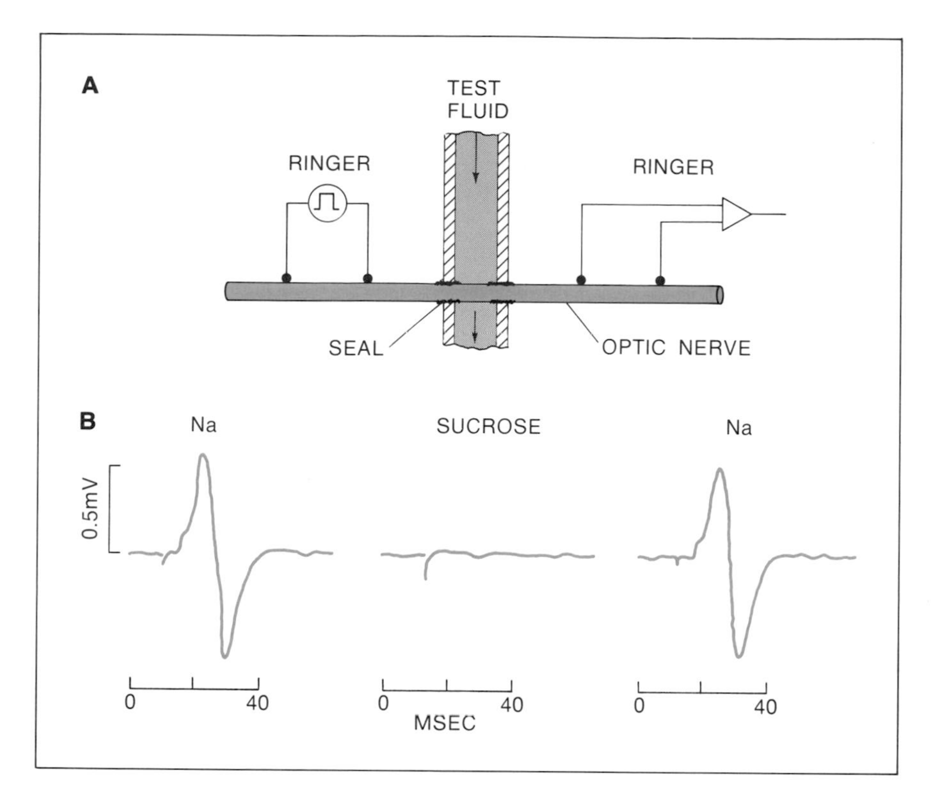

3

RATE OF DIFFUSION THROUGH CLEFTS. Movement of sodium and sucrose through the intercellular clefts of the optic nerve in Necturus. Sodium necessary for impulse conduction is replaced within 12 sec by sucrose perfused through the middle compartment of a chamber shown in A. As a result, conduction through the sucrose-filled region is blocked (middle record in B) and is restored to its original size after sodium is reintroduced (right record). (From Kuffler, Nicholls, and Orkand, 1966)

choline, the action potential becomes smaller (Chapter 5). Increased potassium concentration, on the other hand, reduces the membrane potential. The experimental situation sketched in Figure 3 has been used in the connectives of the leech and in the optic nerve of *Necturus*. The preparation lies in a chamber consisting of three compartments sealed off from one another. The side compartments contain Ringer's fluid and are used for stimulating the nerve and for recording its action potential; the narrow central chamber can be perfused with various test solutions. Figure 3 shows that when the solution perfusing the central chamber is changed to sodium-free Ringer's solution that contains sucrose, conduction rapidly fails in the *Necturus* optic nerve. Within 12 sec after sodium-free fluid reaches the central compartment, all the impulses are blocked. If sodium is introduced again into the perfusion fluid, the conducted potential is fully restored to its control size after 10 sec (Figure 3). The results in the leech connectives are very similar when either sucrose or choline is used to replace sodium.

The above experiments can be interpreted as follows. When sodium is replaced by sucrose, the ion moves out of the extracellular spaces and sucrose takes its place. Block of conduction results because sodium, which is needed for the impulse mechanism, is lost from the immediate environment of the axons. Testing different concentrations of sodium shows that complete block occurs when 60 to 75 percent of the sodium in the central perfusion compartment has been replaced by sucrose. One can therefore conclude that within 12 sec in sodium-free solution, at least 60 percent of the sodium in the optic nerve is exchanged for an equivalent amount of sucrose.

A more accurate measure of the rate of movement through the nervous system can be obtained by recording intracellularly from neurons in leech ganglia.[5] To measure the rate of penetration of potassium, for example, the relation between potassium concentration and the membrane potential of a neuron is first established (Figure 4*B*). From this any membrane potential can be translated into an equivalent potassium concentration. Next, a high concentration of potassium is applied to the ganglion, and the cell becomes depolarized in a few seconds (Figure 4*C*). The equivalent potassium concentration around the neuron at any instant after the potassium has been applied to the bathing fluid can now be estimated. In this way, the influx and efflux of potassium have been plotted. By the same principle, the rates of movement of sodium, sucrose, and choline have been measured, using the size of the action potential as an indicator of concentration. These results have the advantage that a complete diffusion curve can be constructed. The points fall on a reasonably straight line when plotted on a logarithmic scale against time, as would be expected for a diffusion process. The half-time for the exchange of sodium with potassium through the leech ganglion is about 4 sec; for the exchange of sodium and sucrose it is 10 sec.

It is of interest to compare these times with the values predicted on the assumption that simple diffusion is occurring through narrow channels. Knowing the diffusion coefficients for sodium chloride and sucrose, the half-time for diffusion within an individual cleft in the nervous system can be calculated from the formula for linear diffusion.[7]

$$\frac{C}{C_o} = 1 - \frac{2}{\sqrt{\pi}} \int_o^y e^{-y^2} dy$$

In this equation, C_o is the initial concentration and $y = x/2\sqrt{Dt}$, x being the distance (in centimeters) and D the coefficient of diffusion (square centimeters per second). The distance (x), estimated by measuring the length of the clefts (mesaxons), is probably not more than 50μm in a ganglion; in the connectives, it is probably not more than 30μm. Using these values for maximum distance, the half-time for the

[7]Hitchcock, D. I. 1945. In Höber (ed.). *Physical Chemistry of Cells and Tissues.* Blakiston, Philadelphia.

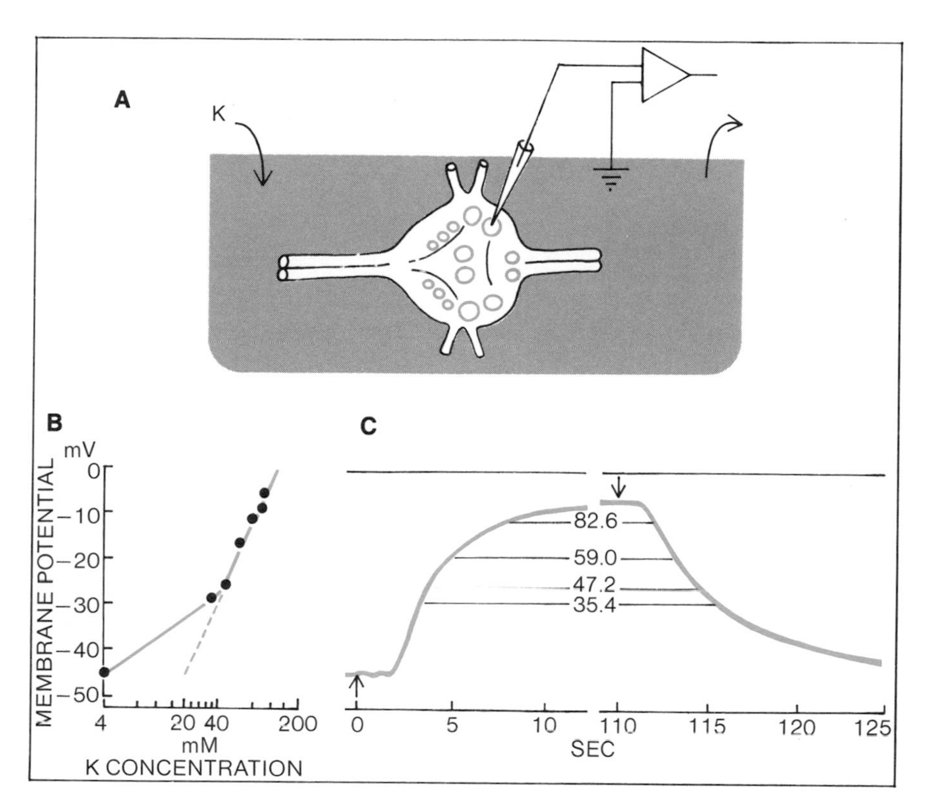

RATE OF DIFFUSION THROUGH A GANGLION. Movement of potassium within a leech ganglion. An intracellular electrode measures the rate of depolarization of a neuron after introduction of a high potassium concentration into the bath, replacing sodium. From the curve in B, membrane depolarization can be directly related to K_o and the rate of movement of potassium through the ganglion can be obtained, as shown in C. The half-time for exchange of sodium and potassium is about 4 sec. The numbers under the curve in C indicate the potassium in mM during the depolarization. About 100 sec was cut out of the record during the gap. (After Nicholls and Kuffler, 1964)

4

exchange of sodium and sucrose is about 3.6 sec in a leech ganglion packet and 1.3 sec in a connective. In each case the half-times measured experimentally were considerably larger (by a factor of over 2) than those predicted from the equation. This is not surprising, since the calculations at best yield only a rough approximation. They do indicate, however, that simple diffusion would probably be rapid enough to account for the rates of movement of sodium, potassium, and sucrose observed in the nervous system.

Exclusion of glia as a pathway for rapid diffusion

Do sodium and sucrose (or choline) exchange by taking a pathway through the glial cells, or around them through the clefts? The initial expectation is that movement through intercellular channels would leave the glial membrane potential unchanged, whereas passage through the cells would inevitably alter it. When glial membrane potentials are measured while isotonic sodium and sucrose move through the nervous system, the potentials remain practically unaffected by the exchange of sodium and sucrose. One can conclude that the electrolytes and sucrose diffuse through the clefts around the glial cells rather than through their cytoplasm.[5,6]

IONIC ENVIRONMENT OF NEURONS IN THE BRAIN

Homeostatic regulation of hydrogen ions in mammalian CSF

Preceding chapters show that the movement of ions and the charges they carry provide the mechanism for signals in nerve cells. It is therefore clear that changes in the ionic environment affect signaling and information processing within the brain. We argue, on teleological grounds, that a "constant" environment must be maintained for constancy of signaling, and that is the reason for a special homeostatic regulation in the brain.

A question that must be decided experimentally is the extent to which changes can be tolerated in the ionic environment of the nervous system without disrupting integrative activity. This problem has been extensively studied by Pappenheimer and his colleagues on the respiratory system of the goat.[8] The composition of the blood plasma and of the CSF was altered in unanesthetized goats whose ventricles were perfused with artificial CSF of known composition and whose blood plasma concentrations were also controlled by intravenous injection of a variety of substances. The respiration of the animal served as a sensitive index, or bioassay, of changes in pH in the vicinity of certain respiratory neurons. These cells are situated in the medulla and are very sensitive to small fluctuations in pH; their activity is immediately detected through changes in the rate of breathing of the animal.

In goats whose body fluid composition was artificially changed, small alterations in the content of H^+, HCO_3, or Cl^- within the ventricles had a large effect on respiration. In contrast, greater changes in the blood plasma had far less effect. These results show that activity in respiratory neurons is related to the fluctuations of concentrations within the CSF rather than in the blood. Even under prolonged modifications of the plasma, such as maintained alkalosis or acidosis, a distinct environment in the brain was maintained. The simplest assumption is that a sharp discontinuity in the concentration profile of H^+ exists between the intercellular fluid around neurons and the blood in the nearby capillaries. It also follows that the differences are maintained by various active processes, such as pumps in the vicinity of the capillaries.

The conclusions derived from the experiments by Pappenheimer and his colleagues would not be affected if it turned out that respiratory neurons send processes (so far not noted) directly into the CSF within the ventricles. Neurons partially protruding into the spinal canal have already been described in reptiles.[9] In any event, there exist direct

[8]Pappenheimer, J. R. 1967. *Harvey Lect.* *61*:71–94.

[9]Vigh, B., Vigh-Teichmann, I., Koritsánszky, S., and Aros, B. 1970. Z. *Zellforsch.* *109*:180–194.

connections to neurons from the CSF via the intercellular clefts (see later).

The mammalian experiments so far mentioned have been extended to amphibians and elasmobranchs (dogfish, shark).[10]

Homeostatic regulation of potassium in intercellular clefts

As explained in Chapter 13, the membrane potential of a glial cell provides a quantitative measure of the potassium concentration in its environment, since it behaves like a perfect potassium electrode. This finding has provided a technique for exploring the composition of fluids in spaces that are only several tens of nanometers wide and are therefore not accessible for conventional chemical analyses. Thus the membrane potentials of glial cells can be used to register maintained changes in the fluid composition in the intercellular clefts. These have been produced by changing the concentration of potassium in the blood plasma of live mudpuppies with intact circulations.

There exists in the brain of *Necturus*, as in mammals and fish, a homeostatic control system.[11] Mudpuppies were kept for long periods in tanks containing increased amounts of potassium; as a result the potassium concentrations in their body fluids increased. Figure 5 illustrates the changes occurring in the CSF when the potassium concentration in the blood had increased up to fivefold. The results demonstrate that the CSF maintained its potassium concentration within the relatively narrow range of 2 mM, in the face of plasma potassium increases of 7 to 8 mM.

Examination of the structural features of the optic nerve of the mudpuppy (Figure 9, Chapter 13) shows clearly that the only separation between the blood within the capillaries and the CSF surrounding them is a single layer of endothelial cells. A sharp gradient must therefore exist in potassium concentration across capillary endothelial cells.

With this background knowledge, we can consider how the potassium concentration changes in intercellular clefts around the neurons during long-term alterations of blood plasma and CSF levels. Do cleft concentrations in the optic nerve follow fluctuations in the CSF or in the

[10]Cserr, H., and Rall, D. P. 1967. *Comp. Biochem. Physiol.* 21:431–434.
[11]Cohen, M. W., Gerschenfeld, H. M., and Kuffler, S. W. 1968. *J. Physiol.* 197:363–380.

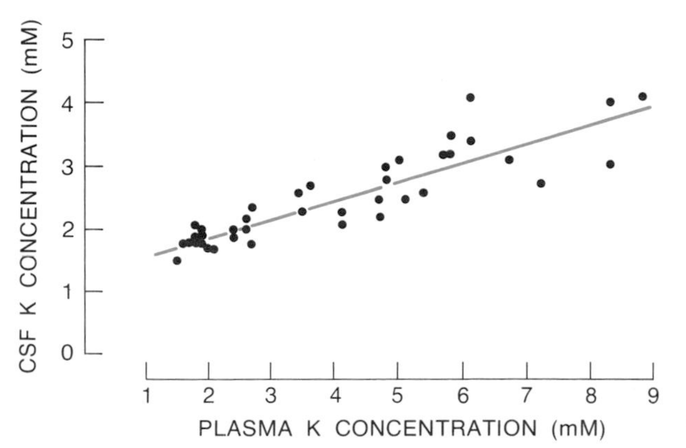

5
POTASSIUM CONCENTRATION in blood plasma and CSF. Determinations in mudpuppies maintained in a high-potassium environment. Each point represents a measurment in a different animal. Potassium in the plasma can be chronically elevated from 1.5 mM to 9 mM, while the CSF maintains its potassium within a range of 2 mM. This shows the existence of a homeostatic mechanism. (From Cohen, Gerschenfeld, and Kuffler, 1968)

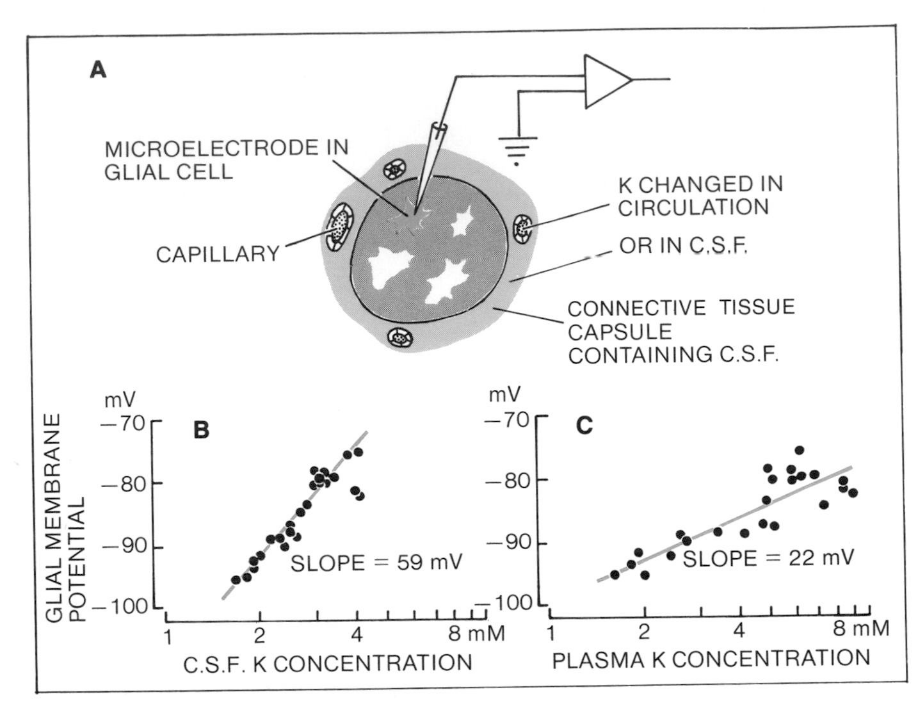

6 **GLIAL CELL AS A POTASSIUM ELECTRODE. Glial membrane potentials measured in mudpuppies with intact circulations. Each point in B and C represents data from one animal maintained in a high-potassium environment. B. The solid line gives the slope predicted by the Nernst equation for a 10-fold change in potassium. The determinations show that potassium within the intercellular clefts follows concentration changes within the CSF. C. In contrast, changes in the potassium in the blood plasma are not reflected by the changes in the potassium within the intercellular clefts. (After Cohen, Gerschenfeld, and Kuffler, 1968)**

plasma within the capillaries? Or do they perhaps maintain an in-between concentration of their own? Once more, the glial membrane potential, behaving like a potassium electrode, provides the answer.

If potassium in the intercellular clefts does faithfully follow fluctuations in the blood plasma, the membrane potential of glial cells should vary, as predicted by the Nernst relation (Chapter 13). The results plotted in Figure 6*C* show that although the glial membrane potential changes when potassium concentrations fluctuate in the blood, the slope is only 22 mV instead of the 59 mV predicted for a 10-fold change in potassium concentration. On the other hand, the relation theoretically predicted by the Nernst equation is faithfully followed when potassium concentration in the CSF is plotted against the membrane potential, as in Figure 6*B*, where the slope is 59 mV.

These experiments show that in the mudpuppy, the potassium concentration in the clefts closely reflects that in the CSF rather than in the blood plasma. Such results support the long-standing view that the principal effect on neurons of fluctuations in ionic concentrations in the blood is indirect; the ionic composition of the CSF must be altered before the effect can be felt by the nerve cells. This shifts the emphasis onto the processes that control the production and composition of the CSF.

The choroid plexus and formation of CSF

Within the brain certain blood vessels protrude into the ventricles to form a rich capillary network, called the choroid plexus. Here the CSF is formed by an active-transport secretory mechanism.[12–14] Figure 7 shows the anatomical arrangement both in the brain parenchyma and in the choroid plexus. The choroid capillaries are lined with a fenestrated endothelium that is structurally different from endothelia in most other regions of the brain (exceptions are certain small areas such as the area postrema and the median eminence). These fenestrated endothelial cells apparently permit molecules, including peroxidase, to leak out. The next cell layer is made up of choroidal epithelial cells. They have the electron microscopic appearance of a secretory epithelium and resemble other active cells in which a regulated exchange occurs (such as those that line the renal tubules). Adjoining choroid epithelial cells are closed off by a circumferential girdle of tight junctions, so that the usual intercellular spaces are obliterated over a continuous beltlike area of contact (Figure 7). This layer of cells has the dual role of (1) preventing proteins from escaping and (2) regulating the newly formed CSF by active transport.[4,14]

Sites other than the choroid plexus also contribute to the composition of CSF. The problem is quantitative, because some leakage is bound to occur from various capillaries into their environment, either directly into the CSF or into the intercellular spaces. These in turn drain into the ventricular and subarachnoid spaces containing CSF (see below). A further contribution to the CSF comes from scattered leakage points which permit blood plasma to enter the CSF directly, bypassing the barriers provided by the choroid epithelial cells. The nascent CSF secreted by choroid plexuses is therefore not identical in composition to the bulk of the CSF, although it dominates the makeup of the latter.

Sites of regulation of the ionic and chemical environment of neurons

In the mammalian brain the differences between CSF and blood plasma are brought about by the endothelium of the capillaries within the brain and by the choroid epithelium. In *Necturus*, physiological results suggest that the clefts communicate freely with the CSF.

The conclusion that a barrier exists at the level of the capillaries is greatly strengthened by tracer studies, similar to those demonstrating that clefts are open for diffusion. In particular, the injection of horseradish peroxidase into the circulation of mice[4,15] and *Necturus*[16] has

---

[12]Ames, A., Higashi, K., and Nesbett, F. B. 1965. *J. Physiol. 181*:506–515.
[13]Cserr, H. 1971. *Physiol. Rev. 51*:273–311.
[14]Wright, E. M. 1972. *J. Physiol. 226*:545–571.
[15]Reese, T. S., and Karnovsky, M. J. 1967. *J. Cell Biol. 34*:207–217.
[16]Bodenheimer, T. S., and Brightman, M. W. 1968. *Am. J. Anat. 122*:249–268.

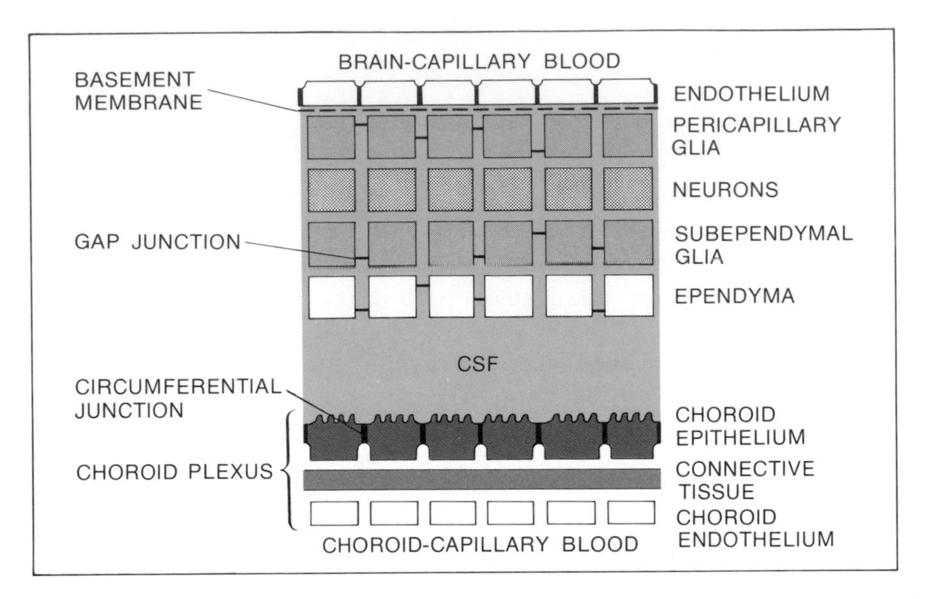

7 **EXCHANGE OF MATERIALS in the brain. Schematic presentation of cells involved in the exchange of materials between blood, CSF and intercellular spaces. Molecules are free to diffuse through the endothelial cell layer lining capillaries in the choroid plexus. They are, however, restrained by circumferential junctions between the choroid epithelial cells which secrete CSF. There are no barriers between the bulk fluid of CSF and the various cell layers such as ependyma, glia, and neurons. The endothelial cells lining brain capillaries are occluded by a circumferential seal, as are the choroid epithelial cells. This prevents free diffusion of molecules out of the blood.**

shown that this enzyme, with a molecular weight of about 43,000, cannot leak out from brain capillaries because these cells are closed off by a continuous belt of tight junctions in the same way as the choroid epithelial cells (Figure 7). However, once materials have reached either the CSF or any of the intercellular spaces, they are free to diffuse throughout the bulk of the brain. Brain capillaries behave differently from those supplying skeletal or heart muscles, which do allow some proteins to escape.[17,18]

In conclusion, the main cellular sites of regulation of the fluid environment reside in the endothelial cells of the capillaries in the brain and in the choroid epithelial cells. Much of the physiological evidence presented above comes from lower vertebrates. There is good structural evidence that homeostasis around neurons and glia is brought about in a similar manner in the brains of the rat and the mouse.

---

[17]Karnovsky, M. J. 1967. *J. Cell Biol. 35*:213–236.
[18]Clementi, F., and Palade, G. E. 1969. *J. Cell Biol. 41*:33–58.

SUGGESTED READING

General reviews

Cserr, H. F. 1971. Physiology of the choroid plexus. *Physiol. Rev. 51*:273–311.

Davson, H. 1972. The Blood-Brain Barrier. In G. H. Bourne (ed.). *The Structure and Function of Nervous Tissue,* Vol. IV. Academic Press, New York, pp. 321–345.

Fleischhauer, K. 1972. Ependyma and Subependymal Layer. In G. H. Bourne (ed.). *The Structure and Function of Nervous Tissue,* Vol. IV. Academic Press, New York, pp. 1–46.

Kuffler, S. W., and Nicholls, J. G. 1966. The physiology of neuroglial cells. *Ergeb. Physiol. 57*:1–90.

Leusen, I. 1972. Regulation of cerebrospinal fluid composition with reference to breathing. *Physiol. Rev. 52*:1–56.

Original papers

Brightman, M. W., Klatzo, I., Olsson, Y., and Reese, T. S. 1970. The blood-brain barrier to proteins under normal and pathological conditions. *J. Neurol. Sci. 10*:215–239.

Brightman, M. W., and Reese, T. S. 1969. Junctions between intimately apposed cell membranes in the vertebrate brain. *J. Cell Biol. 40*:648–677.

Brightman, M. W., Reese, T. R., and Feder, N. 1970. Assessment with the Electronmicroscope of the Permeability to Peroxidase of Cerebral Endothelium in Mice and Sharks. In E. H. Thaysen (ed.). *Capillary Permeability.* Alfred Benzon Symposium II. Munskgaard, Copenhagen.

Cohen, M. W., Gerschenfeld, H. M., and Kuffler, S. W. 1968. Ionic environment of neurons and glial cells in the brain of an amphibian. *J. Physiol. 197*:363–380.

Nicholls, J. G., and Kuffler, S. W. 1964. Extracellular space as a pathway for exchange between blood and neurons in the central nervous system of the leech: Ionic composition of glial cells and neurons. *J. Neurophysiol. 27*:645–671.

Reese, T. S., and Karnovsky, M. J. 1967. Fine structural localization of a blood-brain barrier to exogenous peroxidase, *J. Cell Biol. 34*:207–217.

PART FOUR

SENSORY RECEPTORS

CHAPTER FIFTEEN

HOW SENSORY SIGNALS ARISE AND THEIR CENTRIFUGAL CONTROL

In all sense organs external energy is translated into electrical signals. The receptor structures are specialized to respond to particular stimuli, such as sound, light, touch, or odors. Their properties determine the type of stimulus that can be handled and define the limits of what we can perceive. As examples of how sensory receptors function, two sense organs are compared—stretch receptors in the crayfish and muscle spindles in mammals. Both types of receptors measure the rate and extent of muscle stretch. They function in a basically similar manner and demonstrate the general principles that operate in sensory structures. The terminals of the receptor neurons that are deformed by stretch are inserted into specialized muscle fibers. The crustacean receptors have the great technical advantage of allowing the intracellular recording of events in their nerve endings.

The first electrical signal in nerve terminals in response to extension is a localized depolarization—the generator potential—which is graded and reflects in size and time course the applied stimulus. When the generator potential exceeds the threshold of the sensory axon, impulses propagate toward the central nervous system; their frequency is determined by the size of the generator potential.

The contractile and viscoelastic properties of muscle strands into which sensory terminals insert, combined with the properties of the sensory terminals, determine the character of the sensory discharges; some receptors register preferentially the rate of change and others the maintained phase of stretch.

In addition to responding to passive extension, muscle spindles and crustacean receptors are under the control of the central nervous system. The muscle strands in which the receptor neurons are embedded receive motor innervation that makes them contract; thereby the sensory endings are stretched. This increases the sensitivity to additional stretch or initiates sensory impulses. Further, the crustacean stretch receptor neurons receive direct inhibitory innervation on their sensory terminals that can decrease or silence the sensory discharges in the face of continued stretch-excitation. The principle of centrifugal control, by which the central nervous system adjusts the information it receives, has been observed in a variety of sense organs, including the eye and the ear.

Stretch receptors provide a feedback system that plays an essential role in the fine regulation of muscle movement. Feedback control also operates within the brain itself at different levels in ascending pathways and seems to be part of the general organization of higher centers.

Sensory nerve endings as transducers

The sensory receptors are the gateways through which the outside world enters our minds. Right at the outset they set the stage for all the analysis of sensory events that is subsequently made by the central nervous system. They define the limits of sensitivity and determine the range of stimuli that can be acted upon or perceived. The nerve endings themselves are specialized in both their anatomical and physiological properties to respond preferentially to only one type of external energy. Nevertheless, the stimulus, whatever its form (light, sound waves, temperature changes, or mechanical deformation), always gives rise to an electrical signal that acts as the symbol that can be used by the central nervous system. Furthermore, a great degree of amplification occurs at the receptor level, so that very small amounts of energy provide a trigger to release stored charges that appear in the form of electrical potentials. For example, a few quanta of light trapped by pigment molecules of a receptor in the retina give rise to a photochemical reaction that leads to an electrical potential change, to impulses, and even to a sensation. Similarly, hair cells in the ear that are activated by sound waves respond to small displacements of the order of angstroms.

Despite certain obvious discrepancies, a comparison with familiar mechanical devices such as microphones and photocells is instructive in a number of ways. Sensory receptors in our ears respond to a restricted band width of sound from about 20 to 20,000 cps. Outside this range sound waves produce no response. Our visual system cannot perceive infrared or ultraviolet light because their wavelengths do not activate receptors in the retina.

Animals with appropriately tuned transducers can take account of influences in the external world that we cannot perceive. Though the underlying sensory mechanisms are generally similar in different species, there are wide variations in the spectrum of sensitivity. Thus, dogs can hear higher frequencies and respond to sounds, such as a whistle, that are inaudible to ourselves; certain snakes can detect light in the infrared, while moths respond to ultraviolet light and to odors caused by a few molecules of a specific substance, such as a sex attractant (a pheromone). An involuntary extension of our own range of color perception is not uncommon after removal of a lens that has become cloudy (cataract formation). As a questionable bonus for such inconvenience one becomes aware of the near-ultraviolet wavelengths that are normally filtered by the lens. After removal of a cataract a person is able to read (although poorly) a chart illuminated in the 365-mμ range that is not visible to the normal eye.

Beyond the level of the first electrical signal generated by sensory nerve endings, any simple analogy with conventional electrical hardware ceases to be useful. A microphone or a photocell can be connected by suitable cables to an amplifier with only little distortion of the signals. This is not true in the body, where the primary electrical signals cannot be conducted to the brain by the direct spread of electric currents, owing to the poor conducting properties of the nerve fibers (Chapter 7).

The way in which sensory receptors generate electrical signals in response to external stimuli can be studied conveniently and directly in stretch receptor neurons that register the length of muscles in crustaceans and in vertebrates. The stretch receptors in crayfish, first described by Alexandrowicz,[1] are particularly useful because their cell bodies lie in isolation—not within the central nervous system, but in the periphery, where they can be seen in live preparations (Figure 1). The cell bodies insert their dendrites into a fine muscle strand nearby and send an axon centrally to a segmental ganglion. Stretching the muscle strand deforms the dendrites where the transduction of energy takes place. A microelectrode inserted into the cell body of the neuron is close enough to the dendrites to obtain information about the transduction process and about the impulses initiated in the axon.

The crustacean stretch receptors are analogous to the vertebrate muscle spindles, which are described briefly in Chapter 4, in relation

[1]Alexandrowicz, J. S. 1951. *Q. J. Microsc. Sci.* *92*:163–199.

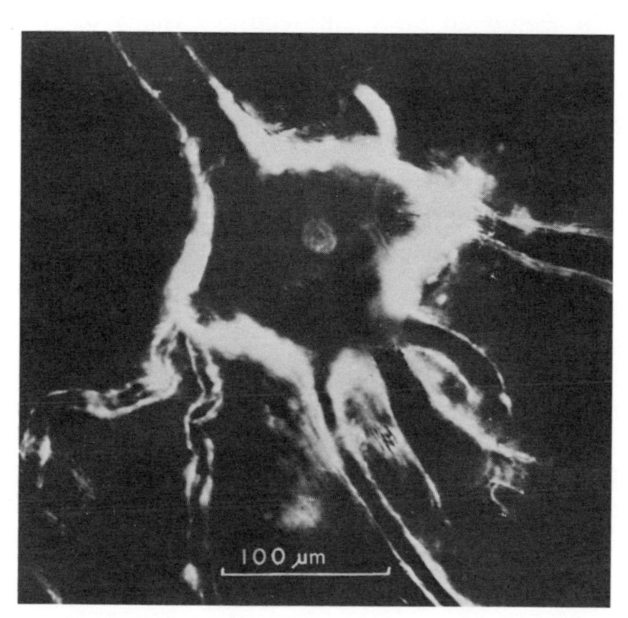

1
LIVING, UNSTAINED STRETCH RECEPTOR NEURON of the lobster viewed with dark-field illumination. Distal portions of six dendrites insert into the receptor muscle, which is not visible. (From Eyzaguirre and Kuffler, 1955)

to the stretch reflex. The adequate stimulus for both types of receptors is mechanical deformation of the sensory nerve terminals produced by stretch of the muscle. These two types of mechanoreceptors serve to illustrate a number of general principles. It will become clear that the apparent simplicity of their function is deceptive. Impinging upon them is an elaborate control system, originating in the central nervous system, which precisely tunes their sensitivity to mechanical stimulation. These sensory cells provide one of the best examples of efferent or centrifugal control in which the central nervous system not merely receives sensory information but reacts to it and modifies the performance of the receptors. Moreover, knowledge of the sensory end organs—their properties, control, and reflex connections—sets the stage for Chapter 16, which deals with integrative mechanisms.

MECHANOELECTRICAL TRANSDUCTION IN STRETCH RECEPTORS

The generator potential

There are two types of crustacean stretch receptor organs with distinct structural and physiological characteristics. Each type of sensory nerve cell has a characteristic appearance, and its dendrites are embedded in a different type of receptor muscle. One neuron responds well to maintained stretch, while the other fires chiefly at the beginning of a stretch (see below). This decrease in response of a sensory nerve to a steady stimulus is called ADAPTATION.

The basic anatomical arrangement of stretch receptors in the crayfish or the lobster is shown schematically in Figures 2 and 5. When the receptor muscle is stretched, the dendrites become deformed and their membrane potential is reduced.[2] This depolarization is the RECEPTOR OR GENERATOR POTENTIAL, a graded, localized event, originating in the dendrites. It reflects, in electrical terms, the intensity and duration of the stretch, as shown in Figure 2*B* with different extensions of the muscle strand. Like synaptic and other localized potentials, the generator potential depends on the passive electrical properties of the membrane and cannot spread far along the neuron without being seriously attenuated. As the stretch is increased, more current flows, until eventually the potential change reaches threshold and an all-or-nothing impulse propagates along the axon toward the central nervous system. Thus, the generator potential in the dendrites depolarizes the cell body and the sensory axons and produces effects analogous to those of positive charges injected through an intracellular microelectrode.

The conducted impulses start in a special region of the axon, called the initial segment or axon hillock, close to the cell body (Figure 2).[3] There the neuronal membrane has a lower threshold for initiation of a regenerative impulse than the cell body, whose dendrites may not conduct impulses at all. Once initiated, the impulses propagate not only toward the central nervous system but also back into the cell body. It

---

[2]Eyzaguirre, C., and Kuffler, S. W. 1955. *J. Gen. Physiol. 39*:87–119.
[3]Edwards, C., and Ottoson, D. 1958. *J. Physiol. 143*:138–148.

appears to be a common property of neurons that impulses are initiated in a special region where the axon emerges (Chapter 16).

As in other conventional receptors, the intensity of the stimulus is expressed in frequency of impulses. The relation between stimulus intensity and impulse frequency is established through the interaction of the maintained generator current in the dendrites and the conductance changes associated with the action potential (Figure 2*B*). At the end of each nerve impulse, the increased potassium conductance that occurs during the recovery phase drives the membrane potential in the hyperpolarizing direction, toward E_K. While the increase in the potassium conductance is transient, the generator current is maintained by the stretch and depolarizes the membrane once more to the firing level. The stronger the current, the sooner the firing level is reached again, and the higher the impulse frequency.

Not all sensory receptors perform as described above. The photoreceptors in the vertebrate retina do not initiate regenerative responses. In crustaceans a mechanoreceptor has been found that depolarizes, as expected, if stretched, but is not capable of generating propagated impulses.[4] Instead, the generator potential spreads passively over a distance of 5 mm to the terminals, which release transmitter onto the second-order cells. A similar situation exists in the eye of the barnacle, in which the next synapse may be several millimeters away.[5] This is a surprisingly long distance, and in order for the passive signal to spread so far, the specific resistance of the membrane must be unusually high, otherwise the potential would fall steeply over a millimeter or so.

[4]Roberts, A., and Bush, B. M. H. 1971. *J. Exp. Biol.* *54*:515–524.
[5]Shaw, S. R. 1972. *J. Physiol.* *220*:145–175.

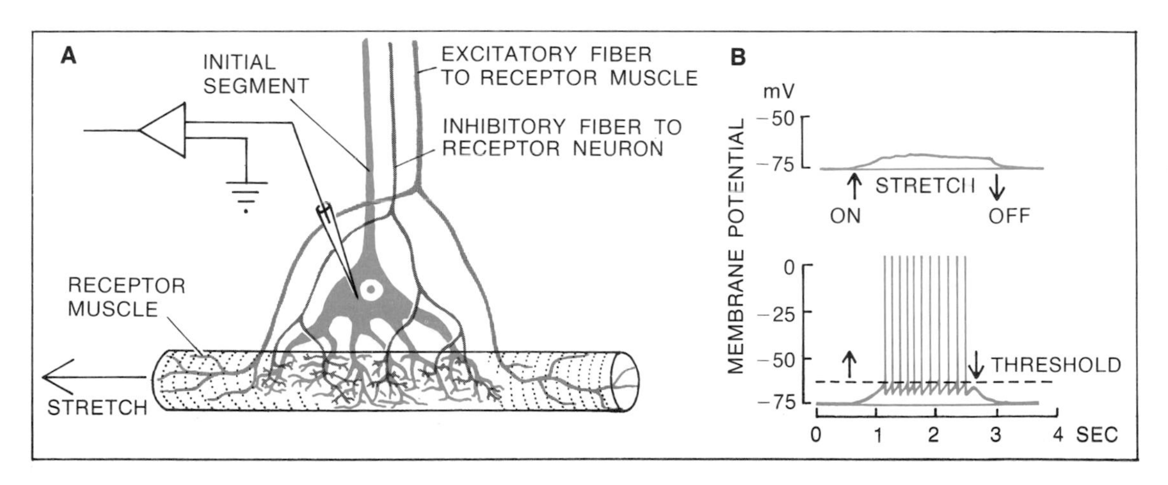

CRUSTACEAN STRETCH RECEPTOR. A. Features of innervation; two additional inhibitory fibers are omitted. B. A weak stretch for about 2 sec causes a subthreshold generator potential. With stronger stretch (lower record), a larger generator potential sets up a series of impulses. (After Eyzaguirre and Kuffler, 1955) 2

Ionic mechanisms underlying the generator potential

The basic mechanism through which deformation leads to a permeability change is not known. Neither are the permeability changes in sensory endings that give rise to the generator potential well understood. One would expect an increased sodium conductance to play a role, as in other excitatory depolarizing potentials. In accord with this idea, the generator potential produced by stretch is reduced when sodium ions are replaced in the bathing fluid.[6] It is interesting that tetrodotoxin does not affect the generator potential.[7] The sodium channels are therefore different from those used for the production of the action potential.

Ions other than sodium that may be involved have not yet been identified. As with other permeability changes, there is a characteristic reversal potential for the generator process that results from the combined equilibrium potentials of the different ions involved. In the stretch receptor, this is probably close to zero potential.[8]

Properties of muscle spindles

In their physiological behavior mammalian muscle spindles (most frequently studied in cats) resemble, in many respects, the more simple crustacean stretch receptors. As in crustaceans, the sensory nerve endings deformed by stretch are associated with specialized muscle fibers. Once again, there are two characteristic types of endings, one of which adapts rapidly and the other slowly. The general outline of the way in which muscle spindles respond was first worked out by Matthews in the early 1930s. For many years his experiments provided one of the best single attempts to describe comprehensively a sensory end organ and its control. Matthews was able to detect impulses in single nerve fibers that arose in individual spindles in frogs and cats. Recordings were made by an oscilloscope he designed for that purpose.[9–11] Since then numerous other studies have revealed the detailed and elaborate specialization of the sensory apparatus and its control.

An essential basis for the physiological experiments is analysis of the structure of the sensory and motor nerves and their terminations within the spindle. Much of this basic information is owed to the systematic tracing of fibers in the laboratories of Barker[12] and Boyd.[13] As in crustaceans, stretch of receptors gives rise to a generator potential, first recorded in spindles of the frog.[14]

Figure 3 illustrates schematically the sensory apparatus of spindles in leg muscles of the cat. The spindle consists essentially of 8 to 10

---

[6]Edwards, C., Terzuolo, C. A., and Washizu, Y. 1963. *J. Neurophysiol. 26:* 948–957.

[7]Nakajima, S., and Onodera, K. 1969. *J. Physiol. 200*:161–185.

[8]Terzuolo, C. A., and Washizu, Y. 1962. *J. Neurophysiol. 25*:56–66.

[9]Matthews, B. H. C. 1931. *J. Physiol. 71*:64–110.

[10]Matthews, B. H. C. 1931. *J. Physiol. 72*:153–174.

[11]Matthews, B. H. C. 1933. *J. Physiol. 78*:1–53.

[12]Barker, D., Stacey, M. J., and Adal, M. N. 1970. *Philos. Trans. R. Soc. Lond. (Biol. Sci.) 258*:315–346.

[13]Boyd, I. A. 1962. *Philos. Trans. R. Soc. Lond.* B *245*:81–136.

[14]Katz, B. 1950. *J. Physiol. 111*:261–282.

modified muscle fibers (called INTRAFUSAL FIBERS) running within a capsule. In the central, or equatorial, region, there is in each fiber a large aggregation of nuclei. Their arrangement provides a basis for the classification of intrafusal BAG or CHAIN FIBERS, depending on whether the nuclei are grouped together in a swollen protuberance or arranged linearly. Two types of sensory neurons innervate each muscle spindle. The large nerve fibers (called GROUP IA AFFERENTS) have a diameter of 12 to 20μm and conduct impulses at velocities of up to 120 m/sec. These are the largest, most rapidly conducting nerve fibers in the mammal; their terminals are coiled around the central parts of both bag and chain fibers to form the primary endings. Smaller sensory nerves

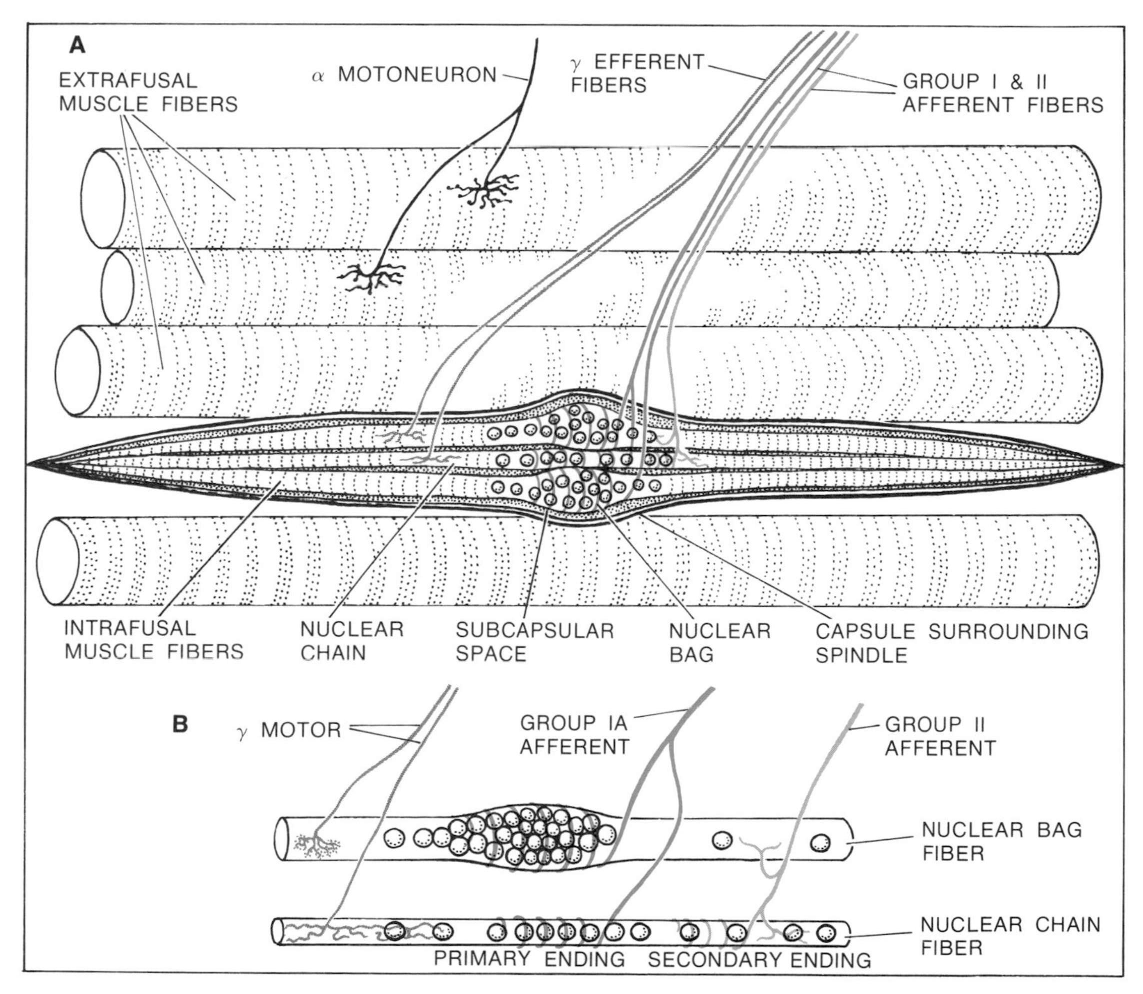

3

MAMMALIAN MUSCLE SPINDLE. A. Scheme of mammalian muscle spindle innervation. The spindle is embedded in the bulk of the muscle, made up of the large extrafusal muscle fibers and supplied by α-motoneurons. B. Diagram of the two intrafusal muscle fiber types and their innervation (see also Figure 5). (B from Matthews, 1964)

(called GROUP II), about 4 to 12μm in diameter, conduct more slowly; their terminals are mainly in the less central region of the chain fibers, where they form the secondary endings. Such a classification according to fiber diameter and conduction velocity provides a convenient approximation for subdividing sensory fibers.[15]

When a rapid stretch is applied to a muscle and thereby to the spindles within it, a burst of impulses arises in both types of sensory fibers. There is, however, a clear difference in the characteristics of the discharges in the two endings. The primary endings, connected to the larger group I axons, are sensitive mainly to the rate of change of stretch. The frequency of discharge is therefore maximal during the dynamic phase, while stretch is increasing, and subsides to a lower steady level while stretch is maintained. The secondary endings connected to the smaller group II fibers behave differently; they are relatively unaffected by the rate of stretch, but are sensitive to the steady level of tension.[16,17] This behavior is illustrated in Figure 4. Both types of nerve fibers contribute to the stretch reflex by exciting motoneurons in the spinal cord (see Chapter 16 and below).

Adaptation in sensory neurons

It is a common experience that the perception of a maintained stimulus tends to fade. One gradually becomes less aware of a constant pressure, an increased temperature on the skin, or contact with clothing or shoes. At least part of such sensory adaptation occurs through a decline in the frequency of firing in the primary receptors during a constant stimulus. This property is common to all receptors despite great differences in the degree and rate of adaptation. For example, certain mammalian mechanoreceptors, called PACINIAN CORPUSCLES, fire only a few impulses at the beginning of a maintained pressure, but can follow oscillatory displacements at high frequencies.[18] Adaptation in this case can be attributed largely to mechanical causes. The nerve terminal is surrounded by a special structure made of a series of re-

---

[15]Lloyd, D. P. C., and Chang, H. T. 1948. *J. Neurophysiol.* *11*:199–207.

[16]Matthews, P. B. C. 1964. *Physiol. Rev.* *44*:219–288.

[17]Bessou, P., and Laporte, Y. 1962. In D. Barker (ed.). *Symposium on Muscle Receptors.* Hong Kong University Press, Hong Kong, pp. 105–119.

[18]Gray, J. A. B. 1959. In J. Field (ed.). *Handbook of Physiology.* Vol. I. American Physiological Society, Washington, D.C., pp. 123–145.

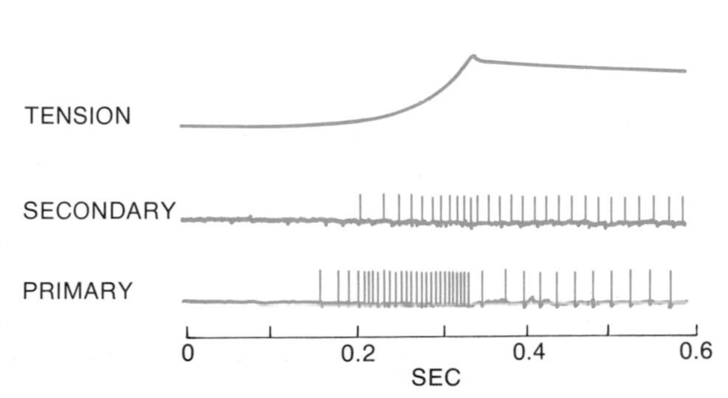

4

DIFFERENCES IN SPINDLE RESPONSES. Recordings from single primary (group I) and secondary (group II) sensory axons orginating in a muscle spindle in the cat. The larger, fast-conducting primary ending greatly increases its response rate during development of tension; afterward, it slowly adapts. The smaller, more slowly conducting secondary ending is little affected by the change in tension but maintains its discharge well during continued tension. (From Jansen and Matthews, 1962)

silient lamellae. It has been observed that after removal of the outer lamellae, the generator potential is better maintained during continued pressure.[19] Conversely, only little adaptation occurs in joint receptors that continually signal information to the central nervous system about the position of our limbs in space. Still other receptors start at a high rate but maintain a lower rate of impulses almost indefinitely.

A number of factors can contribute to the adaptation of sensory nerves. When bright light shines on photoreceptors, the amount of pigment diminishes as it becomes bleached. In addition, however, a neural component in visual adaptation is well established. In mechanoreceptors both their mechanical and their electrical properties can play a part. One example of a mechanical adaptation in the Pacinian corpuscle has already been mentioned. In stretch receptors a contributing factor to adaptation may be the viscoelastic properties of intrafusal muscle fibers or slippage in the attachment of the nerve terminals to the various receptor muscles. During maintained extension, such properties would allow the deformation of the terminals to decrease gradually.

At present it is difficult to assess precisely the relative contribution of mechanical and electrical factors in an individual case. For crustacean stretch receptors, Nakajima and Onodera[20] have shown that a variety of processes occur. The rapidly adapting receptor shows prompt adaptation of its firing rate to steady injected currents. In the slowly adapting receptor, in addition to mechanical factors, trains of impulses lead to an increase in internal sodium concentration and activation of the sodium pump (Chapter 12). The net outward transport of positive charges produces a hyperpolarization, driving the membrane potential away from threshold. With time, a steady stimulus to the slowly adapting neuron becomes less effective in initiating impulses, because in a hyperpolarized cell a greater generator potential is needed to initiate impulses.[21,22] Decreases in firing rate with maintained stimulation can also occur through processes that involve changes in ionic permeability. For example, in certain invertebrate nerve cells repetitive firing causes a steady hyperpolarization through a maintained increase in potassium conductance. In addition, partial inactivation of the sodium mechanism may play a role.

In conclusion, a number of electrical and mechanical factors seem to contribute to adaptation in different sense organs.

CENTRIFUGAL CONTROL OF SENSORY RECEPTORS

The central nervous system not only receives information from sensory receptors but also acts back upon them to modify their responses. The

---

[19] Loewenstein, W. R., and Mendelson, M. 1965. *J. Physiol.* *177*:377–397.
[20] Nakajima, S., and Onodera, K. 1969. *J. Physiol.* *200*:187–204.
[21] Nakajima, S., and Takahashi, K. 1966. *J. Physiol.* *187*:105–127.
[22] Sokolove, P. G., and Cooke, I. M. 1971. *J. Gen. Physiol.* *57*:125–163.

brain, therefore, has the built-in ability to edit, censor, or adjust the flow of information that reaches it. This is usually called a FEEDBACK CONTROL and is executed through pathways leading centrifugally to a variety of peripheral sensory organs. Although centrifugal control has been most fully analyzed in mammalian muscle spindles, it is here discussed first in the crustacean stretch receptor, in which the elements of the system are most easily demonstrated.

Excitatory and inhibitory neural control of stretch receptors

The muscle strands (or receptor muscles) of the crustacean stretch receptors are innervated by excitatory and inhibitory motor axons that can be stimulated electrically. Excitation causes the ends of the muscles to contract more and thereby stretch the centrally located dendrites of the sensory cells. The time course of the generator potential thus mirrors the contraction of the muscle strand; its size, in turn, determines the frequency of the sensory impulses.[2,23] Figure 5*A* shows the principle in a simplified drawing, and for comparison Figure 5*B* presents

[23]Kuffler, S. W. 1954. *J. Neurophysiol.* 17:558–574.

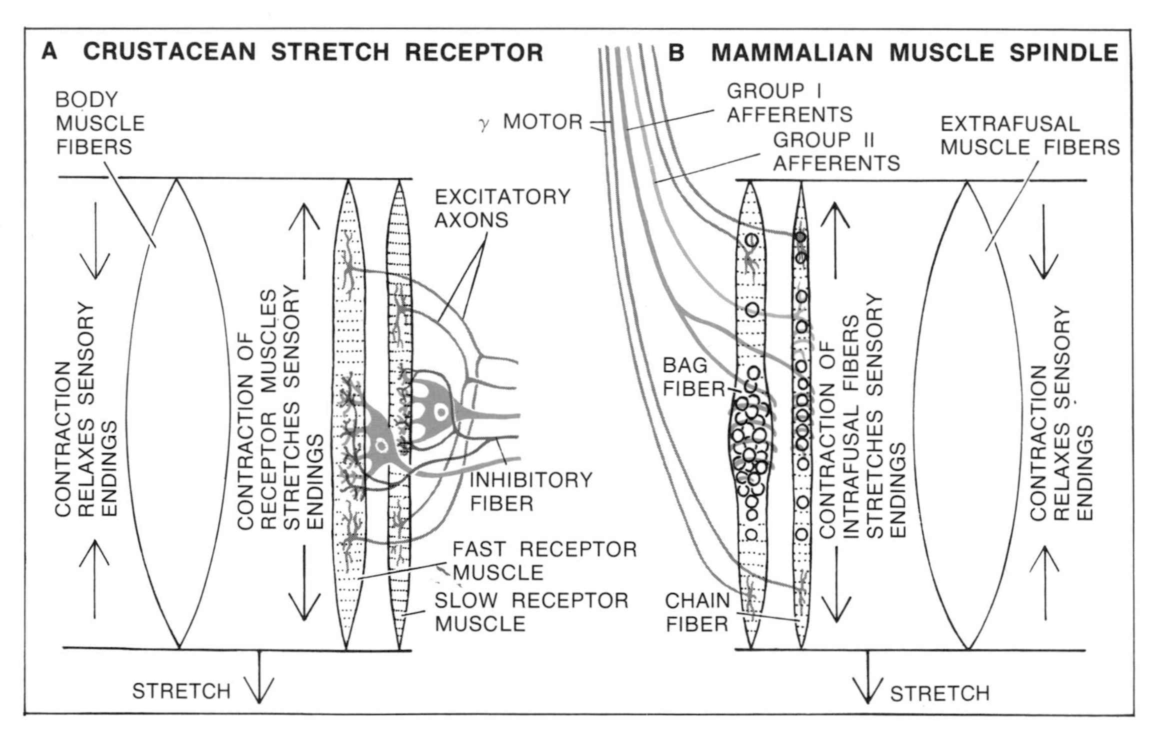

5 **CENTRIFUGAL CONTROL of stretch receptors. A. Excitatory and inhibitory innervation of the crustacean stretch receptor. The slow and fast receptor muscle strands can be made to contract by excitatory motor axons. A second excitatory mechanism is passive stretch of the receptor. The sensory receptor neurons are also innervated by inhibitory fibers that counteract excitation. B. The scheme of excitation is basically similar in the mammalian spindle, which, however, lacks inhibitory innervation. For clarity, the main muscle mass is drawn separately in both preparations. Its contraction reduces stretch on sensory endings.**

a sketch of a mammalian spindle. Many essential features are similar in both systems. In the typical crustacean stretch receptor the sensory neuron is embedded in a muscle strand that contracts when the motor nerves are stimulated. One sensory neuron inserts into a muscle strand that gives graded slow contractions with nerve stimulation, and when stretched the discharges adapt slowly; the dendrites of the other neuron insert into a muscle strand that gives twitchlike contractions, and when stretched the discharges adapt rapidly.

The effects of contractions of fast (twitch) bundles are shown in Figure 6*B* and *C*. The generator potential faithfully reflects the transient rapid individual contractions. Figure 6*D* shows that sensory discharge is greatly accelerated by a contraction in the slow receptor muscle (note different time scales). There exists, therefore, a dual excitatory mechanism for initiating a sensory discharge: (1) passive stretch of the muscle, in which the receptor terminals are embedded, as discussed earlier; and (2) control originating in the central nervous system that makes the receptor muscle itself contract and deforms the endings. Thus, with such a dual control, receptor neurons can be made to signal even when the muscles in which they are embedded are slack (Figure 6*B*, *C*), or their ongoing signals can be accelerated when they are already discharging (Figure 6*D*).

The control of receptor function is capable of even finer modulation than is indicated by the dual excitatory mechanism. The sensory neurons of both types of stretch receptors are innervated by efferent inhibitory fibers.[24] Jansen and his colleagues[25] have shown that as many as three inhibitory axons form synapses on the dendrites and cell bodies of these neurons. An example of inhibitory action is shown in Figure 7, in which only one inhibitory axon is indicated. In the face of a maintained stretch that makes the receptor fire at a rate of 11/sec, a train of inhibitory impulses (at 34/sec) keeps the membrane potential below the threshold level (Figure 7*B*). Once inhibitory stimulation is stopped, the sensory impulses are promptly resumed. In Figure 7*C* the individual contribution of each inhibitory impulse, this time at 21/sec and higher amplification, can be resolved as a separate transient hyperpolarization. At a much greater frequency (150/sec), inhibition practically "clamps" the membrane potential at a steady level (Figure 7*D*). Note that as far as the higher centers are concerned it makes no difference whether the stimulus is 34 or 150/sec, because both frequencies suppress all the sensory signals. However, a weaker inhibition can be more readily overruled or canceled by stronger excitation.

The stretch receptor provides an example of multiple excitatory and inhibitory actions characteristic of neurons within the central nervous system. The depolarizing generator action drives the membrane potential beyond its firing level, while the inhibitory synaptic action tends to

---

[24]Kuffler, S. W., and Eyzaguirre, C. 1955. *J. Gen. Physiol.* *39*:155–184.

[25]Jansen, J. K. S., Njå, A., Ormstad, K., and Walloe, L. 1971. *Acta Physiol. Scand.* *81*:273–285.

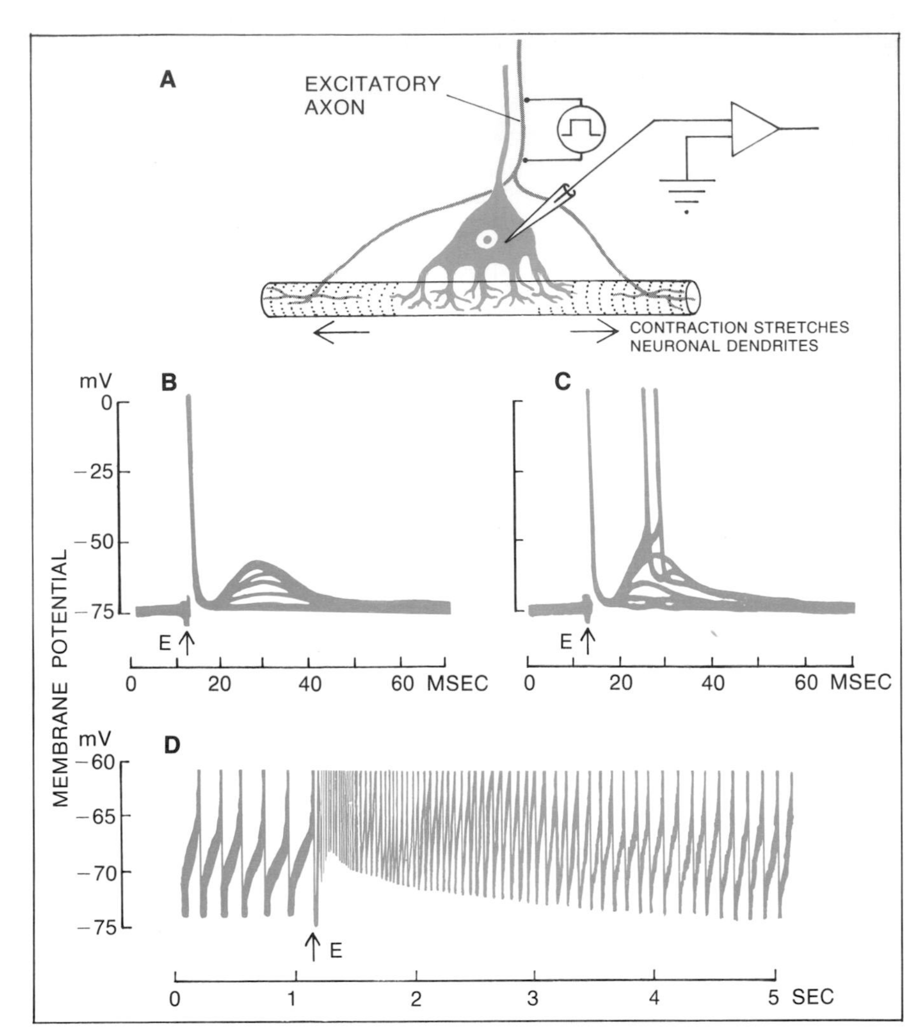

6 **CENTRIFUGAL EXCITATORY CONTROL OF SENSORY IMPULSES.** **A. Excitatory stimulation of the motor nerve to the receptor muscle while registering potential changes in terminals and cell body of the receptor neuron. B. Successive superimposed records during stimulation at 4/sec of a fast receptor muscle. The time course of the contraction is reflected in the subthreshold generator potentials. C. With stimulation at 10/sec, the contractions build up until threshold is reached and sensory impulses are set up. D. A slow receptor muscle gives a maintained sensory discharge that is accelerated by two closely spaced motor stimuli (arrow). The duration of the increased discharge rate reflects the time course of contraction. Note different amplifications and time scales. In B and C the large rapid deflections preceding the generator potentials are from the sensory axon which happened to be excited by the excitatory stimulus. In D only the lower portions of impulses are seen.**

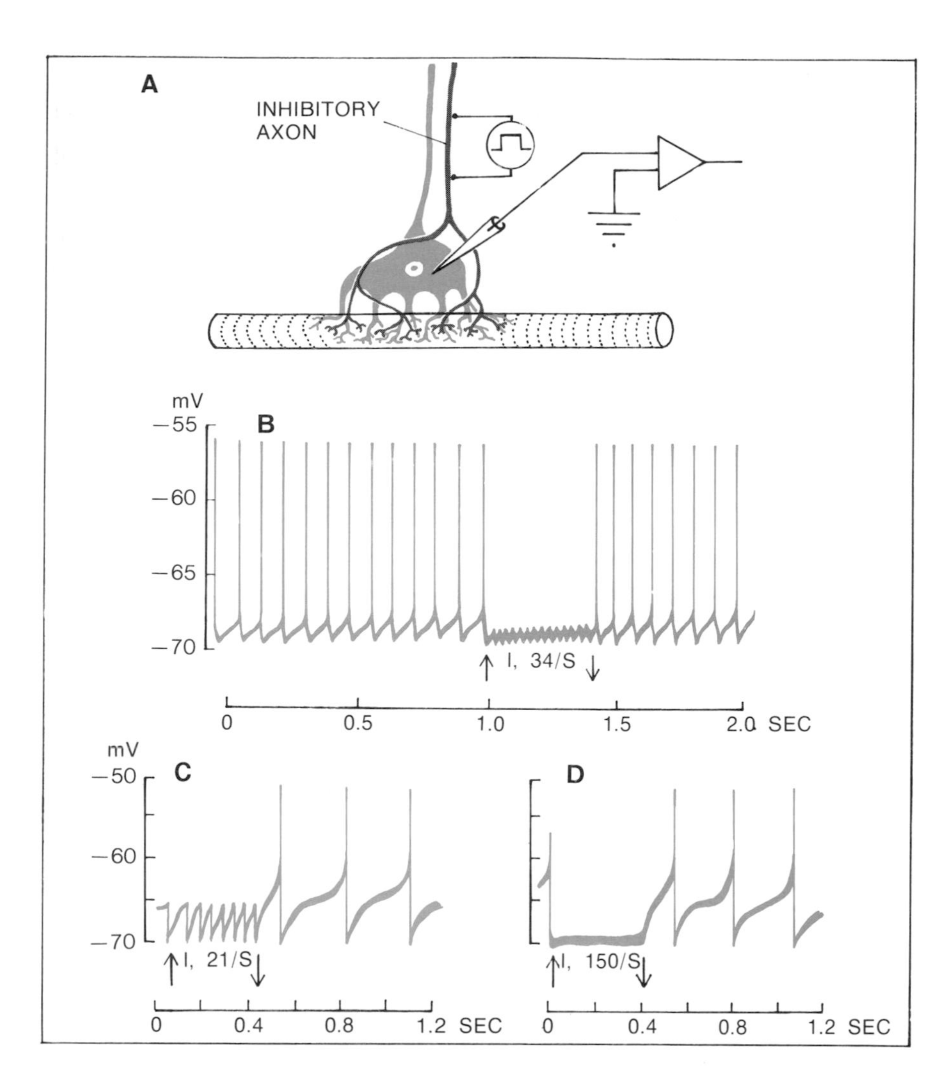

INHIBITORY CONTROL OF SENSORY DISCHARGES. A. Stimulating electrodes on inhibitory axon. B. Maintained sensory signals are suppressed for the duration of inhibitory stimulation. C. At higher amplification each inhibitory impulse at 21/sec is seen to transiently hyperpolarize the membrane potential. D. High-frequency inhibitory impulses keep the membrane potential steady in the face of continued stretch. (From Kuffler and Eyzaguirre, 1955) 7

keep the membrane potential below threshold. The balance of these two competing influences determines whether the membrane potential of a receptor neuron is "set" above or below the critical threshold level, so that the cell keeps discharging or remains quiescent. The effectiveness of synaptic inhibitory modulation depends, as expected, on the frequency of the inhibitory impulses. But, in addition, the crustacean receptor in this example has a choice of three neurons with different pathways. Thus, the largest inhibitory axon exerts a more powerful inhibitory effect than the two smaller ones, which require a greater frequency of stimuli to produce an equivalent reduction in the rate of sensory discharges. They all act by the same ionic mechanism, producing conductance increases for the same ions, predominantly for chloride; they also use the same transmitter, γ-aminobutyric acid (GABA; Chapter 11).[26]

On closer scrutiny, then, the information that leaves the simple crustacean stretch receptors is highly controlled and adjustable over a wide range. The signals are determined by the excitatory action of graded amounts of external stretch, by contraction (shortening) of the receptor muscles, and by inhibitory synapses that apply a variable negative bias. These processes interact and exert their influence on the initial segment of the axon, where impulses start. The end result of impulses in the stretch receptors is to initiate a well-defined reflex by activating motoneurons within the central nervous system.[27]

Centrifugal control of muscle spindles

The motor control of mammalian muscle spindles is similar to that of crustacean stretch receptors. Stretch of the muscle in which they are embedded (Figures 3 and 4) gives rise to sensory discharges, and as explained below the same happens when intrafusal muscle fibers are stimulated. The original experiments by B. H. C. Matthews[11] suggested that stimulation of the motor nerves innervating a muscle can produce two effects. First, contraction of the main mass of ordinary muscles causes a cessation of the afferent discharge arising in muscle spindles (Figure 10). This occurs because the spindles lie in parallel with the contracting fibers, so that their shortening reduces the tension on the sensory endings (Figure 5*B*). Second, stronger motor nerve stimulation frequently adds a burst of sensory impulses which he suspected to be caused by stimulation of small diameter motor fibers, initiating contraction of the intrafusal muscle fibers.

In 1945, Leksell,[28] a neurosurgeon, studied the efferent nerve supply to skeletal muscles that contained a distinct group of small motor nerves (2 to 8μm in diameter), originally described by Eccles and Sherrington.[29] Stimulation of these fibers, now called FUSIMOTOR or γ-FIBERS, caused no increase in muscle tension but did increase the sensory activity originating in muscles. Confirming these ideas, the role of fusimotor fibers was soon firmly established on various limb muscles of the cat. The procedure was to record in the dorsal root the activity in an individual afferent fiber coming from a spindle in an anesthetized cat while stimulating a fusimotor fiber in the ventral root going to the same spindle (Figure 8).[30] Such stimulation caused an increase in the

---

[26]Hagiwara, S., Kusano, K., and Saito, S. 1960. *J. Neurophysiol.* *23*:505–515.

[27]Fields, H. L., Evoy, W. H., and Kennedy, D. 1967. *J. Neurophysiol.* *30*: 859–874.

[28]Leksell, L. 1945. *Acta Physiol. Scand.* *10* (Suppl. 31): 1–84.

[29]Eccles, J. C., and Sherrington, C. S. 1930. *Proc. R. Soc. Lond.* B *106*:326–357.

[30]Kuffler, S. W., Hunt, C. C., and Quilliam, J. P. 1951. *J. Neurophysiol.* *14*: 29–54.

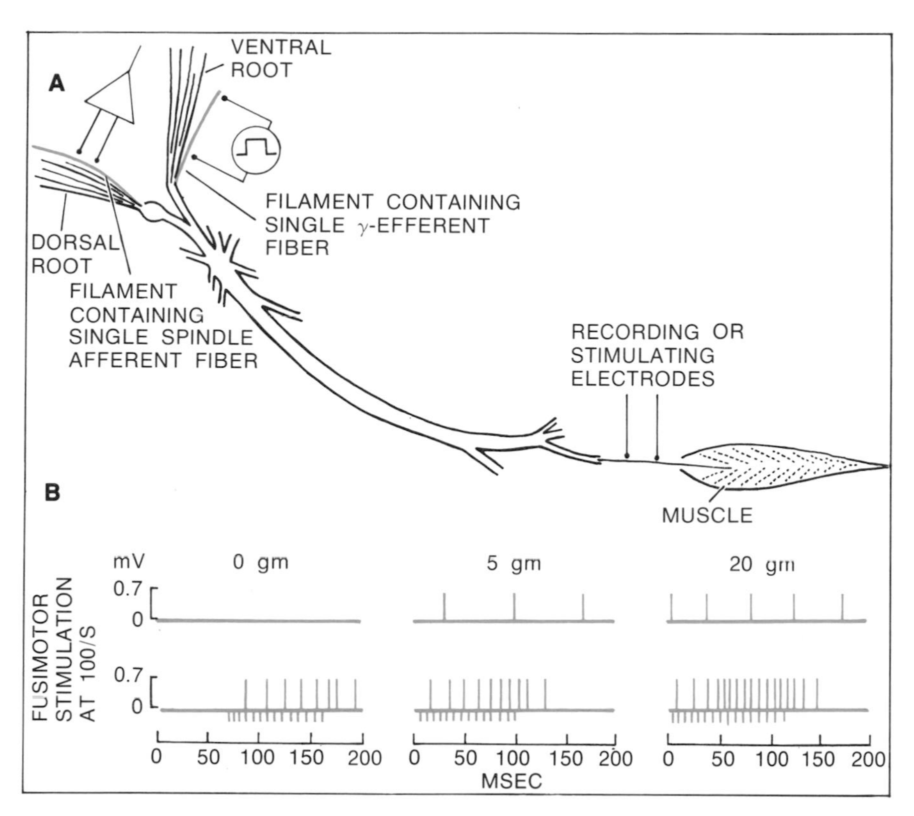

8

CENTRIFUGAL CONTROL OF MUSCLE SPINDLE. A. Recordings from a single dorsal root axon arising in a muscle spindle of an anesthetized cat; a single fusimotor (γ) axon in the ventral root innervating the same spindle is stimulated. B. Upper records show sensory discharges when the muscle is slack (0-g tension) or is lightly stretched (5- and 20-g tension). A brief train of 14 to 15 stimuli at 100/sec either initiates sensory discharges (lower left) or accelerates them. Extrafusal muscle fibers remain inactive. (After Kuffler, Hunt, and Quilliam, 1951)

frequency of the sensory spindle discharge without visible muscle contraction or impulses in skeletal muscle fibers. Trains of impulses in the fusimotor neurons initiated a sensory discharge even in the absence of passive stretch in a completely slack muscle (0 g tension in Figure 8*B*). If the initial load (or stretch) caused a background discharge, this was accelerated during a train of γ-impulses.

Further studies on the effects of intrafusal contraction in the cat have established the existence of two types of efferent fibers supplying mammalian spindles.[16,31] One produces an increase in the static discharge; the other affects mainly the dynamic component (Figure 9). It seemed likely that this result could be explained by assuming that there are two kinds of fusimotor fibers: the DYNAMIC (γ_d) FIBERS supplying the

[31]Jansen, J. K. S., and Matthews, P. B. C. 1962. *J. Physiol.* *161*:357–378.

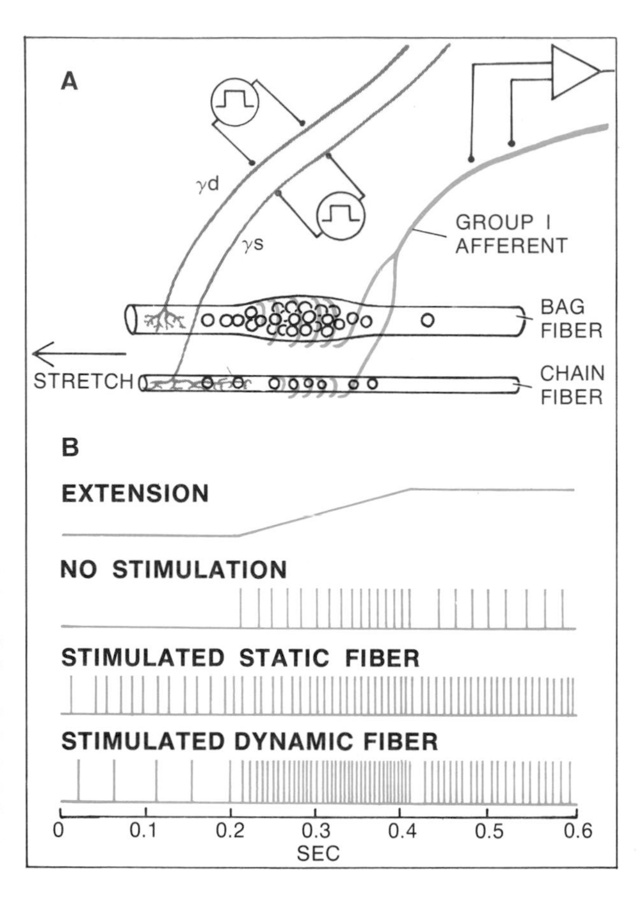

9

SEPARATE INNERVATION of bag and chain fibers. A. Scheme according to Matthews (1964). B. Different effects of stimulating two fusimotor fibers on the discharge arising in a primary (group I) sensory ending of a muscle spindle in the cat. During extension the sensory fiber responds with an increased frequency. Stimulation of the static fusimotor fiber (γ_s), starting near the beginning of the second record, causes a steady increase in the rate of sensory discharges. Stimulation of the dynamic fusimotor fiber (γ_d) causes a larger increase in sensory discharge during stretch. (After Crowe and Matthews, 1964)

nuclear bag fibers and the STATIC EFFERENT FIBERS (γ_s) innervating the chain fibers. More recent experiments by Barker, Laporte, and colleagues have shown, however, that this is probably an oversimplification.[32,33] In several instances the same fusimotor fiber may innervate both bag and chain fibers.

So far the basic similarities between stretch receptors have been stressed. There are, of course, differences as well, even between closely related species. For example, in the cat centrifugal innervation of intrafusal muscle fibers is, with few exceptions, separate from the innervation of extrafusal muscle fibers that make up the bulk of skeletal muscles. In frogs and toads a distinction of this sort is not found. A large motoneuron causing muscle twitches will at the same time innervate intrafusal muscle fibers and make them contract.[34,35] In the frog and toad there exists a rigid coupling of these two functions. This arrangement apparently reflects a less finely tuned efferent control by the central nervous system.

---

[32]Barker, D., Emonet-Dénand, F., Laporte, Y., Proske, U., and Stacey, M. J. 1973. *J. Physiol.* *230*:405–427.

[33]Brown, M. C., and Butler, R. G. 1973. *J. Physiol.* *233*:553–573.

[34]Katz, B. 1949. *J. Exp. Biol.* *26*:201–217.

[35]Eyzaguirre, C. 1957. *J. Neurophysiol.* *20*:523–542.

Recently, other preparations have been found that offer special advantages for studying spindles. In the tail of the cat and the rat, and in snake muscles, spindles can be dissected out and maintained in vitro.[36] The sensory nerve endings and the intrafusal fibers have been viewed directly by Hunt and Ottoson[37] and their colleagues. This has provided a good opportunity for precise measurements of the displacement produced by stretch or contraction and for a correlation with sensory discharges.

A NOTE ON THE CLASSIFICATION OF NERVE FIBERS IN MAMMALS

It is convenient to distinguish among different nerves according to their diameter and conduction velocity. If one stimulates a nerve bundle electrically at one end and records from it some distance away, one sees a series of potential peaks on the oscilloscope tracing. These are the result of dispersion of nerve impulses that travel at different speeds and therefore arrive at the recording electrodes at different times. On the basis of their electrical properties, the fibers in mammalian nerves were subdivided into groups named A, B, and C. Next, the myelinated, rapidly conducting A fibers were grouped into four subdivisions designated α, β, γ, and δ, in order of decreasing conduction velocity. The motor efferent fibers that supply the intrafusal muscles correspond to the γ group, but the term "fusimotor," introduced at a later stage, is meant to include any axon that innervates intrafusal fibers. When, in subsequent years, extensive recordings were made from group A sensory nerves supplying muscle, these nerves were classified according to a different convention as group I, group II, and group III.[15] Group Ia fibers correspond to Aα and form the primary endings on muscle spindles; group Ib fibers, discussed below, have endings that respond to stretch of receptors called "tendon organs"; group II fibers correspond to Aβ and form secondary endings. The still smaller myelinated fibers of group III are not discussed here.

Stretch receptors and the control of movement

What is the role of crustacean stretch receptors and muscle spindles? With the detailed, but still growing, knowledge about the peripheral mechanisms that govern the behavior of these sense organs, the search is shifting more and more to their central connections and the control that higher centers exercise. The magnitude of the task of working out the central organization becomes apparent: about one third of all the motor nerves that leave through the ventral roots is concerned with the control of muscle spindles. The fusimotor neurons together with their cell bodies in the spinal cord and attached connections constitute a cell system that is impressively large in its mass and numbers. It is therefore not surprising that knowledge of the actual workings of these neurons is incomplete and that discussion is still couched in terms of "general outlines" and "basic roles" rather than the desired details of

[36]Fukami, Y., and Hunt, C. C. 1970. *J. Neurophysiol.* 33:9–27.
[37]Hunt, C. C., and Ottoson, D. 1975. *J. Physiol.* 252:259–281.

the wiring diagram for function that would give a comprehensive view. Besides, spindles are just one part of a larger interconnected sensory system and must be considered together with other peripheral sensory components from the muscles themselves, the joints, and the skin.

For many years it was thought that sensory information from muscle spindles was concerned solely with reflex regulation of muscles and did not reach the cerebral cortex. Position sense was attributed entirely to receptors situated in joints. Recently, however, it has been shown that information from spindles does reach the cerebral cortex and consciousness: patients and volunteers have reported sensations of movements or stiffness when muscles were vibrated or stretched after all sensation had been abolished in joint receptors.[38]

Muscle spindles are sensing elements that provide information about the state of muscles, their actual length, and the rate at which their length is changing. The skeletal muscles that do the work are directly regulated by groups of large motoneurons (usually called α-MOTONEURONS) in the ventral part of the spinal cord. All the neural apparatus that influences movement must converge on these cells. Considering numbers once more, this is an apparatus in which one third of all the motoneurons, the γ- or fusimotor neurons, is concerned with regulation rather than execution of movement. Apparently, the more finely honed the movement, the more nervous machinery is devoted to its control. For example, the muscles of our hands are more densely supplied with spindles than are the larger limb muscles or the diaphragm.

One role played by the efferent innervation of intrafusal muscle fibers in the spindle becomes apparent when sensory discharges during muscle movement are considered. As the muscle contracts, the tension on the sensory element is reduced and the spindle is unloaded (Figure 10). During the shortening, the rate of sensory discharges is reduced or the signals stop. Therefore, in the absence of efferent γ- or fusimotor impulses to the intrafusal muscle elements, the spindle would temporarily send fewer impulses, or if the muscle stayed in a shortened state, sensory signals would cease altogether. In such a situation stimulation of the γ-efferents can shorten intrafusal muscle elements, take up the slack, and restore the tension on the sensory terminals (Figure 10*C*).[39] One basic part played by efferent nerves is to ADJUST THE SENSITIVITY OF THE MEASURING INSTRUMENT, the spindles, so that they can perform over a wide range of muscle length. Spindles, therefore, can be made to maintain their discharge frequencies even when the external stretch on them is reduced during contraction of muscles (Figures 10 and 12).

Spinal connections of stretch receptors

The simplified diagram in Figure 11 outlines connections used for the stretch reflex and for reciprocal organization of extensor and flexor muscles in the cat. It is the simplest of many pathways and represents

---

[38]Goodwin, G. M., McCloskey, D. I., and Matthews, P. B. C. 1972. *Brain 95:* 705–748.

[39]Hunt, C. C., and Kuffler, S. W. 1951. *J. Physiol. 113:*283–297.

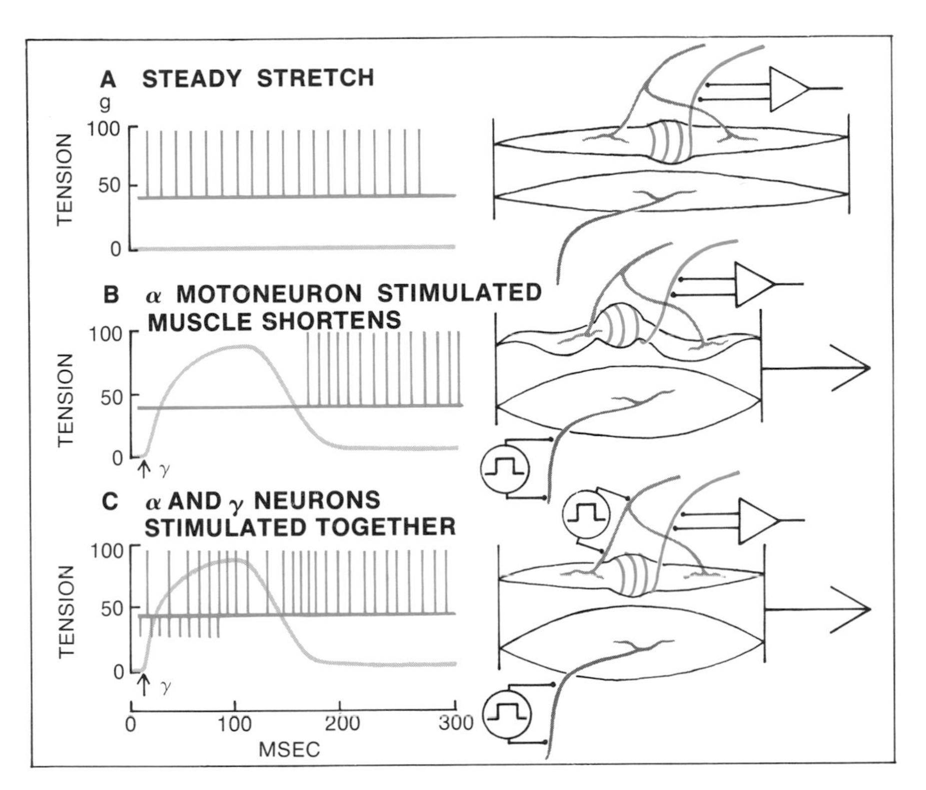

EFFECT OF FUSIMOTOR STIMULATION during contraction. Sensory discharges from a muscle spindle in the leg of a cat, recorded from an afferent fiber in the dorsal root. A. During steady stretch the discharges are maintained. Tension is registered on the lower trace. B. During stimulation of the large α-motoneurons, the muscle shortens and the spindle ceases to fire. C. If the α-motoneurons and the γ-efferent (fusimotor) fiber are stimulated together, the spindle maintains its discharges during the muscle shortening. (After Hunt and Kuffler, 1951)

10

only a small part of what is known about spinal reflexes. Assume that the sketch in Figure 11 illustrates an extensor muscle—for example, a muscle that straightens the leg. If the muscle is stretched, or when the tendon is tapped, as occurs in the knee-jerk reflex (Chapter 4), the primary sensory endings are deformed and initiate impulses; a somewhat stronger pull also brings in the secondary spindle endings. In each case the afferent frequency is highest while the muscle is extended. The first event is a monosynaptic excitation of the α-motoneurons going back to the muscle that had been extended; discharges from the smaller sensory fibers (secondary group II, not shown) reinforce the effect shortly thereafter, both monosynaptically and by way of interneurons.[40,41] The reflex contraction of the extensor muscle is

---

[40]Matthews, P. B. C. 1969. *J. Physiol.* *204*:365–393.
[41]Kirkwood, P. A., and Sears, T. A. 1974. *Nature* *252*:243–244.

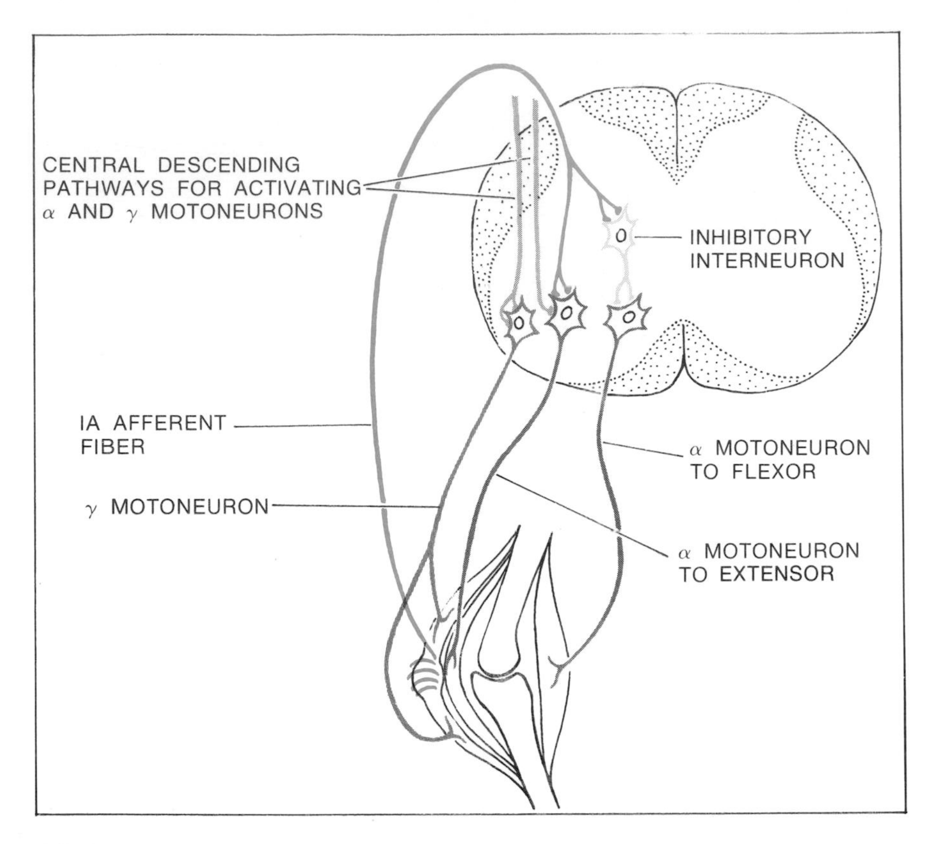

11 **CONNECTIONS FOR STRETCH REFLEX. Sketch of two antagonistic muscles, an extensor and a flexor, together with some of their connections in the spinal cord that are activated through sensory discharges from a spindle. When the extensor muscle is briefly stretched the α-motoneuron is excited monosynaptically (the stretch reflex) while the flexor is inhibited via interneurons (reciprocal innervation). Thereby antagonistic muscles cooperate. Central descending pathways activate α-motoneurons to initiate voluntary movements and excite γ-motoneurons to maintain sensory discharges during contraction (see text).**

further helped by a simultaneous inhibition of the α-motoneurons that innervate the antagonistic flexors. This principle of one group of muscles being excited while its antagonists are inhibited is owed to Sherrington, who called it RECIPROCAL INNERVATION.

If, instead of a brief tap on the tendon of an extensor muscle, a steady pull is exerted, the muscle contracts in response to discharges from primary afferents (marked Ia), the load is taken off the spindles, and thereby the drive on the α-motoneurons is reduced. This, in turn, leads to a renewed lengthening of the muscle and a renewed sensory acceleration of spindle afferent discharges. Such an alternate lengthening and shortening, an oscillation, is in phase with the fluctuations of the discharges from spindles. If activation of muscle spindles is now interposed at the appropriate time through the γ-fibers, the fusimotor discharges can reduce the slack on the muscle spindles during muscle shortening (Figure 10).

For the sake of simplicity a number of the known pathways and connections are omitted from Figure 11. These include the group II afferents within the spindle, which play a part in the excitatory drive for the stretch reflex, and the connections of various interneurons within the spinal cord that have been worked out in detail by Lundberg and his colleagues.[42,43] Further, there exist other sensory receptors within muscles that play a role in the stretch reflex. The most important of these are usually found near the tendon-muscle junctions and are called the TENDON ORGANS; they lie in series with the contracting regular skeletal muscle fibers and are therefore made to discharge impulses by both passive stretch and contraction of muscle fibers.[44] They have no centrifugal control. Their axons, classified Ib, activate interneurons that inhibit the α-motoneurons innervating the muscle from which the axons of the tendon organs arise.

A number of schemes have assigned to spindles roles in movement and in the maintenance of posture. For example, fusimotor fibers can in theory be used to initiate reflex contractions by producing sensory discharges that in turn excite the α-motoneurons. This is an attractive hypothesis from an engineering and a teleological point of view.[45] Probably the greatest weight of evidence, obtained in cats, suggests that the γ-efferent system is used chiefly to maintain the synaptic drive during muscular contraction and at different muscle lengths. In terms of feedback loops, the occurrence of α- and γ-coactivation could be considered as a servoassistance mechanism for controlling movement.[46]

A favorable situation for demonstrating a role for γ-efferent fibers in normal movement has been found in respiratory muscles. Sears,[47] von Euler,[48] and their colleagues have recorded the discharges in afferent fibers from spindles in the intercostal muscles used for inspiration and expiration. Figure 12 shows that under normal conditions the afferent discharge from an inspiratory muscle is greatest during inspiration. This at first is surprising, because during inspiration that muscle actually shortens (Figure 12*A*). A simple explanation is that the fusimotor fibers are being excited together with the α-motoneurons to the same muscle and thereby overcome the slack created by contraction. This is confirmed in Figure 12*B*, which shows the responses of the afferent fiber after the γ-efferents to the inspiratory muscle were paralyzed by

---

[42]Lundberg, A. 1970. In P. Andersen and J. K. S. Jansen (eds.). *Excitatory Synaptic Mechanisms.* Universitetsforaget, Oslo, pp. 333–340.

[43]Jankowska, E., and Lindström, S. 1972. *J. Physiol.* *226*:805–823.

[44]Houk, J., and Henneman, E. 1967. *J. Neurophysiol.* *30*:466–481.

[45]Merton, P. A. 1953. In G. E. W. Wolstenholme (ed.). *The Spinal Cord.* Churchill, London, pp. 247–255.

[46]Matthews, P. B. C. 1972. *Mammalian Muscle Receptors and Their Central Actions.* Edward Arnold, London.

[47]Sears, T. A. 1964. *J. Physiol.* *174*:295–315.

[48]Critchlow, V., and von Euler, C. 1963. *J. Physiol.* *168*:820–847.

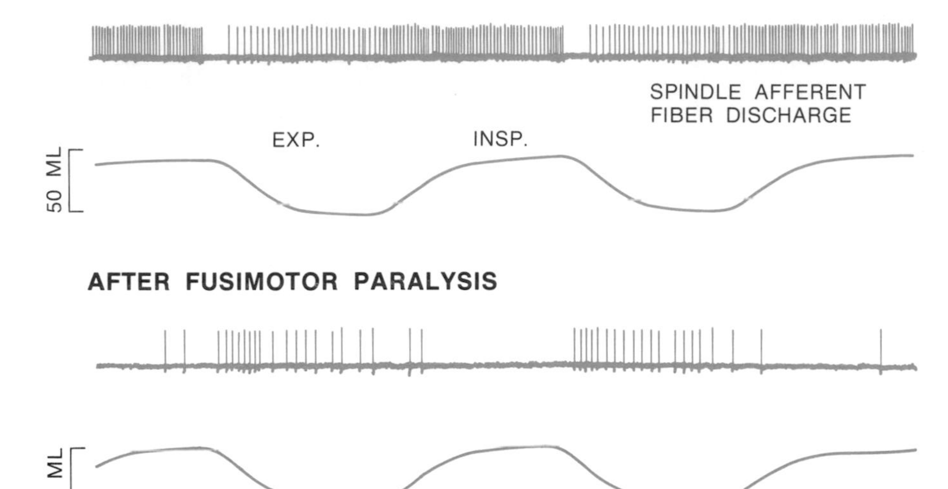

12 **REFLEX CONTROL OF SPINDLE ACTIVITY. Evidence in the cat for the simultaneous efferent activation of the motor innervation of an intercostal respiratory muscle and of its muscle spindles during breathing. Normally (A) the sensory discharge frequency from an inspiratory muscle is highest during inspiration, even though the muscle is shortening. After the fusimotor fibers have been selectively blocked by procaine (B), the spindle behaves passively and its discharges cease during inspiration when the muscle contracts. This is the response expected for a muscle spindle afferent fiber in the absence of γ-control. Upper traces register spindle afferent discharge, lower traces register respiration. (After Critchlow and von Euler, 1963)**

applying the local anesthetic procaine to the intercostal nerve. This procedure blocks the relatively small diameter γ-fibers, but not the larger α-fibers, allowing the extrafusal muscle fibers to contract as before. Under these conditions the afferent discharge occurs, as expected, only during expiration while the inspiratory muscle is being stretched. Other examples of the parallel activation of γ- and α-motoneurons are provided by experiments on walking movements in the cat and finger movements in man.[49,50]

Clearly, the efferent fusimotor fibers emerging from the spinal cord play an essential role in controlling and modulating the reflexes used in maintaining posture and in executing smooth movements. Clinical studies, particularly those on injuries to dorsal roots, tend to support this view.

---

[49]Serverin, F. V., Orlovskii, G. N., and Shik, M. L. 1967. *Biophysics 12*:575–586.
[50]Vallbo, A. B. 1971. *J. Physiol. 218*:405–431.

OTHER RECEPTORS AND CONTROL OF ASCENDING PATHWAYS

Spindles and stretch receptors are convenient models of sensory receptors because the mechanisms underlying their signaling are similar in many sense organs. Thus, depolarizing generator potentials have been recorded and studied in detail from various mechanoreceptors, including the vibration-sensitive Pacinian corpuscles (see above), and from olfactory chemoreceptors.[51] Among the photoreceptors, an interesting difference exists between vertebrates and invertebrates: in invertebrate eyes, such as the eyes of horseshoe crabs, barnacles, and crayfish, illumination causes a conductance increase that leads to a depolarization; in vertebrate eyes, photoreceptors become hyperpolarized as a result of a conductance decrease for sodium (Chapter 2).

While the principle of centrifugal control has been discussed for spindles and stretch receptors only, similar mechanisms also operate in other sense organs. For example, an inhibitory efferent innervation to the first-order sensory cells in the auditory system has long been known,[52] and the cellular basis of this is now being worked out in detail, including the synaptic innervation of the hair cells.

A control system, with excitatory as well as inhibitory centrifugal fibers, has been demonstrated in the lateral line organs of fish.[53] The comparative chemistry is unexpectedly and interestingly different in cells of the lateral line organs. An unusual transmitter action found here is that efferent nerves appear to liberate acetylcholine (ACh) as an inhibitory transmitter, while γ-aminobutyric acid may be the excitatory transmitter used for initiating sensory discharges (Chapter 11).[54]

Within the central nervous system itself, the distinction between afferent sensory and efferent or motor becomes at times difficult. The visual system offers a good example of progressive feedback control that occurs in successive stages as visual information ascends to higher centers. First, the amount of light allowed to fall on the receptors is regulated by the size of the pupil. The properties and functional significance of this feedback loop have been worked out in detail. Feedback also goes on within the retina: receptors influence horizontal cells and these then act back upon the receptors and modify their responses to light. In addition, as mentioned in Chapter 2 in birds there exists an efferent control from the brain to the eye.[55] The isthmo-optic nucleus sends into the retina centrifugal fibers, probably ending on amacrine cells, that modulate the discharges set up by light stimulation; thereby the higher centers are able to alter the responses from individual ganglion cells.[56] In the cat, descending fibers run from the visual cortex to

[51]Ottoson, D. 1971. In L. M. Beidler (ed.). *Handbook of Sensory Physiology*, Vol. IV, *Olfaction*. Springer, New York, pp. 95–131.

[52]Fex, J. 1962. *Acta Physiol. Scand. 55* (Suppl. 189): 1–68.

[53]Flock, A., and Russell, I. J. 1973. *J. Physiol. 235*:591–605.

[54]Flock, A., and Lam, D. M. K. 1974. *Nature 249*:142–144.

[55]Cowan, W. M. 1970. *Br. Med. Bull. 26*:112–118.

[56]Miles, F. A. 1972. *Brain Res. 48*:115–129.

the colliculi, again modifying the responses of neurons to visual stimuli. A tract from the cortex to the geniculate nucleus is also known, but no role has yet been assigned to it; and a corticofugal innervation to the olfactory bulb of the rat has been described.[57]

In the somatosensory system, centrifugal fibers descend from the cortex and end in the first relay station for touch and pressure neurons in the medulla oblongata; these are the gracile and cuneate nuclei. The next relay in the ventrobasal nucleus of the thalamus also is supplied by fibers from the cortex.[58]

Centrifugal control has been mainly considered in peripheral sense organs because the principal mechanisms can be most easily demonstrated and understood in these examples. At the same time it bears emphasizing that corticofugal pathways may have a much wider role that is as yet insufficiently formulated and therefore remains obscure. The observations that ascending pathways can be suppressed at different levels suggest that efferent inhibition may be used to prevent unwelcome or distracting information from reaching certain parts of the brain, for example, during attention. Of particular interest is the possibility of suppressing or "gating" pathways that convey pain.[59] Efferent control could also serve the opposite purpose—to enhance the flow of afferent information, thereby lowering the threshold for sensory stimuli. Many of these schemes can now be tested with available methods. An intriguing but fanciful idea is the use of efferent pathways to stir up subcortical centers for recall of stored information.

Centrifugal control, therefore, seems to be part of the general organization of our higher centers that continuously edit and transform information at various stages of processing.

SUGGESTED READING

General reviews

Fex, J. 1974. Neural Excitatory Processes of the Inner Ear. In W. D. Keidel and W. D. Neff (eds.). *Handbook of Sensory Physiology*, Vol. V, *Auditory System*. Springer, New York, pp. 585–646.

Granit, R. (ed.). 1966. *Muscle Afferents and Motor Control*. Wiley, New York.

Hunt, C. C. 1974. The Physiology of Muscle Receptors. In C. C. Hunt (ed.). *Handbook of Sensory Physiology*, Vol. III, *Muscle Receptors*. Springer, New York, pp. 191–234.

Kuffler, S. W. 1960. Excitation and inhibition in single nerve cells. *Harvey Lect. Academic Press*, New York, pp. 176–218.

Matthews, P. B. C. 1972. *Mammalian Muscle Receptors and their Central Actions*. Edward Arnold, London. (An authoritative, critical, and compre-

---

[57]Price, J. L., and Powell, T. P. S. 1970. *J. Anat.* *107*:215–237.

[58]Towe, A. L. 1973. In A. Iggo (ed.). *Handbook of Sensory Physiology*, Vol. II, *Somatosensory System*. Springer, New York, pp. 700–718.

[59]Wall, P. D. 1973. In A. Iggo (ed.). *Handbook of Sensory Physiology*, Vol. II, *Somatosensory System*. Springer, New York, pp. 253–270.

hensive review of the structure, properties, and functions of muscle spindles.)

Stein, R. B. 1974. Peripheral control of movement. *Physiol. Rev. 54*:215–243.

Original papers

Critchlow, V., and von Euler, C. 1963. Intercostal muscle spindle activity and its γ motor control. *J. Physiol. 168*:820–847.

Eyzaguirre, C., and Kuffler, S. W. 1955. Process of excitation in the dendrites and in the soma of single isolated sensory nerve cells of the lobster and crayfish. *J. Gen. Physiol. 39*:87–119.

Houk, J., and Henneman, E. 1967. Responses of Golgi tendon organs to active contractions of the soleus muscle of the cat. *J. Neurophysiol. 30*:466–481.

Jansen, J. K. S., and Matthews, P. B. C. 1962. The central control of the dynamic response of muscle spindle receptors. *J. Physiol. 161*:357–378.

Kuffler, S. W., Hunt, C. C., and Quilliam, J. P. 1951. Function of medullated small-nerve fibers in mammalian ventral roots: Efferent muscle spindle innervation. *J. Neurophysiol. 14*:29–54.

Matthews, B. H. C. 1933. Nerve endings in a mammalian muscle. *J. Physiol. 78*:1–53. (The classical paper describing the physiological properties of muscle spindles in the cat.)

Nakajima, S., and Onodera, K. 1969. Adaptation of the generator potential in the crayfish stretch receptors under constant length and constant tension. *J. Physiol. 200*:187–204.

Sears, T. A. 1964. Efferent discharges in alpha and fusimotor fibers of intercostal nerves of the cat. *J. Physiol. 174*:295–315.

PART FIVE

HOW NERVE CELLS TRANSFORM INFORMATION

INTEGRATION BY INDIVIDUAL NEURONS

The central nervous system is faced incessantly with the task of making decisions on the basis of information about the outside world provided by sensory end organs situated in skin, muscles, internal organs, eyes, ears, and nose. At any one instant incoming signals from these diverse sources bombard the brain, some tending to reinforce and others to counteract each other.

The mechanism by which the various types of information are taken into account and assigned priorities is called integration. It is carried out by the nervous system as a whole, so that actions appropriate to the particular set of circumstances can be taken by the animal. It may not be stretching the analogy too far to picture individual cells engaged in a similar task. Each cell must perform an integrating function according to its particular position in the central nervous system. Examples already mentioned include ganglion cells in the retina and simple, complex, and hypercomplex cells in the visual cortex. Each sifts and sorts out signals as visual processing progresses (Chapters 2 and 3).

To look at integration from a different angle, consider the consequences of an act such as stroking an area of skin. This may give rise to a scratching movement by one hand, while other groups of muscles perform different movements to make the limbs flex or extend to maintain steady posture. A smooth, coordinated movement has been initiated in the face of all the other external and internal stimuli to which the body is being subjected. This reflex may, however, be overridden at any instant by another stronger input arising at a different place in the skin or through some other sensory system. In our own brains an action of this type involves the confluence of literally millions of conflicting messages toward a single decisive action.

The question, then, is: how are such diverse influences handled by the brain? We cannot answer this question with regard to the total behavior of an animal, but we can go a long way toward understanding how the response of a single cell with known properties comes about. Out of the medley of incoming signals, every neuron makes its own synthesis and comes up with one piece of information that is typically its own. In other words, the cell does not simply hand the message on, but transforms it.

Chapter 16 describes the manner in which individual cells integrate, combining excitation and inhibition into new instructions. To demonstrate these processes in action, three familiar cells are chosen: the crustacean muscle fiber, the Mauthner cell in the goldfish, and the mammalian spinal motoneuron. One can hardly imagine a more diverse set of excitable cells; yet, they all exemplify similar basic principles of summed synaptic action. Chapter 17 carries the analysis on to a description of individual cells and their relation to behavior in a simple invertebrate, the medicinal leech. There are many invertebrate nervous systems in which reflexes and integrative mechanisms can be studied at the cellular level. Each type of animal offers certain advantages and disadvantages. Instead of providing a comprehensive review, we have again chosen to follow in some detail one example, the leech nervous system, chiefly because of our familiarity with the preparation.

CHAPTER SIXTEEN

TRANSFORMATION OF INFORMATION BY SYNAPTIC ACTION IN INDIVIDUAL NEURONS

Our higher centers continually receive and integrate information arising in a great variety of sources on the surface of the body and in the internal organs. A typical central neuron faces a task similar to that of the brain as a whole. It is the target of converging excitatory and inhibitory signals whose information it synthesizes before taking action of its own in the form of a new nerve impulse. This integrative activity of individual neurons—the combining of a variety of information—is centered on synaptic transmission. Its universality is illustrated by samples from three diverse cell systems—in crustaceans, fish, and mammals.

A simple example is the opener muscle of the claw in the lobster and crayfish. All the muscle fibers are innervated at multiple sites by two nerve fibers only, one excitatory and the other inhibitory. With this innervation the muscle is able to execute finely graded movements. On an individual muscle fiber the interaction of inhibitory and excitatory synaptic potentials determines the level of the total postsynaptic potential and thereby the level of contraction. Besides its postsynaptic chemical action, the inhibitory axon also acts presynaptically and can reduce the amount of transmitter released by the excitatory terminal.

The Mauthner cell in the medulla of the goldfish is a large neuron with two dendrites, each several hundred microns long. Impulses arising in the neuron are controlled by five different types of synaptic actions: (1) excitatory chemical synapses, (2) inhibitory chemical synapses, (3) excitatory electrical synapses, (4) inhibitory electrical synapses, and (5) presynaptic chemical inhibition occurring on some of the excitatory nerve terminals. The large dendrites of the Mauthner neuron do not give regenerative impulses. The cell is, therefore, able to combine or integrate through its cable properties the diverse converging synaptic influences. All these are channeled to the crucial site, the axon hillock (axon cap) region, where conducted impulses are initiated.

The third example of integration, the spinal motoneuron, integrates the excitatory and inhibitory action of many thousands of chemically transmitting synapses. Here also presynaptic inhibition can selectively eliminate or reduce the action of certain pathways that reach the motoneuron. As in the Mauthner cell the integrating action of the neuron is focused on the initial axon segment where impulses are initiated.

The synapse considered in greatest detail so far is the vertebrate myoneural junction; it differs in an important way from the majority of synapses within the central nervous system. Each skeletal muscle fiber receives its excitatory input from only one or sometimes two motor nerve fibers, and under normal conditions each impulse in the presynaptic nerve gives rise to an impulse in the muscle. The situation differs in neurons within the brain. They are, as a rule, supplied by many converging axons, some excitatory and others inhibitory. Each presynaptic impulse usually releases only a few quanta of transmitter, producing a small subthreshold effect. These synaptic potentials, like other graded localized potentials, can sum with each other either to reinforce or to inhibit. The main point of this chapter is to show that in order for a standard cell in the brain to discharge, the collective input from many other cells is required. The integration of synaptic events is crucial for information processing in general, since the rigidly determined all-or-none impulses cannot sum. It is at synapses that the flexibility of the nervous system resides.

The three preparations illustrated in Figure 1 have many features in common. Each is subjected to convergent excitatory and inhibitory influences, and each, directly or indirectly, gives rise to movement. The crustacean muscle contracts, the large Mauthner cell in the goldfish medulla activates motoneurons, and the motoneuron in the spinal cord innervates muscle fibers. However, while one crustacean muscle fiber receives just two or a few incoming fibers, the motoneuron and the Mauthner cell receive many hundreds. Furthermore, a new dimension is added through the placement of synaptic terminals on specific regions of the cell. Despite these differences, the underlying mechanism of integration used to produce smooth, graded movements is highly similar in the three types of cells.

CRUSTACEAN MYONEURAL SYNAPSES

The nerve-muscle junctions in the lobster and the crayfish provide a preparation that is impressively simple, yet shows the basic elements that nerve cells in vertebrate brains use for integration. The opener muscle of the claw is supplied by two axons only, each forming chemical synapses—one excitatory and the other inhibitory (Chapter 11). Yet the contraction of the entire muscle, composed of many individual muscle fibers, can be finely graded by these two axons. Both axons branch in unison (Figure 1) and distribute themselves over the surface of each muscle fiber, forming multiple synapses close to each other

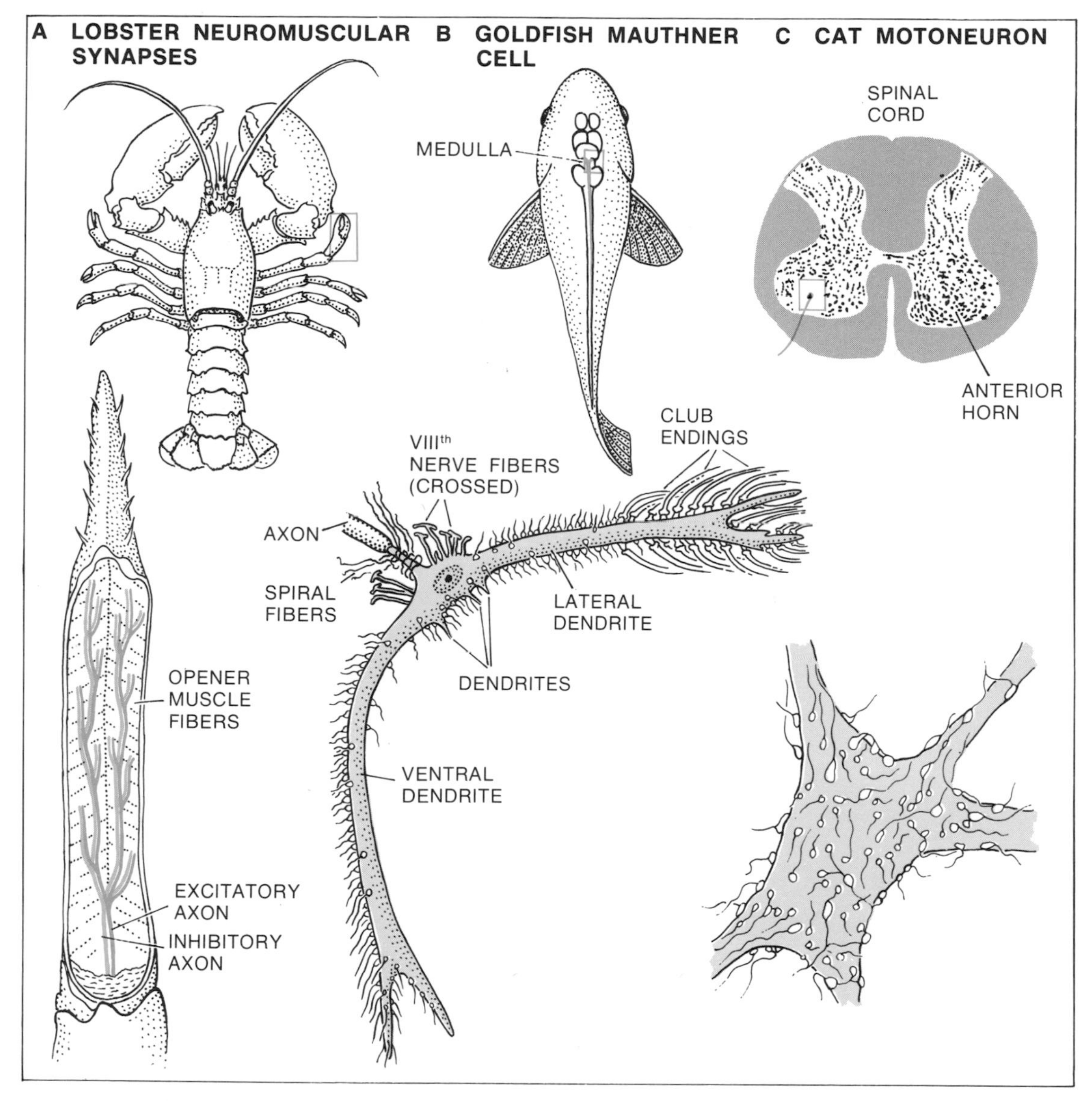

SYNAPSES IN LOBSTER, FISH AND CAT. Synaptic arrangements in (A) a lobster (or crayfish) muscle, (B) a large neuron in the brain of a goldfish, and (C) a motoneuron in the spinal cord of a cat. The crustacean muscle is supplied by only one excitatory and one inhibitory axon, each of which forms scattered synapses on the muscle surface; the typical motoneuron receives thousands of synapses from hundreds of other neurons making excitatory and inhibitory connections; in the Mauthner cell some of the incoming nerve fibers with distinct function form synapses in specific regions of the cell. In the muscle and the motoneuron chemical synaptic excitation and chemical post- and presynaptic inhibition have been studied. The Mauthner cell receives, in addition, electrical excitatory and inhibitory synapses. (B after Bodian, 1942)

1

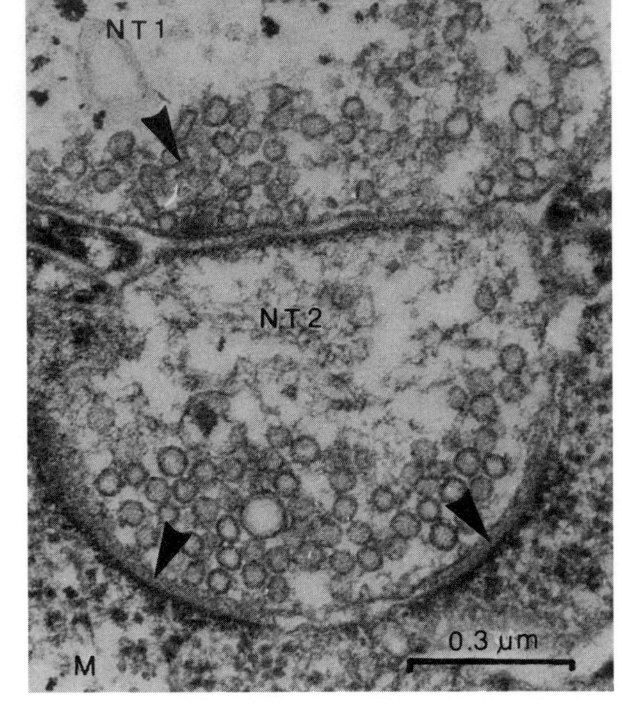

2

CHEMICAL SYNAPSES in crustaceans. A nerve terminal (NT2) forms chemical synapses (arrows), presumably excitatory, with a muscle fiber (M) in an opener muscle of the crayfish. The nerve terminal itself receives a chemical synaptic contact (NT1) which is, therefore, presynaptic and inhibitory. (From Lang, Atwood, and Morin, 1972)

(Figure 2).[1–3] In this way, synaptic excitation and inhibition can be exerted at many spots almost simultaneously.

If such a muscle worked like human skeletal muscles, in which each muscle fiber gives a regenerative conducted response and a rapid twitch, the entire muscle would contract maximally with each motor nerve impulse; it would behave as one motor unit (see below). To achieve versatility and compensate for the deficiency in numbers of nerves, each muscle fiber in the opener of the claw gives graded local contractions, the amplitude being determined by the size of the synaptic potentials. Their amplitude in turn depends on two processes: (1) TEMPORAL FACILITATION, in which each successive synaptic potential becomes larger than the last; and (2) TEMPORAL SUMMATION, which occurs when potentials arrive in quick enough succession so that each adds its effect to the preceding one.

When the muscle surface is explored in detail, by means of iontophoretic application of γ-aminobutyric acid (GABA) and glutamate (a candidate for the excitatory transmitter), the membrane is seen to be a mosaic of spots sensitive to one or the other of the two substances only in the circumscribed region of the terminals (Chapter 10). In this manner the sites of excitatory and inhibitory synapses are determined.[4]

---

[1]Wiersma, C. A. G., and Ripley, S. H. 1952. *Physiol. Comp. Oecol.* 2:391–405.
[2]Atwood, H. L., and Morin, W. A. 1970. *J. Ultrastruct. Res.* 32:351–369.
[3]Lang, F., Atwood, H. L., and Morin, W. A. 1971. *Z. Zellforsch.* 127:189–200.
[4]Takeuchi, A., and Takeuchi, N. 1965. *J. Physiol.* 177:225–238.

Through its cable properties, the postsynaptic muscle membrane shows SPATIAL SUMMATION of all the discrete distributed synaptic potentials, so that in practice the entire cell changes its potential uniformly.

In the opener of the claw, then, two types of chemically transmitting synapses engage in opposing excitatory and inhibitory actions. These determine the state of the postsynaptic muscle membrane potential in a finely graded manner. The contraction, in turn, reflects the level of the membrane potential. An example of separate and joint synaptic excitation and inhibition is shown in Figure 3.

Integration at the crustacean neuromuscular junction displays a further type of synaptic interaction seen in the central nervous system

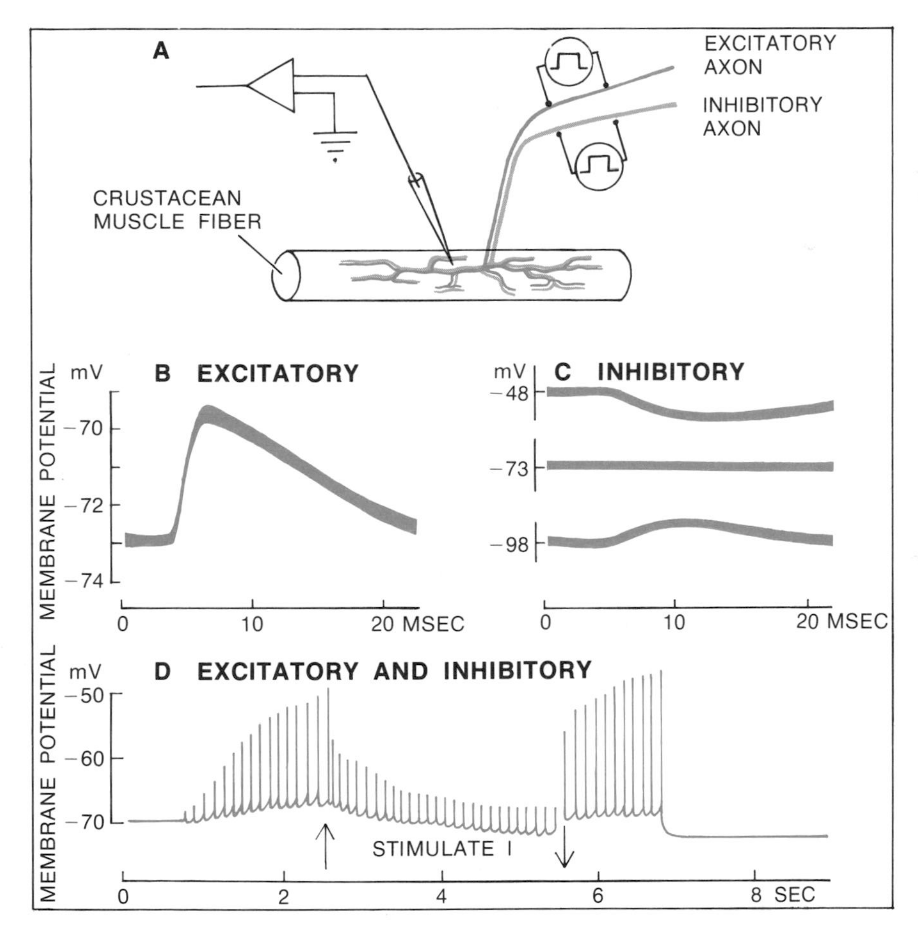

3 **CHEMICAL EXCITATION AND INHIBITION in crustacean muscle. A. Excitatory and inhibitory axons can be stimulated separately or together. B. An excitatory synaptic potential in the opener of the claw at a membrane potential of −73 mV. C. Stimulation of the inhibitory axon causes no potential change at −73 mV, but hyperpolarizes at −48 mV and depolarizes at −98 mV. D. Simultaneous excitatory and inhibitory stimulation at 10/sec (between arrows) greatly reduces the excitatory potentials. (C from Fatt and Katz, 1953; D from Atwood and Bittner, 1971)**

of vertebrates—presynaptic inhibition (Chapter 8). Synapses formed by the inhibitory nerve on the excitatory terminals (Figure 2) reduce the number of quanta liberated by excitatory terminals in response to each impulse. The timing is critical. The inhibitory transmitter, in this case γ-aminobutyric acid, must be released just before the action potential reaches the excitatory terminals.[5] The significance of presynaptic interactions for integration is more apparent in the next type of preparation, the Mauthner cell (see below).

To sum up, the crustacean neuromuscular junction shows great versatility in its integrating action by making use of chemical synaptic excitation and inhibition, facilitation, temporal and spatial summation, and presynaptic inhibition. All these mechanisms are also used by higher nervous systems for the same basic purpose.

SYNAPTIC INTERACTIONS AT THE MAUTHNER CELL

In the medulla of teleost fish, on each side of the midline, there is an unusually large, peculiarly shaped neuron—the Mauthner cell (Figure 1).[6] It has attracted the attention of many investigators, not only because of its size but also because of the structural diversity of its synaptic contacts. The Mauthner cells represent a miniature nervous system that combines all the synaptic mechanisms known at present. In addition to chemically mediated excitation and inhibition of the type seen at crustacean myoneural junctions, the Mauthner cell is excited and inhibited by electrical synapses.

Another attractive feature of Mauthner cells is the well-defined geometry of the synaptic inputs. In the crustacean muscle, knowledge of the particular distribution of synapses is not crucial for understanding integration. In contrast, in the Mauthner cell the various types of synapses occupy characteristic positions on specific parts of the neuronal surface and have different degrees of effectiveness. The regularity of the anatomical pattern in cell after cell in animals of the same species has greatly facilitated finding the typical structures for a detailed morphological analysis and for a correlation with physiological performance.

Structure and function of Mauthner cells

The distribution of synapses on the goldfish Mauthner cell is shown in Figures 1 and 4. The neuron lies about 1 mm beneath the surface of the medulla and cannot be seen in live preparations. In a comprehensive, detailed series of experiments, Furshpan and Furukawa[7] mapped the synaptic connections by using a coordinate system that enabled them to place electrodes within different regions of the Mauthner cell and accurately onto specific parts of its surface. They made extracellular and intracellular recordings, studied the membrane properties of the cells, and determined which structures gave rise to the various synaptic potentials. Later on this knowledge was used in micropharmacological

---

[5]Dudel, J., and Kuffler, S. W. 1961. *J. Physiol.* *155*:543–562.
[6]Bodian, D. 1937. *J. Comp. Neurol.* *68*:117–159.
[7]Furshpan, E. J., and Furukawa, T. 1962. *J. Neurophysiol.* *25*:732–771.

tests in which drugs or transmitters were applied to known selected areas of the cell surface.[8]

The massive lateral and ventral dendrites can be several hundred microns long. The initial portion of the axon, close to the cell body, is the AXON HILLOCK, and it is here (as in the crustacean stretch receptor neuron—Chapter 15—and spinal motoneuron) that impulses are initiated. This region plays a critical role in the integrative mechanisms of the cell, since the dendrites and cell body do not generate action potentials (see later). The segment of nerve between the axon hillock near the cell body and the starting point of the myelinated axon is surrounded by a wrapping of glial cells into which a number of fine axons penetrate to form a spiraling network. Some of the axons emerge again and later on make synapses on the nearby cell body. This structure is the site of electrical inhibition and is called the AXON CAP.

The large axon (up to 50μm in diameter) of each Mauthner neuron crosses to the opposite side, runs down along the spinal cord, and makes synapses with numerous motoneurons. Each time the Mauthner neuron discharges a single impulse, it causes a rapid, vigorous contraction of one side of the tail by synchronously activating motoneurons in the spinal cord. This escape reaction can be readily seen when one taps the side of a tank containing a number of fish; the fish execute a sudden tail flip that propels them sideways.

Synaptic inputs to Mauthner cells

Each neuron receives numerous fibers that come directly or indirectly from the eighth nerve on the two sides and from the other Mauthner cell on the opposite side of the medulla. The excitatory inputs are derived from two groups of eighth nerve fibers (Figure 4). The larger fibers have diameters of 10 to 15μm and are particularly conspicuous; they originate on the same side of the animal in a part of the labyrinth, a sensory structure in the inner ear concerned with somatic proprioception. The sensory axons run directly to the distal portion of the lateral dendrite of the Mauthner cell, where they form the so-called "club" endings (Figure 5). Electron microscopy shows that gap junctions occur in the contact area of the club endings with the dendrite;[9,10] as mentioned in Chapter 8, these junctions, where the pre- and postsynaptic membranes are in close apposition, are the sites of electrical excitatory synapses. Experiments described below indicate that the current generated by impulses in the large axons spreads directly into the lateral dendrites of the Mauthner cell. The smaller eighth nerve fibers responsible for chemical excitation probably end diffusely on the lateral dendrite.

The principal inhibitory inputs to the Mauthner cell come from the contralateral eighth nerve, from the contralateral Mauthner cell, and from the same Mauthner cell by way of recurrent collaterals. All these

---

[8]Diamond, J. 1968. *J. Physiol.* *194*:669–723.

[9]Robertson, J. D. 1963. *Cell. Biol.* *19*:201–221.

[10]Robertson, J. D., Bodenheimer, T. S., and Stage, D. E. 1963. *J. Cell Biol.* *19*: 159–199.

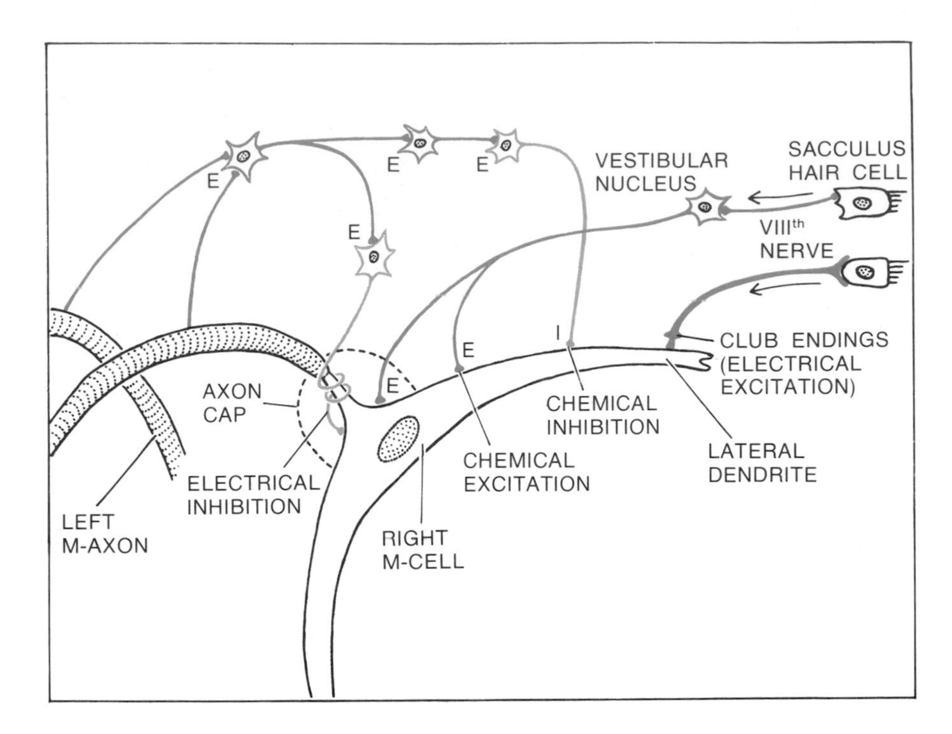

4 **SCHEME OF MAUTHNER CELL SYNAPSES. The eighth nerve on the same side provides excitatory electrical synapses on the distal portion of the lateral dendrite and excitatory chemical synapses elsewhere on the cell. The inhibitory input comes mainly from the Mauthner cell and the eighth nerve on the other side and by way of axon collaterals from the same side. Probably these pathways involve interneurons. Electrical inhibition is confined to the axon cap region, and presynaptic chemical inhibition is seen in the large axons in the distal portion of the lateral dendrites. (After Furukawa, 1966)**

pathways probably involve interneurons. They give rise to three distinct forms of inhibition: (1) conventional chemical inhibitory potentials, similar to those in crustacean muscle, evoked by stimulation of the contralateral eighth nerve; (2) electrical inhibition produced by current flow in the region of the axon cap by stimulating the contralateral eighth nerve and the Mauthner cells; and (3) presynaptic inhibition occurring on the endings of eighth nerve fibers where they synapse upon the lateral dendrite of the Mauthner cell.[11]

Analysis of synaptic mechanisms: excitatory electrical and chemical transmission

If the large axons in an ipsilateral eighth nerve are stimulated electrically and the responses are recorded with an intracellular electrode in the distal portion of the lateral dendrite, a potential of up to 50 mV appears with a latency of about 0.1 msec. This synaptic delay is so short that the transmission must be electrical (Figure 6*B*, potential 1). The size of the potential declines as the recording electrode is in-

[11]Furukawa, T., Fukami, Y., and Asada, Y. 1965. *J. Neurophysiol.* *26*:759–774.

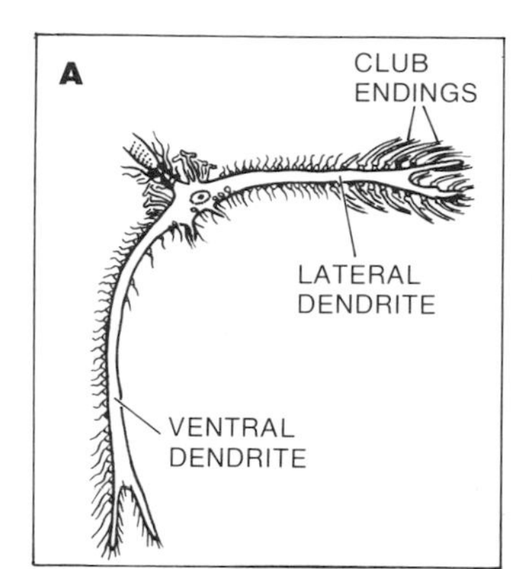

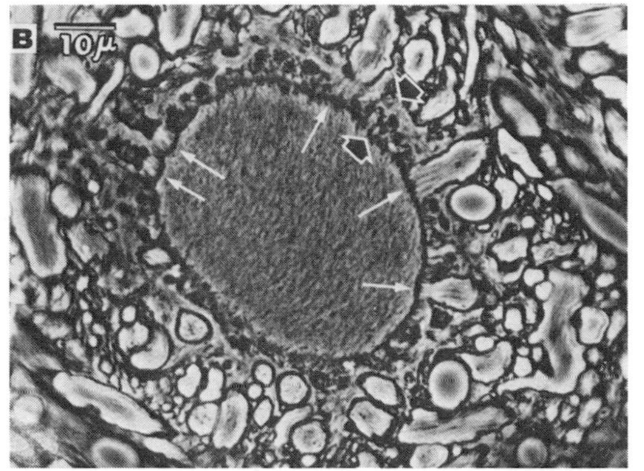

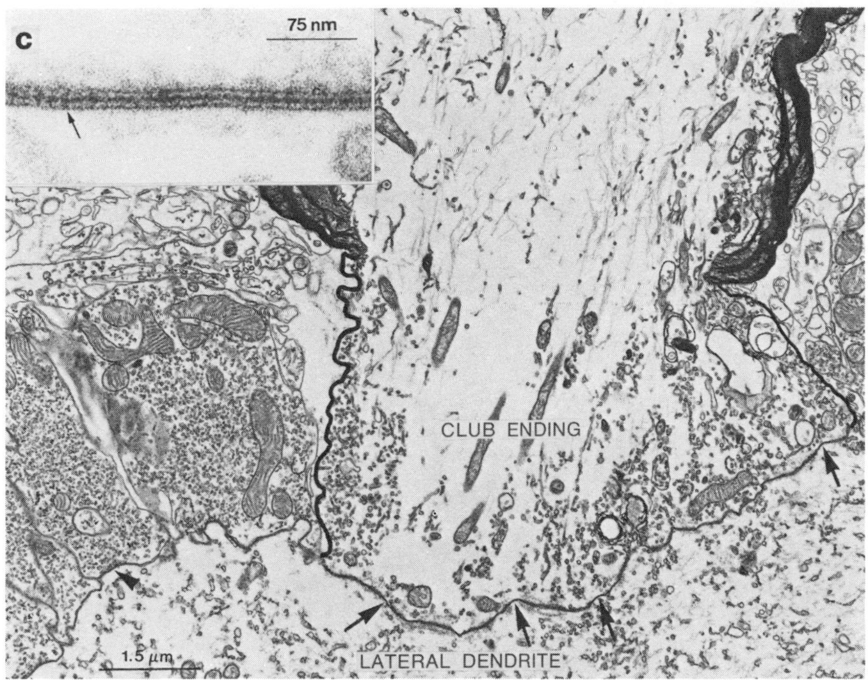

5

CHEMICAL AND ELECTRICAL SYNAPSES on the same neuron. A. Scheme of the Mauthner cell. The electrically transmitting club endings lie on the distal portion of the lateral dendrite. B. Histological view of several large club endings (thin arrows) in a transverse section through the distal portion of the lateral dendrite. Region of chemical synapses appears between thick arrows. C. Electron micrograph of a club ending on the lateral dendrite. Arrows indicate special contact areas. One of these is shown at high magnification in the inset and is typical of electrical (gap) junctions, which are traversed by crossbridges (small arrow). The lateral boundaries of the unmyelinated portion of the club ending have been outlined in ink. To the left are smaller axons forming chemical synapses; two of these are marked by arrow heads. (A from Bodian, 1942; B from Robertson, Bodenheimer, and Stage, 1963; C courtesy of Y. Nakajima, see also Nakajima, 1974; other details on analogous chemical and electrical synapses in mammals are given in Figure 7, Chapter 8)

serted at different distances away from the club endings toward the cell body. This potential, decreasing with distance, is therefore a non-regenerative, excitatory postsynaptic event. The spatial distribution of potentials also indicates that the site of electrical coupling is close to the end of the lateral dendrite.[12]

When the ipsilateral eighth nerve is stimulated more strongly, so that its smaller diameter axons are excited, an additional delayed post-synaptic potential can be recorded in the Mauthner cell after a latency of less than 1 msec (Figure 6*B*, potential 2). Judged by the usual criteria (Chapter 8), this delayed excitatory synaptic potential is derived from chemically transmitting synapses, probably by way of interneurons. Either of these synaptic potentials can lead to conducted impulses in the Mauthner axon (see below).

Chemical inhibitory synaptic transmission

Stimulation of the eighth nerve, which excites the Mauthner cell on the same side, inhibits the contralateral Mauthner cell. This, together with the electrical inhibitory circuit through axon collaterals of each

[12]Furshpan, E. J. 1964. *Science 144*:878–880.

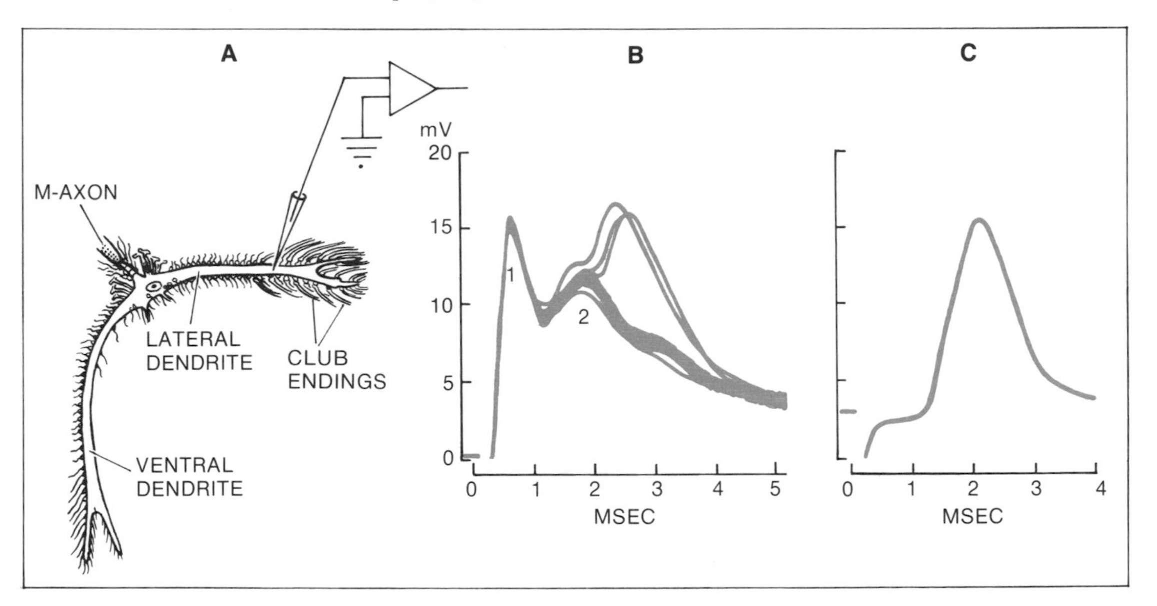

6 **EXCITATORY ELECTRICAL AND CHEMICAL TRANSMISSION in a Mauthner cell. The eighth nerve is stimulated on the same side (Figure 4). A. Recording from the distal portion of the lateral dendrite. B. Ten to twelve superimposed records. The first response (1) appears with a negligible delay and is due to electrical synaptic transmission. The second, later synaptic response (2) is chemical; on three of the sweeps it gives rise to impulses in the axon hillock area at a distance of over 300μm from the recording site. The impulses therefore appear small. C. A single impulse generated by direct electrical stimulation of the Mauthner axon. It also appears relatively small because it spreads passively to the recording electrode in the lateral dendrite, which does not give regenerative responses. (From Diamond, 1968)**

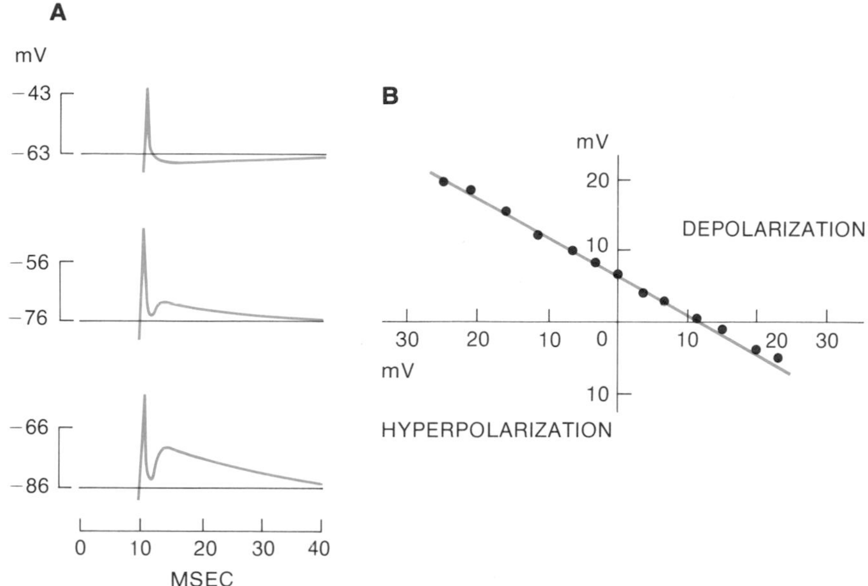

7

CHEMICAL INHIBITORY SYNAPTIC POTENTIALS in a Mauthner cell that was impaled with two microelectrodes, one for recording, the other for passing current through the cell, thereby displacing the membrane potential to various levels (as indicated). A. Direct stimulation of the Mauthner axon produces first an impulse, followed by a slower inhibitory potential. It depolarizes at the resting potential of −76 mV, increases further at −86 mV, while at −63 mV the potential reverses its polarity. B. The peak amplitudes of the inhibitory potential (ordinate) are plotted at different membrane levels (abscissa). The reversal potential occurs close to 10 mV depolarization. (After Furukawa and Furshpan, 1963)

Mauthner cell (see below), prevents simultaneous contraction in tail muscles on both sides of the animal. The sketch of connections in Figure 4 indicates the consequences of ipsilateral and of contralateral stimulation of the Mauthner cells.

The chemical inhibition produced by these pathways is similar to other inhibitory processes already discussed in detail. An example of the inhibitory potential and its reversal is shown in Figure 7. The chemical transmitter increases the conductance principally for chloride ions. Normally, the reversal potential for chloride is close to the resting potential. During the inhibitory transmitter action, the effect of all types of excitatory stimuli, chemical or electrical, is reduced. A convenient test consists of recording from a dendrite while an antidromic nerve impulse conducts from the spinal cord into the axon hillock region. During an inhibitory potential, the impulse height recorded in the dendrite is reduced, and this decrease is a measure of the shunt—that is, of the inhibitory conductance in the cell body and dendrite region.[7]

Electrical inhibition at the axon cap

The fibers that give rise to electrical inhibition run into the axon cap region that is surrounded by a network of glial cells (Figure 1). The fibers arise from the eighth nerve and can be activated by stimulat-

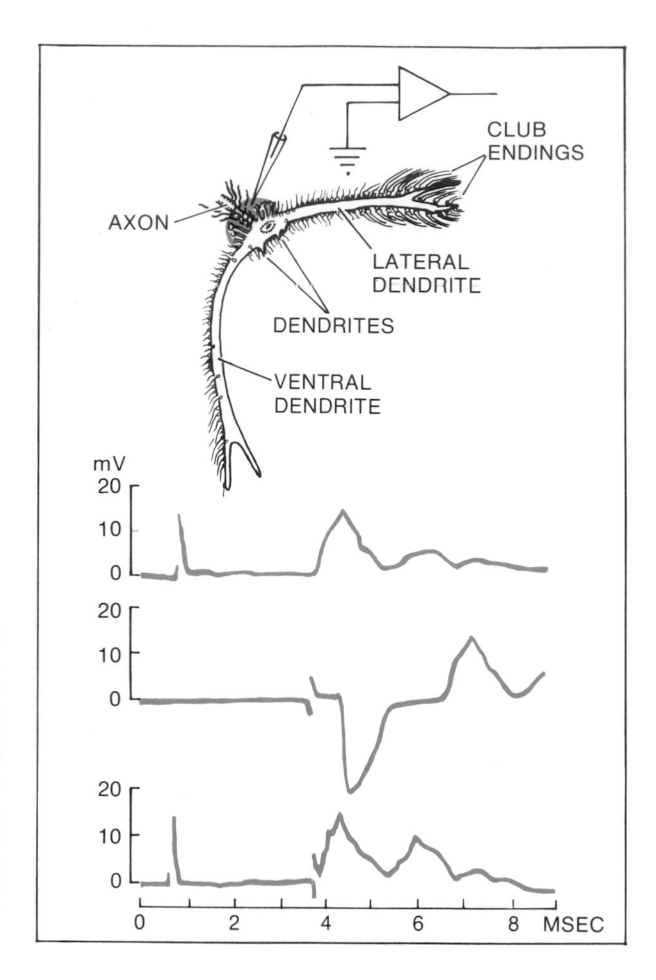

8

ELECTRICAL INHIBITION caused by a hyperpolarization recorded with an electrode within the axon cap region just outside the axon. In record 1 the opposite Mauthner cell is stimulated; it causes hyperpolarization of almost 15 mV (upward deflection). In record 2 the Mauthner cell axon is stimulated directly, the first downward deflection being the axonal impulse conducting into the axon cap segment. If both Mauthner axons are stimulated (record 3), the axonal impulse is blocked by the hyperpolarization. (From Furukawa, 1966)

ing either Mauthner cell (Figure 8). After a brief delay, an extracellular electrode placed accurately in the cap region registers a potential that lasts for about 1 msec.[13] The potential change is highly localized, and its sign indicates an increase in external positivity of about 15 mV. This restricted focal hyperpolarization is generated by the current flow in fibers that penetrate the axon cap; it can be faithfully mimicked by current passed directly into the axon cap through an extracellular microelectrode. The injected currents and the potentials they create are confined to the axon cap region; therefore, they are not recorded as a potential difference between an intracellular electrode placed inside the cell body and a second indifferent electrode at a distance from the cap region.

The hyperpolarization at the axon cap occurs at a strategically important site, because this is the region where impulses are initiated. While it persists, the threshold for excitation is increased in the initial axon segment and conduction can be blocked. The nature of this effect is interesting because it is one of the few known physiological examples

[13]Furukawa, T., and Furshpan, E. J. 1963. *J. Neurophysiol.* *26*:140–176.

of inhibition produced by extrinsic electrical currents.[14] We do not know how the geometry of the glial packing is arranged to bring about a distribution of current flow in such a way that the extracellular spaces around the initial segment become hyperpolarized; nor do we know the physiological properties of the glial cells in the axon cap. In this context it is useful to recall another instance in which glial cells confine current flow to a restricted region: oligodendrocytes make the myelin that channels current flow in nerve fibers through the nodes of Ranvier, thereby ensuring saltatory conduction (Chapter 13).

Presynaptic inhibition at Mauthner cell synapses

There is experimental evidence for a third type of inhibition at the Mauthner cell. The reduction in the size of excitatory potentials observed during inhibition is too great to be accounted for by increases in conductance of the Mauthner cell. This may be explained by the observation that the action potentials in eighth nerve fibers are reduced in amplitude during inhibition.[11] The decrease apparently is brought about by an inhibitory action on the eighth nerve axons themselves, making them less effective in evoking excitatory synaptic potentials in the motoneuron. The significance of this presynaptic inhibition is that one input to the postsynaptic cell can be selectively depressed without reducing the excitability of the cell as a whole (Chapter 8).

INTEGRATION BY THE SPINAL MOTONEURON

The motoneuron is the most studied mammalian nerve cell, and it has a pivotal role in all movements we perform.[15] Each motoneuron innervates a group of muscle fibers and together with them forms a functional unit called the MOTOR UNIT. When a motor nerve fiber discharges, all its muscle fibers contract. The motor unit is, therefore, the basic component of normal movement. This is so because individual mammalian skeletal muscle fibers give maximal nongraded contractions in response to motor nerve impulses. (There is evidence for an exception to this rule in some of the fibers in muscles that control eye movement.) In muscles that produce fine movements, such as those of the fingers, each motoneuron innervates fewer fibers than in a larger muscle used to maintain posture. The smoothness and precision of our movements are brought about by varying the number and timing of motor units brought into play. The rapid action of each motor unit is not apparent when the whole muscle contracts, because the individual contributions are asynchronous and are smoothed out by the elastic properties of muscles. By keeping track of motor unit discharges in muscles, one obtains a picture of the activity pattern of motoneurons in the spinal cord, even without recording from them. This technique, called electromyography, is a useful tool in clinical and physiological work.

Sherrington called the spinal motoneuron the FINAL COMMON PATH because all the neural influences that have to do with movement or

[14]Korn, H., and Faber, D. S. 1975. *J. Neurophysiol.* *38*:452–471.
[15]Eccles, J. C. 1964. *The Physiology of Synapses.* Springer, Berlin.

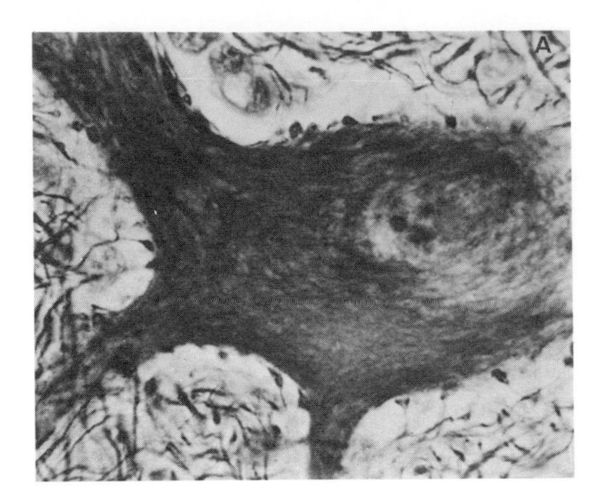
A

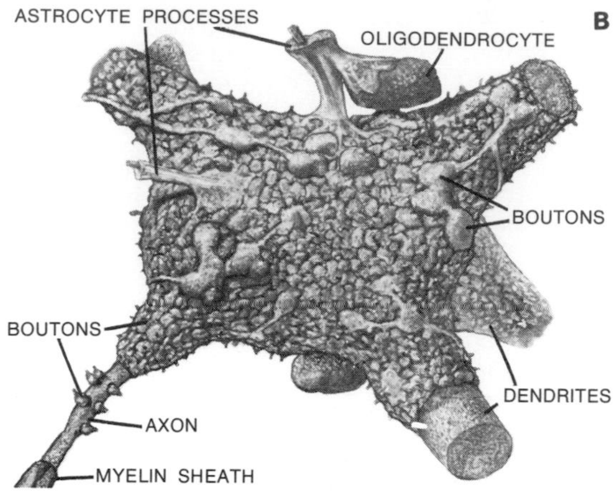
B
ASTROCYTE PROCESSES
OLIGODENDROCYTE
BOUTONS
BOUTONS
DENDRITES
AXON
MYELIN SHEATH

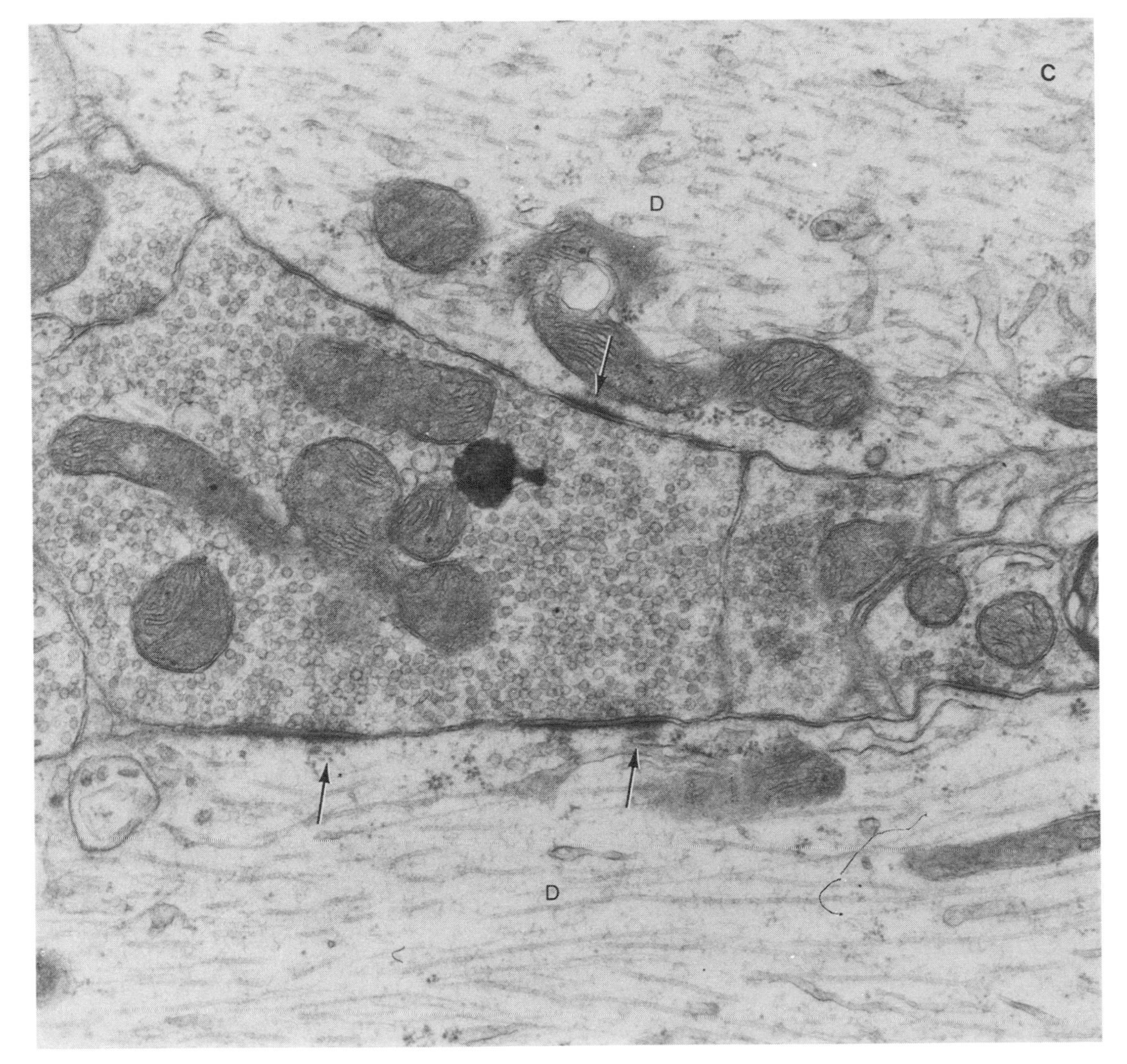
C
D
D

posture converge on it. It is, therefore, not surprising that the average motoneuron has many connections on it, up to about 10,000 synapses by some estimates, providing information derived from all over the body and from higher centers.

At the cellular level, the fine structure of the synapses[16] and the mechanisms acting on motoneurons resemble those found in crustaceans and in Mauthner cells. Once again, chemical postsynaptic excitation, inhibition, and presynaptic inhibition interact in determining the end result of converging signals. As in the Mauthner cell, there is good evidence that the impulse originates in one particular region of the cell, the axon hillock. Figure 9*A* shows a stained section through a motoneuron with numerous synaptic contacts in the relatively thin section that goes through the cell body. The real density of synaptic boutons covering the cell surface is more apparent in the drawing based on electron microscopic studies (Figure 9*B*).

Through extensive work, pioneered by Eccles and his colleagues, much is now known from intracellular recordings about the mechanisms of synaptic transmission at this cell and the interaction of excitatory and inhibitory synapses (Figure 10).[15] In the context of integration it is likely that since there are so many endings, each one would contribute only little. Thus, Kuno has shown that individual group Ia fibers originating in muscle spindles release only a few quanta with each impulse.[17] Excitation therefore depends upon a roughly simultaneous activation by many converging fibers. Among other features of the synaptic inputs to motoneurons is the finding by Henneman and his coworkers that each motoneuron that supplies a limb muscle in the cat receives inputs from all the group Ia afferent fibers arising in spindles of the same muscle; a single fiber may contribute to the innervation of as many as 300 motoneurons.[18] These workers have also shown a clear correlation between the size of various motoneurons and the sequence in which they are recruited to take part in the stretch reflex. According

[16]McLaughlin, B. J. 1972. *J. Comp. Neurol.* *144*:429–460.
[17]Kuno, M. 1971. *Physiol. Rev.* *51*:647–678.
[18]Mendell, L. M., and Henneman, E. 1971. *J. Neurophysiol.* *34*:171–187.

9
SYNAPSES ON MAMMALIAN MOTONEURONS. A. Silver-stained section of a spinal motoneuron. Axons and their terminal synaptic boutons converging on the dendrites and cell body are stained black. Their diameters are several μm. B. Drawing based on an electron-microscopic study of the distribution of synaptic boutons covering the surface of a motoneuron cell body. C. Several nerve terminals are apposed to two dendrites (D) of a motoneuron. Three chemical synapses are marked by arrows. (A, unpublished photograph by the late F. deCastro; B from Poritsky, 1969; C from Peters, Palay, and Webster, 1976)

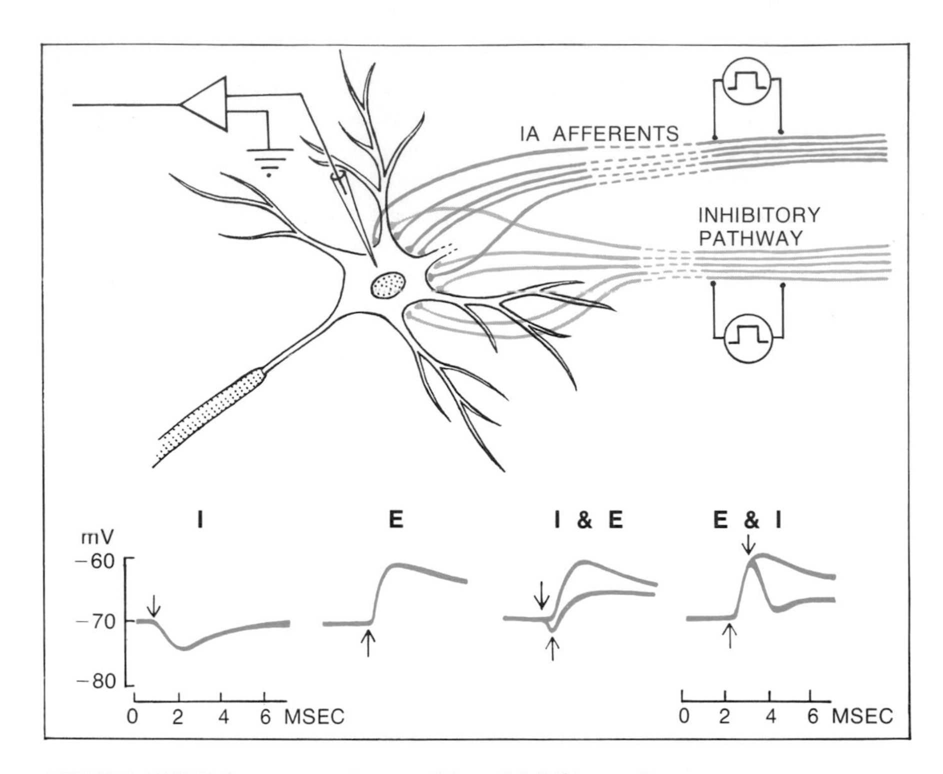

10 **INTERACTION between excitatory (E) and inhibitory (I) synaptic potentials in a motoneuron of the cat. The sum of the two effects determines whether an impulse is initiated at the initial segment of the axon. (After Curtis and Eccles, 1959)**

to this SIZE PRINCIPLE, small motoneurons are activated before larger ones.[19]

In numerous studies the spatial aspects of integration have been correlated with the structure and membrane properties of motoneurons.[20,21] Motoneurons have been injected with the dye procion yellow and their dimensions reconstructed. These findings, combined with electrical recordings from different sites in the cell, allow reasonable estimates of the decline of a synaptic potential as it spreads passively from dendrite to cell body.[22]

As in the crayfish and Mauthner cell synapses, inhibition is brought about by an increase in chloride conductance and by presynaptic inhibition upon the endings of group Ia afferent fibers originating in muscle spindles (Chapter 8). In addition, recurrent inhibition feeds back into the cell as a result of its own activity.[15] The circuit for this proc-

[19]Henneman, E., Somjen, G., and Carpenter, D. O. 1965. *J. Neurophysiol. 28:* 560–580.

[20]Burke, R. E., and Ten Bruggencate, G. 1971. *J. Physiol. 212:*1–20.

[21]Jack, J. J. B., and Redman, S. J. 1971. *J. Physiol. 215:*321–352.

[22]Barrett, J. N., and Crill, W. E. 1974. *J. Physiol. 239:*301–324, 325–345.

ess involves an interneuron called the Renshaw cell (Chapter 11). The motoneuron releases acetylcholine (ACh) at its terminals with muscle fibers, and there is evidence that it does the same at the synapses it makes with interneurons in the spinal cord.

Still lacking for this cell is more detailed knowledge of certain important anatomical and physiological characteristics: the sites of termination of the various excitatory and inhibitory inputs are not known, and information about the membrane properties of the dendrites is incomplete. As a result one cannot predict what effect a particular synapse seen in the microscope will have on the integrative activity of the cell. Nevertheless, there is more information about this cell than about any other in the mammalian central nervous system. Furthermore, its integration plainly depends on the same principles that govern integration in crustacean muscle and the Mauthner cell.

COMPARISON OF INTEGRATION IN THE THREE CELL TYPES

The following paragraphs compare the various factors that contribute to the summation of different synaptic inputs on the three cell types: crustacean muscle, Mauthner cell, and spinal motoneuron. The Mauthner cell and the spinal motoneuron have in common a single restricted site for impulse initiation, the axon hillock. The membrane potential of this part of the cell determines the outcome of all that goes on in the cell body and the dendrites—the conflict or interplay between excitation and inhibition. A similar situation exists in the stretch receptor neuron (Chapter 15). The feature of making the initial axon segment the focal point of integrative activity has important consequences and can be most easily discussed in detail for the Mauthner cell. A similar analysis in the motoneuron is more difficult because, as mentioned before, the specific distribution of inhibitory and excitatory synapses on dendrites and cell body is unknown.

The integrative activity on the cell body of the Mauthner cell and its dendrites can be summarized roughly as follows, with the help of the sketch in Figure 4. When the ipsilateral eighth nerve fibers are stimulated, excitatory synaptic potentials arise at the distal end of the lateral dendrite. From there they spread passively to the axon hillock up to 300μm away, decreasing by more than 50 percent through the cable properties of the cell. If at this point the depolarization reaches threshold, an impulse and a tail flip follow. Chemical inhibition tends to prevent firing in two ways: (1) hyperpolarization spreads passively to the axon hillock driving the potential away from threshold, and (2) the increase in conductance, by reducing the length constant of the dendrite, attenuates the spread of excitatory depolarization and causes it to decrease more steeply. In other words, excitatory current spread is curtailed before it can reach the critical axon hillock region. Clearly, what matters is not simply the presence of a synapse: the SITES OF THE SYNAPSES ON A NEURON play a key role in establishing their effectiveness. For example, the inhibition would be less useful if it occurred at the tip of the dendrite or if the excitatory synapses were close to the axon hillock; similarly, electrical inhibition blocks impulse initiation most effectively if it occurs in the critical area where impulses start. Finally, through presynaptic inhibition, the input from the ipsilateral eighth nerve can be reduced without influencing the ability of the cell to respond to other excitatory synapses on its surface.

In contrast to the motoneuron and Mauthner cell, spatial considerations of the distribution of inhibitory and excitatory synapses on muscle fibers in crustaceans are less important. Since the excitatory and inhibitory neurons in the extensor make synapses close together, graded doses of excitatory and inhibitory action can counteract each other rather uniformly over the whole surface of the muscle fiber. Thus, with synapses scattered over its surface, the muscle fiber behaves evenly along its length, because the electrical spread of current between adjacent synapses blurs the distinction between active synaptic spots and the intervening inactive extrasynaptic regions.

Timing of the arrival of impulses is very important at the crustacean myoneural junction. For presynaptic inhibition to work, the inhibitory impulse is maximally effective if it arrives at the motor terminal just before the excitatory impulse, whose release of transmitter it reduces. A critical feature of timing, already mentioned, occurs in the goldfish. To cause a flip of the tail one Mauthner cell must be inhibited electrically when the other one discharges.

The presence of an electrical rather than a chemical synapse in the goldfish, representing a gain of less than 1 msec, can make a great difference considering the speed with which a fish must react when it takes evasive action. As yet, however, there is no complete scheme for sorting out the special functional significance of the fish's varied synaptic assembly for movement.

An important principle that bears on problems of development (Chapters 18 and 19) emerges from the preceding discussion of integration. Clearly, for all the various types of synaptic interactions to work, neurons must make highly precise and specific connections with the appropriate postsynaptic targets. Furthermore, during development, synapses are formed in the appropriate regions of those cells on particular dendrites or regions of the cell body.

SUGGESTED READING

General reviews

Atwood, H. L. 1972. Crustacean Muscle. In G. H. Bourne (ed.). *The Structure and Function of Muscle*, Vol. I. Academic Press, New York, pp. 421–489.

Diamond, J. The Mauthner Cell. In W. S. Hoar and D. J. Randall (eds.). *Fish Physiology*, Vol. V. Academic Press, New York, pp. 265–346.

Eccles, J. C. 1964. *The Physiology of Synapses*. Springer, Berlin.

Henneman, E. 1974. Organization of the Spinal Cord. In V. B. Mountcastle (ed.). *Medical Physiology*, Vol. I. Mosby, St. Louis, pp. 636–650.

Shepherd, G. M. 1974. *The Synaptic Organization of the Brain.* Oxford University Press, New York.

Original papers

Barrett, J. N., and Crill, W. E. 1974. Specific membrane properties of cat motoneurones. *J. Physiol. 239*:301–324.

Barrett, J. N., and Crill, W. E. 1974. Influence of dendrite location and membrane properties on the effectiveness of synapses on cat motoneurones. *J. Physiol. 239*:325–345.

Furshpan, E. J., and Furukawa, T. 1962. Intracellular and extracellular responses of the several regions of the Mauthner cell of the goldfish. *J. Neurophysiol. 25*:732–771.

Furukawa, T., and Furshpan, E. J. 1963. Two inhibitory mechanisms in the Mauthner neurons of goldfish. *J. Neurophysiol. 26*:140–176.

Takeuchi, A., and Takeuchi, N. 1965. Localized action of gamma-aminobutyric acid on the crayfish muscle. *J. Physiol. 177*:225–238.

CHAPTER SEVENTEEN

A SIMPLE NERVOUS SYSTEM: THE LEECH

Invertebrates perform various tasks that appear complex; yet, they use only a relatively small number of nerve cells. This makes them useful for the study of some problems that at present seem too difficult to be approached in the mammalian brain. In particular, simple reflexes can be studied in terms of individual cells that give rise to them and the various mechanisms of signaling. At a higher level of integration it is possible to show how elementary units of behavior are combined into coordinated movements of the animal as a whole.

The central nervous system of the leech is used to illustrate this approach. The segmental ganglia in the leech resemble each other, each containing only about 350 nerve cells, which can be seen under a dissecting microscope and impaled by microelectrodes. Individual sensory and motor cells have been identified, their synaptic connections traced, and the fields they innervate in the periphery are accurately defined. This information is an essential requirement for tracing the normal wiring diagram of the nervous system and for studying how the properties of neurons change with repeated use.

In ganglia of the leech one can correlate the performance of the chemical and electrical synapses between sensory and motor nerve cells with reflex movements. For example, when impulses are initiated at their normal frequencies in sensory neurons by natural stimulation of the skin, certain chemical synapses show sequentially both facilitation and depression. The reflexes mediated by cells connected through such synapses reflect this sequence in motor acts that are readily observed. In contrast, transmission mediated by cells with electrical connections shows little variability under similar conditions. The sodium pump is another neural mechanism that can be related to functional performance in reflexes. Under physiological conditions of use, sensory neurons become hyperpolarized, and this influences their subsequent responses to sensory stimuli. The nervous system of the leech also lends itself well to following the formation of specific connections and the mechanisms by which use or disuse can bring about changes in the properties of synapses.

Rationale for studying invertebrate central nervous systems

Throughout the previous chapters discussion is oriented toward problems of higher nervous systems. In the case of cellular mechanisms, it is quite satisfactory to take examples from a wide variety of species. The squid axon, for example, provides an excellent model of nearly universal validity for the conduction of impulses; cells of other invertebrates illustrate the mechanisms of permeability sequences found in synaptic inhibition and excitation in vertebrates. Similarly, the neurochemical studies in ganglia of lobsters have direct relevance for neurons in the mammalian brain (Chapter 11).

The reasons for choosing invertebrate preparations are usually technical. Certain problems can be solved more easily in invertebrate nervous systems which tend to survive in isolation; in addition, if their cells are large and readily accessible, they can be easily recognized and studied with electrical recording methods and biochemical techniques.

The relevance of the nervous systems of invertebrates, however, extends beyond basic mechanisms of conduction and synaptic transmission. These preparations also offer advantages for the exploration of some aspects of complex behavior that seem too difficult to handle in higher vertebrates. Along these lines, Kandel,[1] Tauc,[2] Strumwasser,[3] and others have broken new ground by using cell groups in the sea snail *Aplysia* to analyze simple behavioral responses by tracing the neural circuits involved in their performance. Some of these responses are performed by relatively few neurons, while analogous responses in mammals would require many thousands of neurons. For example, in one series of experiments Kandel and his colleagues[4] studied the synaptic mechanisms and the individual cells associated with the habituation of a reflex; this is a process by which a sensory stimulus, when repeated at intervals, evokes progressively weaker reflex responses in an animal. Similarly, Kennedy and his colleagues[5] traced the neural circuits for coordinated elementary units of behavior, such as escape reactions and swimming, in crayfish. Other invertebrate central nervous systems (of insects, for example) have also been used; each has its own set of advantages for approaching specific problems at the cellular level.

We have singled out for fuller discussion the nervous system of the leech, mentioned in Chapter 13 in relation to the properties of neuroglial cells. This compact nervous system contains relatively few cells arranged in an orderly manner; yet, it provides a useful model for the study of certain problems of neural organization found in the brains of higher animals. At the same time the ganglia in the leech use synaptic mechanisms similar to those that have been identified in vertebrates. Figures 1 and 2 show the principal features of the

[1]Kandel, E. R., and Spenser, W. A. 1968. *Physiol. Rev. 48*:65–134.

[2]Tauc, L. 1967. *Physiol. Rev. 47*:521–593.

[3]Strumwasser, F. 1973. *Physiologist 16*:9–42.

[4]Carew, T. J., Pinsker, H. M., and Kandel, E. R. 1972. *Science 175*:451–454.

[5]Kennedy, D. 1974. In F. O. Schmitt and F. G. Worden (eds.). *The Neurosciences: Third Study Program.* MIT Press, Cambridge, Mass., pp. 378–388.

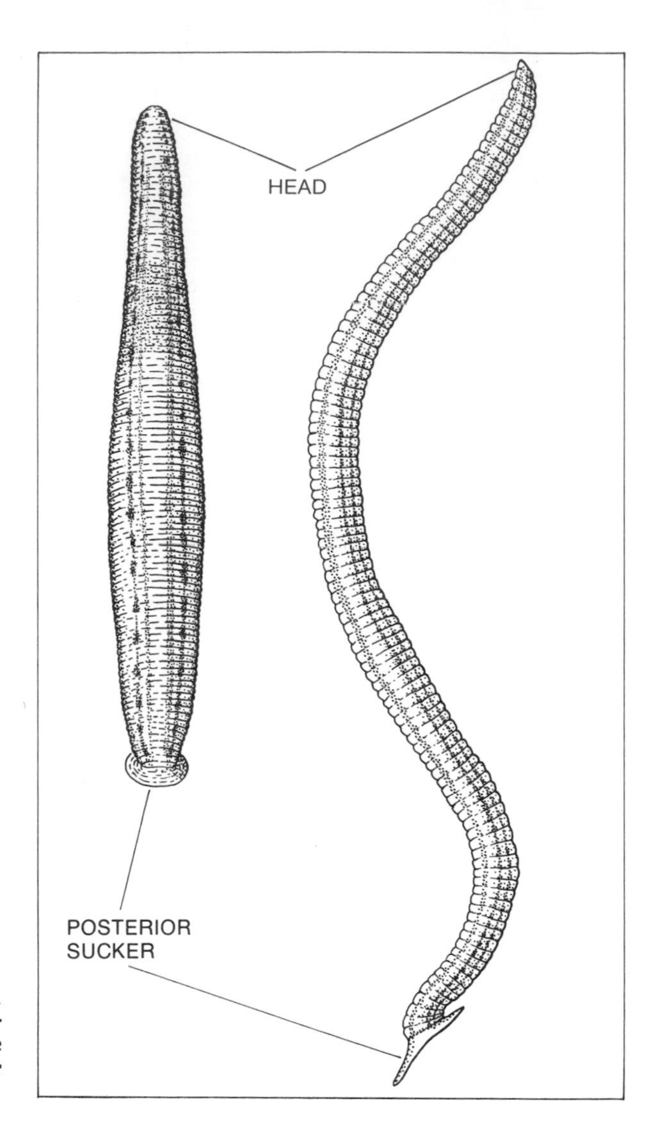

1
THE LEECH, *Hirudo medicinalis*, has a segmented body with a sucker at each end. The animal can measure 5 in. in length after feeding.

animal and of its central nervous system, which consists of a chain of stereotyped ganglia.

Working at the level of single cells, one can identify the individual neurons that mediate reflexes, trace their synaptic connections, and establish their characteristic properties—whether they are chemical or electrical, excitatory or inhibitory. This detailed knowledge of the wiring diagram sets the stage for studying how neural components act in concert to produce coordinated elements of behavior. As a next step, one can investigate how the properties of neurons and synapses change as a result of repeated natural sensory stimuli, and how these changes are reflected in the performance of the animal. For example, repetitive electrical stimulation of neurons may lead to prolonged hyperpolarization through the activity of an electrogenic sodium pump (Chapter 12). In the leech the same effect can be produced by natural stimulation of

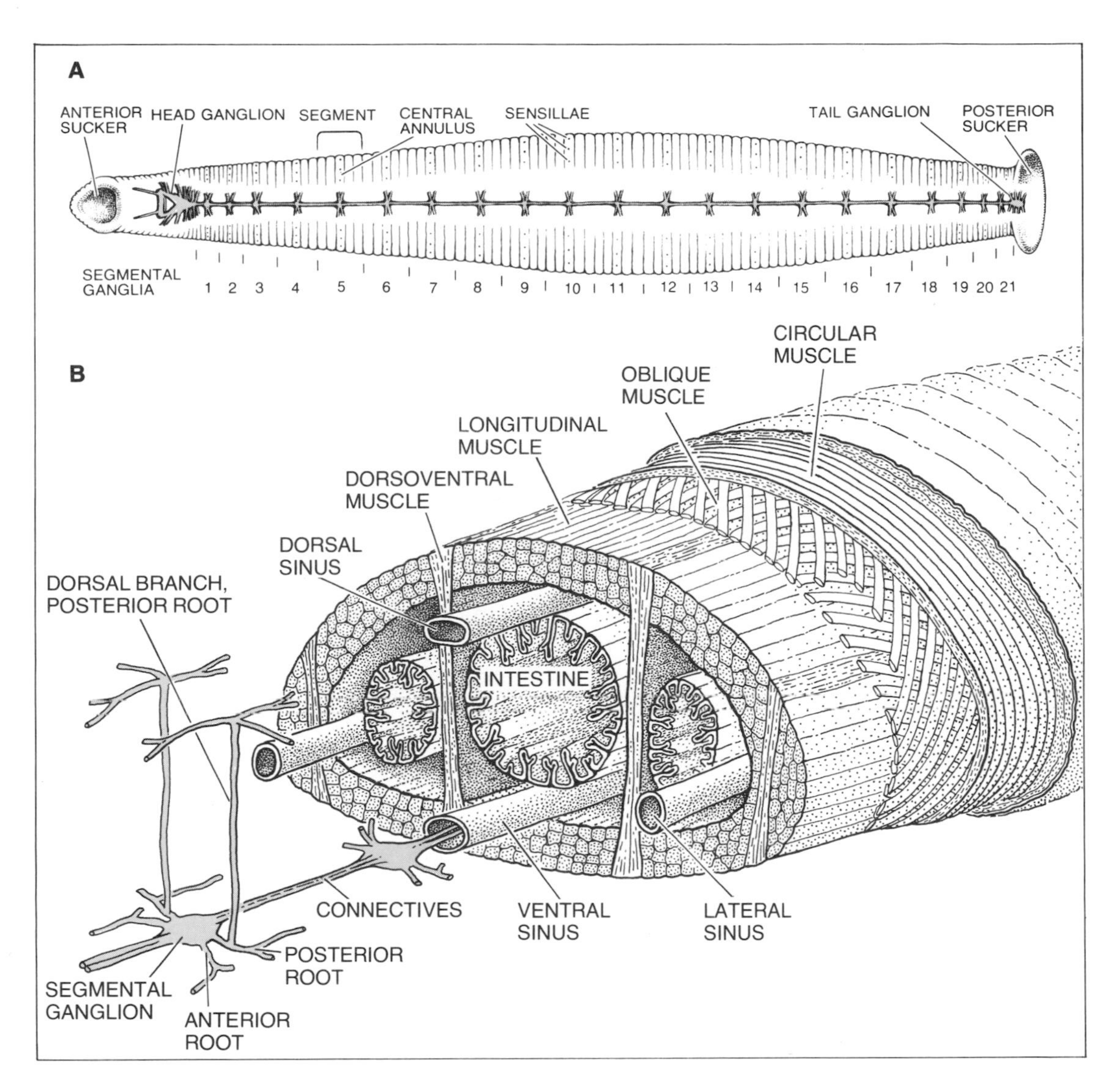

CNS OF THE LEECH. A. The central nervous system of the leech consists of a chain of 21 segmental ganglia, a head ganglion, and a tail ganglion. Over most of the body five circumferential annuli make up each segment; the central annulus is marked by sensory end organs responding to light (the sensilla). B. The nerve cord lies in the ventral part of the body within a blood sinus. Ganglia which are linked to each other by bundles of axons (the connectives), innervate the body wall by paired roots. The muscles are arranged in three principal layers: circular, oblique, and longitudinal. In addition, there are dorsoventral muscles that flatten the animal and fibers immediately under the skin that raise it into ridges.

2

sensory neurons, and the consequences for signaling in reflex pathways can be examined. The advantage of using natural stimuli is that only in this way is one assured that impulses are arising in the appropriate fibers at similar frequencies as in the intact animal.

Semiautonomous units: the ganglia

The principal advantages of the leech accrue from the simplicity of its body plan, which is reflected in the structure of its nervous system (Figures 2 and 3). Both body and nervous system are rigorously segmented and consist of a number of repeating units (SEGMENTS) that are similar throughout the length of the animal. Since the animal has no limbs, its behavior consists of a relatively simple repertoire of movements performed by layered groups of muscles. Each segment is innervated by a ganglion which is similar to the others within one animal and to those in other animals. Even the specialized head and tail "brains" consist of fused ganglia where many characteristic features of segmental ganglia are still recognized.

Each ganglion contains only about 350 nerve cells, which have distinctive shapes, sizes, positions, and branching patterns.[6] A ganglion innervates a well-defined territory of the body by way of paired axon bundles (ROOTS), and it communicates with neighboring and distant parts of the nervous system through another set of bundles (CONNECTIVES). Integration thus occurs in a succession of clear-cut steps: (1) each segmental ganglion receives information from a circumscribed body segment whose performance it directly regulates, (2) neighboring ganglia influence each other by direct interconnections, and (3) the coordinated operation of the whole nerve cord and the animal is governed by the brains at each end of the animal. The distinct segmental subdivisions can be studied on their own or together.

ANALYSIS OF REFLEXES MEDIATED BY INDIVIDUAL NEURONS

Sensory and motor cells in leech ganglia

When one strokes, presses, or pinches the skin of a leech, a sequence of reflexes follows. One segment or more shortens abruptly, and the skin becomes raised into a series of distinct ridges. Subsequently, the animal swims away or executes writhing movements.

As a first step in defining the pathways, the synaptic mechanisms, and the changes that occur in time, the individual cells responsible for producing the reflexes must be identified.

In living leech ganglia one can reliably identify the sensory and motor cells according to their shapes, sizes, positions, and electrical characteristics.[7] For example, the 14 cells labeled *T*, *P*, and *N* in Figure 3 are all sensory and represent three sensory modalities. Each cell responds selectively to touch, pressure, or noxious mechanical stimulation of the skin. Representative intracellular records are also shown in Figure 3. The impulses of T cells are always similar to, but smaller and briefer than, those in P or N cells. With practice, it is usually enough

---

[6]Coggeshall, R. E., and Fawcett, D. W. 1964. *J. Neurophysiol.* *27*:229–289.
[7]Nicholls, J. G., and Baylor, D. A. 1968. *J. Neurophysiol.* *31*:740–756.

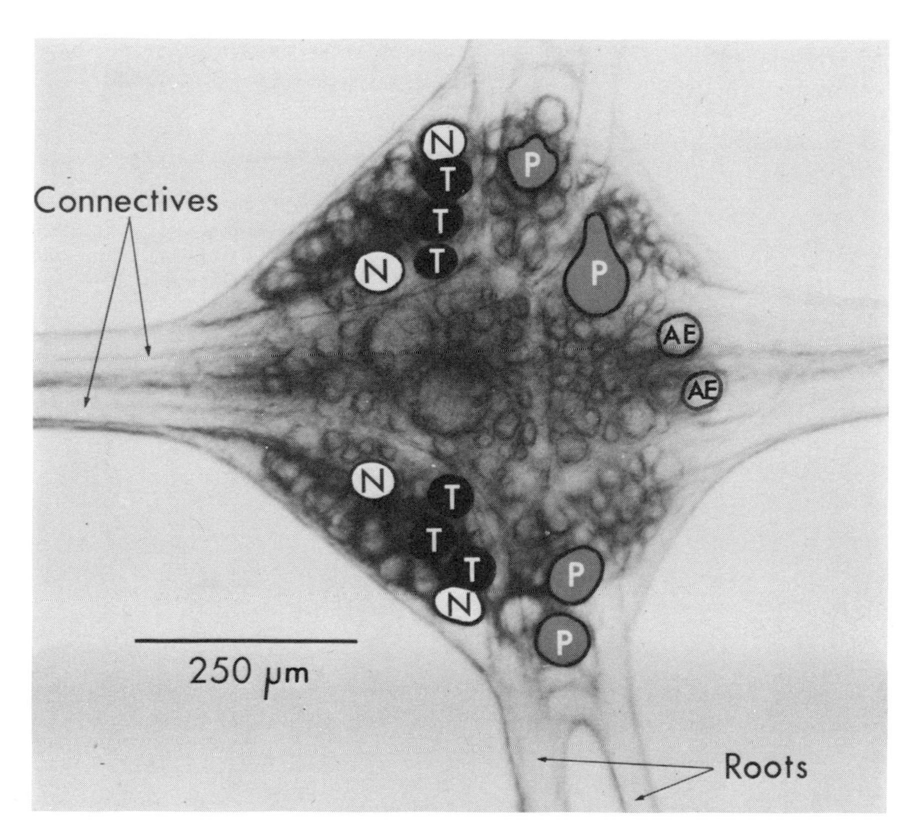

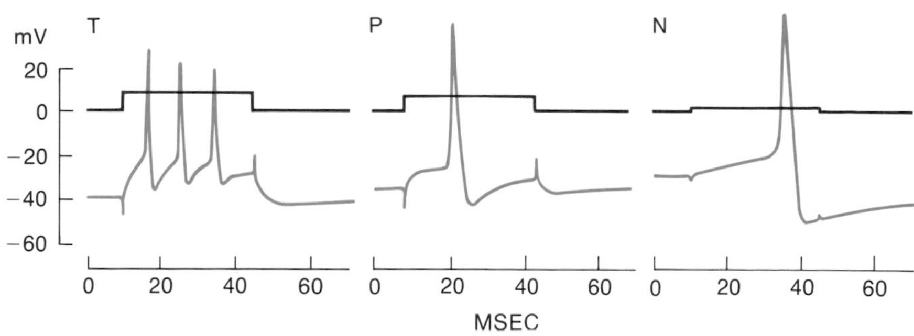

3

VENTRAL VIEW OF A SEGMENTAL GANGLION. Individual cells are clearly recognized. The three sensory cells responding to touch (T) and the pairs of cell types responding to pressure (P) or noxious (N) mechanical stimulation of the skin are labeled. Each type of cell gives distinctive action potentials, as shown by the traces below. Impulses in T cells are briefer and smaller than those in P or N cells. Current injected into cells through the microelectrode is monitored on the upper traces. The cells outlined in the posterior part of the ganglion are the annulus erector (AE) motoneurons. (After Nicholls and Baylor, 1968)

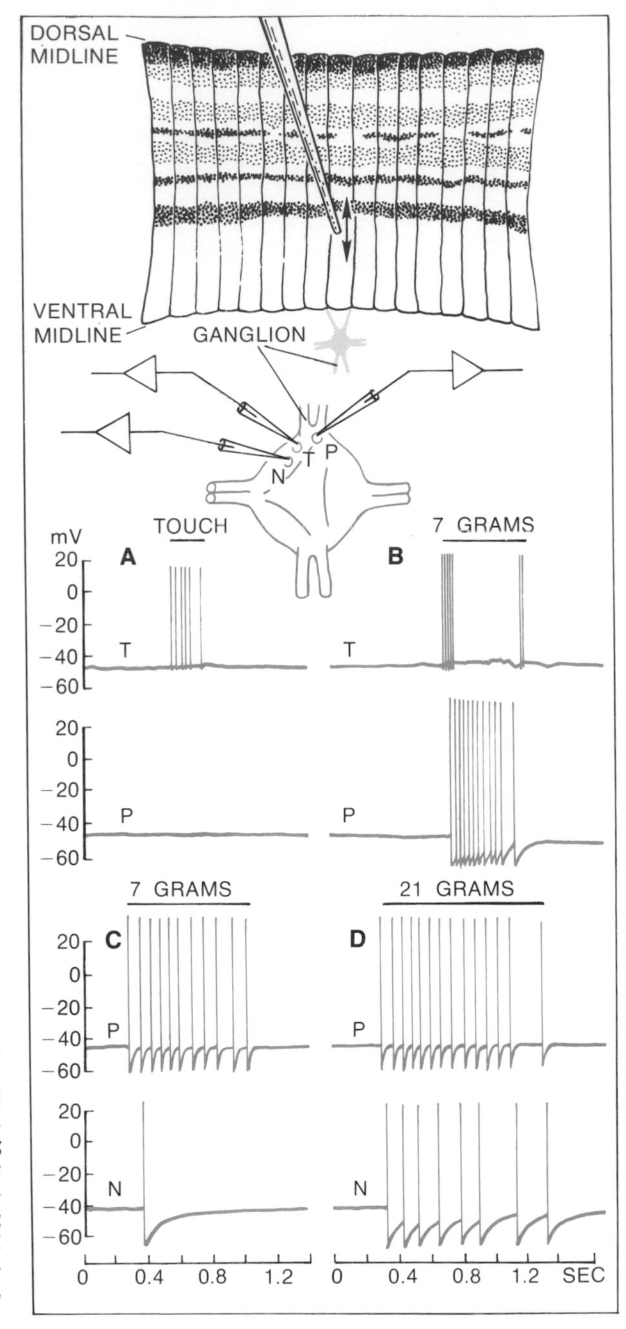

4

RESPONSES TO SKIN STIMULATION illustrated by intracellular records of T, P, and N sensory cells. The preparation consists of a piece of skin and the ganglion that innervates it. Cells are activated by touching or pressing their receptive fields in the skin. (A) A T cell responds to light touching which is not strong enough to stimulate the P cell. (B) Stronger maintained pressure evokes a prolonged discharge from the P cell and a rapidly adapting "on" and "off" response from the T cell. (C and D) Still stronger pressure is needed to activate the N cell. (After Nicholls and Baylor, 1968)

just to look at a single action potential to be certain which type of cell has been impaled.

Figure 4 illustrates the response of the sensory cells to various forms of cutaneous stimuli. The T cells give transient responses to light touch of the skin surface or even to eddies in the solution bathing the skin. They adapt rapidly to a maintained step indentation and usually cease firing within a fraction of a second. As expected, a tactile stimulus moved back and forth over the skin gives rise to a maintained discharge whose frequency can be graded by the rate at which the stimulus moves or by the frequency at which the skin is indented at a point. The P cells respond only to a marked deformation of the skin. Their discharge is slowly adapting and lasts 10 to 20 sec or more during maintained pressure. Again, the frequency is graded with the extent of the indentation. Light touch is ineffective in activating P cells. The N cells require still stronger mechanical stimuli. The threshold has not been determined precisely, but the stimulus that gives the highest frequency and best maintained discharge is a radical deformation produced by pinching the skin with blunt forceps or scratching it with a pin. The N cells, like the P cells, are slowly adapting and often continue to fire after the stimulus has been removed.

The modalities and responses of these sensory neurons resemble those of mechanoreceptors in the human skin, which also distinguish between touch, pressure, and noxious or painful stimuli. Similar types of cells have also been found in the central nervous systems of the sea snail[8] and the lamprey.[9]

A number of experiments indicate that these 14 T, P, and N cells in the leech are true first-order sensory cells rather than second- or third-order neurons driven indirectly by sensory cells in the periphery. They are the principal cells conveying sensory information about touch, pressure, and noxious mechanical stimuli to the segmental ganglia. Each of the 14 sensory cells innervates a clearly defined area of the skin and responds only to stimuli applied within one of these circumscribed receptive fields. The boundaries of a field can be identified by landmarks, such as segmentation or the coloring of skin, so that one can predict reliably which cells will fire when a particular area is touched, pressed, or pinched.

Comparable experiments have been made to identify the motor cells and their fields of innervation.[10] The leech performs only a limited repertoire of simple movements. These include shortening its body in response to cutaneous stimuli, swimming, twisting, and walking like an inchworm by using its suckers. The muscles of the leech are controlled by excitatory and inhibitory motor nerve cells in the segmental ganglia. So far 17 pairs of motoneurons that meet these criteria have been identified. These 34 cells probably constitute most of the motoneurons in the leech ganglia.

Only two motor cells are considered here in detail, to exemplify simple reflex pathways. Upon these two cells converge sensory cells of different modalities. One cell is the large longitudinal motoneuron (labeled *L* in Figure 7). Its axon runs to the body wall on the opposite

---

[8]Byrne, J., Costelluci, V., and Kandel, E. R. 1974. *J. Neurophysiol.* *37*:1041–1064.

[9]Martin, A. R., and Wickelgren, W. O. 1971. *J. Physiol.* *212*:65–83.

[10]Stuart, A. E. 1970. *J. Physiol.* *209*:627–646.

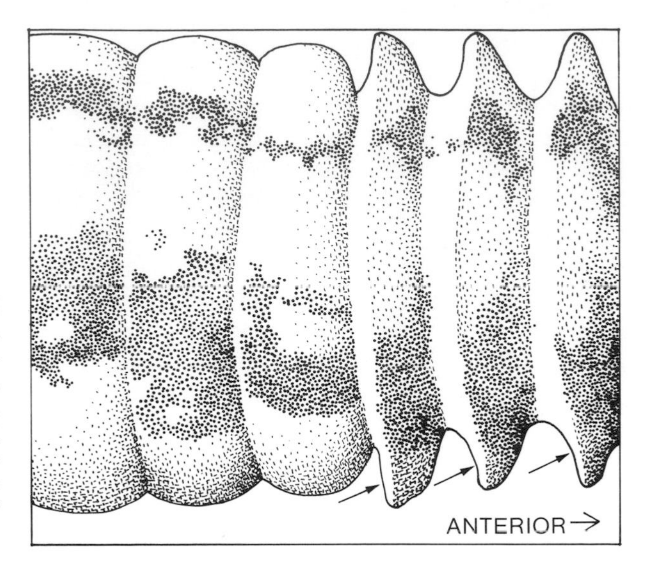

5

ERECTION OF ANNULI by the AE motor neuron. The AE cells are situated in the posterior part of the ganglion, one on each side (Figures 3 and 8). Impulses in an AE cell cause the skin on the other side of the animal to be raised into ridges (marked by arrows). The effect is so obvious that one can tell when this cell is firing by observing the animal's skin. (After Stuart, 1970)

side of the animal, where it innervates longitudinal muscle fibers of the segment to produce a rapid shortening. The other motor cell is the annulus erector (labeled *AE* in Figure 8), which causes the skin to be raised into ridges. The effect of this cell's action is so obvious that one knows it has been stimulated simply by observing the animal (Figure 5). For both AE and L cells there is evidence that the transmitter is acetylcholine (ACh).

Here, then, is a system in which one can infer which sensory cells are activated when a mechanical stimulus is applied to a particular area of skin and which motor cells are firing when the animal performs a movement.

Orderly distribution of synapses

In the leech, as in other invertebrates, the synapses between neurons are situated not on the cell bodies but on fine processes within a central region of the ganglion (the NEUROPIL).[6] The neuropil is highly complex and resembles that in the vertebrate brain. The synaptic potentials generated in the neuropil cannot be recorded directly by microelectrodes because the processes are too small. They do, however, spread into the nearby cell body where excitatory and inhibitory potentials several millivolts in amplitude can be recorded. Similarly, currents injected into the cell body can influence synaptic potentials and the release of transmitter.

Despite its apparent complexity the neuropil is organized in an orderly manner: this is known from the constancy of connections between identified cells (see below). Another clear sign of orderliness is the characteristic configuration of the branching patterns of sensory and motor cells.[11,12] For each type of cell the fingerprint is similar from ganglion to ganglion. An example of the typical ramifications of a touch cell is shown in Figure 6A. This neuron was injected with horse-

[11]Muller, K. J., and McMahan, U. J. 1976. *Proc. R. Soc. Lond.* B.
[12]Purves, D., and McMahan, U. J. 1972. *J. Cell Biol.* *55*:205–220.

radish peroxidase which penetrates into all the processes. The enzyme can also be detected in electron micrographs, as in Figure 6B in which two characteristic chemical synapses are marked by arrows. In this manner, then, one can identify the synaptic input to at least some of the synapses in the neuropil. Other chemical synapses, formed by a neuron that does not contain horseradish peroxidase, also appear in Figure 6.

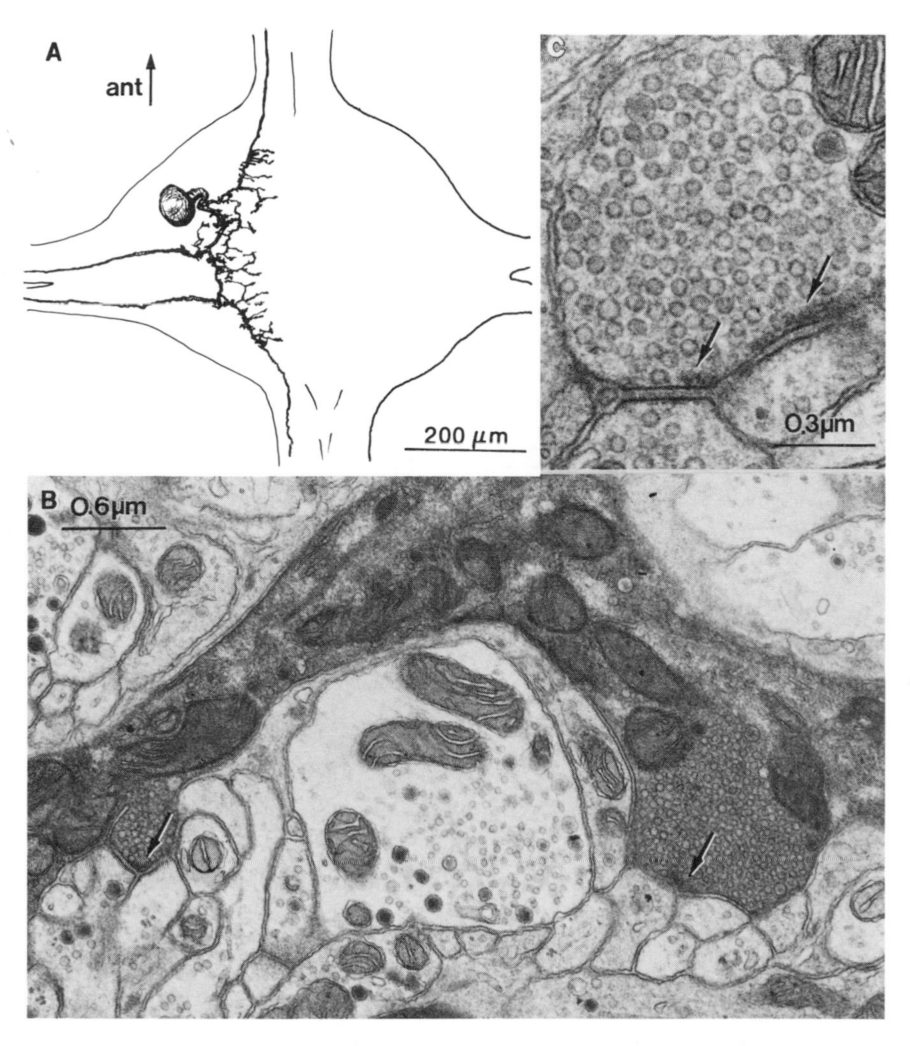

6

SHAPES AND SYNAPSES OF LEECH NEURONS. A. Arborization of a touch cell after injection with horseradish peroxidase. The cell body sends a single process to the neuropil where all the synaptic connections of the ganglion are made. Axons run to neighboring ganglia through connectives and to the body wall through roots. Small processes within the neuropil form synaptic contacts, as shown in the micrograph of B in which the terminals of the injected cell can be identified by the dark reaction product. C shows an unidentified chemical synapse. (From Muller and McMahan, 1976)

Performance of chemical and electrical synapses

The T, P, and N sensory cells all converge on the L motoneuron and thereby influence the longitudinal musculature of the body wall. A remarkable feature is that the mechanism of transmission onto the L motor cell is characteristically and consistently different for each of the sensory cells.[13] The N cells act through chemical synapses, the T cells through rectifying electrical synapses, and the P cells by a combination of both mechanisms. These simple pathways, therefore, exemplify the different forms of signaling encountered in the nervous systems of higher animals. Several lines of evidence indicate that the connections of T, P, and N sensory cells with the L motoneuron are direct—that there are no unknown intermediary cells. This is an important point, because only if each constituent and its properties are known can one pinpoint the sites at which any interesting modifications in signaling take place. The arrangement also offers an opportunity to dissect functionally the contribution that may be characteristic for a particular mode of synaptic transmission.

The mechanisms of synaptic transmission between the three types of sensory cells and the large longitudinal motoneuron have been distinguished from one another by (1) observing the latency of the synaptic potential in the motoneuron; (2) bathing the ganglion in high concentrations of magnesium ions, which block chemical synaptic transmission in leech ganglia; and (3) changing the membrane potential of the pre- and postsynaptic neurons.

For simplicity, the initial analysis of signaling mechanisms is usually made by observing the effects of single impulses. There are indications, however, that under normal conditions stimulation of the skin causes trains of impulses while the animal is swimming in water or moving on a surface. As a second step, therefore, the performance of the motoneurons is tested by sustained sensory stimulation, to differentiate between the effects of chemical and electrical synapses on reflexes.

From the material presented in Chapter 8 on the characteristics of transmission processes, the general expectation is that electrical synapses will remain relatively stable in their performance under a variety of conditions, while chemical synapses will be much more labile. This is borne out by the effects of activating N or T cells repetitively.[14] For example, when an N cell fires in response to noxious stimulation, the synaptic potentials recorded in the L cell during a train first increase (facilitation) and then decrease (depression). In sharp contrast is the synaptic effect of the touch neuron (the T cell) that makes an electrical synapse. With repeated stimulation under the same conditions, the synaptic potentials it evokes in the L cell remain unchanged (Figure 7).

Another instructive example is the effect of cooling. Leeches frequently live at low temperatures in ponds or are kept in refrigerators for long periods. In isolated preparations cooled to 2° C, electrical transmission from T cell to L cell is not appreciably changed. The

[13]Nicholls, J. G., and Purves, D. 1970. *J. Physiol.* *209*:647–667.
[14]Nicholls, J. G., and Purves, D. 1972. *J. Physiol.* *225*:637–656.

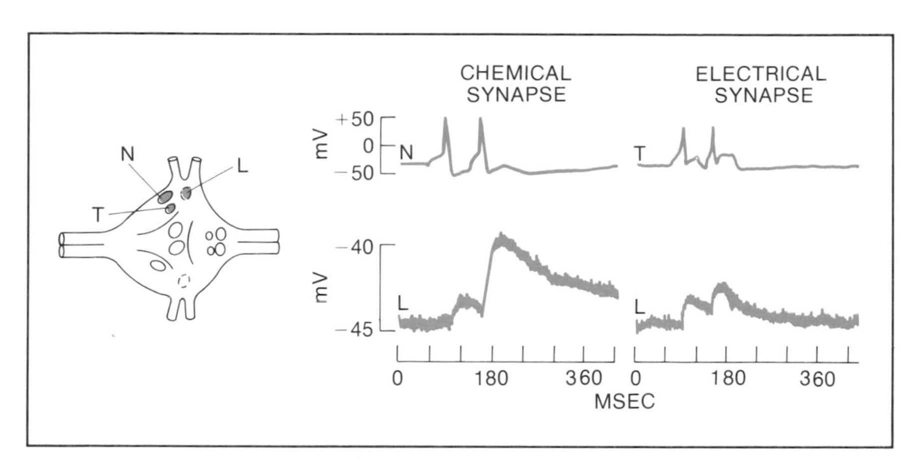

7

CHEMICAL AND ELECTRICAL TRANSMISSION. An N cell is stimulated twice in succession and at the same time its impulses are recorded (upper records). At the chemical synapse between N and L cells facilitation occurs so that a second impulse leads to a larger synaptic potential (bottom left). In contrast, two potentials evoked by T-cell impulses in an L cell cause similar postsynaptic potentials with double or multiple stimulation. This is typical of electrical synapses. (After Nicholls and Purves, 1972)

chemical transmission by the N cell, however, becomes practically ineffective even at low rates of stimulation (1/sec) for 10 sec.

Stimulation of the pressure cell produces a different situation. The P cells form mixed electrical-chemical synapses whose components can be separated by the simple expedient of cooling the ganglion. Thus, at 2° C the latency of the chemical component becomes longer, causing a "notch" to be visible. Eventually this chemical component drops out, while the electrical synaptic potential remains.

Such observations raise the question whether the variable effectiveness of different chemical synapses may not be part of a functional design that permits sequential activation of different postsynaptic structures. A single presynaptic neuron would then activate first one and then the other postsynaptic cell. This type of differential effect would adequately explain how pressing the skin of a leech leads first to a shortening reflex and then, after a delay, to erection of the annuli.

In leech ganglia the N and P sensory cells make chemical synapses not only on the L cell but also on the AE motor cell that raises the skin into ridges.[15] The potential recorded in the AE cell is considerably smaller than that in the L cell (Figure 8). With trains of impulses in a single sensory cell, the synaptic potentials in the AE and L motoneurons undergo phases of facilitation and depression; the facilitation, however, is characteristically greater and longer lasting at synapses upon the AE motoneuron (Figure 8). A number of experiments suggest that the differences in synaptic transmission can be accounted for by

---

[15]Muller, K. J., and Nicholls, J. G. 1974. *J. Physiol.* *338*:357–369.

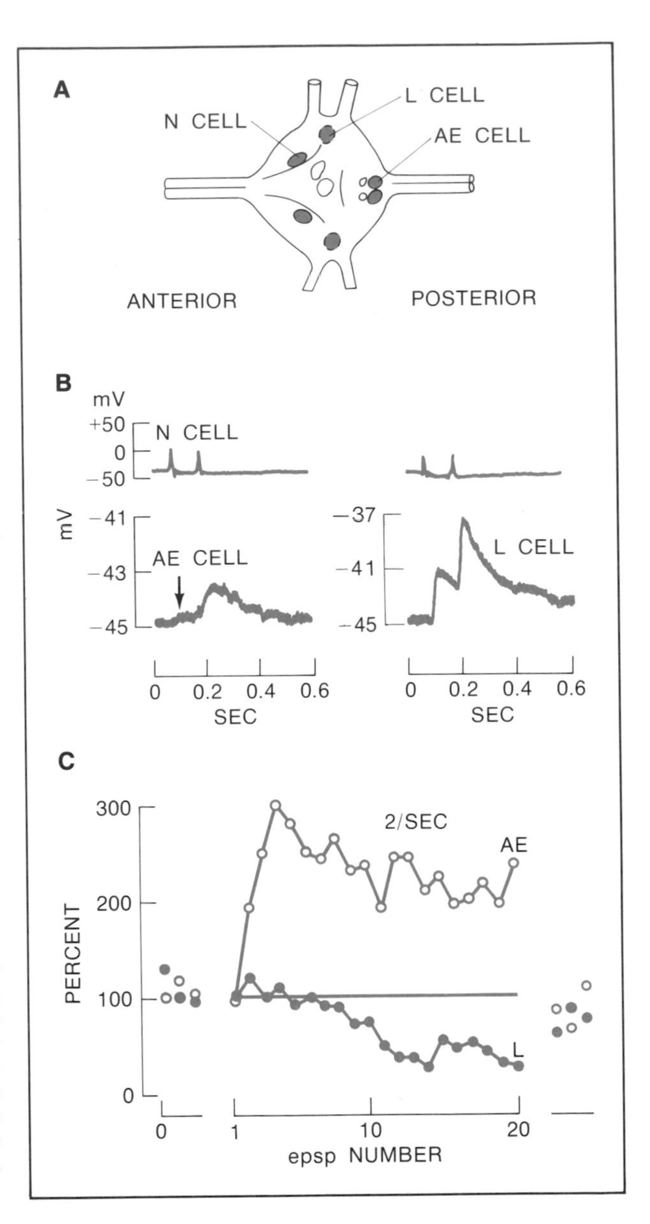

8

CHARACTERISTICS OF TRANSMITTER RELEASE at different synapses made by a presynaptic neuron. A. An N cell is stimulated and responses are recorded in L and AE cells. B. Facilitation is greater at the N-AE synapse. (The small first synaptic potential in the AE cell is marked by an arrow.) C. When the N cell is stimulated at 2/sec the synaptic potentials in the AE cell are facilitated to more than double their original size, while those in the L cell decrease in amplitude. The abscissa indicates the number of the synaptic potential recorded in the train. The ordinate gives the proportional height of the synaptic potentials compared with the average value before the train (100 percent). (After Muller and Nicholls, 1974)

variations in the amount of transmitter released at presynaptic N-cell terminals, rather than by differences in the postsynaptic cells.

Similar observations have been made at crustacean neuromuscular synapses, where it was shown that not all synapses made by a single motor fiber behave in the same way.[16,17] Some of the synapses made on one muscle fiber cause pronounced facilitation, while synapses on another muscle fiber by the same motoneuron show much less facilitation. The difference apparently lies in the presynaptic nerve terminals

[16]Atwood, H. L., and Bittner, G. D. 1971. *J. Neurophysiol.* *34*:157–170.
[17]Frank, E. 1973. *J. Physiol.* *233*:635–658.

rather than the muscle fibers, in that some terminals release more transmitter than others.

A satisfactory aspect of these studies is the general agreement between the properties of synaptic transmission and the behavioral reactions that occur in response to mechanical stimuli. In the whole animal, or in one of its isolated segments, two reflexes follow different time courses: the shortening of the body wall occurs abruptly and is poorly maintained, while the annuli become erect more slowly and stay erect longer. This correlates well with the synaptic potentials that facilitate at different rates and trigger the two events in sequence.

A further difference between synapses made by the same P or N cell on the two motoneurons is brought out by the effect of cooling. While in cooled ganglia transmission fails rapidly between P and L cells, the synapse between P and AE cells continues to function at low temperatures and even shows facilitation. This also is reflected, as expected, in the reflex responses of the animal in the cold.

It will be of interest to learn how extensively other animals use such differential effects in setting up a sequence of reactions to a maintained stimulus. So far there is little information about analogous situations in the central nervous systems of vertebrates. One example is provided by afferent fibers from muscle spindles that have characteristically different effects at different synapses. One branch causes a large, powerful excitatory potential in the neurons of Clarke's column, while another evokes only a small subthreshold potential in spinal motoneurons.[18] It is not yet known how these synapses behave with trains of impulses.

Modification of signaling in sensory neurons by repetitive firing

Repeated natural stimulation of the skin by stroking or pressing influences the performance of sensory neurons, quite apart from the effects on chemical synapses described above. These effects can persist for many seconds or even minutes after the end of stimulation. They are brought about by two distinct mechanisms: (1) the activity of an electrogenic sodium pump, described in Chapter 12; and (2) a prolonged increase in potassium permeability which allows the membrane potential to increase towards E_K, the equilibrium potential for potassium (Chapter 5).[19] The conductance change is greatly enhanced if the concentration of calcium is high while the cell is firing. Leech sensory cells therefore provide an opportunity for studying how a process like an ionic pump or a conductance change alters signaling in a reflex pathway.

As a consequence of pump and conductance mechanisms, sensory cells in the leech, in common with crustacean stretch receptors, snail neurons, and unmyelinated mammalian nerve fibers, become hyperpolarized by trains of impulses (Chapter 12). Interesting features of this hyperpolarization are its large amplitude and long duration. For example, natural stimulation of a T, P, or N cell by stroking or pressing the skin can induce an increase in the membrane potential of as much

---

[18]Kuno, M., and Miyahara, J. T. 1968. *J. Neurophysiol.* *31*:624–638.
[19]Jansen, J. K. S., and Nicholls, J. G. 1973. *J. Physiol.* *229*:635–655.

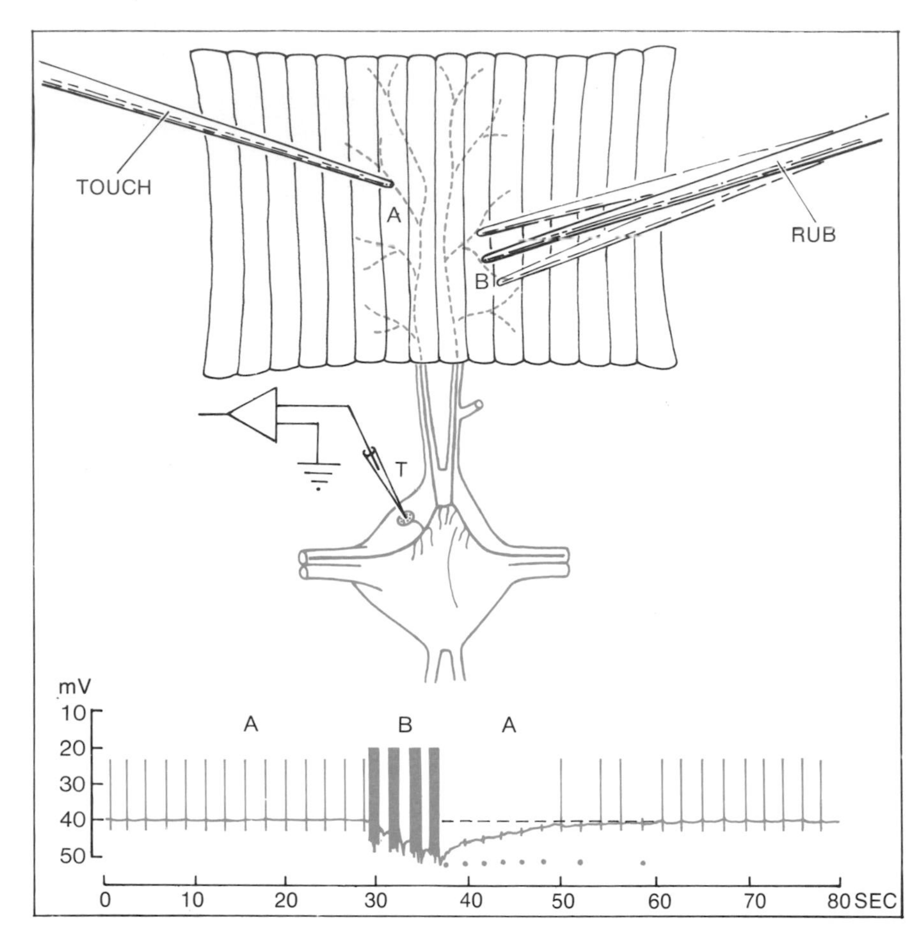

9 **CHANGE IN SENSITIVTY TO TOUCH as a result of sustained impulse activity in a T neuron. The two axons of the T cell run in the roots to innervate different areas of skin. Each light touch applied once every 2 sec at A initiates an impulse that is recorded in the cell body. Stroking the skin at B, 1 mm away from A, at intervals evokes four trains of impulses. During the hyperpolarization that follows the trains, the same light touches at A fail to set up impulses for 10 sec or more (marked by dots). As the hyperpolarization declines, the stimuli at A once again become effective. (After Van Essen, 1973)**

as 20 mV that gradually declines over a period of up to 15 min. Figure 9 shows an example produced by natural stimulation of a T cell. The hyperpolarization is not restricted to the cell body, but also occurs in its peripheral axon and in fine processes within the neuropil.

The precise proportion in which pump activity and conductance changes contribute to hyperpolarization is hard to assess. The predominant mechanism in T cells is the pump and in N cells the increase in potassium conductance; in P cells both mechanisms contribute to about the same extent.

The effects of brief trains of impulses have been most thoroughly studied in T cells.[20] After a small area of the skin of a leech has been stroked with a light stylus for a few seconds, stronger mechanical or electrical stimulation is required to initiate subsequent impulses. This increase in threshold occurs throughout the whole of the receptive field of the T cell (Figure 9). Wherever impulses travel in this cell, orthodromically to the ganglion or antidromically back to the skin, they leave behind a hyperpolarization. With more prolonged stroking of the skin and sustained impulse activity, the threshold of a T axon can increase to such an extent that impulses fail to propagate at regions of low safety factor, such as branch points. Conduction block occurs first at the periphery, so that not all the action potentials initiated in the skin actually reach the ganglion. Later, as a result of protracted firing, impulses can also fail to conduct at branch points within the central nervous system where the safety factor for propagation is low. These activity-evoked effects can be mimicked by artifically hyperpolarizing a cell through a microelectrode. Conversely, the consequences can be reversed by depolarizing currents.

One result of the hyperpolarization, then, is to reduce impulse traffic right at its origin at the receptor level, partly by raising the threshold and partly by blocking conduction in peripheral branches; subsequently, communication within the ganglion and with the neighboring ganglia may become impaired. Another effect is on integration by the sensory cells that themselves receive excitatory and inhibitory inputs from other T cells within the same ganglion.

At present numerous unexplained problems remain. For example, it is not known to what extent the hyperpolarization of terminals influences the amount of transmitter they release. Neither is it clear whether conduction block at branch points plays a part in the normal course of events in an intact animal's nervous system.[21]

Higher levels of integration

One aim of studies on an invertebrate like the leech is to analyze how complex behavioral acts are built up from simple, elementary reflexes. Smooth, coordinated movements of the animal as a whole are produced through interaction of individual ganglia with their neighbors, with distant ganglia, and with the head and tail brains.

It is not surprising that the connections of cells between ganglia exhibit the same orderliness and precision as those within a ganglion. Thus P and N sensory cells in one ganglion make direct excitatory connections on the L cell in the preceding and following ganglia.[22] The synaptic potentials are, however, smaller. As a result, pressing the skin causes a localized shortening which is strongest within the stimulated segment and spreads decrementally to neighboring segments.

A good example of a higher level of interaction is the coordinated

---

[20]Van Essen, D. C. 1973. *J. Physiol.* *230*:509–534.

[21]Barron, D. H., and Matthews, B. H. C. 1935. *J. Physiol.* *85*:73–103.

[22]Jansen, J. K. S., Muller, K. J., and Nicholls, J. G. 1974. *J. Physiol.* *242*:289–305.

sequential contraction and relaxation of segments during swimming. This complex movement has been approached at the cellular level by Stent, Kristan, and their colleagues.[23] The individual muscle groups that participate, and the motor cells that control them, have now been identified and some of the central connections traced. The key problems are (1) to analyze the pattern of synaptic interconnections and the mechanisms that allow the waves of contraction to travel repeatedly from head to tail along the flattened body, and (2) to explain the source of the rhythm and the factors that modulate it.

The head brain is not essential for swimming movements, which can occur even in a few segments of the animal. At present, evidence suggests that the basic rhythm is established by synaptic interactions within ganglia and that the role of peripheral receptors is to trigger, enhance, depress, or halt the swimming. In this respect the leech resembles other invertebrates, such as the crayfish, the locust, and the cricket, in which central motor "programs" involving a small number of individual cells have been shown to control complex patterns of coordinated movements.[24] There are also similarities to the performance of coordinated movements in vertebrates. Complex acts such as respiration or walking in decerebrate cats are built up on the basis of specific connections discussed in Chapter 16. Although millions of neurons rather than hundreds or thousands are involved in cats, the underlying principles resemble those in invertebrates.

Relevance for understanding more complex systems

This chapter, devoted to the leech, illustrates the varied uses of a simple preparation for exploring integrative mechanisms at the cellular level. Although analysis of a brain composed of so few cells may seem to present a well-defined, finite problem, many essential questions relating to the anatomy, chemistry, and physiology of the nerve cells and synapses remain to be explored. Some questions can be answered simply by doing the appropriate experiments; others must await technical advances. One hope lies in the use of organ culture in which neurons can be studied in a controlled environment over long periods.[25] Another promising approach is to study the nerve cells in leech embryos with the expectation of seeing how the nervous system becomes wired up during development. In addition, it seems possible that the leech nervous system may provide clues to the mechanisms by which use, disuse, or abnormal use can alter the properties of synapses and thereby modify the performance of the animal. Finally, as always, one expects a simple system to provide a stepping stone toward an understanding of complex nervous systems. We suspect that Nietzsche[26] anticipated our problems when he wrote:

---

[23]Kristan, W. B. Jr., Stent, G. S., and Ort, C. A. 1974 *J. Comp. Physiol.* *94*:155–176.

[24]Wilson, D. M. 1966. In *Nervous Hormonal Mechanisms of Integration. Symp. Soc. Exp. Biol. 20*:199–228. Academic Press, New York.

[25]Miyazaki, S., Nicholls, J. G., and Wallace, B. G. 1976. In *The Synapse. Cold Spring Harbor Symp. Quant. Biol. 40*:483–493.

[26]Nietzsche, F. *Thus Spake Zarathustra.*

"Then you must be a scientist whose field is the leech," said Zarathustra, "and you must pursue the leech to its last rock-bottom, you conscientious man!"

"Oh Zarathustra," answered the man, "that would be an enormity. How could I take up such a huge task! What I am the master and connoisseur of is the *brain* of the leech: that is *my* field! And it is a whole universe!"

SUGGESTED READING

General reviews

Bullock, T. H., and Horridge, G. A. 1965. *Structure and Function in the Nervous Systems of Invertebrates.* W. H. Freeman, San Francisco.

Carew, T. J., and Kandel, E. R. 1974. Synaptic Analysis of the Interrelationships Between Behavioral Modifications in Aplysia. In M. V. L. Bennett, (ed.). Raven Press, New York, pp. 339–383. *Synaptic Transmission and Neuronal Interaction.* Society of General Physiologists Series, Vol. 28.

Kennedy, D. 1971. Crayfish interneurons. *Physiologist 14*:5–30.

Nicholls, J. G., and Baylor, D. A. 1969. The specificity and functional role of individual cells in a simple nervous system. *Endeavour 103*:3–7.

Original papers

Jansen, J. K. S., and Nicholls, J. G. 1973. Conductance changes, an electrogenic pump and the hyperpolarization of leech neurones following impulses. *J. Physiol. 229*:635–655.

Kristan, J. B. Jr., and Stent, G. S. 1976. Peripheral feedback in the leech swimming rhythm. In *The Synapse. Cold Spring Harbor Symp. Quant. Biol. 40*:663–674.

Muller, K. J., and McMahan, U. J. 1976. The shapes of sensory and motor neurons and the distribution of their synapses in ganglia of the leech: A study using intracellular injection of horseradish peroxidase. *Proc. R. Soc. Lond. B* (in press).

Muller, K. J., and Nicholls, J. G. 1974. Different properties of synapses between a single sensory neurone and two different motor cells in the leech C.N.S. *J. Physiol. 338*:357–369.

Nicholls, J. G., and Baylor, D. A. 1968. Specific modalities and receptive fields of sensory neurons in the CNS of the leech. *J. Neurophysiol. 31*:740–756.

Nicholls, J. G., and Purves, D. 1972. A comparison of chemical and electrical synaptic transmission between single sensory cells and a motoneurone in the central nervous system of the leech. *J. Physiol. 225*:637–656.

Stuart, A. E. 1970. Physiological and morphological properties of motoneurones in the central nervous system of the leech. *J. Physiol. 209*:627–646.

Van Essen, D. C. 1973. The contribution of membrane hyperpolarization to adaptation and conduction block in sensory neurones of the leech. *J. Physiol. 230*:509–534.

PART SIX
NATURE AND NURTURE

IN SEARCH OF THE RULES OF FORMATION, MAINTENANCE, AND ALTERATIONS OF NEURAL CONNECTIONS

One of the most striking features of the nervous system is the high degree of precision with which nerve cells are connected to each other and to different tissues in the periphery, such as skeletal muscle and skin. This orderliness of the connections made during development is a necessary prerequisite for all the integrative mechanisms described so far.

The nervous system appears constructed as if each neuron had built into it an awareness of its proper place in the system. During development the neuron grows toward its target, ignores some cells, selects others, and makes permanent contact not just anywhere on a cell but with a specified part of it. Conversely, neurons behave as if they were aware when they have received their proper connections. When they lose their synapses, they show their recognition in various ways. For example, denervated neurons or muscle fibers develop supersensitivity to the chemical transmitter, owing to the appearance in their membranes of new receptor proteins. At times denervated cells provoke sprouting of processes from their neighbors. Even in the absence of denervation, simply as a result of changed patterns of activity, the effectiveness of synapses may change and new connections may possibly be formed.

One of the tasks at present is to sharpen and widen our general recognition of these phenomena and, above all, to find out about the processes responsible for such behavior of cells. We therefore present selected examples that substantiate these statements and point to instances where the rules are relaxed, as when cells become able to accept foreign nerves.

ORDERLINESS OF CONNECTIONS

Four examples illustrate the precise design of the architecture of the nervous system. First, Chapter 15 mentions the stretch reflex that arises as the result of impulses in the sensory nerve cell that innervates a muscle spindle. Its cell body, which lies in a dorsal root ganglion, sends some of its processes into the periphery to appropriate regions of the intrafusal muscle fiber. It also sends processes centrally to search out and make synapses exclusively on those motoneurons that innervate the same skeletal muscle from which the sensory neuron arises. Other branches run in the dorsal columns to end in a localized region of the dorsal column nuclei, and still other branches end on various interneurons. A second example (Chapter 2) is provided by individual neurons in the visual cortex that selectively recognize a vertically oriented light bar shone into their receptive field. This is possible because the inputs are derived from selected lower order neurons, some of which are excitatory and others inhibitory. A third case of specificity is the Mauthner cell (Chapter 16), which initiates a rapid tail flip in the goldfish. Well-defined regions of its cell body and dendrites receive different inputs from the two auditory nerves and from the other Mauthner cell. One consistently finds certain recognizable synapses in the expected places, synapses that must be of the correct type—excitatory or inhibitory, electrical or chemical. Finally, a good example is the distribution of synapses made on Purkinje cells in the cerebellum, worked out by Ramón y Cajal, Eccles, Szentagothai, Ito, Llinás, and Palay. These large neurons (Chapter 1) provide the only known output from the cerebellum and are therefore the end stations for the integrative activity of all the other cells. They may receive more than 100,000 synapses that end on appropriate parts of the neuron. Thus, the climbing fibers terminate on smooth dendrites and the basket cells on cell bodies and axons, while the granule cells make synapses with spiny processes of Purkinje neurons.

A number of questions naturally arise when one considers the above examples. What cellular mechanisms enable one neuron to select another out of myriad choices, to grow toward it and form synapses? Are both cells specified, or does the arrival of one determine the fate of the other? As for the precision of the wiring, how much variability is there in the connections between certain cells in different animals? What directs the systematic growth of nerve fibers along their well-defined paths? The answer to these questions influences thinking about the genetic blueprint for wiring up a brain containing 10^{10} to 10^{12} cells with a much smaller number of genes, 10^6 or less.

EXPERIMENTAL APPROACHES

One approach to these problems is to study the FORMATION OF CELLS DURING DEVELOPMENT. Using embryonic and newborn animals, Sidman and Rakic have followed cells as they migrate to their final destinations and form connections. This involves painstaking reconstruction of the positions, shapes, and distribution of various cell types at different stages of development. Layered structures with distinctive cells organized in a regular manner—the cerebellum, hippocampus, and cortex—lend themselves best to this sort of study. To follow cell lineage, one can label cell nuclei with tritiated thymidine. The label becomes incorporated into DNA only when a cell is in the process of dividing; in subsequent divisions the radioactivity is rapidly diluted out and lost. By labeling at different stages of development, one can establish by autoradiography the time when a cell underwent its last division (the birthday of a cell), where it arises, and into what type of cell it develops.

In addition, Rakic and Sidman have exploited genetic techniques and mutant animals.[1,2] If one class of cells fails to develop, one can ask what happens to the cells that would normally form synapses on them. Other promising approaches, which we merely mention, make use of the methods that have proved successful for molecular biology, the GENETIC DISSECTION of relatively simple nervous systems, including the brains of the fruit fly, *Drosophila*,[3] of a small crustacean, *Daphnia*,[4] and of various worms.[5] Because of the short life cycle of such animals, mutants can be selected and cloned with a view to determining how the presence or absence of a single gene influences the

---

[1]Sidman, R. L., and Rakic, P. 1973. *Brain Res. 62*:1–35.

[2]Rakič, P., and Sidman, R. L. 1973. *J. Comp. Neurol. 152*:103–132, 133–162.

[3]Benzer, S. 1971. *J.A.M.A. 218*:1015–1022.

[4]LoPresti, V., Macagno, E. R., and Levinthal, C. 1973. *Proc. Natl. Acad. Sci. U.S.A. 70*:433–437.

[5]Ward, S., Thomson, N., White, J. G., and Brenner, S. 1975. *J. Comp. Neurol. 160*: 313–338.

behavior of an animal and the architecture of its nervous system.

Lately, the formation of connections has been studied successfully in vitro, using TISSUE CULTURE techniques.[6] Several types of nerve cells have been used, including neurons obtained from autonomic and dorsal root ganglia and spinal cord.[7–9] When dissociated nerve cells are cultured together with cardiac or skeletal muscle fibers, functioning nerve-muscle junctions have been grown. As the synapses develop, the chemistry of the transmitters and the associated enzymes undergoes changes. Dissociated embryonic cells can also form aggregates that resemble the tissue of origin. In cultures of cells from different regions of the brain one can observe that retinal, hippocampal, cerebellar, or cortical neurons form separate characteristic arrangements. Moreover, there is evidence for specific proteins that promote the aggregation.[10]

A major difficulty for culture studies on neurons compared with other tissues in the body is that neurons do not divide. A recent development is the introduction of NEURAL TUMOR CELLS, which are able to divide and can be cloned.[6] In such neuroblastoma cultures the formation of functional synapses has not yet been observed. In the long run, tissue culture methods promise to be most useful for determining the chemical requirements for neurons to form contacts and to manipulate experimentally the conditions under which contacts are made.

An alternative path to the study of the formation of connections is to make use of REGENERATION, the fact that the processes of mature nerve cells can grow back to form synapses after they have been severed by a lesion. For example, in lower vertebrates such as frogs or fish, one can cut the optic nerve and allow the animal to recover. After a few weeks, the fibers grow back into the brain to reform connections that restore vision. In mammals, although no new neurons can be formed in the central nervous system, and tracts do not regenerate, there is evidence that existing fibers do sprout. In any event, mammalian peripheral axons do grow back and innervate skeletal muscles and sensory end organs in the skin. This approach will gain in value if it turns out that the ground rules are similar during early development and regeneration or repair.

A different but related question concerns the STABILITY OF SYNAPSES once they have been formed. How is the effectiveness of transmission influenced by use, disuse, or inappropriate use? There are many examples to show that synapses can be changed in a number of ways, ranging from an increase in efficiency to complete block. On the short time scale, one impulse arriving at the chemical synapse can produce a residual change, so that the next impulse liberates more transmitter. This facilitation may last for seconds or minutes and has been studied in detail at many types of synapses. But there are also longer term changes that occur over days, weeks, or months. Among the most remarkable of these are the changes in the visual system produced by closing one eye or by producing a squint in a kitten. Subtle alterations of the sensory input disturb performance and disrupt pathways that had previously been effective.

Another important factor determining neural organization is CELL DEATH. In the embryo there is a considerable redundancy of nerve cells, such as those that supply limb muscles in amphibians or neuronal assemblies in the avian brain. Before connections become permanent, a sorting

---

[6]Nelson, P. G. 1975. *Physiol. Rev. 55*:1–61.

[7]Peterson, E. R., and Crain, S. M. 1972. *Exp. Neurol. 36*:136–159.

[8]Fischbach, G. D. 1972. *Devel. Biol. 28*: 407–429.

[9]Fischbach, G. D., and Dichter, M. A. 1974. *Devel. Biol. 37*:100–116.

[10]Garber, B. B., and Moscona, A. A. 1972. *Devel. Biol. 27*:217–234.

out process occurs according to a predestined plan, only the fittest surviving. The disappearance of cells, however, is a continuing process in the adult brain. Support for such an assertion rests on counts of nerve cell nuclei, but hard numbers are difficult to come by in normal individuals.[11–13] Conversely, the relative number of glial cells increases. Such a constant attrition of neurons goes generally without special notice, perhaps because we regard the consequences as part of our normal aging process. One faces here the serious difficulty of diagnosing or defining "normal" and what the statement implies. For example (Chapter 15), about one third of the spinal motoneuron pool is solely concerned with centrifugal control of sensory discharges rather than direct execution of movement. How many cells, and which ones, would we miss if they gradually dropped out? Only after various injuries resulting from trauma, disease, or senility do clear-cut symptoms appear of such magnitude that we can correlate them with confidence to specific lesions. Even with large cell deficits, as produced by some tumors in the frontal lobes, diagnosis may require considerable skill. To some extent, therefore, one can maintain that we possess a great redundancy of nerve cells that automatically take care of continued neuron death.

The scope of all the problems relating to development, synapse formation, neural specificity, and changes in efficiency is too great for a comprehensive discussion. Many aspects are covered in detail elsewhere.[14] The following two chapters deal with a few selected experimental approaches: (1) the use of skeletal muscles and autonomic nerve cells to study the effects of denervation and prolonged inactivity and the factors required for reinnervation; (2) the formation of nerve connections in the brain after lesions; (3) the properties and isolation of a nerve growth factor that influences the growth of sympathetic nerve fibers; and (4) the effects of sensory deprivation and a mutation in the visual system of the cat.

---

[11]Landmesser, L., and Pilar, G. 1974. *J. Physiol. 241*:715–736, 737–749.

[12]Cowan, W. M. 1973. In M. Rockstein (ed.). *Development and Aging in the Nervous System.* Academic Press, New York.

[13]Brody, H. 1955. *J. Comp. Neurol. 102*: 511–556.

[14]Jacobson, M. 1970. *Developmental Neurobiology.* Holt, New York.

CHAPTER EIGHTEEN

SPECIFICITY OF NEURONAL CONNECTIONS

How are the orderly and precise synaptic connections of the nervous system formed during development? These questions of neural specificity are discussed in the framework of two diverse experimental approaches: (1) denervation and regrowth of nerve fibers and (2) the consequences of a genetic abnormality in the wiring of the cat's visual system.

Neurons in the heart of the frog or vertebrate skeletal muscle fibers, deprived of their synapses by denervation, develop new chemoreceptors and an increased sensitivity to the transmitter acetylcholine. Instead of being concentrated around the synapses, additional acetylcholine receptors appear over the entire neuronal and muscle surface, accounting for the supersensitivity. Direct electrical stimulation of supersensitive muscles causes the chemosensitivity to shrink back to the original end plate area. The distribution of chemoreceptors in the muscle membrane can be controlled by the level of muscle activity; additional unknown factors may also play a part. Supersensitivity goes hand in hand with the ability of muscle fibers to accept innervation, but no causal relation has been established. Regenerating nerves usually grow back to their appropriate muscle fibers and neurons. When muscles are denervated, however, even foreign nerves will form connections.

A more direct experimental demonstration of neural specificity can be obtained in amphibians and fish in which nerve fibers grow back to their specific targets; thus, regenerating optic nerve fibers can reform connections so that function is restored. Still more specificity is exemplified in the central nervous system of the leech, where an individual nerve cell can grow back and selectively restore functional connections with identified target cells.

The closest approach to an identified substance that promotes the growth of certain nerve cells and their processes is a nerve growth factor. This is a protein that acts selectively on sympathetic neurons in the autonomic nervous system and dorsal root ganglion cells. Antibodies to nerve growth factor injected into immature mice can lead to failure of development of the sympathetic nervous system.

In the visual system of the Siamese cat, genetically determined errors of connections occur: during development some optic nerve fibers take the wrong pathway at the optic chiasm and terminate to make synapses with inappropriate groups of neurons in the lateral geniculate nucleus. These cells in turn supply inappropriate cells in the visual cortex. The results indicate that specificity of connections need not be absolute.

DENERVATION AND FORMATION OF CONNECTIONS

The denervated muscle membrane

Neuromuscular synapses have provided a useful model for mechanisms of synaptic transmission between neurons in the higher centers. Similarly, changes occurring in denervated muscles are relevant for thinking about the appearance and disappearance of neural connections in general.[1]

Some phenomena in skeletal muscles after severance of their nerve supply were originally described toward the end of the last century; they were usually noted in muscle fibers that could be readily seen, such as those in the tongue.[1] After a variable period, individual muscle fibers start to exhibit spontaneous, asynchronous contractions called fibrillation. The onset of fibrillation may be as early as 2 to 5 days after denervation in rats, guinea pigs, or rabbits, or well over a week in monkeys and humans. Fibrillation occurs through changes in the muscle membrane and is not initiated by acetylcholine (ACh).[2]

Before or at the start of fibrillation, mammalian muscle fibers become SUPERSENSITIVE to a variety of chemicals. This means that the concentration of a substance required to produce depolarization, or shortening of a muscle, is reduced by a factor of several hundred to a thousand. For example, a denervated mammalian skeletal muscle is about 1000 times more sensitive to its transmitter, ACh, than is a normally innervated one if the ACh is either applied directly in the bathing fluid or injected into an artery supplying the muscle.[3] The increase in chemosensitivity is not restricted to the physiological transmitter, ACh, but occurs for a wide variety of chemical substances and even makes the muscle more sensitive to stretch or pressure.[4] Characteristic changes also occur in the passive electrical properties of denervated muscle fibers in frogs and mammals. The membrane resistance becomes approximately doubled owing to a decrease in the resting potassium conductance.[5] This effect, however, is far too small to account for the observed increases in sensitivity to ACh. The action potentials in denervated muscles are also changed in that they are more resistant to tetrodotoxin, the puffer fish poison that blocks sodium channels (Chapter

---

[1]Cannon, W. B., and Rosenblueth, A. 1949. *The Supersensitivity of Denervated Structures: Law of Denervation.* Macmillan, New York.

[2]Purves, D., and Sakmann, B. 1974. *J. Physiol.* *239*:125–153.

[3]Brown, G. L. 1937. *J. Physiol.* *89*:438–461.

[4]Kuffler, S. W. 1943. *J. Neurophysiol.* *6*:99–110.

[5]Nicholls, J. G. 1956. *J. Physiol.* *131*:1–12.

8); thus impulses can still be generated, with calcium substituting for sodium.[6]

Other changes that occur in denervated muscle, such as the gradual atrophy or wasting and the many chemical changes that follow,[7,8] are not discussed here.

Formation of new ACh receptors after denervation or prolonged inactivity

Supersensitivity is explained by an altered distribution of ACh receptors in denervated muscles. This has been demonstrated by applying ACh locally to part of the muscle surface by electrophoretic release from an extracellular micropipette while recording the membrane potential with an intracellular microelectrode. As explained in Chapters 8 and 10, in normal frog, snake, and mammalian muscle the end plate region, where the nerve makes synapses, is sensitive to ACh, while the rest of the muscle membrane has a very low sensitivity.

When ACh is applied locally to denervated muscles, the results are very different from normal. The chemosensitivity increases day by day after a nerve to a mammalian muscle is cut, until by about 7 days the surface of the muscle is almost uniformly sensitive to ACh (Figure 1).[9] In frog muscle, the changes are relatively small and take many weeks to develop.[10]

The new receptors in denervated muscles are in many respects similar to those that are present normally. ACh still increases the permeability of the membrane to both potassium and sodium ions, but not chloride,[11] and its action is blocked by curare and the snake venom α-bungarotoxin (Chapter 10). There are, however, certain differences in the properties of the junctional and new extrajunctional receptors (see below).

How does section of a nerve lead to the appearance of new receptors? Is it simply inactivity of the muscle or is there some additional mechanism? Lømo and Rosenthal[12] investigated this problem by blocking conduction in rat nerves by a local anesthetic or by diphtheria toxin. The substances were applied by means of a cuff to a short length of the nerve some distance from the muscle. With this technique the muscles became completely inactive, since motor impulses failed to conduct past the cuff. Occasional test stimulation of the nerve distal to the block produced a twitch of the muscle as usual, and miniature end plate potentials still occurred normally, showing that synaptic transmission was basically intact. And yet after 7 days of nerve block the muscle had become supersensitive (Figure 2).

Other experiments have shown that long-term application of curare or of α-bungarotoxin also leads to the appearance of new receptors.[13] All these results show that supersensitivity can be produced without denervation.

---

[6]Harris, J. B., and Thesleff, S. 1971. *Acta Physiol. Scand.* *83*:382–388.
[7]Gutmann, E. 1962. *Rev. Can. Biol.* *21*:353–365.
[8]Guth, L. 1968. *Physiol. Rev.* *48*:645–687.
[9]Axelsson, J., and Thesleff, S. 1959. *J. Physiol.* *147*:178–193.
[10]Miledi, R. 1960. *J. Physiol.* *151*:1–23.
[11]Jenkinson, D. H., and Nicholls, J. G. 1961. *J. Physiol.* *159*:111–127.
[12]Lømo, T., and Rosenthal, J. 1972. *J. Physiol.* *221*:493–513.
[13]Berg, D. K., and Hall, Z. W. 1975. *J. Physiol.* *244*:659–676.

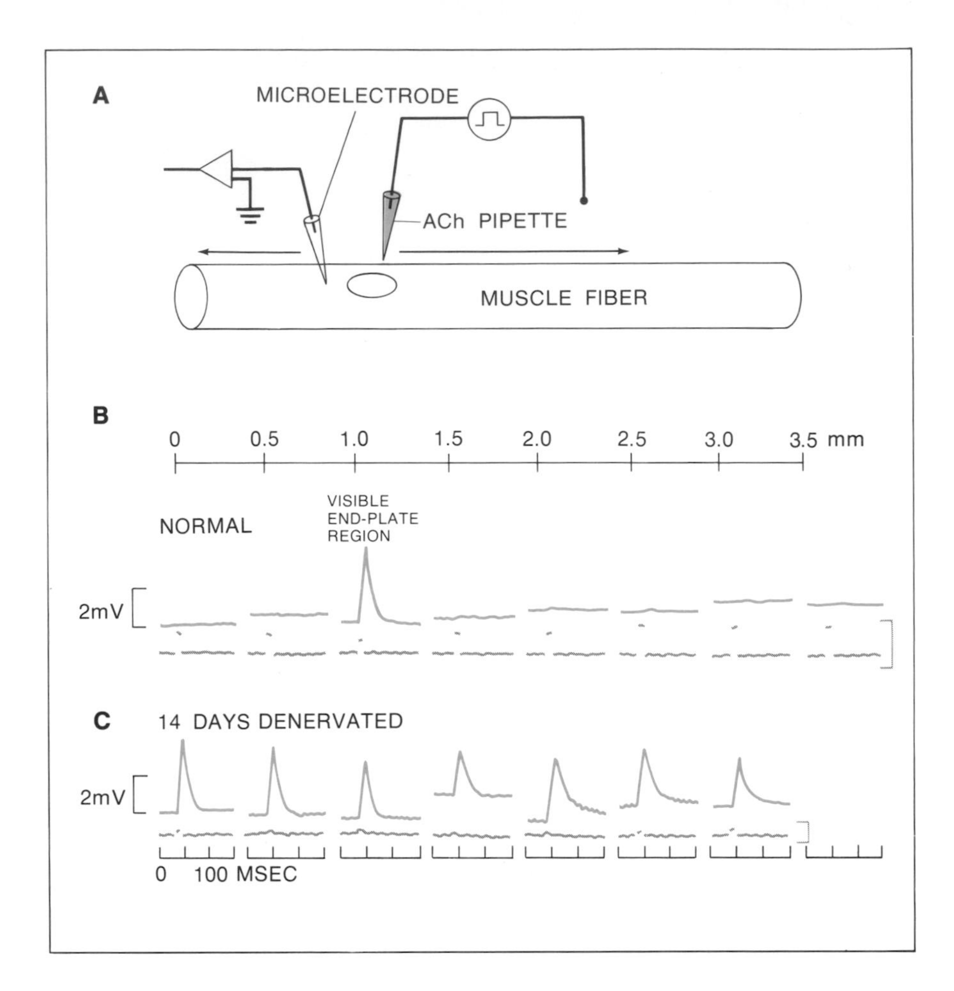

1 **NEW ACh RECEPTORS appear in a muscle of the cat after denervation. A. ACh-filled pipette is moved to different positions where it releases ACh onto the surface of a muscle fiber. B. Pulses of ACh are applied to a muscle fiber with intact innervation. A response is seen only in the vicinity of the end plate. C. After 14 days of denervation, muscle fibers respond to ACh along their entire length. (After Axelsson and Thesleff, 1959)**

The role of activity itself as an important factor in controlling supersensitivity was further shown in other experiments in which supersensitive denervated muscles in the rat were stimulated directly through electrodes permanently implanted around the muscle. Repetitive direct stimulation of muscles over several days caused the sensitive area to become restricted, so that once again only the synaptic region was sensitive to ACh (Figure 3).[12] The frequency of stimulation and the interval of quiescence were seen to be important variables in the development or reversal of supersensitivity. This explains the apparent inconsistency that denervated mammalian muscle fibers develop supersensitivity in spite of the ongoing contractions associated with fibrillation. Sampling the activity of individual fibers showed that fibrillation is cyclical, periods of activity alternating with inactivity. The level of spontaneous activity is, however, below that required to reverse the effects of denervation on the distribution of ACh receptors.[14]

There remains the question whether nerves provide the muscle with some products other than transmitters that keep the muscles normal. Effects on the muscle that cannot be interpreted simply in terms of activity have been produced by the drug colchicine. Colchicine has long been used as a remedy for gout; it is known to disrupt axoplasmic flow (Chapter 11) and to break up neurotubules found in nerves. It causes the development of supersensitivity in muscles without paralysis, so that nerve conduction and synaptic transmission are still effective and use of the muscle appears normal.[15] At present, it is not known how

[14]Purves, D., and Sakmann, B. 1974. *J. Physiol.* *237*:157–182.
[15]Cangiano, A. 1973. *Brain Res.* *58*:255–259.

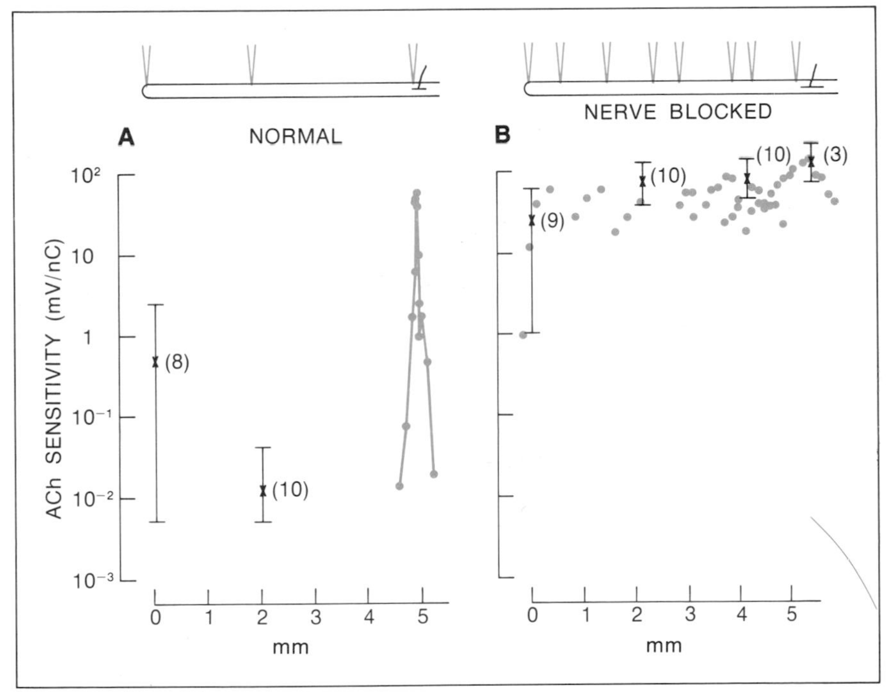

NEW ACh RECEPTORS in a muscle after block of nerve conduction. The nerve to the rat soleus muscle was blocked for 7 days by a local anesthetic. Neuromuscular transmission still functions if the nerve is stimulated below the block. A. In the normal muscle the ACh sensitivity is restricted to the end-plate region (near the 5 mm position). B. In a muscle fiber whose nerve was blocked for 7 days the ACh sensitivity is distributed over the entire muscle fiber surface. Sensitivity is expressed numerically in millivolts per nanocoulomb. The crosses and bars represent the mean and range of sensitivities in adjacent muscle fibers. (From Lømo and Rosenthal, 1972) 2

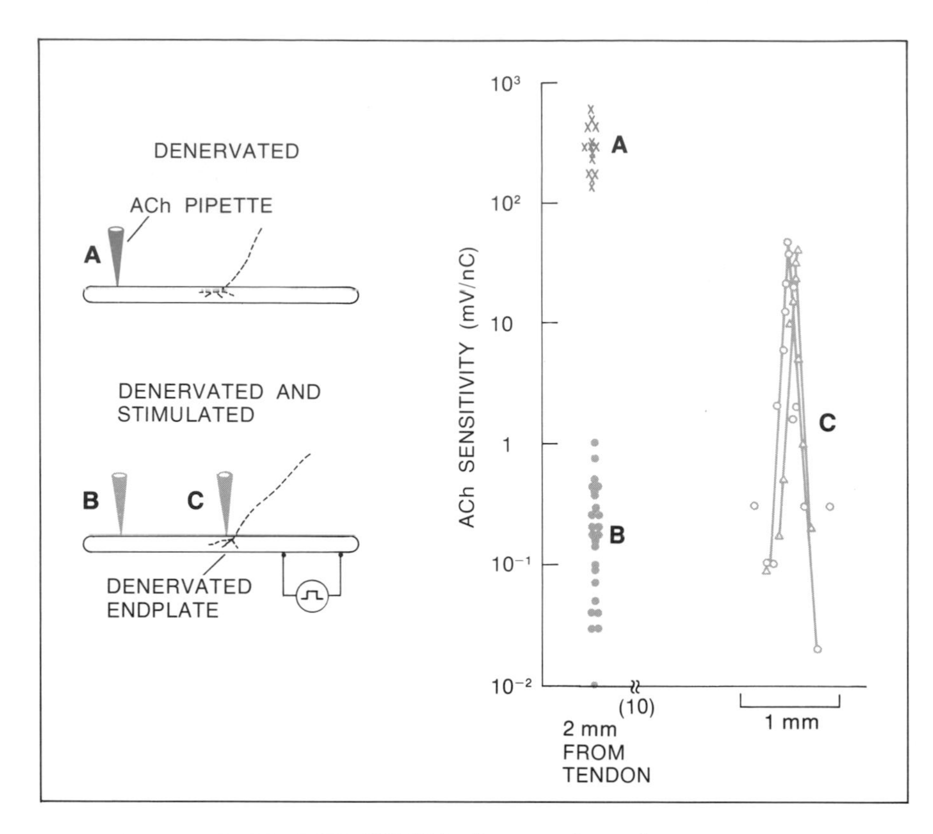

3 **REVERSAL OF SUPERSENSITIVITY in denervated muscle of the rat by direct stimulation of the muscle fibers. A. Increased ACh sensitivity in the nerve-free portion of a muscle fiber after 14 days of denervation. B. Sensitivity in the nerve-free region of a muscle that had been denervated for 7 days without stimulation and then was intermittently stimulated for another 7 days. This reversed the denervation supersensitivity. C. ACh sensitivity in two stimulated fibers of the same muscle near their denervated end plate regions. The high sensitivity is confined to this region in the stimulated muscle. (After Lømo and Rosenthal, 1972)**

colchicine produces its effects. A direct action on the muscle is a possibility, since electrical stimulation does not prevent colchicine-induced supersensitivity.[16]

Other experiments on frogs have shown that if fibers in the sartorius muscle are deprived of some of their multiple end plates, supersensitivity develops in the denervated portions of the muscle fibers; yet these fibers have kept contracting all along.[10] This whole area of work is in an interesting state of flux and needs more direct experimental evidence that will allow assessment of the relative importance of activity

---

[16]Lømo, T. and Westgaard, R. H. 1976. In *The Synapse. Cold Spring Harbor Symp. Quant. Biol.* 40:263–274.

Synthesis and turnover of new receptors in supersensitive muscles

and possible "chemical" or "trophic" factors for the control of receptor distribution.[17]

Many questions arise about the ACh receptors that appear in supersensitive muscles. Are they synthesized de novo or uncovered in new regions of membrane? Or do they perhaps migrate over the surface from the end plate where they are normally aggregated? These considerations lead to more general problems of how the mechanisms for synthesizing and turning over receptor proteins are controlled.

Experiments by Katz and Miledi have shown that the end plate region is not required for the appearance of new receptors.[18] Frog muscles were cut in two (Figure 4); fragments, which had never been innervated, separated from the end plate, survived and developed increased sensitivity to ACh. In this case the receptors could not have originated in the synaptic region, which had been physically separated.

---

[17]Harris, A. J. 1974. *Annu. Rev. Physiol.* *36*:251–305.
[18]Katz, B., and Miledi, R. 1964. *J. Physiol.* *170*:389–396.

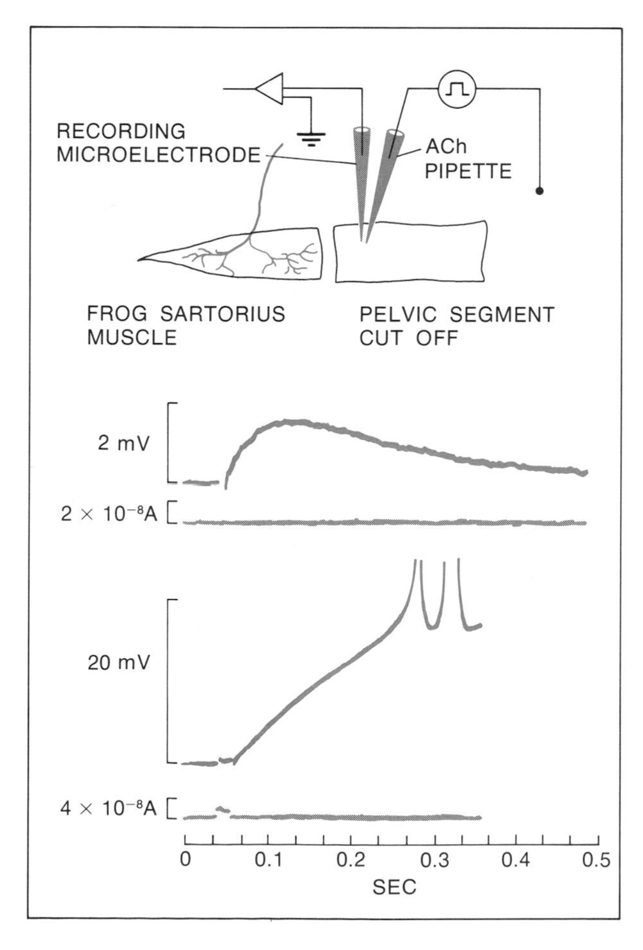

4

DEVELOPMENT OF NEW RECEPTORS in nerve-free isolated segment of a muscle in the frog. The pelvic muscle fibers had been severed from the rest of the sartorius muscle for 10 days. Iontophoretic pulses depolarize and in the lower trace initiate muscle action potentials. (After Katz and Miledi, 1964)

A valuable technique for studying the development of new receptors is to label them with radioactive α-bungarotoxin, which binds strongly and with a high degree of specificity. The rat diaphragm muscle is a convenient preparation because the end plates are restricted to a narrow zone that can be dissected out. The method used by several workers has been to bathe normal and denervated muscles in toxin and compare toxin binding to end plate and end plate-free areas. As expected, the amount and distribution of the labeled toxin were changed after denervation. Estimates of the binding sites at postsynaptic areas of the muscle are of the order of 10^4/sq μm compared with less than 10/sq μm at end-plate-free areas.[19–21] After denervation, however, receptor sites in the extrasynaptic regions increase to about 10^3/sq μm. The increase in numbers of receptors and development of supersensitivity have also been observed in muscles maintained outside the body in organ culture. In such isolated muscles substances that block protein synthesis (for example, actinomycin and puromycin) prevent supersensitivity and the formation of new receptors.[22]

As mentioned, extrasynaptic receptors behave similarly to synaptic ones in a limited number of physiological tests. Biochemical tests show several differences: newly formed extrasynaptic receptor proteins turn over in denervated muscles at a faster rate than those at the region of the original end plate;[23] they differ slightly in their affinity to certain pharmacological agents, such as curare and α-bungarotoxin; and they have a different isoelectric point, that is, they migrate differently in an electrical field.[24] Experiments of this type open up the possibility of actually studying the biochemical steps involved in the synthesis of receptors and their regulation.

Development of extrasynaptic receptors in nerve cells

Our knowledge of how nerve cells behave if some of their synaptic connections have degenerated while others are maintained is still restricted, largely because of technical difficulties. Loss of connections is a common occurrence in the brain after injury or in disease, and there is no question that neurons in the central nervous system undergo changes when part of their synaptic input is destroyed. One of the consequences is an altered response to injected drugs.[1]

Recently, it has been possible to study certain nerve cells in as detailed a manner as is described above for muscles. The alterations of the neuronal surface membrane after loss of synapses have been studied in parasympathetic ganglion cells that innervate the heart of the frog. These nerve cells can be seen in the transparent interatrial septum and, like skeletal muscles, are highly sensitive to the transmitter ACh only

---

[19]Barnard, E. A., Wieckowski, J., and Chiu, T. H. 1971. *Nature* 234:207–209.
[20]Hartzell, H. C., and Fambrough, D. M. 1972. *J. Gen. Physiol.* 60:248–262.
[21]Berg, D. K., Kelly, R. B., Sargent, P. B., Williamson, P., and Hall, Z. W. 1972. *Proc. Natl. Acad. Sci. U.S.A.* 69:147–151.
[22]Fambrough, D. M. 1970. *Science* 168:372–373.
[23]Berg, D. K., and Hall, Z. W. 1975. *J. Physiol.* 244:659–676.
[24]Brockes, J. P., and Hall, Z. W. 1975. *Biochemistry* 14:2100–2106.

at selected spots on their surfaces, immediately under the presynaptic terminals (Figure 5*A*; see also Chapter 10).

In studies on denervation the two vagus nerves to the heart are cut and the frog is left to recover.[25] Synaptic transmission between vagal nerve terminals and ganglion cells fails rapidly at room temperature starting on the second day after denervation. At the same time the chemosensitivity of the neuronal surface membrane starts to increase and is fully developed within 4 to 5 days. Thus, ACh applied iontophoretically causes a membrane depolarization wherever it is applied (Figure 5*B*). In this respect the sensitivity of the neurons differs drastically from normal, while the membrane potentials remain in the normal range and the cells are able to fire impulses. Normal chemosensitivity reappears if the original nerve is allowed to grow back into the heart. As in muscles (see below), the supersensitive area becomes restricted once more to the vicinity of the synapses.

SUPERSENSITIVITY AND REINNERVATION OF MUSCLES

Some clues about the possible significance of supersensitivity come from studying the changes a muscle undergoes in the process of rein-

[25]Kuffler, S. W., Dennis, M. J., and Harris, A. J. 1971. *Proc. R. Soc. Lond.* B. *177*:555–563.

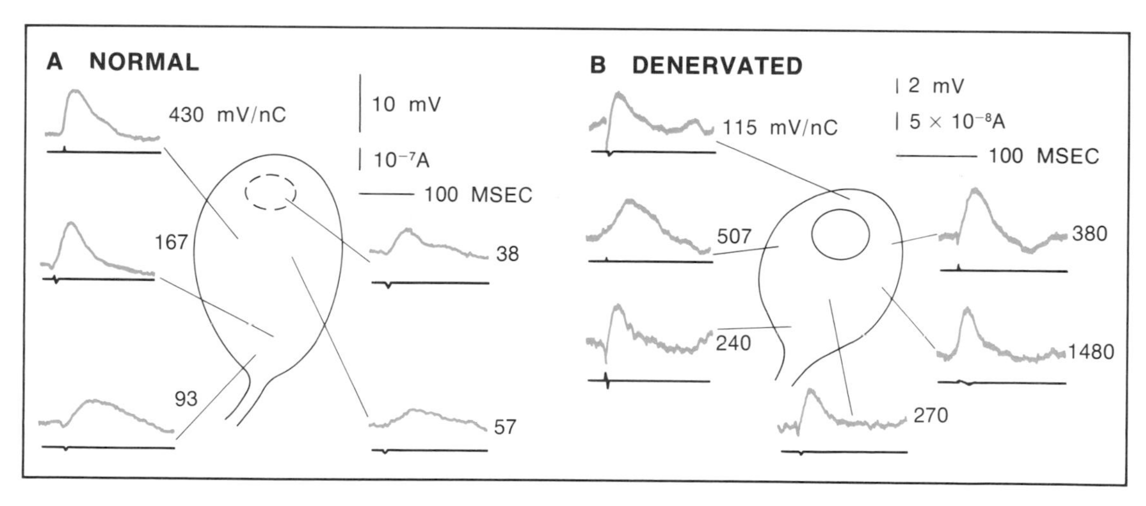

DEVELOPMENT OF NEW RECEPTORS in parasympathetic nerve cells in the heart of the frog after denervation. A. In the normal neuron the high ACh sensitivity is confined to synaptic regions. The large numbers indicate high sensitivity (expressed in millivolts per nanocoulomb). When ACh is applied to the extrasynaptic region, more must be released to have an effect. Such responses rise relatively slowly because ACh has to diffuse to a nearby sensitive synaptic spot. B. After 21 days of denervation, the sensitivity of the neuronal surface is high wherever ACh is released. (After Kuffler, Dennis, and Harris, 1971) 5

nervation. Normal skeletal muscle fibers in amphibians and mammals can be innervated by more than one nerve fiber.[26] A fully innervated muscle fiber cannot, however, be innervated by an additional accessory nerve. Thus, if a cut motor nerve is placed on an innervated muscle, it will not "take" to form additional new end plates on the muscle fibers. In contrast, nerve fibers do grow out and reinnervate a denervated or injured muscle.

What are the conditions that enable a denervated muscle to accept a nerve? Several tests have been made to see whether a correlation exists between increased chemosensitivity and synapse formation. In one such experiment Miledi used the frog sartorius in which the muscle fibers had been cut.[27] The pieces of muscle fibers that were separated from their end plate regions (Figure 4) became sensitive to ACh along their length. When a cut motor nerve was placed in apposition to these fragments, they became innervated. This is a remarkable result. It means that nerve fibers grow out to a muscle fiber and form synapses in a region that was never normally innervated. Perhaps even more remarkable is the observation that the nerve induces in the postsynaptic muscle fiber the formation of a subsynaptic specialization that this section of muscle fiber never possessed. In particular, newly grown subsynaptic folds develop. In related experiments muscles that had been cut into small fragments became reconstituted and reinnervated with normal end plates at which synaptic transmission occurred.[28]

Other investigations have shown that application of botulinum toxin produces neuromuscular block not by destroying nerve terminals but by preventing them from releasing transmitter (Chapter 9). Despite the presence of the intact-looking nerve terminals, the muscle membrane develops supersensitivity in previously extrasynaptic areas[29,30] and accepts additional new innervation. Similarly, after a muscle in the rat is made supersensitive as a result of blocking impulse transmission in the nerve, a foreign nerve is able to form additional synapses. After the block is removed from the first nerve, an individual muscle fiber may contract in response to simulation of each of the two nerves (Figure 6).[31] Conversely, when a denervated muscle is stimulated directly, the ability to accept new innervation is lost together with the supersensitivity.

Is it a normal prerequisite that the muscle be supersensitive for innervation to occur? If this were so, one might expect muscles to be supersensitive when they become innervated for the first time in the fetus. This was shown in fetal rat muscles, whose fibers are sensitive to ACh along their length. After innervation, the ACh-sensitive area

---

[26]Hunt, C. C., and Kuffler, S. W. *1954. J. Physiol. 126*:293–303.
[27]Miledi, R. 1962. *Nature 193*:281–282.
[28]Bennett, M. R., Florin, T., and Woog, R. 1974. *J. Physiol. 238*:79–92.
[29]Thesleff, S. 1960. *J. Physiol. 151*:598–607.
[30]Fex, S., Sonessin, B., Thesleff, S., and Zelená, J. 1966. *J. Physiol. 184*:872–882.
[31]Jansen, J. K. S., Lømo, T., Nicolaysen, K., and Westgaard, R. H. 1973. *Science 181*:559–561.

shrinks over a period of about 2 weeks and becomes restricted to a region around the end plate.[32] A somewhat analogous situation exists in developing muscle fibers in tissue culture as they become innervated.[33]

Both initial innervation and reinnervation therefore occur when the

[32]Diamond, J., and Miledi, R. 1962. *J. Physiol.* *162*:393–408.

[33]Steinbach, J. H., Harris, A. J., Patrick, J., Schubert, D., and Heineman, S. 1973. *J. Gen. Physiol* *62*:255–270.

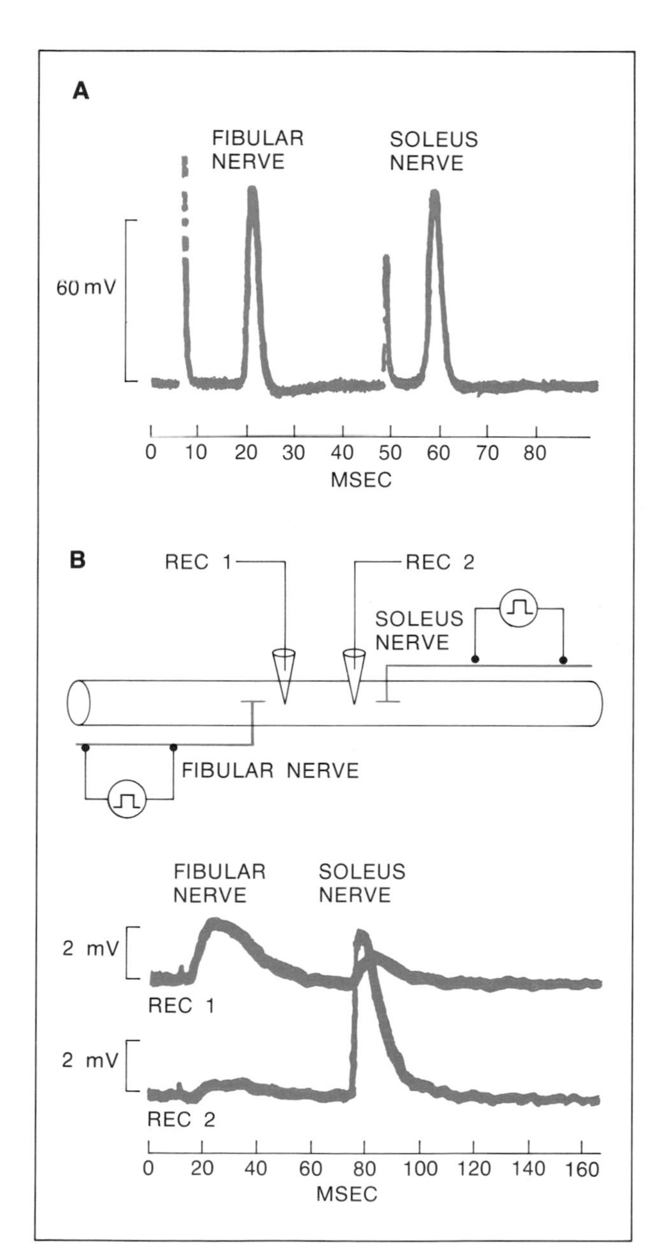

6

DUAL INNERVATION of a single muscle fiber in the soleus of the rat by its original nerve and by a transplanted foreign nerve (the fibular nerve). A. Both nerves initiate conducted action potentials in the same muscle fiber recorded with an intracellular electrode. B. Recordings with two microelectrodes (Rec 1, Rec 2) from one muscle fiber after transmission block by an increased magnesium concentration. Synaptic potentials only are caused by the original and foreign nerve stimulus. The synaptic potentials are relatively large near the point of innervation (see drawing) and become attenuated with distance. (After Jansen, Lømo, Nicolaysen, and Westgaard, 1973)

muscle fibers are supersensitive. However, supersensitivity itself may be merely an outward expression of an unknown process that makes the muscle receptive for innervation. Several experiments indicate that synapse formation is not dependent upon the receptor itself, or at least not on that part of it to which α-bungarotoxin or curare binds. Thus, reinnervation still occurs in denervated rat and toad muscles in the presence of α-bungarotoxin and curare.[34,35]

Reinnervation of skeletal muscle by foreign nerves

Under special circumstances foreign or anomalous extra nerves can be made to form functional synapses with denervated skeletal muscles. Such experiments raise a number of questions regarding the specificity of synapse formation and the way in which nerve and muscle cells influence each other. What properties must the alien nerve have to be accepted? Does it alter the property of the muscle? Is the nerve itself altered as a result of innervating the wrong muscle?

Some observations on these questions date back to 1904, when Langley and Anderson showed that muscles of the cat could become innervated by cholinergic preganglionic sympathetic fibers[36] which normally make synapses in ganglia. Similarly, formation of synaptic connections between vagal nerve fibers and the superior cervical ganglion has also been reported.[37]

Recently, the properties of synapses formed by vagus nerves on the frog sartorius muscle have been studied in detail by Landmesser.[38,39] The procedure was to transplant the denervated muscle to the thoracic region of the frog, where the cut gastric vagus nerve was then sutured to it. Within 50 days, stimulation of the vagal nerve produced synaptic potentials and visible contractions of the muscle. The synaptic potentials were different from those evoked by stimulation of a normally innervated sartorius muscle. They resembled those observed in the multiply innervated skeletal "slow" muscle fibers of the frog whose terminals make synapses at many widely distributed spots on the surface rather than at one or two discrete motor end plates. In addition, the individual synaptic potentials were small and facilitated with repetitive firing. There was no evidence to suggest that the properties of the nerve were changed or that the abnormal type of innervation had altered the electrical characteristics of the muscle.

In contrast to these experiments, the properties of certain other muscles in frogs become markedly changed with foreign innervation. For example, the slow muscle fibers in the frog (see above) are quite distinctive. Apart from their diffuse innervation they differ in their fine structure and cannot give regenerative impulses or twitches.[40] After denervation the slow fibers can become reinnervated by nerves that

---

[34]Van Essen, D., and Jansen, J. K. S. 1974. *Acta Physiol. Scand.* *91*:571–573.
[35]Cohen, M. W. 1972. *Brain Res.* *41*:457–463.
[36]Langley, J. N., and Anderson, H. K. 1904. *J. Physiol.* *31*:365–391.
[37]Vera, C. L., Vial, J. D., and Luco, J. V. 1957. *J. Neurophysiol.* *20*:365–373.
[38]Landmesser, L. 1971. *J. Physiol.* *213*:707–725.
[39]Landmesser, L. 1972. *J. Physiol.* *220*:243–256.
[40]Kuffler, S. W., and Vaughan Williams, E. M. 1953. *J. Physiol.* *121*:289–317.

normally innervate the twitch muscle fibers at discrete end plates. Under these conditions, the slow fibers change and give conducted action potentials and twitches.[41]

Cross-innervation and double innervation of mammalian muscles

Eccles, Buller, Close, and their colleagues have cut and interchanged the nerves to rapidly and more slowly contracting muscles in kittens and rats. After they have become reinnervated by the inappropriate nerves, the slow muscles become faster and the fast ones, slower.[42–44] (It should be emphasized that both types of mammalian muscle fibers give conducted action potentials and therefore differ from the slow fibers in the frogs.) These results show, therefore, that the types of nerves that innervate a muscle can influence its properties. One cannot say as yet how this action of the nerve is brought about. It seems that a major factor is the pattern of impulses and contraction, since the motoneurons innervating slow and fast muscle fibers tend to fire at different frequencies. The type of use to which a muscle is subjected apparently influences its physiological performance and its structure.

The finding that muscles can accept additional, foreign innervation has led to speculation about the possible preference for one type of nerve over another. One idea is that when a muscle is innervated by a foreign nerve as well as its normal one, the inappropriate additional synapses cease to function while still retaining their morphological appearance.[45] However, recent results obtained in both mammalian and fish muscles indicate that transplanted nerves can be as effective as the original ones in innervating fibers. Further, muscle fibers with dual innervation, one original the other foreign, can both function simultaneously (Figure 6).[31,46,47]

One clear example of the rejection of "inappropriate" innervation is provided by observations on neonatal rat muscles. In this situation an individual fiber is supplied by a number of axons, each forming an effective synapse. Over the first 2 weeks or so after birth, axons lose their connections until the picture resembles that in the adult, with just one axon supplying an end plate.[48] These experiments raise a number of interesting problems: one concerns the mechanism by which the muscle rejects some fibers while allowing one fiber to remain. There is evidence that the total number of axons in peripheral nerves and motoneurons in the spinal cord decreases during this period;[49,50] it is therefore possible that only those cells survive that make effective

[41]Miledi, R., Stefani, E., and Steinbach, A. B. 1971. *J. Physiol.* *217*:737–754.

[42]Buller, A. J. 1970. *Endeavour* *29*:107–111.

[43]Buller, A. J., Eccles, J. C., and Eccles, R. M. 1960. *J. Physiol.* *150*:399–416.

[44]Close, R. I. 1972. *Physiol. Rev.* *52*:129–197.

[45]Mark, R. F. 1974. *Br. Med. Bull.* *30*:122–125.

[46]Frank, E., Jansen, J. K. S., Lømo, T., and Westgaard, R. H. 1975. *J. Physiol.* *247*:725–743.

[47]Scott, S. A. 1975. *Science* *189*:644–646.

[48]Redfern, P. A. 1970. *J. Physiol.* *209*:701–709.

[49]Hughes, A., and Edgar, M. 1972. *J. Embryol. Exp. Morphol.* *27*:389–412.

[50]Prestige, M. C. 1967. *J. Embryol. Exp. Morphol.* *17*:453–471.

and permanent synaptic connections with their target muscle cells and that the others die. A similar series of events occurs during the development of the ciliary ganglion of the chick: nerve cells that fail to make effective synapses with their targets die.[51]

NERVE GROWTH FACTOR: WHAT INDUCES GROWTH?

There are no answers yet to the question of what makes a nerve select a certain muscle and what in the muscle attracts a nerve. There has, however, been support for many years for the idea that certain substances in tissues can attract neurons. For example, numerous experiments have shown that transplanting an extra leg onto the back of a tadpole, a lizard, or a newt causes the outgrowth of nerve fibers from the central nervous system.[52] This idea has received a considerable boost from the work of Levi-Montalcini and her colleagues,[53] who have found a factor that selectively influences the growth of sympathetic and sensory neurons. These studies have helped in approaching the problems raised in this chapter, and the course of the investigations also illustrates the manner in which much research progresses in the hands of perceptive investigators. The search for the growth factor is a remarkable sequence of coincidences—false but profitable leads, extraordinary and apparently fortunate choices, all leading to an important development in the area of the study of nerve growth.

Following up the idea that there must be substances in transplanted limbs capable of attracting nerve fibers, it was reasonable to test the effect of rapidly growing tissues on the growth of neurons. The initial experiments were made by implanting onto chick embryos a connective tissue tumor (sarcoma) obtained from mice. On the side where the sarcoma had been implanted there was a profuse outgrowth of sensory and sympathetic nerve fibers from the embryo into the tumor. To show that the effect was caused by a humoral factor, sarcomas were grafted onto the chorioallantoic membrane, a tissue that surrounds the embryo. The only communication between the embryo and the tumor was indirect, but once again the dorsal root ganglia and sympathetic neurons on the side of the implant grew profusely.[54]

Next, it was shown that a similar dramatic effect is produced by sarcoma cells on tissue-cultured chick ganglia. The active factor in the sarcoma initially appeared to be a nucleoprotein. To see if nucleic acids were essential components of the growth-promoting factor, tumors were incubated with a snake venom whose action would hydrolyze the nucleic acids and thereby render the tumor fraction inactive. With venom present, however, the growth, far from being inactivated, was further increased. In fact, the control experiment of add-

[51]Landmesser, L., and Pilar, G. 1974. *J. Physiol.* *241*:737–749.

[52]Hamburger, V. 1939. *Physiol. Zool.* *12*:268–284.

[53]Levi-Montalcini, R. 1964. *Science* *143*:105–110.

[54]See Levi-Montalcini, R., and Angeletti, P. U. 1968. *Physiol. Rev.* *48*:534–569, for references to earlier work.

ing snake venom without the sarcoma extract surprisingly revealed that the venom itself was a far richer source of growth factor than the sarcoma.[55] This in turn gave rise to the speculation that, since venom is secreted by the salivary gland, possibly salivary glands from other animals might also contain a similar factor.

The animal selected was the mouse.[55,56] It was fortunate that adult male mice were chosen, because the salivary glands of female or immature mice contain far less growth factor than do those of other animals that have since been tried. Extracts of salivary glands of adult male mice are potent in causing the growth of cultured sympathetic ganglion cells (Figure 7*A*). In this regard the functional role of salivary glands in the animal is not clear, especially since removal of these glands from young animals has minor effects on nerve growth.

The substance extracted from snake venom and salivary glands of mice has been called NERVE GROWTH FACTOR (NGF). It has been shown to be a protein made up of three types of subunits; the amino acid sequence has been worked out in the β-subunit, which is the biologically active part.[57] In immature animals it causes an increase in the growth of sympathetic and dorsal root ganglion cells. The cells increase not only in size but also in number.

It is possible to argue that everything described so far is a purely pharmacological phenomenon—the effect of an agent which, while extractable from salivary glands, plays no part in the normal maturation or growth of neurons. This was put to a test by producing a specific antibody to nerve growth factor by injecting it into rabbits.[58] When the serum was injected into newborn mice, they failed to develop normal sympathetic nervous systems (Figure 7*B*). The parasympathetic nervous system was apparently not affected, and the dorsal root ganglia were only slightly smaller than normal. The animals lived normally but responded poorly to stress conditions. In adults, the antibody was much less effective. One cannot say at present whether there is an effect on the central nervous system.

The protein therefore seems to play a part in normal maturation and growth. It is not yet known whether other factors act in the central nervous system, but a growth protein has been found for epithelia.

The observations described above have set the stage for an analysis of the mechanism of action of the nerve growth factor by a number of groups, including those of Levi-Montalcini and of Shooter.[59] Their studies have explored such questions as the fraction of the protein which is most effective in inducing growth, the receptors on the mem-

[55]Cohen, S. 1959. *J. Biol. Chem.* *234*:1129–1137.

[56]Cohen, S. 1960. *Proc. Natl. Acad. Sci. U.S.A.* *46*:302–311.

[57]Angeletti, R. H., Mercanti, D., and Bradshaw, R. A. 1973. *Biochemistry* *12*: 90–100, 100–115.

[58]Levi-Montalcini, R., and Cohen, S. 1960. *Ann. N.Y. Acad. Sci.* *85*:324–341.

[59]Herrup, K., and Shooter, E. M. 1973. *Proc. Natl. Acad. Sci. U.S.A.* *70*: 3884–3888.

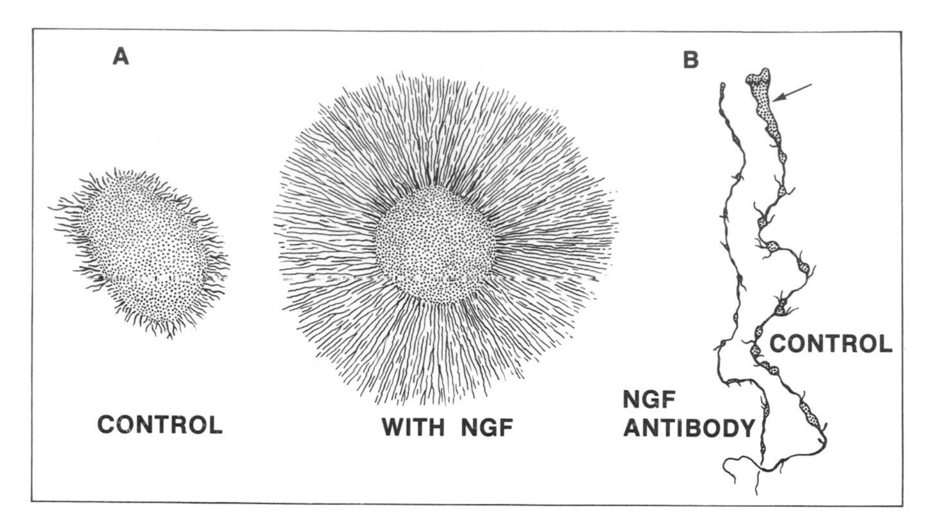

7 **EFFECT OF NERVE GROWTH FACTOR on neuron growth in 7-day-old sensory ganglion from a chick embryo kept in a culture medium for 24 hours. A. Ganglion in control medium is to the left. To the right is a ganglion in a medium supplemented with nerve growth factor from the salivary gland of a male mouse. It shows profuse growth. B. The thoracic sympathetic chain of ganglia from a control animal (mouse) is to the right. Arrow points to the stellate ganglion. To the left is a ganglion chain, much smaller in size, from a mouse injected 5 days after birth with the antiserum to nerve growth factor. (A after Levi-Montalcini, 1964; B after Levi-Montalcini and Cohen, 1960)**

brane that interact with nerve growth factor, and the metabolic events that subsequently occur. In some respects actions of nerve growth factor resemble those of insulin. The discovery of nerve growth factor offers hope for determining at a molecular level the site of action, the nature and distribution of the receptors, and the mechanisms by which a nerve cell is induced to grow. However, as Levi-Montalcini has recently stated:[60] "The NGF has still not found its place in the everyday broadening panorama of neuroscience. . . . It has disclosed only a few of its traits and keeps us wondering where it is heading for."

REINNERVATION IN THE CENTRAL NERVOUS SYSTEM

How do nerve fibers find their targets?

There is evidence that when a motor nerve grows back into a muscle, the axons generally find the sites of the former end plates. Similarly, in denervated mammalian skin sensory axons grow back to reinnervate the original end organs that remain.[61]

---

[60]Levi-Montalcini, R. 1975. In *The Neurosciences: Paths of Discovery.* F. G. Worden, J. P. Swaze, and G. Adelman (eds.). MIT Press, Cambridge, Mass., pp. 245–265.

[61]Burgess, P. R., English, K. B., Horch, K. W., and Stensaas, L. J. 1974. *J. Physiol.* *236*:57–82.

The studies with clearest results on the formation of connections during reinnervation have been done on lower vertebrates and invertebrates. Even after extensive lesions to the brain, functions can be restored through the regrowth of axons. For example, it was shown by Stone,[62] Sperry,[63,64] and their colleagues that if the optic nerve is cut in a frog or a salamander, fibers grow back to the appropriate region of the brain (the tectum). There they form synapses, and eventually the animal is able to see once again. There is much evidence that this regrowth is orderly and that neurons reform their original connections with a high degree of precision. In Sperry's original experiments the optic nerve of a frog was cut and the eye was rotated through 180°. About 3 weeks later, the nerve had regenerated and the frog could see again, but it behaved as though its vision was inverted. Thus, all its movements directed toward objects, as when the frog struck at a fly, were 180° out of phase. The simplest interpretation of this and other similar behavioral experiments is that fibers had grown back from the inverted retina to their original destinations in the tectum. This was confirmed by Gaze[65] and Jacobson,[66] using physiological techniques.

A further important observation is that the animal never learned to correct its mistakes: frogs with rotated eyes continued to strike downward at a fly held up in the air for as long as they lived after the operation. There was therefore no evidence from these experiments that synaptic connections could be reorganized through the stimulus of making behavioral errors. In this respect the behavior of frogs is different from that of higher animals; a man or a monkey can soon compensate for the effects of an inverting prism placed on the eyes (this does not imply that higher centers have reorganized their connections).

Anatomical evidence in the goldfish has provided confirmation that optic nerve fibers can grow back to their original destinations. In normal fish, each point on the retina projects onto a particular part of the tectum, producing there an orderly and stereotyped map. During regeneration of the optic nerve, groups of fibers can be observed to course directly to the appropriate region of the tectum, following a similar route to that taken during development.[67] We have no good clues to the sorts of mechanisms involved in the multiple choices the nerve fiber makes as it grows and forms connections. As for the question of reorganization of central connections, there are some experiments in which half the tectum in a goldfish was removed and as a result the entire retina now projected onto the remaining portion of the tectum. This suggests that the connections may not be specified in an inflexible

---

[62]Stone, L. S., and Zaur, I. S. 1940. *J. Exp. Zool. 85*:243–269.

[63]Sperry, R. W. 1944. *J. Neurophysiol. 7*:57–69.

[64]Sperry, R. W. 1945. *J. Neurophysiol. 8*:15–28.

[65]Gaze, R. M. 1970. *The Formation of Nerve Connections.* Academic Press, New York.

[66]Jacobson, M. 1970. *Developmental Neurobiology.* Holt, New York.

[67]Attardi, D. G., and Sperry, R. W. 1963. *Exp. Neurol. 7*:46–64.

manner and that the central nervous system in the goldfish has some latitude to reorganize itself.[68]

It should be emphasized that the interpretation of maps obtained by recording electrically is not altogether clear. The recordings so far have all been made with relatively large external electrodes from numbers of neurons activated more or less synchronously. It is not known for certain whether the signals recorded arise from impulses in nerve fibers along their course, from terminals, or from postsynaptic cells. One cannot therefore be sure whether the map represents the actual pattern of synaptic connections that have been reformed (that is, the end points) or indicates the general area into which regenerating axons grow.

Accuracy of regeneration of individual neurons

The results obtained in vertebrates suggest that regenerating cells grow to predestined targets whenever possible. The degree of selectivity is not yet established, and it would be of interest to know how cells reconnect in a system where individual neurons rather than whole populations can be examined. A convenient preparation for a study of the precision of regenerating nerve fibers is the leech central nervous system (Chapter 17), where individual cells can be recognized without ambiguity and their connections in normal animals traced.

To observe regeneration in the leech, the procedure is to sever the axons that link two ganglia and test whether the connections become reestablished.[69,70] For example, an individual sensory cell in one ganglion is known to initiate a synaptic potential in a specific motor cell in the next ganglion. Such a connection between identified neurons can, in fact, be successfully reestablished. In other experiments individual sensory neurons have been shown to reinnervate the appropriate patch of skin after their processes have been severed.[71]

The results in the leech, obtained by physiological techniques, demonstrate that a cell can discriminate among many targets in the complex neuropil so as to interact once more with a particular neuron in preference to others. In one instance there is anatomical evidence for accurate regeneration. A single fiber, about 5μm in diameter, runs along the whole length of the leech nerve cord. It can be clearly recognized in cross sections through the connectives, since it is the largest of all the axons. If the nerve cord is cut or crushed, the two regenerating portions of the axons search out each other's cut ends and fuse, so that transmission becomes reestablished.[72] There is also evidence for such a process in the peripheral nervous system of crustaceans.[73] We do not know whether repair by end-to-end fusion also occurs in vertebrates.

---

[68]Yoon, M. 1972. *Exp. Neurol. 37*:451–462.

[69]Baylor, D. A., and Nicholls, J. G. 1971. *Nature 232*:268–270.

[70]Jansen, J. K. S., and Nicholls, J. G. 1972. *Proc. Natl. Acad. Sci. U.S.A. 69*: 636–639.

[71]Van Essen, D., and Jansen, J. K. S. 1976. In *The Synapse. Cold Spring Harbor Symp. Quant. Biol. 40*:495–502.

[72]Frank, E., Jansen, J. K. S., and Rinvik, E. 1975. *J. Comp. Neurol. 159*:1–14.

[73]Hoy, R., Bittner, G. D., and Kennedy, D. 1967. *Science 156*:251–252.

Other invertebrates, such as the cricket, provide additional examples of regeneration with a high degree of precision.[74] In these creatures it has been shown anatomically and physiologically that mechanoreceptor fibers regenerate to form functional connections on the appropriate neurons within the central nervous system. These experiments also provide evidence for reorganization of the central connections when regeneration is not allowed to occur after removal of parts of the sensory apparatus. The neurons within the centers that had lost their normal outputs now receive innervation from elsewhere.

Sprouting by neurons in the mammalian CNS

It has been generally believed that sprouting in the adult mammalian central nervous system is quite restricted, largely because transection of tracts is not followed by regeneration and restitution of function. In recent years, however, it has become apparent that uninjured neurons within the mammalian brain can form new processes following injuries to other cells that innervate the same target area. The design of the experiments is to examine well-defined layered structures in the brain, such as the hippocampus, the colliculi, or the visual cortex.[75–78] A lesion is then made in a bundle of axons known to end in a circumscribed area of the tissue. After the terminals have degenerated, tests are made to see whether new growth has occurred into the denervated area from nearby regions. Examples come from work in which lesions made in the hippocampus or colliculi resulted in orderly reinnervation by neurons whose field of innervation had expanded. Such expansion can occur even in the absence of degeneration that follows surgical intervention. Chapter 19 describes how in the newborn monkey, closure of one eye during a critical period causes shrinkage of the cortical area supplied by that eye. This is accompanied by a corresponding increase in the area innervated and functionally activated by the normal eye.

ABNORMAL CONNECTIONS IN THE MAMMALIAN CENTRAL NERVOUS SYSTEM: THE SIAMESE CAT

From the preceding descriptions of regeneration and repair two general features emerge: (1) cells find their way to become connected to certain specific targets, and (2) despite this ability to discriminate there is some flexibility—it seems as though neurons can accept inappropriate or abnormal inputs under certain conditions. Both principles are clearly exemplified in an interesting abnormality of the visual pathways that occurs in Siamese cats. The account that follows is based largely on the work of Guillery,[79] Hubel and Wiesel,[80] and their colleagues.

[74]Palka, J., and Edwards, J. S. 1974. *Proc. R. Soc. Lond.* B *185*:105–121.

[75]Chow, K. L., Mathers, L. H., and Spear, P. D. 1973, *J. Comp. Neurol. 151*: 307–322.

[76]Lund, R. D., and Lund, J. S. 1973. *Exp. Neurol. 40*:377–390.

[77]Raisman, G., and Field, P. M. 1973. *Brain Res. 50*:241–264.

[78]Steward, O., Cotman, C. W., and Lynch, G. S. 1973. *Exp. Brain Res. 18*: 396–414.

[79]Guillery, R. W., and Kaas, J. H. 1971. *J. Comp. Neurol. 143*:73–100.

[80]Hubel, D. H., and Wiesel, T. N. 1971. *J. Physiol. 218*:33–62.

In Siamese cats certain optic nerve fibers fail to grow along their usual pathways during development. An outward sign of this disorder is that the animals are frequently cross-eyed. Figure 8 shows the connections between the retina and the lateral geniculate nucleus in normal (wild type) and Siamese (mutant) cats.

In the normal cat the fibers arising in the lateral part of the retina project to the middle layer of the lateral geniculate nucleus on the same side of the animal (Figure 8; see also Chapter 2). The upper and lower layers of the geniculate are supplied by fibers that arise from the medial half of the opposite eye and cross at the chiasm (shown in gray in Figure 8*A*). The points to be noted here are that: (1) the eyes project to different layers, and (2) only half of each retina projects to each lateral geniculate nucleus. This means that the animal's left field of vision is represented in the right lateral geniculate nucleus, and vice versa.

In the Siamese cat, during development a number of fibers cross at the chiasm instead of staying on the same side of the animal. In Figure 8*B* these aberrant fibers are seen to originate from ganglion cells close to the midline in the lateral part of the retina and to make connections in the middle layer of the "wrong" lateral geniculate nucleus. On both sides of the animal these abnormally crossed fibers take similar courses. Therefore, in the mutant animal a group of neurons in the middle layer of the geniculate receives an inappropriate input because a group of optic nerve fibers has crossed abnormally.

A number of intriguing problems arise. The optic nerve fibers that have taken a wrong course terminate in a foreign, unusual region of the brain. At the same time the cells they would have supplied have been deprived of their "expected" inputs. This means that certain cells receive information from unaccustomed regions of the visual field and from the wrong eye. What is the effect of such an initial mistake in wiring on the orderly map of the retina in the higher centers? The results can most simply be explained by assuming that the main genetic abnormality in Siamese cats is confined to groups of ganglion cells, while other neuronal elements in the geniculate and higher centers are genetically normal. Following the fate of the aberrant fibers provides ideas about the capacity of the central nervous system for dealing with abnormal connections.

Having crossed onto the wrong side, upon what particular cells of the middle layer of the lateral geniculate nucleus do the aberrant fibers end? They occupy just the region they would have gone to in the OTHER geniculate in a normal animal had they not crossed. It is as though the geniculate cells that had not received their usual input now accept abnormal, but corresponding, fibers coming from the opposite eye (Figure 8*B*). The eye that supplies the fibers is on the wrong side of the animal, but once crossed the fibers take the place normally reserved for the corresponding fibers from the correct side. So far, then, abnormal fibers are accommodated in the "correct" layer of the lateral geniculate nucleus by those cells left "vacant" because their expected axons did not arrive.

The next stage is the pathway from the lateral geniculate nucleus to the visual cortex, and here a new problem arises. In normal animals input from corresponding points in the visual field of the two eyes converges on cortical neurons in a precise and systematic manner, thus providing the basis for binocular interaction (Chapter 2). Further, the visual cortex on one side receives information only about the opposite half of the visual world. But in the Siamese cat the cortex receives an unusual projection from visual fields on the SAME SIDE of the animal by way of crossed fibers.

How is the map of the cortex to be arranged and which cells act as targets for the wrong incoming fibers in a Siamese cat? The surpris-

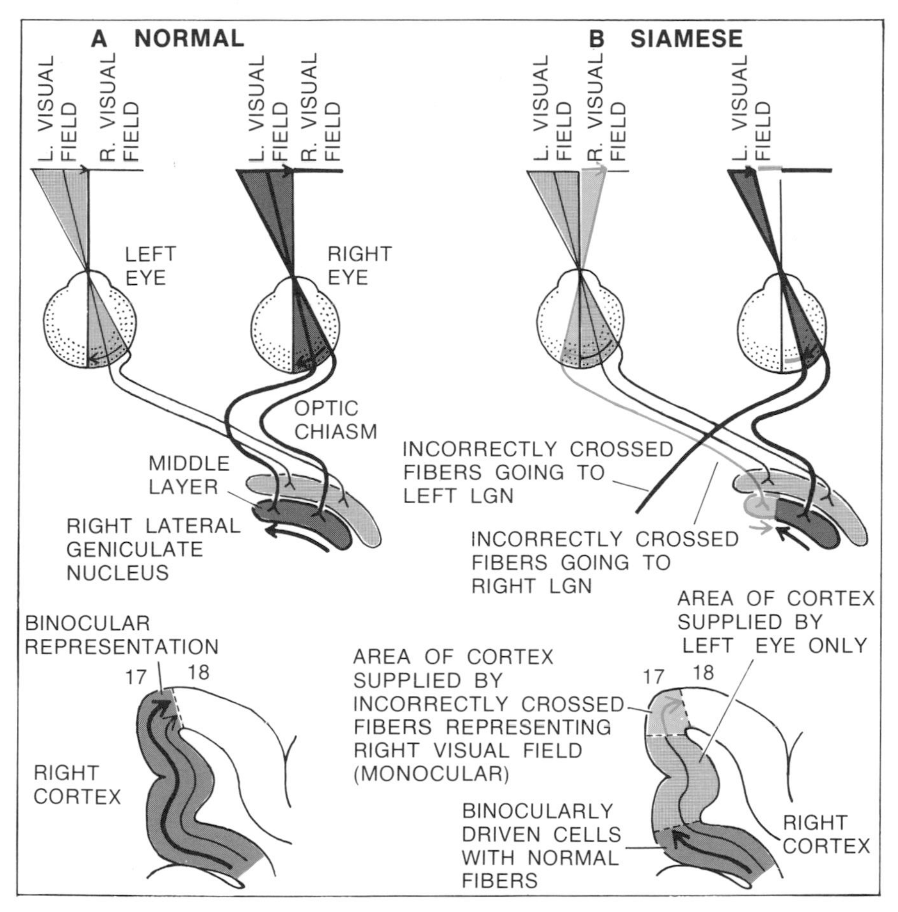

8

ABNORMAL CONNECTIONS in the Siamese cat between retina, lateral geniculate nucleus (LGN) and cortex. A. Normal control. Each eye projects to a different layer of the LGN. B. Siamese cat. Some optic nerve bundles from each eye cross to the LGN on the wrong side. The cortical projections of the abnormally crossed fibers (from the left retina) represent part of the right visual field rather than the (normal) left visual field. Further, the cells are driven from one eye only. The representation of the abnormal projection remains orderly. Additional complex variants of cortical reorganization are omitted. (After Hubel and Wiesel, 1971)

ing answer is shown in Figure 8*B*. The geniculate neurons that receive the abnormally crossed fibers end for the most part in a special region. In the normal cat, cells in the same region of the cortex respond to appropriately oriented illumination of BOTH eyes, and the receptive fields, as stated before, are located in the visual field on the opposite side of the animal. In the type of Siamese cat discussed here, however, the receptive fields of cells in this special region of the cortex have two abnormal features: (1) they are not binocularly driven, but receive projections from only one eye, and (2) their receptive fields are situated in the visual field on the same side of the animal instead of on the opposite side.

The complete array of connections in the cortex exemplifies still further variants from the normal, not described here. Nevertheless, the general conclusion to be drawn from this series of observations is that neurons in the Siamese cat can receive quite abnormal inputs. This is shown by the presence of a specific region of visual cortex devoted to the ORDERLY representation of the ABNORMAL portion of the visual field. At the same time some of the ground rules for normal cell connections are still followed. Thus the rules for organizing the remainder of the geniculocortical connections stay normal; logic and order prevail, rather than complete scrambling.

The genetic aspects of the defect in development are of considerable interest. They have been explored in a detailed, systematic study by Guillery,[81] who examined a large variety of animals, including rats, minks, mice, guinea pigs, Himalayan rabbits, ferrets, and even one white tiger. In all these species the neurological defect goes hand in hand with a pigment deficit. A direct, but not invariable, connection is with the albino gene. The absence of pigment in the eye is somehow associated with the abnormal course taken by the optic nerve fibers of certain ganglion cells. Here, then, is an example of an intriguing genetic defect, involving in its most obvious form the color of an animal but also causing errors in the specificity of connections.

GENERAL CONSIDERATIONS OF NEURAL SPECIFICITY

The question of neural specificity—how nerve cells find their appropriate targets—is just one facet of a general property of the regulation of cell growth. Other more mundane cells in the body also appear to know in what direction to grow and when to stop. The restitution of skin tissue after partial removal, the appropriate closing of a wound, regrowth of an injured organ such as the liver to its proper size are related phenomena. Just as a denervated muscle seems to attract nerve fibers, so does a transplanted muscle attract new capillaries that suddenly grow into it, splitting off from an adjoining vascular bed. But how can we find out the precise mechanisms whereby nerve cells become connected and aggregated into tissues?

---

[81]Guillery, R. W. 1974. *Sci. Am.* *230*:44–54.

Perhaps a rather commonplace analogy may be encouraging. Let us assume that we are ignorant about the workings and design of the postal system. A chapter from a book on the nervous system, without its illustrations, is posted in Boston addressed to Tokyo, where it arrives a few days later. How does it get there? The writer knows only the closest letterbox and is unaware even of the post office in his district. The postal worker who empties the letterbox knows the post office; there the clerk who handles the mail may not know where Tokyo is but does know how to direct the package to the airport, and so on, to the right country, city, street, building, and eventually the correct person. If this were not enough, the illustrations that complement the chapter are posted by separate mail from Stanford to the same destination, where they arrive almost simultaneously with the chapter from Boston.

Some aspects of neural specificity may not be too different. Suppose that a nerve cell in the left retina finds its specified target in the lateral geniculate nucleus and that a geniculate neuron finds the proper address in the visual cortex among the millions of other inhabitants of that particular visual area. And another nerve cell in the right eye also reaches its appropriate destination, so that eventually connections are made in the visual cortex at about the same time in order to complement the information sent from the left eye.

The comforting feature of this lengthy analogy is that the problem seems altogether baffling at first sight. Yet, one can solve the postal puzzle by following the mail step by step to its destination. This would reveal some of the logic and design of postal organization (albeit without disclosing the identity of the designer). At any one step, only a limited number of instructions are followed and a limited number of mechanisms operate. It would not necessarily be profitable to ask how the letterwriter knows the way to the right room and person in Tokyo, or what attracts the letter to the postal employee, the right franking machine, or airplane; equally, it may not be useful at this stage to ask how the neuron in the right eye finds its way to deliver its message to the correct column in the cortex to arrive in time for the rendezvous with the signals originating in the neuron from the left eye.

It seems that the problem of neural specificity can be broken down into analyzable parts. Perhaps the best tools at our disposal at present, in addition to those discussed in previous chapters, are a combination of genetic, tissue culture, and developmental approaches at the cellular level.

SUGGESTED READING

Items marked with an asterisk (*) are reprinted in Cooke, I., and Lipkin, M., Jr. 1972. *Cellular Neurophysiology.* Holt, New York.

General reviews

Guth, L. 1968. "Trophic" influences of nerve. *Physiol. Rev. 48*:645–687.

Harris, A. J. 1974. Inductive functions of the nervous system. *Annu. Rev. Physiol. 36*:251–305.

Jacobson, M. 1970. *Developmental Neurobiology.* Holt, New York.

Nelson, P. G. 1975. Nerve and muscle cells in culture. *Physiol. Rev. 55*:1–61.

The Synapse. 1976. *Cold Spring Harbor Symp. Quant. Biol.* 40.

Original papers

DENERVATION SUPERSENSITIVITY

*Axelsson, J., and Thesleff, S. 1959. A study of supersensitivity in denervated mammalian skeletal muscle. *J. Physiol. 147*:178–193.

Frank, E., Jansen, J. K. S., Lømo, T., and Westgaard, R. H. 1975. The interaction between foreign and original nerves innervating the soleus muscle of rats. *J. Physiol. 247*:725–743.

Hartzell, H. C., and Fambrough, D. M. 1972. Acetylcholine receptors: Distribution and extrajunctional density in rat diaphragm after denervation correlated with acetylcholine sensitivity. *J. Gen. Physiol. 60*:248–262.

Katz, B., and Miledi, R. 1964. The development of acetylcholine sensitivity in nerve-free segments of skeletal muscle. *J. Physiol. 170*:389–396.

Kuffler, S. W., Dennis, M. J., and Harris, A. J. 1971. The development of chemosensitivity in extrasynaptic areas of the neuronal surface after denervation of parasympathetic ganglion cells in the heart of the frog. *Proc. R. Soc. Lond.* B *167*:555–563.

Landmesser, L. 1972. Pharmacological properties, cholinesterase activity and anatomy of nerve-muscle functions in vagus-innervated frog sartorius. *J. Physiol. 220*:243–256.

Lømo, T., and Rosenthal, J. 1972. Control of ACh sensitivity by muscle activity in the rat. *J. Physiol. 221*:493–513.

NEURONAL SPROUTING AND SPECIFICITY

Attardi, G., and Sperry, R. W. 1963. Preferential selection of central pathways by regenerating optic fibers. *Exp. Neurol. 7*:46–64.

Jansen, J. K. S., and Nicholls, J. G. 1972. Regeneration and changes in synaptic connections between individual nerve cells in the central nervous system of the leech. *Proc. Natl. Acad. Sci. U.S.A. 69*:636–639.

Raisman, G., and Field, P. M. 1973. A quantitative investigation of the development of collateral innervation after partial deafferentation of the septal nuclei. *Brain Res. 50*:241–264.

Yoon, M. 1972. Transposition of the visual projection from the nasal hemiretina onto the foreign rostral zone of the optic tectum in goldfish. *Exp. Neurol. 37*:451–462.

NERVE GROWTH FACTOR

Levi-Montalcini, R. 1975. NGF: An Uncharted Route. In *The Neurosciences: Paths of Discovery.* F. G. Worden, J. P. Swaze, and G. Adelman (eds.). MIT Press, Cambridge, Mass., pp. 245–265.

Levi-Montalcini, R., and Angeletti, P. U. 1968. Nerve growth factor. *Physiol. Rev. 48*:534–569.

Perez-Polo, J. R., Bamburg, J. R., De Jong, W. W. W., Straus, D., Baker, M., and Shooter, E. M. 1972. Nerve Growth Factors of the Mouse Submaxillary Gland. In *Nerve Growth Factor and Its Antiserum.* E. Zaimis and J. Knight (eds.). London, Athlone Press, pp. 19–34.

SIAMESE CAT

Guillery, R. W., and Kaas, J. H. 1971. A study of normal and congenitally abnormal retinogeniculate projections in cats. *J. Comp. Neurol.* *143*:73–100.

Hubel, D. H., and Wiesel. T. N. 1971. Aberrant visual projections in the Siamese cat. *J. Physiol.* *218*:33–62.

CHAPTER NINETEEN

GENETIC AND ENVIRONMENTAL INFLUENCES IN THE MAMMALIAN VISUAL SYSTEM

An approach has been made in the mammalian visual system to the questions of the relative importance of genetic factors and of the environment for the establishment and proper performance of synaptic interactions. In newborn, visually naive kittens and monkeys many features of the neuronal organization are already present. Cells in the retina, the lateral geniculate nucleus, and the visual cortex can respond to stimuli in much the same way as those in adult animals. The neurons, however, are far more susceptible to change and can be irreversibly affected by inappropriate use.

In the cat, closure of the lids of one eye during the first 3 months of life leads to blindness in that eye. The abnormality occurs chiefly at the level of the cortex; although geniculate cells are still driven by the eye that had been closed, the great majority of cortical cells are not. The other eye develops and functions normally. The period of greatest susceptibility in cats occurs at about 4 to 6 weeks of age, when lid closure for 3 to 4 days produces serious damage. Lid closure in adult cats has no effect. In monkeys the period of greatest sensitivity to visual deprivation begins at birth and extends over the next 6 to 8 weeks, followed by a period of lower sensitivity lasting for about 1 year. Closure of the lids of one eye during this period leads not only to impairment of vision but also to a shrinkage of the cortical dominance columns supplied by that eye. A corresponding increase occurs in the width of the columns of the normal eye.

The organization of the cortex can also be disrupted by an abnormal sensory input. A demonstration is provided by the effects produced by a squint that results from cutting extraocular muscles. Under such conditions each eye is exposed to the normal amount of visual input and only the fixation of the two eyes upon objects is altered. Yet, the way cortical cells are driven by the two eyes is changed in kittens or immature monkeys that have been made to squint. The cells have normal receptive field properties but only a few are driven by both eyes; instead, one eye or the other is effective on its own. The subtle functional difference produced by a squint is therefore enough to disrupt synaptic interactions that had been present at birth. Since there is no disuse, it appears that impulse traffic in convergent pathways must continue in an appropriately balanced manner for the normal functional organization to be maintained.

The direct cause of such changes is not clear. At the same time these results have a wide significance for considering a variety of aspects of development in the central nervous system. Other sensory systems and higher functions may also have critical periods during which their performance can be sharpened by appropriate use or irreversibly damaged by disuse or inappropriate use.

We have emphasized repeatedly the constancy of the wiring that is necessary for the nervous system to function properly. It is also clear that development continues after birth for varying periods in different animals; at the cellular level we do not know whether synaptic connections are immutably fixed even in the adult. For example, kittens are born with their eyes closed. If the lids are opened and light is shone into an eye, the pupil constricts, although the animal had not previously been exposed to light and appears to be completely blind. By 14 days, the kitten shows evidence of vision, and thereafter begins to recognize objects and patterns. When kittens are brought up in darkness instead of their normal environment, they become blind, but the pupillary reflex continues to function.[1] It is as though there were a hierarchy of susceptibility with "hard" and "soft" wiring in different parts of the brain.

Behavioral experiments support the view that simpler, more basic reflexes resist change and are altered only by drastic experimental procedures such as lesions; in contrast, higher functions may fail to develop, are far more susceptible, and can be modified by subtle changes in the environment.[2]

Changes in the performance of the nervous system raise a number of questions. What are the relative contributions of genetic factors and experience, summed up by the phrase "nature and nurture"? To what extent are the neuronal circuits required for vision already present and ready to work at birth? What effect has light falling into the eyes on their development? Does a kitten or a monkey brought up in darkness become blind because the connections fail to develop or because connections that had originally been there have either withered away or become ineffective? The visual system offers great advantages for approaching directly questions such as these because the relay stations are accessible and the background of light and natural stimulation can readily be altered.

---

[1]Riesen, A. H., and Aarons, L. 1959. *J. Comp. Physiol. Psychol.* 52:142–149.
[2]Held, R., and Bauer, J. A. 1974. *Brain Res.* 71:265–271.

As in previous parts of this book, we have chosen to describe one system without presenting a full review of the field of sensory deprivation. And within the visual system we again emphasize the pathways from retina to cortex in cat and monkey. For our purposes it is convenient to focus largely on work that follows logically from the material presented in Part One.

THE VISUAL SYSTEM IN NEWBORN KITTENS AND MONKEYS

A good deal is known about the organization of the visual connections that underlie perception in the adult cat and monkey (Part One). Thus a simple cell in the cat or monkey cortex selectively "recognizes" one well-defined type of visual stimulus, such as a narrow bar of light, oriented vertically, in a particular region of the visual field of either eye. It is natural to wonder whether cells of this type are already present in the newborn animal or whether visual experience or learning is required so that a random set of preexisting connections is reformed or at least modified for such a specific task. For technical reasons it is difficult to record from cells in newborn animals. Most experiments on kittens were made during the first 3 weeks after birth. To prevent form vision the lids were sutured or the cornea was covered by a translucent occluder.[3] Similarly, visually naive monkeys were produced by suturing the animals' lids immediately or several days after birth; in some instances monkeys were delivered by Caesarean section for later examination, care being taken to avoid exposure to light.[4]

A newborn monkey appears visually alert and is able to fixate. In contrast a newborn kitten whose lids have been opened by surgery is behaviorally blind. Nevertheless, many of the features seen in adults are already present in the performance of cortical neurons in both animals. For example, recordings made from individual cells in area 17 of the visual cortex show that the cells are not driven by diffuse illumination of the eyes. As in a mature animal, they fire best when light or dark bars with a particular orientation are shone onto a particular region of the retina of either eye. The responses of the receptive fields are also organized into antagonistic "on" and "off" areas that are similar in the two eyes.

In newborn animals the responses of cortical cells are different in certain respects from those of the adult (see below). In particular, the discharges tend to be weaker and some cells are unresponsive. The principal point to be made here, however, is that certain features of the basic wiring are already established. One useful test has been to compare newborn and adult animals for binocular representation and binocular dominance. As explained in Chapter 3, in the adult cat and monkey most of the cortical neurons have receptive fields in both eyes. Although the receptive fields are similar in the two eyes, some

---

[3]Hubel, D. H., and Wiesel, T. N. 1963. *J. Neurophysiol.* *26*:994–1002.
[4]Wiesel, T. N., and Hubel, D. H. 1974. *J. Comp. Neurol.* *158*:307–318.

cortical cells are driven better by one eye, some by the other, and some equally well by both. About 20 percent of all the 1116 cells in Figure 1*A* from visually normal monkeys are driven solely by one eye and about the same percentage by the other. The degree of dominance can be conveniently expressed in a histogram by grouping neurons into seven categories according to the discharge frequency with which they respond to stimulation of one or the other eye. In the visually inexperienced kitten and two-day-old monkey the histogram appears rather normal, the majority of cells responding to appropriate illumination of either eye, as shown in Figure 1*B*, *C*. As in the adult (Chapter 3) the cells are grouped in the cortex according to stimulus preference. With oblique penetrations (Figure 2), the preferred orientation changes in a regular sequence as the electrode moves through the cortex.[4] Moreover, in many instances the "tuning" curve or range of orientations in animals lacking prior visual experience[5] cannot be distinguished from that in adults. These observations therefore indicate that kittens and monkeys are born with their intricate connections for orientation selectivity arranged and in place.

[5]Sherk, H., and Stryker, M. P. 1976. *J. Neurophysiol.* 39:63–70.

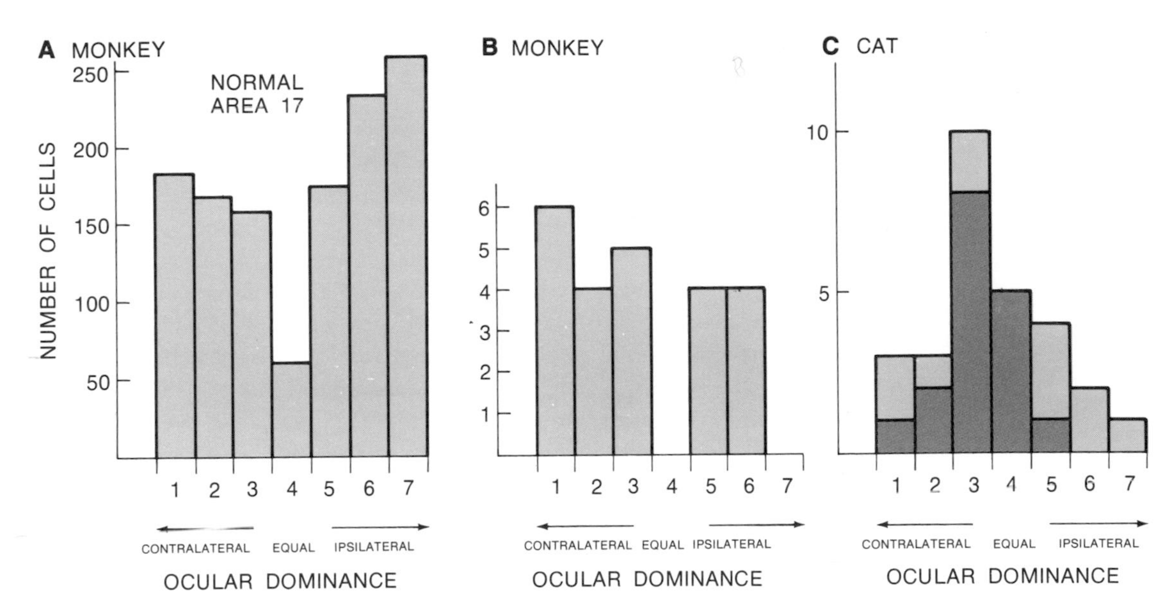

1

OCULAR DOMINANCE DISTRIBUTION in the visual cortex of normal monkeys. A. Cells in groups 1 and 7 of the histogram are driven by one eye only (ipsi- or contralateral). All other cells have an input from both eyes. In groups 2,3 and 5,6 one eye predominated; in group 4 both had a roughly equal influence. B. A similar ocular dominance distribution in a 2-day-old monkey. C. Histogram of ocular dominance distribution in a 20-day-old kitten (light color) with normal visual experience and two kittens (8 and 16 days old) without previous visual exposure (dark color). (A,B from Wiesel and Hubel, 1974; C from Hubel and Wiesel, 1963b)

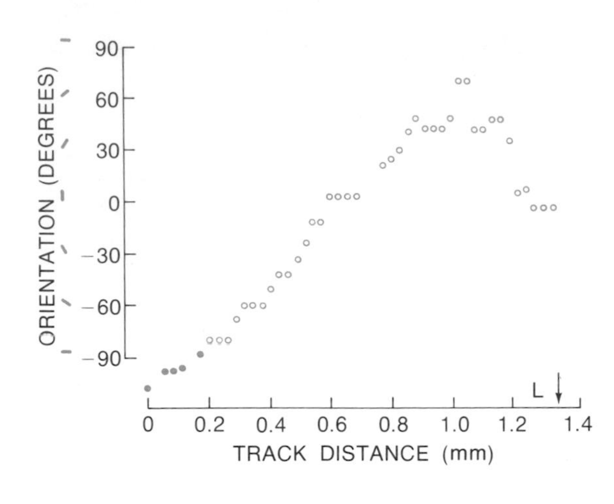

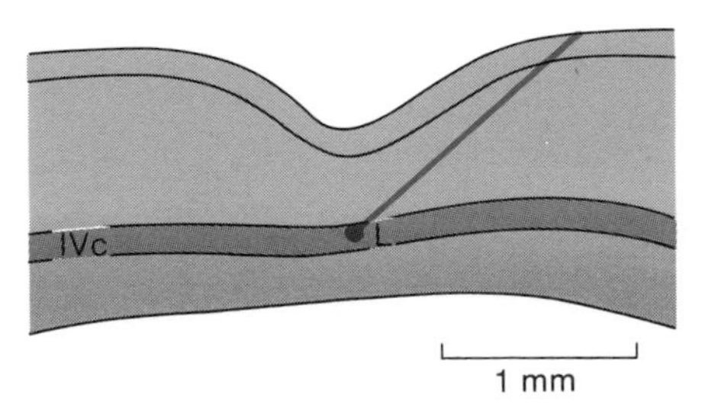

2

AXIS ORIENTATION COLUMNS in the absence of visual experience. Top, axis orientation of receptive fields encountered by an electrode during an oblique penetration through the right cortex of a 17-day-old baby monkey whose eyes had been sutured closed on the second day after birth. The circle at L (bottom) marks a lesion made at the end of the electrode track in layer IV. The orientation of receptive fields changes progressively as orientation columns are traversed, indicating that normally organized orientation columns are present in the visually naive animal. Closed circles in graph are from the ipsilateral eye, and the open circles from the other eye. (From Wiesel and Hubel, 1974)

The quantitative aspects of these questions remain to be settled. Are some cells still unspecified, others perhaps specified, and still others in the final state? On the whole it is apparent that the cortex and its performance in the newborn naive animal is not fully developed, as has been emphasized by Barlow, Blakemore, and Pettigrew.[6] As expected, there also are differences in different species of animals. For example, a newborn monkey's visual performance is better than that of a newborn kitten whose eyes remain closed for about 10 days.

The findings in immature animals come as no great surprise. Although the development of the cortex is susceptible to environmental changes, one would be rather surprised if the basic outline of neural organization with its orderly and intricate visual connections were shaped by the vagaries of the visual environment.

EFFECTS OF ABNORMAL EXPERIENCE

This section describes three stages of experiments, mostly by Hubel and Wiesel who deprived animals of normal visual stimuli and studied the effects on the physiological responses of nerve cells in the visual system after (1) closing the lids of one or both eyes, (2) preventing form vision but not access of light to the eye, and (3) leaving light and form vision intact but producing an artificial stabismus (squint) in one eye. These

[6]See Blakemore, C. 1974. *Br. Med. Bull.* *30*:152–157.

procedures cause abnormalities in function; they are also followed by anatomical changes whose finer details are now starting to emerge.

Cortical cells after monocular deprivation in kittens

When the lids of one eye were sutured during the first week of life, kittens developed normally and used their unoperated eye. However, at the end of 1 to 3 months, when the operated eye was opened and the normal one closed, it was clear that the animals were practically blind in the operated eye. They would bump into objects and fall off tables.[7] There is no gross evidence of a defect within such eyes; pupillary reflexes appear normal and so does the electroretinogram which serves as an average index of the electrical activity of the eye (Chapter 13). Records made from retinal ganglion cells in these animals showed no obvious changes in their responses.[8] Similarly, the receptive fields and physiological properties of the cells in the lateral geniculate nucleus appeared normal in deprived animals (see below). There were, however, striking changes in the responses of cortical cells.

When electrical recordings were made in the visual cortex, only a few of the cells could be driven by the eye that had been closed. Figure 3 shows the ocular dominance histogram obtained from 199

---

[7]Wiesel, T. N., and Hubel, D. H. 1963. *J. Neurophysiol.* *26*:1003–1017.
[8]Sherman, S. M., and Stone, J. 1973. *Brain Res.* *60*:224–230.

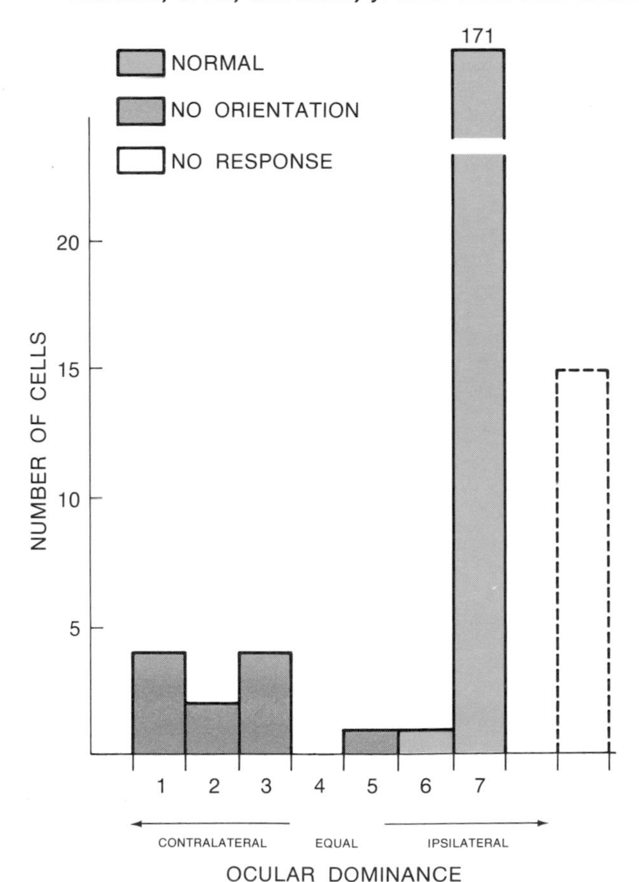

3

DAMAGE PRODUCED BY CLOSURE of one eyelid. Ocular dominance distribution in 8- to 14-week-old kittens without visual experience in the right eye. Only 13 out of 199 cells responded to stimulation from the deprived eye; all but one of the neurons had abnormal receptive fields. The cells from the eye that had not been closed gave normal responses (light color). The column with interrupted lines represents cells that did not respond to either eye. (From Wiesel and Hubel, 1965a)

cells examined in five kittens raised with closure of an eye for 2 to 3 months. Only 13 cells responded to stimulation of the deprived eye and nearly all of those had abnormal receptive fields. More recently, similar results have been obtained in newborn monkeys in which one eye was closed for varying periods.

Period of susceptibility to lid closure and the extent of recovery

When the lids of one eye are closed in a cat over 4 months of age, none of the abnormal consequences are seen. In an adult cat, even if an eye is closed for over a year, the cells in the cortex continue to be driven normally by both eyes, with the normal ocular dominance histogram.

The period during which susceptibility in kittens is highest has been narrowed down to the fourth and fifth weeks after birth. During the first 3 weeks or so of life, eye closure has little effect. This is not surprising since the kittens' eyes are normally closed at first. But, abruptly, during weeks 4 and 5, sensitivity increases. Closure at that age for as little as 3 to 4 days leads to a sharp decline in the number of cells that can be driven by the deprived eye.[9]

An experiment in which littermates are compared is shown in Figure 4. In this example, 6- and 8-day closures starting at the age

[9]Hubel, D. H., and Wiesel, T. N. 1970. *J. Physiol.* *206*:419–436.

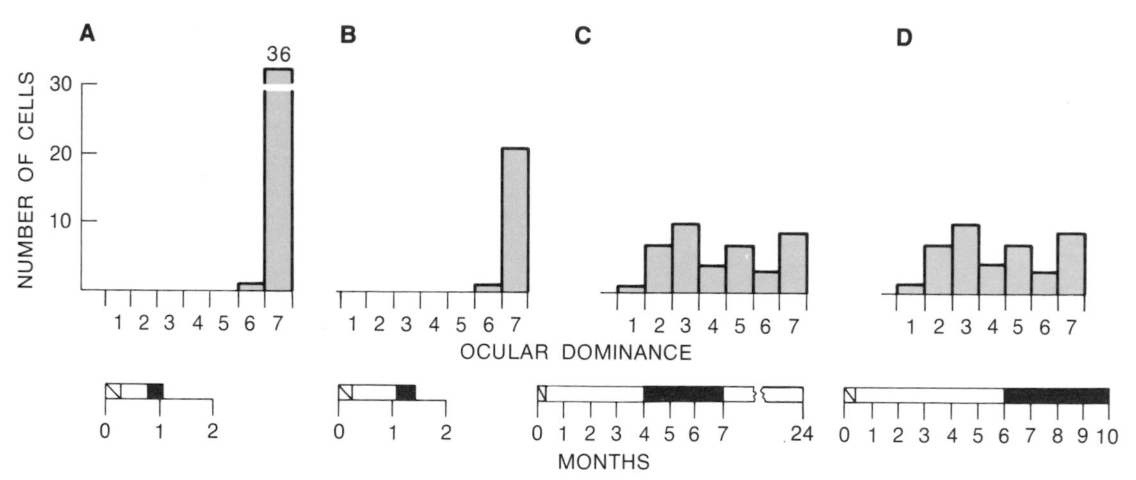

4 **CRITICAL PERIOD in kittens. Histograms of ocular dominance distribution in the visual cortex in kittens that are littermates. A. The right eye was sutured shut for 6 days (age 23 to 29 days). B. The right eye was closed from age 30 to 39 days. In each animal only one cell was weakly influenced by the temporarily deprived eye. The damage was about as great as eye closure for 3 months or longer. C. The right eye was open for first 4 months then sutured for 3 months, then opened again. Recordings were made at age of 2 years. D. The eye was open for the first 6 months, then closed for 4 months. Ocular dominance was determined at 10 months of age. Ocular dominance distribution appeared normal for both eyes. The black segment below the abscissa indicates closure period. (From Hubel and Wiesel, 1970a)**

of 23 and 30 days (Figure 4*A*, *B*) caused about as great an effect as 3 months of monocular deprivation from birth. The susceptibility to lid closure declines after the critical period has passed and eventually disappears at about 3 to 4 months (Figure 4*C*, *D*).

Except in special circumstances, little recovery occurs after the damage has been inflicted. In one experiment an eye was closed for 3 months and then opened. At this time the other eye was closed for 1 year, during which period the animal had to use the previously closed eye. Finally, both eyes were kept open for almost 5 years more (Figure 5). At the end of this period recordings were made from the cerebral cortex and it was found that almost no cells could be driven by the eye that had been first closed five and a half years earlier. The animal still appeared to be behaviorally blind in this eye, while the other eye was indistinguishable from normal. In related experiments Blakemore and Van Sluyters[10] obtained significant recovery in kittens. Their procedure was to suture one eye closed from birth until 5 weeks of age. (This would lead to blindness in that eye.) Next, at the height of the critical period after closure for 5 weeks, the deprived eye was opened and the normal control eye sutured. Nine weeks later the initially deprived eye was dominant in such kittens with reversed eye suturing; the eye closed at 5 weeks was ineffective in driving cortical

[10]Blakemore, C., and Van Sluyters, R. C. 1974. *J. Physiol.* 237:195–216.

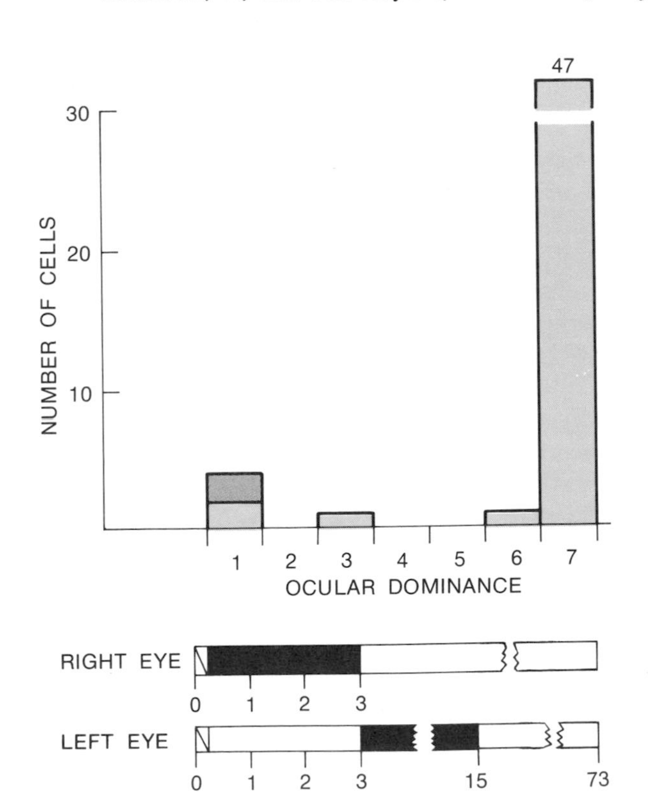

5
PERMANENT DAMAGE AFTER LID CLOSURE. Ocular dominance histogram in a cat whose right eye was closed for the first 3 months, then opened and the left eye closed for 1 year. From then on both eyes remained open for almost 5 years more. At 6 years of age the ocular dominance distribution, recorded in the left visual cortex, was highly abnormal. The eye that had been closed 5½ years earlier was able to drive only a few neurons in the visual cortex. The black portions in the lower bars indicate times of closure of one or the other eye. (From Hubel and Wiesel, 1970a)

cells. Reversed suturing performed later (for example, at 14 weeks) was ineffective in restoring vision to the initially deprived eye.

The susceptibility to lid closure has been studied recently in immature monkeys. Suturing the lid of one eye may lead to blindness. The duration of the critical period has not been explored so extensively as in the cat. It seems, however, that the vulnerability to deprivation is greatest during the first 6 to 8 weeks.

The susceptibility of kittens and monkeys during early life is reminiscent of some clinical observations made in man. It has long been known that removal of a clouded or opaque lens (cataract) can lead to a restoration of vision, even though the patient has been blind for many years. In contrast, a cataract that develops in a baby can lead to blindness without the possibility of recovery after surgery.[11]

Relative importance of diffuse light and form vision for maintaining normal responses

The results described so far indicate that although modifications and refinements do occur in the first few months, the principal features of the synaptic architecture are already present in the newborn animal. If, however, one eye is not normally used, its power wanes and it ceases to be effective. These far-reaching changes are produced by the relatively minor procedure of sewing the lids, without cutting any nerves.

What is the important condition for maintaining and developing proper visual responses? Is diffuse light adequate? Lid closure reduces light that reaches the retina but does not exclude it. One would suspect, therefore, that diffuse light alone would not keep an eye functioning normally. Instead of closing eyelids, a series of experiments were made in which a plastic occluder (like frosted glass or a Ping-pong ball) was placed over the cornea of a newborn kitten, preventing form vision but not exluding light by more than 2 log units. In one experiment a thin sheet of opaque tissue, the nictitating membrane, was sewn across one eye, blurring vision, with even less attenuation of the light.[7]

All these cats were still functionally blind in the deprived eye. Furthermore, the cortical cells were no longer driven by the deprived eye. Neither retinal nor geniculate responses were noticeably changed under such conditions. Thus, FORM VISION, rather than the presence of light, is an important stimulus required to preserve the normal responses of cortical cells already present in the visual cortex at birth. Surprisingly, however, even form vision alone may not be adequate to maintain complete normality (below).

Morphological changes in the lateral geniculate nucleus after visual deprivation

Cells in the lateral geniculate nucleus of the cat are arranged in three layers, each predominantly supplied by one or the other eye (Chapter 2). In the same kittens that showed marked abnormalities in the cortex after lid closure, the geniculate cells seemed to behave normally. The cells in the appropriate layers responded with "on" or "off" discharges to small spots of light shone into the deprived or the normal eye, and no clear-cut differences from normal firing patterns

---

[11]Von Senden, M. 1960. *Space and Sight: The Perception of Space and Shape in the Congenitally Blind Before and After Operation.* Free Press, Glencoe, Ill.

could be observed. Nevertheless, it was shown that marked changes in morphology occurred after lid closure of an eye: the cells were noticeably smaller in the layers supplied by the deprived eye (layer A_1 in the lateral geniculate body on the same side of the animal and layers A and B on the contralateral side).[12] The cell bodies were only about half as large as those in the normal layers, the reduction in size depending on the duration of lid closure.

In rabbits a similar reduction has been seen in the cell size of the lateral geniculate nucleus. This effect could be largely reversed by the procedure of reverse suturing of the lids, discussed earlier for cortical neurons of the cat.[13]

Occlusion with a plastic diffuser or with the nictitating membrane admitted much more light than did lid closure and produced less shrinkage. It is possible that light intensity as well as the duration of diffuse illumination is a stimulus for geniculate cells to grow. It seems surprising that cells in the lateral geniculate show relatively obvious morphological changes but little significant physiological deficit. This may be due in part to a selective disappearance of one particular cell type, such as the Y or fast-adapting neurons[14] (Chapter 2). Competition between the pathways serving the two eyes seems to be another factor controlling the size of cells.[15]

Morphological changes in the cortex after visual deprivation

The morphological consequences of eye closure are not so obvious in the cortex as in the geniculate. In mice reared in the dark a number of changes have been noted by several workers. The changes include reduction in the number of spines in apical dendrites of pyramidal cells.[16] Significantly, the deficit is particularly conspicuous in layer IV of the striate cortex where geniculate fibers terminate.

More recently changes in the cortex of 18-month-old monkeys have been studied after (1) removal of one eye at birth and (2) closure of one eye. Striking changes in ocular dominance columns were demonstrated by two different methods. One is the reduced silver staining method[17] and the other is use of autoradiography to follow the migration along visual pathways of radioactive material injected into the eye (Chapter 3). With both techniques there occurred a marked reduction in the width of the ocular dominance columns that received their input from the eye that had been either removed or occluded. The columns with input from the normal eye, however, showed a corresponding increase in width. This expansion and shrinkage of ocular dominance columns is demonstrated in a striking manner in Figure 6 in

---

[12]Wiesel, T. N., and Hubel, D. H. 1963. *J. Neurophysiol. 26*:978–993.

[13]Chow, K. L., and Stewart, D. L. 1972. *Exp. Neurol. 34*:409–433.

[14]Sherman, S. M., Hoffmann, K.-P., and Stone, J. 1972. *J. Neurophysiol. 35*: 532–541.

[15]Guillery, R. W., and Stelzner, D. J. 1970. *J. Comp. Neurol. 139*:413–422.

[16]Valverde, F. 1970. In *Contemporary Research Methods in Neuroanatomy.* W. J. H. Nauta and S. O. E. Ebbeson (eds.). Springer, New York, pp. 12–31.

[17]LeVay, S., Hubel, D. H., and Wiesel, T. N. 1975. *J. Comp. Neurol. 159*: 559–575.

which the normal columns can be compared with columns in animals in which one eye had been closed during the first 18 months. The drastic change indicates that geniculate nerve terminals activated by the normal eye have taken over some of the territory of their weaker, visually deprived neighbors. The expansion was also confirmed physiologically by recording from layer IV where the geniculate fibers terminate.[18]

One explanation for the expansion and shrinkage of columns is that sprouting has occurred by the geniculate axon terminals from the normal eye. (For evidence for sprouting in the mammalian brain see Chapter 18.) This interpretation assumes the existence in the newborn monkey

---

[18]Hubel, D. H., Wiesel, T. N., and LeVay, S. 1976. In *The Synapse. Cold Spring Harbor Symp. Quant. Biol. 40.*

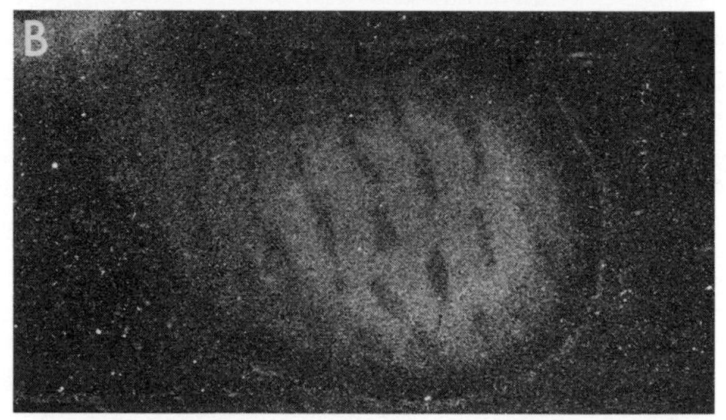

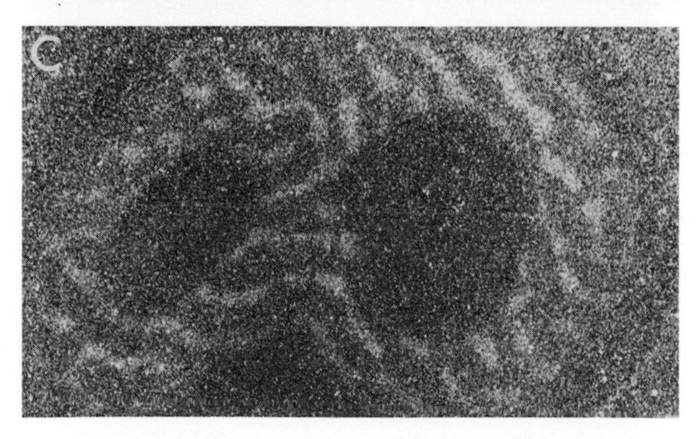

6

OCULAR DOMINANCE COLUMNS AFTER CLOSURE OF ONE EYE. A. Normal control, adult rhesus monkey. The right eye had been injected with a radioactive proline-fucose mixture 10 days before a section was made tangential to the exposed dome-shaped primary visual cortex of the right hemisphere. The section passes through layer V, which is seen as a dark oval area near the center of the figure; just outside this is layer IV, seen as a ring of finger-like alternating dark and light processes. With dark-field illumination the radioactivity in the geniculate axon terminals in layer IV appears as fine white granules, forming the light stripes (columns) that correspond to the injected eye. The dark intervening bands correspond to the other eye. Roughly 9 sets of stripes are shown. B. A similar section in an 18-month-old monkey whose right eye had been closed at age of 2 weeks. Proline-fucose injected into normal left eye. Here the plane of section grazes layer IV, which is seen as an oval. The white label again demonstrates the columns whose input is derived from the intact eye. The columns, however, are larger than normal and alternate with narrowed columns, seen as dark gaps, supplied by the eye whose lids had been closed. C. A section from a 6-month-old monkey whose right eye had been closed at 3 weeks. Here the right eye was injected and the section made 2 weeks later. The plane of section passes through a dimpled region of cortex, grazing layer V twice, forming a pair of dark ovals; these are surrounded by layer IV, producing an 8-like figure. The labelled columns are now the shrunken ones, alternating with dark bands that are widened. The abnormality is less than in B, either because the eye was closed later, or because the time it remained closed was shorter. (From Hubel, Wiesel and Le Vay, 1976)

of ocular dominance columns that is not presently established. Some evidence actually suggests that such columns do not develop fully for several weeks after birth. If that is the case, the expanded columns of Figure 6*B,C* could be due to pre-existing overlapping terminals from the two eyes which were maintained by the normal eye but lost by the visually deprived eye. This interpretation would be consistent with the idea that during the critical period fibers from the two eyes compete for cortical connections. Normally, the balance is fairly equal. However, if one eye is deprived or if the eyes see discordant images, the equilibrium shifts and with it the pattern of connections.

Morphological changes in mouse brain following removal of whiskers

An interesting, well-defined change in morphology in a different area of cortex has been described by Woolsey and van der Loos.[19,20] In the somatosensory area in the mouse they noticed characteristically shaped aggregates of cells. Viewed from the surface, the group of cells appears arranged in rings. In sections through the depth of the cortex the distribution has the shape of barrels, or circular cylinders, with less dense cores. The barrels are neatly aligned in regular rows. At first this seemed puzzling, since similar groupings are not seen elsewhere in the mouse cortex. By counting the barrels and examining their distribution in the sensory cortex, a good correlation was found with the number of whiskers on the animal's face: each barrel corresponded to one vibrissa on the opposite side of the face. Moreover, as in the visual system, the morphology of cell groupings can be markedly and permanently altered by the early history. When a vibrissa is removed in a newborn mouse, the corresponding barrel cannot be found in the adult brain. Such questions as the presence or absence of barrels in newborn mice, the critical period, the role of sensory experience, and the effects of disuse have not yet been explored. Nevertheless, this finding provides a further example of flexibility of neuronal architecture in a very different system.

REQUIREMENTS FOR MAINTENANCE OF FUNCTIONING CONNECTIONS IN THE VISUAL SYSTEM

At this stage one might be tempted to conclude that loss of activity in the visual pathways is the main factor that tends to disrupt normal responses of cortical neurons. After all, cortical cells are not driven by diffuse illumination, but by shapes and forms. The following discussion shows that some of the causes must be far more subtle. To maintain normal responses, even form vision is not enough. There must occur in addition a special interaction between the two eyes whose manner of working we cannot yet explain.

Binocular lid closure

The first clue that loss of visually evoked activity cannot on its own account for the changed performance of neurons is shown by the following experiments. Both eyes in monkeys, newborn or delivered by

[19]Woolsey, T. A., and van der Loos, H. 1970. *Brain Res.* *17*:205–242.

[20]Van der Loos, H., and Woolsey, T. A. 1973. *Science* *179*:395–398.

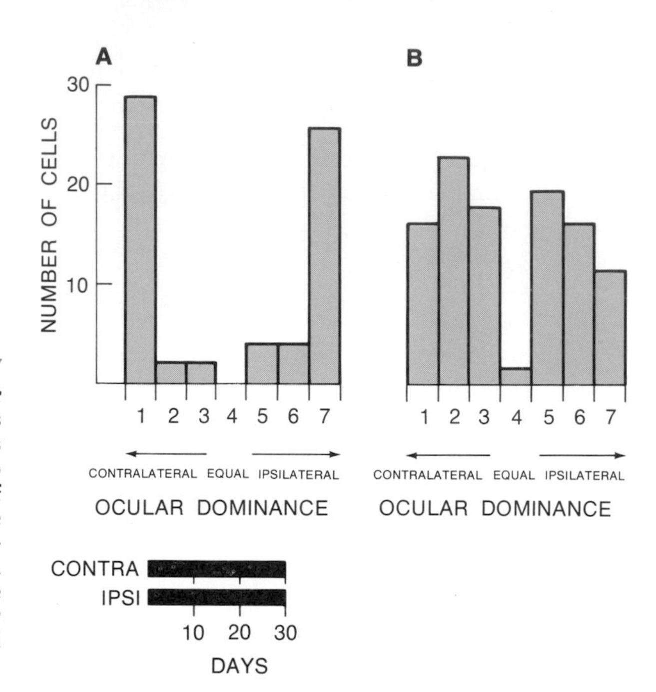

7

OCULAR DOMINANCE HISTOGRAMS after closure of both eyes at birth. A. A monkey was delivered by Caesarean section and recordings were made at 30 days of age. In contrast to the effects of monocular deprivation, each of the two deprived eyes could drive cells in the visual cortex. The receptive fields appeared normal, except that relatively few cells were driven by both eyes. Black area at the bottom of the histogram indicates closure time. B. Ocular dominance histogram from a normal 21-day-old monkey. (From Wiesel and Hubel, 1974)

Caesarean section, were closed.[4] From the preceding discussion one would expect that cells in the cortex would subsequently be driven by neither eye. Surprisingly, however, after binocular closure for 17 days or longer, most cortical cells could still be driven by appropriate illumination; the receptive fields of simple, complex, and hypercomplex cells appeared largely normal. Moreover, the columnar organization for orientation was similar to that in controls (Figure 2). The principal abnormality was that a substantial fraction of the cells could not be driven binocularly (Figure 7). In addition, some cells could not be driven at all and others did not require specifically oriented stimuli. The conclusion one can draw from these experiments is that some, but not all of the ill effects expected from closing one eye are reduced or averted by closing both eyes. Once more one might theorize that the two different pathways from the two eyes are somehow competing for representation in cortical cells and with one eye closed the contest becomes unequal.

Effects of artificial squint

All the abnormal effects described in the preceding paragraphs were produced by suturing eyelids or using translucent diffusers, implicating loss of form vision. Following the clue that in children squint (strabismus) can produce severe loss of vision or blindness, Hubel and Wiesel produced artificial squint in cats and monkeys by cutting an eye muscle.[4,21] The optical axis of the eye is thereby deflected from normal. Under such conditions illumination and pattern stimulation for each eye remain unchanged.

The experiment at first seemed disappointing, because after several months vision in both eyes of the operated kittens appeared normal,

[21]Hubel, D. H., and Wiesel, T. N. 1965. *J. Neurophysiol.* *28*:1041–1059.

and Hubel and Wiesel were about to abandon a laborious set of experiments (personal communication). Nevertheless, they recorded from cortical cells and obtained the following surprising results. Individual cortical cells had normal receptive fields and responded briskly to precisely oriented stimuli. But almost every cell responded only to one eye; some were driven exclusively by the ipsilateral eye and others by the contralateral, but very few were driven equally well by both. The cells were, as usual, grouped in columns with respect to eye preference and field axis orientation. As expected, no atrophy occurred in the lateral geniculate body. The almost complete lack of binocular representation on cortical cells is shown in a histogram (Figure 8*B*) constructed from pooled results in four cats. Similar observations were made in monkeys with artificial squint or binocular lid closure.[4,22]

Squint provides an example in which all the usual parameters of light are normal—the amount of illumination and form and pattern stimuli. The only apparent change consists in a failure of the images on the two retinas to superimpose. Because the cortex in such an animal

[22]See also Baker, F. H., Grigg, P., and van Noorden, G. K. 1974. *Brain Res.* *66*:185–208.

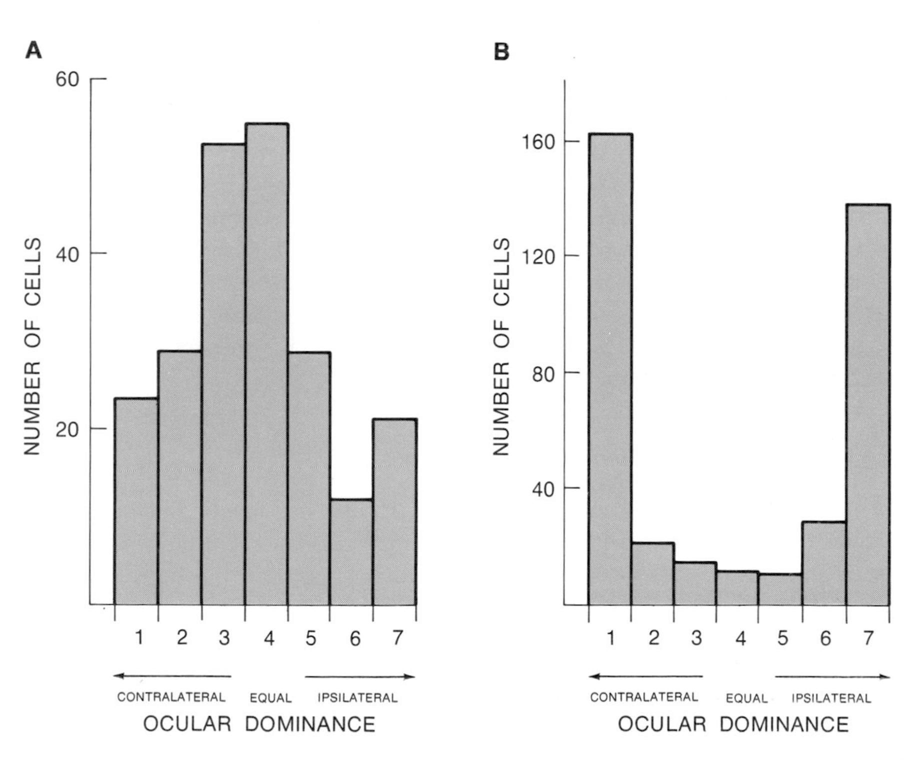

8

EFFECT OF SQUINT in kittens. A. Ocular dominance in normal cats. B. Ocular dominance in four cats in which an artificial squint had been produced. Binocular representation largely disappeared and most cells were driven by one or the other eye. (From Hubel and Wiesel, 1965b)

is rich in responsive cells and columns, it seems unlikely that the large percentage of cells that had originally been driven by both eyes could have dropped out.

We have no detailed structural analyses of the possible lesions that may explain the loss of binocular connections with squint. The factor that seems important for maintaining the normal effectiveness of connections from the lateral geniculate body to the cortex is some form of congruity of input from the two eyes. It is as though the homologous receptive fields in both eyes must be in register with, and superimposable on, each other, and excitation must be simultaneous. The following experiments further support this idea. During the first 3 months or longer the eyes of a kitten were occluded with one plastic occluder that was switched on alternate days from one eye to the other, so that the two eyes received the same total experience, but at different times.[21] Once again the result was the same as in the squint experiment: cells were driven predominantly by either one eye or the other, but not by both. The maintenance of normal connections, therefore, depends not only on the amount of impulse traffic but also on the normal interrelations between activity in the different incoming fibers.

In a series of related studies kittens were raised in various abnormal surroundings and then examined. For example, kittens were made to wear goggles marked by horizontal or vertical stripes[23] or were kept in an environment with striped walls, so that they were exposed to only one kind of orientation.[24] On examination of the cortical neurons it seemed as if the great majority of cells could respond only to light that had an orientation similar to that prevalent in the environment; that is, they needed horizontal or vertical stimuli. Two interpretations of these observations have been proposed: (1) cortical cells could change their orientation preference and (2) only those cells responded specifically that had been appropriately stimulated; the remaining cells that had not received appropriate experience lost their orientation specificity. To what extent such plastic changes occur and which mechanisms account for them is under study in several laboratories.[25]

SENSORY DEPRIVATION IN EARLY LIFE

The effects produced by altered sensory inputs in kittens and immature monkeys have a number of important implications for our understanding of the nervous system. At the level of synaptic mechanisms it is not at all clear how use, disuse, and inappropriate use of the visual pathways can alter their effectiveness in such a drastic and permanent manner. Are we dealing, singly or in combination, with a loss of synapses and therefore with an actual loss of connections, or with a change in effectiveness of physically persisting connections? An addi-

---

[23] Hirsch, H. V. B., and Spinelli, D. N. 1971. *Exp. Brain Res.* 12:509–527.
[24] Blakemore, C., and Cooper, G. F. 1970. *Nature* 228:477–478.
[25] Stryker, M. P., and Sherk, H. 1975. *Science* 190:904–906.

tional factor is the possible formation of new connections, such as may happen during expansion of cortical dominance columns (Figure 6).

At the level of behavior, the demonstration of a critical period of vulnerability to deprivation or to abnormal experience is not new. In the recent experiments, however, abnormalities in physiological behavior in signaling have been pinned down to relays in the cortex and not significantly to lower levels. There exists a wealth of literature reporting other complex behavioral processes in a variety of animals that show periods of susceptibility. Imprinting is one example.

Lorenz has shown that birds will follow any moving object presented during the first day after hatching, as if it were their mother.[26] In higher animals, for example in dogs, behavioral studies indicate that if they are handled by humans during a critical period of 4 to 8 weeks after birth, they are far more tractable and tame than animals that have been isolated from human contact.[27] The critical period in an animal's development may possibly represent a time during which a significant sharpening of senses or faculties occurs.

It is tempting to speculate about the effects of deprivation on higher functions in man. One can imagine, as Hubel has said:[28]

> Perhaps the most exciting possibility for the future is the extension of this type of work to other systems besides sensory. Experimental psychologists and psychiatrists both emphasize the importance of early experience on subsequent behavior patterns—could it be that deprivation of social contacts or the existence of other abnormal emotional situations early in life may lead to a deterioration or distortion of connections in some yet unexplored parts of the brain?

To find a physiological basis for such behavioral problems seems a distant but not impossible goal.

CONCLUDING REMARKS

At the beginning of this book we asked how far the cellular approach can take us in understanding the nervous system.

From our knowledge of signaling and the physical properties of neurons we can form a picture of how groups of nerve cells in the brain integrate diverse signals and put together information. Many areas of the brain seem accessible to these approaches, but we still do not understand the basic mechanisms that underlie changes in membrane permeability and active transport. For the solution of these problems it now appears that an important step will be a recognition and characterization of the membrane components that regulate transport and permeabilities, again at the cellular and subcellular levels.

We have witnessed in the past two decades promising steps that

---

[26]Lorenz, K. 1970. *Studies in Animal and Human Behavior*, Vols. I, II. Harvard University Press, Cambridge, Mass.

[27]Fuller, J. L. 1967. *Science 158*:1645–1652.

[28]Hubel, D. H. 1967. *Physiologist 10*:17–45.

lead to a correlation of known groups of cells and specific signals with certain experiences during perception. The analysis of what happens to the information that is originally impressed upon the surface of the retina and then ascends stepwise from one group of cells to the next will undoubtedly be extended. So far, in following and deciphering visual clues we have progressed into the brain through only seven synaptic relays.

For further unraveling of neural organization a more thorough recognition of the chemical architecture of cell assemblies, discussed in Chapter 11, may prove most valuable. Nevertheless, we must ask how useful the cellular approach will continue to be as we go still further, and at what stage the analysis will have to be enlarged. To interpret successfully the activity of large populations of cells, new methods are needed.

Although the field is by no means neglected, we have no good approach as yet for investigating the cellular basis of memory, learning, consciousness, or even how a simple act of movement is initiated. But perhaps we should feel encouraged by the vivid awareness of our many glaring deficiencies and by our ability to define many areas of ignorance. Consider the question of how continued activity of nerve cells, particularly during growth and development, modifies their ability to influence other nerve cells. We can now at least point to specific processes in the visual cortex where certain groups of cells lose their synaptic drive from one eye that has been inappropriately used during a critical early period of life. In other words, some cortical neurons will be drastically changed for the rest of their lives because they have been used in the "wrong" way. But we do not know how normal impulse traffic acts to maintain and further develop anatomical connections between cells that are present at birth.

One could maintain that what is most needed is an understanding of the rules that govern the assembly of nerve cells in the first place. What is the ground plan according to which the instructions contained in the genes are translated into normal wiring? With such knowledge one might distinguish between genetically determined deviations and changes or abnormalities that are superimposed by the environment in the course of later experience.

Nonetheless, despite the enormous gaps in our knowledge, one can take some satisfaction by looking back about two decades and recalling how impossible it seemed then to think of mechanisms by which nerve cells could recognize even a simple shape such as a corner or the letter *L*.

SUGGESTED READING

Items marked with an asterisk (*) are reprinted in Cooke, I., and Lipkin, M., Jr. 1972. *Cellular Neurophysiology*. Holt, New York.

General review

Hubel, D. H. 1967. Effects of distortion of sensory input on the visual system of kittens. *Physiologist 10*:17–45.

Original papers

Guillery, R. W., and Kaas, J. H. 1974. The effects of monocular lid suture upon the development of the visual cortex in squirrels (*Sciureus carolinensis*). *J. Comp. Neurol. 154*:443–452.

Hubel, D. H., and Wiesel, T. N. 1963. Receptive fields of cells in striate cortex of very young, visually inexperienced kittens. *J. Neurophysiol. 26*:994–1002.

Hubel, D. H., and Wiesel, T. N. 1965. Binocular interaction in striate cortex of kittens reared with artificial squint. *J. Neurophysiol. 28*:1041–1059.

Hubel, D. H., and Wiesel, T. N. 1970. The period of susceptibility to the physiological effects of unilateral eye closure in kittens. *J. Physiol. 206*: 419–436.

*Wiesel, T. N., and Hubel, D. H. 1963. Single-cell responses in striate cortex of kittens deprived of vision in one eye. *J. Neurophysiol. 26*:1003–1017.

Wiesel, T. N., and Hubel, D. H. 1965. Comparison of the effects of unilateral and bilateral eye closure on cortical unit responses in kittens. *J. Neurophysiol. 28*:1029–1040.

Wiesel, T. N., and Hubel, D. H. 1965. Extent of recovery from the effects of visual deprivation in kittens. *J. Neurophysiol. 28*:1060–1072.

Wiesel, T. N., and Hubel, D. H. 1974. Ordered arrangement of orientation columns in monkeys lacking visual experience. *J. Comp. Neurol. 158*: 307–318.

APPENDIX

CURRENT FLOW IN ELECTRICAL CIRCUITS

A few basic concepts are required to understand the electrical circuits used in this presentation. An especially clear and lively treatment is found in Rogers's book.[1] For our purposes it is sufficient to describe the properties of various circuit elements and explain how they work when connected together in ways that correspond to the circuits described for nerves. The numerical examples in Figures 3 to 5 provide useful insights into the behavior of nerve membranes that are not obvious unless one is already familiar with circuitry.

Terms and units describing electric currents

The difficulties encountered on first reading accounts of electricity often stem from the apparently abstract nature of the forces and energy involved. It is therefore reassuring to realize that many of the original pioneers must have been faced with similar problems, since the terms devised in the last century actually refer to the movement of water or heat. The words "current," "flow," "potential," "resistance," and "capacity" all apply to hydraulic models and to electricity in a manner that is quite analogous. It is somewhat ironic that nowadays an engineer faced with a complex problem in hydraulics may use electrical circuits to work out the solution.

Consideration of the movement of water through a simple hydraulic circuit illustrates how the terms apply to both systems. In Figure 1 the amount of water flowing through the pipes in unit time is the same all around the circuit. An identical flow rate is measured at points *a*, *b*, and *c*. Similarly, an electric current flows all around the circuit. Narrower tubes offer a relatively higher resistance to the flow of water, and thinner wires act in the same way for electric current. Keeping the water currents or electric currents flowing around the circuit requires a source of energy in the form of a pump or a battery. In the case of electricity, the flow along a wire consists of the movement of electrons; the more electrons that flow past each second, the larger the current. The choice of units in which to express the rate of flow is largely a matter of convenience; one can measure the water flowing through a

[1]Rogers, E. M. 1960. *Physics for the Enquiring Mind.* Princeton University Press, Princeton, N.J.

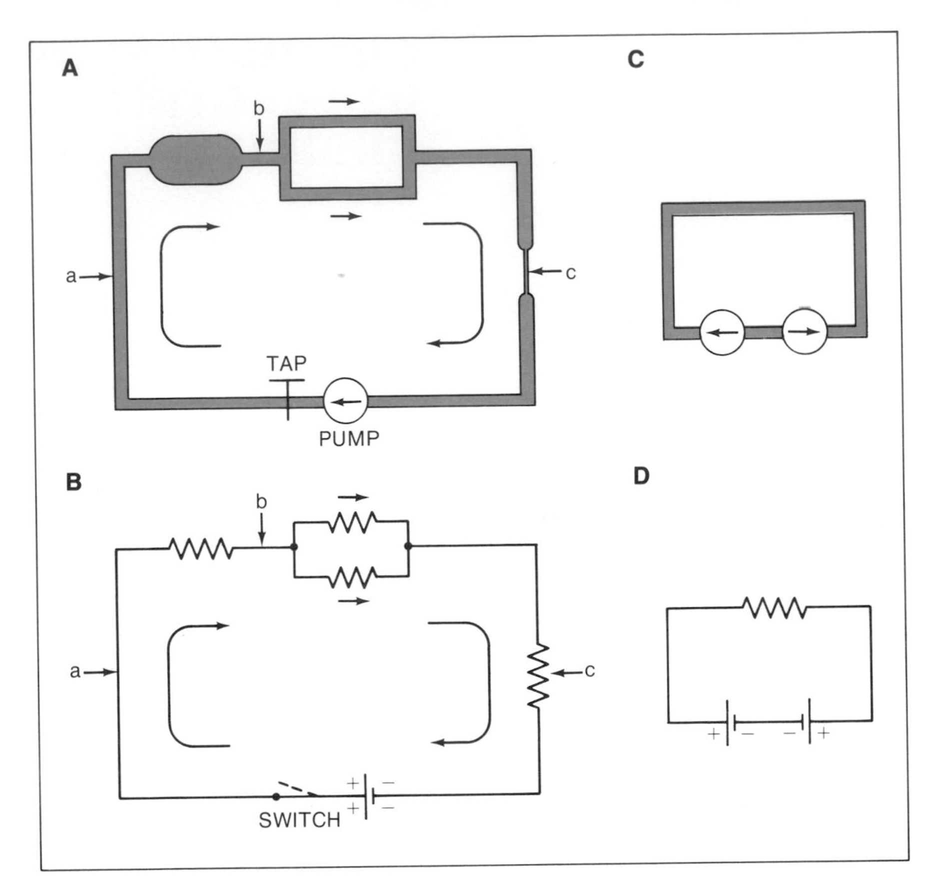

1 **HYDRAULIC AND ELECTRICAL CIRCUITS. A, B. Analogous circuits for the flow of water and electric current. When currents flow in opposite directions, they sum algebraically (C, D).**

pipe in gallons per second, for example, although in some other situations molecules per second might be a more suitable unit. Similarly, electric current is measured in COULOMBS per second instead of electrons per second, the coulomb being a multiple unit, equivalent to 6×10^{18} electrons. (Like other electrical units, the coulomb is named eponymously.) The term AMPERE (abbreviated A) is the same as coulombs per second and is a measure of current flow; current is usually designated in equations as I. Pursuing further the comparison of the flow of electricity and of water along a tube, the net flow of current can move in one direction or the other, but not both at once (Figure 1). If the currents are equal and opposite in the two directions, there is no net flow. Current is in fact a vector quantity, with a direction and magnitude. It is therefore convenient to denote the direction of flow by labeling flow in one direction as positive and in the other as negative.

What do the terms POSITIVE and NEGATIVE mean with regard to electricity? Here the analogy with hydraulic systems does not help.

Before the nature of electricity was understood, it was clear that currents flowing in opposite directions could produce very different effects on chemicals in solution. Thus positive and negative do not simply represent directions of flow with exactly equivalent actions. This is readily illustrated by considering the effects of passing current through two copper wires that dip into a solution of copper sulfate. Copper ions move in the direction of the current, being expelled into the solution from the positive wire and deposited from the solution onto the negative one; sulfate ions move the opposite way. The units that measure charge and current, coulombs and amperes, mentioned earlier, are based on this property. For example in a copper plating bath 1 A deposits 0.329×10^{-6} kg/sec.

Only later was it discovered that the electric currents in metal wires consist of negatively charged electrons moving toward the positive polarity. In a circuit such as that shown in Figure 3, the current marked by the arrow, by convention, moves from positive to negative. In fact, within the wires only negative charges (electrons) are free to move, and they do so in the opposite direction. It does not matter for the present purpose, looking at circuit diagrams, whether negative charges are thought of as moving one way or positive charges the other way, and we shall only refer to current moving from positive to negative. In salt solutions, both positively and negatively charged ions do move in both directions, as in the solution of copper sulfate.

To explain the energy sources for current flow and the meaning of electrical POTENTIAL, the hydraulic analogy is again useful. The flow of fluid depicted in Figure 1 depends on the difference in pressure, the direction of flow being from high to low. No net movement occurs between two parts of the circuit at the same pressure. A difference in pressure can be maintained between connected points only by the expenditure of energy. Similar considerations apply to electric currents that flow under the influence of difference in potential. No current flows between two points at the same potential. In the fluid system the source of energy that creates the pressure is the pump, and in the electrical circuits considered here it is a BATTERY, a device in which chemical work has been done to separate positive and negative electrical charges. The stored energy is released when the external circuit is completed. The energy is measured in JOULES (J), which can also be expressed as ergs or foot-pounds. Potential, charge (q), and energy are related as follows: $1 \text{ V} = 1 \text{ J}/q$. Thus the VOLTAGE is an indication of the amount of energy delivered by each coulomb. A VOLTMETER, which measures potential differences, corresponds to a pressure gauge; it indicates how much energy each coulomb carries. An AMMETER, which measures current, indicates how many coulombs flow past each second.

In a circuit such as that of Figure 3, the current is driven by the difference in potential between the two poles of the battery. A larger potential difference provides more energy for each charge and a larger flow of current. The potentials of the two batteries arranged in series, as in Figure 5, sum.

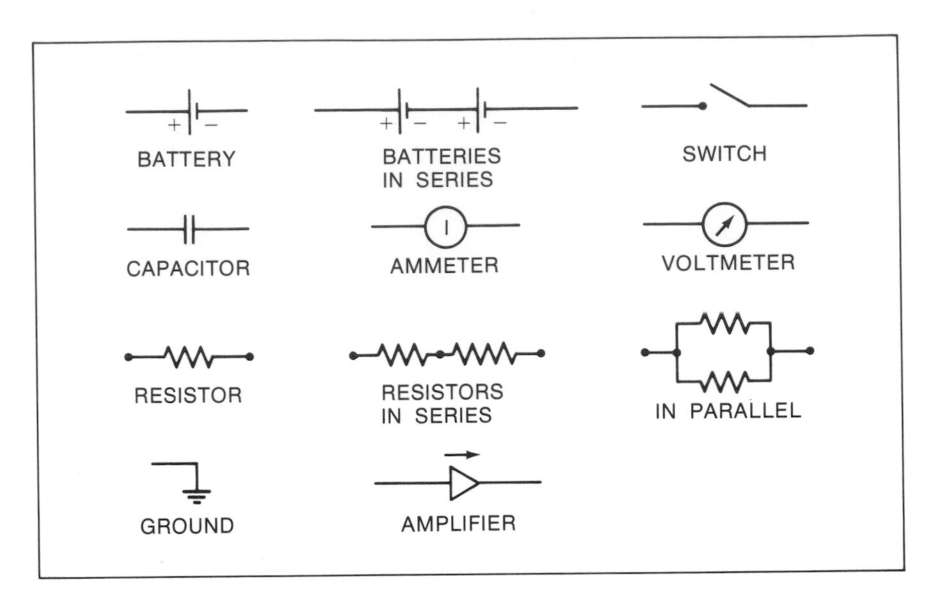

2 **SYMBOLS IN ELECTRIC CIRCUITRY.**

Ohm's law and electrical resistance

So far it has been shown that the current (I) is proportional to the potential difference, just as the flow of water is proportional to the pressure. The constant that determines the ratio of pressure difference to flow rate is an inherent characteristic of the pipes, their RESISTANCE. More pressure is needed to produce the same flow rate in a narrower tube. The same situation applies in the wires or salt solutions that conduct electricity. Smaller wires and more dilute solutions have higher resistances because fewer electrons or ions are available for carrying current. In the 1820s Ohm discovered the law relating current and potential difference for a wire: $I = V/R$, where I is current, V is the potential difference between the two ends of the wire, and R is its resistance measured in OHMS (Ω). The reciprocal of resistance measures the "ease" with which current flows. It is called CONDUCTANCE and is indicated by $g = 1/R$; the units are mho's. The symbols used in this section are shown in Figure 2.

Significance of Ohm's law for understanding circuits

Ohm's law holds whenever the graph of current against potential is a straight line. In any circuit or part of a circuit for which this is true, one of the variables can be estimated if the other two are known. For example:

1. We can pass a known current through a nerve membrane, measure the change in potential, and estimate the resistance of the membrane.
2. If we measure the potential difference produced by an applied current and know the resistance, we can estimate how much current is flowing through the membrane.
3. If we pass a known current through the measured membrane resistance, we can estimate the change in potential.

Two additional simple but important laws (Kirchoff's laws) should be mentioned: (1) the algebraic sum of all the currents flowing toward a junction is zero (this is merely a statement that charge is neither created nor destroyed); (2) the algebraic sum of all the battery voltages is equal to the algebraic sum of all the *IR* drops in a circuit loop (this is equivalent to conservation of energy).

We can now turn to an analysis of the circuits in Figures 3 and 4, which are needed to construct a model of the membrane. Figure 3 shows a battery (*V*) of 10 V connected to a resistance (*R*) of 10Ω. The switch (*S*) opens or closes the circuit and thereby interrupts or establishes current flow. The potential difference between the two ends of *R* is 10 V. The current (*I*) measured by the ammeter is therefore 1 A. Suppose *R* is replaced by two resistors, R_1 and R_2, each of 10Ω, which are IN SERIES. The total resistance is now 20Ω, and $I = 0.5$ A. The potential difference between *a* and *c* is still 10 V (the whole battery), but between *a* and *b* it is only 5 V (=0.5 A × 10Ω). Note also that *a* is more positive than *b*, which is more positive than *c*.

How does the current flow if the circuit of Figure 3 is extended by adding another resistance, R_3, also 10Ω, in parallel (Figure 4)? Some flows through R_1 and some through R_3. The potential difference across both these resistors is the same 10 V. We can now solve for the current flowing through R_1 and R_3, employing Ohm's law:

$$I_{R_1} = \frac{V}{R_1} = \frac{10}{10} = 1\ \text{A}$$

$$I_{R_3} = \frac{V}{R_3} = \frac{10}{10} = 1\ \text{A}$$

The ammeter (*I*) therefore measures 2 A. We can conclude that resistors in parallel sum as their reciprocals. Together, their effective

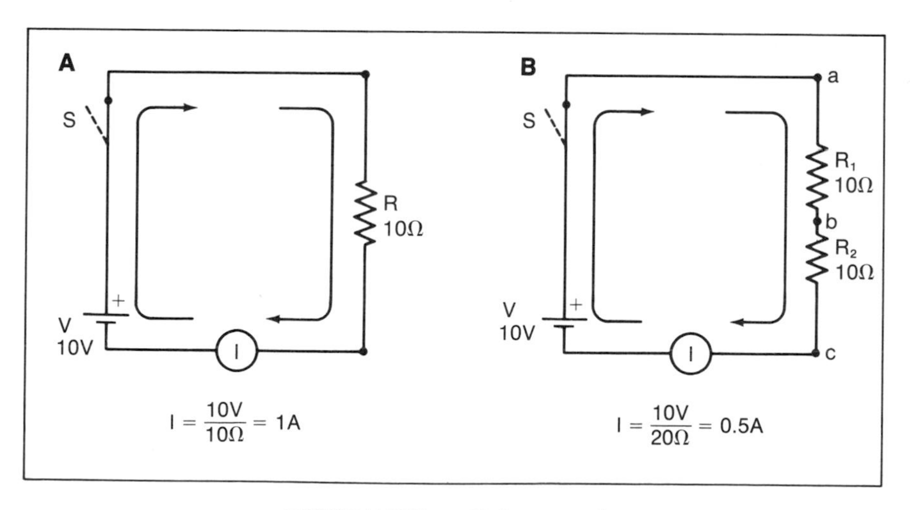

3 **OHM'S LAW applied to simple circuits. In A the current I = 10 V/10 Ω. B is a voltage divider; the current is 0.5 A and the drop across each resistor is 5 V.**

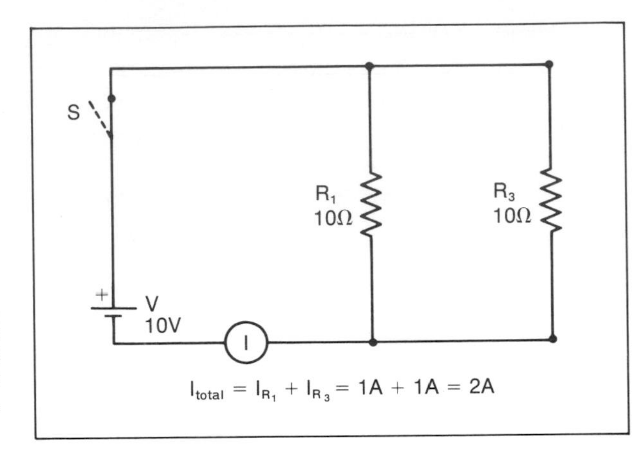

4

PARALLEL RESISTORS. With R_1 and R_3 in parallel, the current is 2 A and the voltage drop across each resistor is 10 V. In parallel, resistors add reciprocally.

resistance is 5Ω. In the hydraulic model two pipes alongside each other act similarly by offering less resistance and enabling more flow to occur with the same head of pressure. To summarize, in series resistors add simply:

$$R_{total} = R_1 + R_2 + R_3 \text{ etc.}$$

In parallel they add reciprocally:

$$\frac{1}{R_{total}} = \frac{1}{R_1} + \frac{1}{R_2} + \frac{1}{R_3} \text{ etc.}$$

Thus in parallel the conductances, being reciprocal resistances, add:

$$g_{total} = g_1 + g_2 + g_3$$

Applying Ohm's law to the membrane model

Figure 5 shows a circuit similar to that used for the nerve membrane. Notice that the two batteries are both driving current around the circuit in the same direction, and that resistances R_1 and R_2 are in series, as in Figure 3. What is the potential difference between points *b* and *d*? We can determine the current flowing in the circuit. The potential difference between points *a* and *c* is 150 V, *a* being positive. The total resistance of this circuit is 100Ω. The current is therefore 1.5 A. When 1.5 A flows across 10Ω, a potential difference of 15 V is produced, *a* being positive with respect to *b*. The potential difference between "outside" and "inside" is therefore 100 V − 15 V = +85 V. We can solve in the same way for R_2 and get the same result: 1.5 A across 90Ω produces 135 V, *b* being positive with respect to *c*. The potential difference between "outside" and "inside" is therefore +135 − 50 = +85 V. This has to be so since the potential difference between two points must have a unique value.

In Figure 5*B*, R_1 and R_2 have been exchanged. The current is still 1.5 A. Now the potential drop along $R_2 = 90 \times 1.5 = 135$ V, *a* being positive. Consequently, the differences between "inside" and "outside" is 100 − 135 = −35 V, OUTSIDE NEGATIVE; the same figure can be calculated from the current through R_2. This simple circuit shows how THE POTENTIAL ACROSS A MEMBRANE CAN CHANGE AS A RESULT OF RE-

SISTANCE CHANGES WHILE THE BATTERIES REMAIN CONSTANT. A simple equation derived from Ohm's law which relates the potential between b and d to the batteries and resistors in the circuits of Figure 5 is:

$$V = \frac{\frac{V_1}{R_1} + \frac{V_2}{R_2}}{\frac{1}{R_1} + \frac{1}{R_2}} = \frac{\frac{R_2}{R_1} V_1 + V_2}{\frac{R_2}{R_1} + 1}$$

Electrical capacitance and time constant

Capacitors introduce the time element into the consideration of current flow and changes in potential. In all the circuits described so far, closing or opening the switch produces instantaneous and simultaneous changes in current and potential (Figure 3). Electrical capacitors can hold or store electrical charges and thereby prevent them from flowing through the circuit. A capacitor consists of two conducting plates (usually of metal) separated by an insulator (air or paper). The larger the plates and the closer together they lie, the greater the CAPACITY for storing and separating charges. Electrons cannot flow directly from one plate to the other because of the insulator. The better the insulator, the larger the dielectric constant and the capacity. Nevertheless, a brief momentary surge of current occurs when the switch (S_1) in Figure 6*A* is closed, as indicated by the arrow. Positive charges accumulate at one plate and are drawn away from the other. (Remember that although currents in wires in fact consist of negatively charged elec-

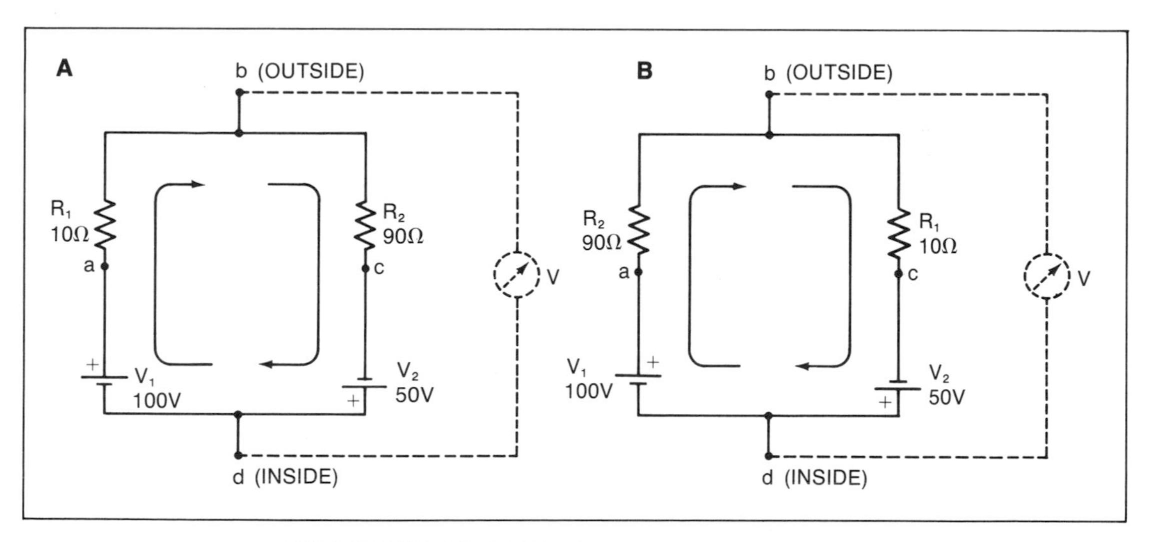

5

ANALOGOUS CIRCUITS FOR NERVE MEMBRANES. In A and B the resistors R_1 and R_2 are switched around; otherwise the circuits are the same. The batteries V_1 and V_2 are in series. In A, b is positive with respect to d by 85 V; in B it is negative (−35 V). These circuits illustrate how changes in resistance or conductance can give rise to changes in potential even though the batteries remain constant. The terms "outside" and "inside" are used by analogy for the models of nerve membranes (Part Two).

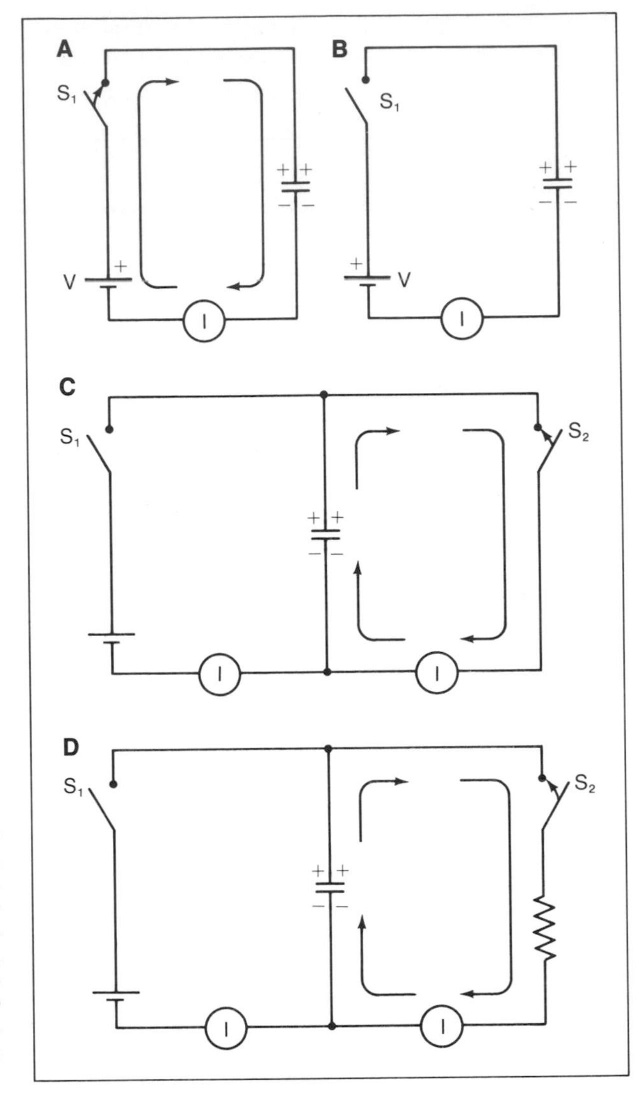

6

CAPACITORS in electrical circuits. A, B, and C are idealized circuits having no resistance. When S_1 is closed in A, the capacitor is instantaneously charged to a voltage (V). The potential remains constant if S_1 is then opened (B). In C, S_2 is closed, allowing the capacitor to discharge instantaneously. In D the capacitor discharges through the resistor (R). The maximum current (rate of removal of charge) is V/R.

trons moving, one can still speak of "positive charges" going in the opposite direction.) Once the full potential has been built up across the capacitor, no further current flows. If the battery is removed from the circuit (by opening the switch) (Figure 6*B*), the potential remains steady. This demonstrates the ability of the capacitor to STORE charges. The relation between potential (V), charges in coulombs (q), and capacity (C) is given by $V = q/C$; larger capacitors store more charges for a given potential difference between the two plates. The unit of capacitance is the farad (F), which represents the capacity to separate (and store) 1 coulomb of charge for each volt between the plates. When S_2 (Figure 6*C*) is closed, the charges flow off the plates, positive being attracted to negative, as usual. This constitutes a brief current.

In the circuits of Figure 6*A*, *B* and *C* time plays no part because with no resistance ($R = 0$) in the idealized circuit, the current is infinitely large and also infinitely brief for a given potential ($I = V/0$). In practice, every circuit has a finite resistance that limits current flow. Figure 6*D* shows a circuit in which the capacitor discharges through a resistor instead of a short circuit. The larger the resistance, the fewer the number of coulombs that can be drawn off in each second—that is, the smaller the current. Consequently, the potential difference between the plates of the condenser decreases slowly instead of instantaneously. The time for discharge varies directly with both R and C.

Another example is shown in Figure 7, where the resistor is in series with the capacitor. Here the charging up process cannot be achieved instantaneously, again because the resistor limits the current flow that deposits charges onto the plates of the capacitor. As soon as the capacitor is fully charged to the same voltage as the battery, no further current flows.

These examples show that current flows "across" a capacitor only when the potential is changing:

$$I_C = C\frac{dV}{dt}$$

V_C denotes the voltage across the capacitor. When this is fully charged or discharged and V_C is steady, no current flows. It has "infinite resistance" for a steady potential difference, but low "resistance" for changing potential difference. Indeed the extent to which a capacitor impedes the current varies inversely with the frequency of a time-varying voltage. Figure 7*B* and *E* show the currents flowing through a resistor and capacitor in parallel.

The properties of a capacitor in an electrical circuit can be illustrated by a slightly more elaborate hydraulic analogy. In Figure 7*C* the capacitor is represented by an elastic diaphragm that forms a partition in a water-filled chamber. When the pump begins to generate pressure, the diaphragm bulges until it is as fully distended as it is going to be for the particular pressure applied (pressure being equivalent to potential). There is a brief surge of water current, but only while the diaphragm is being distended. If, later, the pressure is suddenly released, the diaphragm recoils, producing flow in the opposite direction. If a tube is placed alongside, as in Figure 7*D*, some water flows through the tube and some is used to expand the diaphragm. If the resistance of the pipe is high, the pressure difference at its two ends is larger for a given flow. In this case, the distention of the diaphragm is greater and takes longer to achieve. Similarly, if the capacity of the cylinder is larger, more water is diverted from the pipe. Again it takes longer to achieve the steady state in which the diaphragm is steady and water is flowing at a constant rate through the pipe. (Note that the water flows at its final rate through the pipe only once the diaphragm has stopped being distended.) Thus the characteristic time of the system to achieve a steady state (its TIME CONSTANT) varies with the product of resistance and capacity.

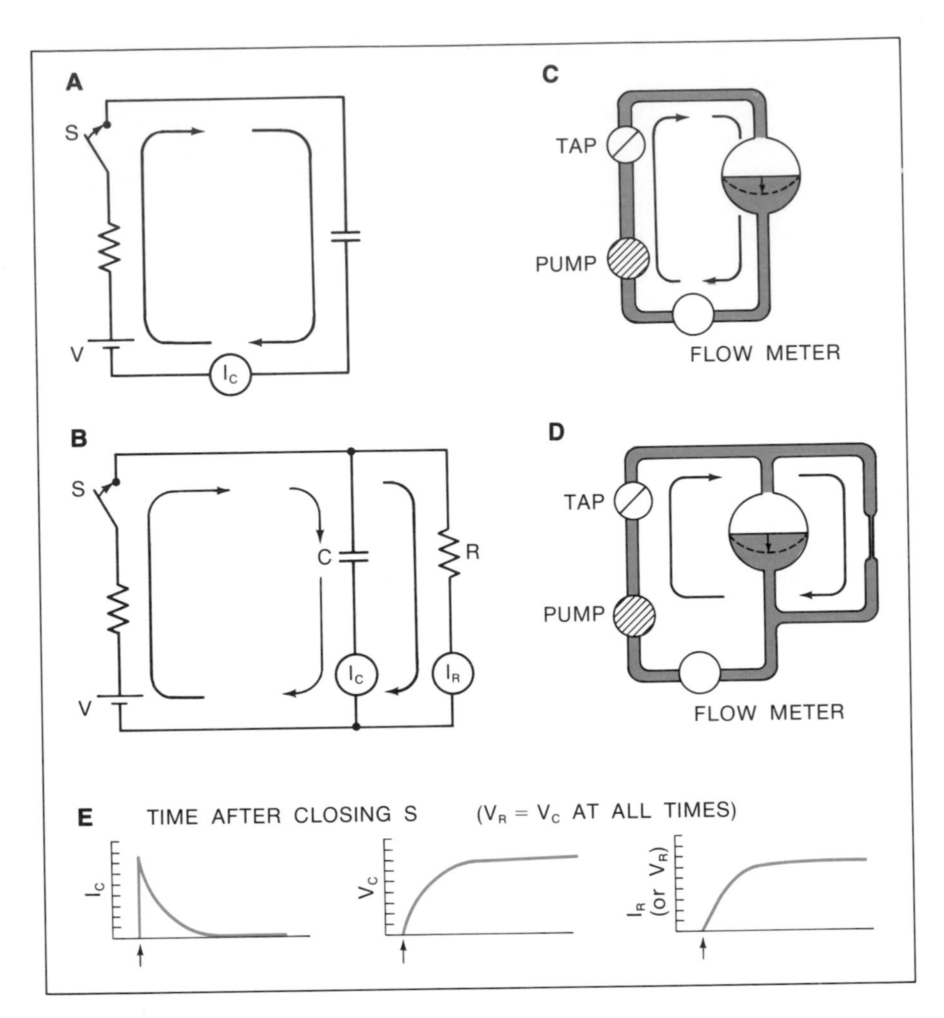

7 **CHARGING OF CAPACITOR. In A the capacitor is charged at a rate limited by the resistor. In B the time depends on both resistors in the circuit. E shows the capacitative current (I_C) and the current flowing through the resistor (R). This reaches its final value only when the capacitor is fully charged. At all times $V_C = V_R$. C and D are hydraulic analogies.**

When capacitors are arranged in parallel, the total CAPACITANCE increases. The explanation is shown in Figure 8*A*. Capacitors C_1 and C_2 both take up charges and therefore more charges are stored for a given potential difference. This is the situation in nerve, where capacitors distributed along the length produce progressively greater slowing in the time course of localized potentials. In contrast, the capacitance DECREASES when capacitors are arranged in series (Figure 8*B*). The potential difference between the two plates of each of the capacitors is smaller and fewer charges can be held. Such a situation occurs in myelin, where several layers of membrane fuse to form a thick multilayered

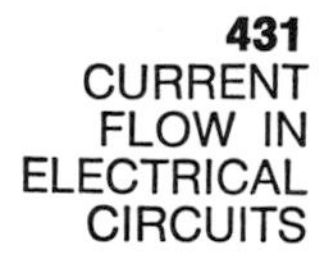

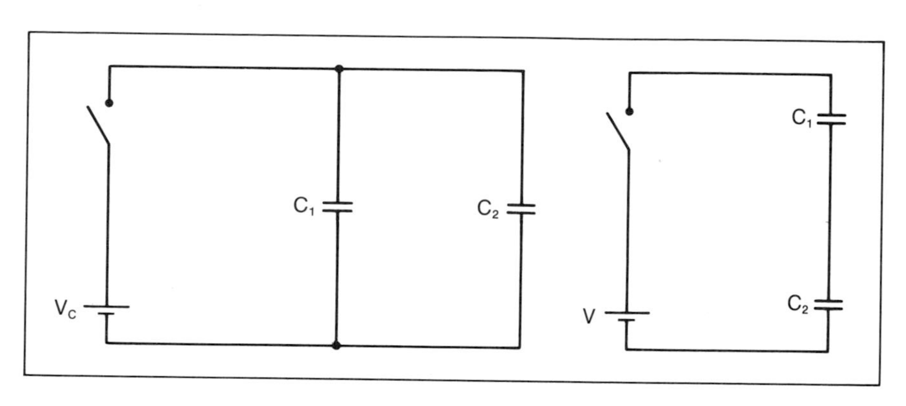

CAPACITORS IN PARALLEL (left) sum; those in series (right) add reciprocally.

8

sheath that has a lower capacity than a single membrane. Hence in parallel:

$$C_{total} = C_1 + C_2 + C_3$$

while in series

$$\frac{1}{C_{total}} = \frac{1}{C_1} + \frac{1}{C_2} + \frac{1}{C_3}.$$

GLOSSARY

The definitions below apply to the terms used in the context of this book. Thus, **excitation, adaptation,** and **inhibition** all have additional meanings that are not included.

acetylcholine, ACh $CH_3—CO—O—CH_2—CH_2—N—(CH_3)_3$. Transmitter liberated by vertebrate motoneurons, preganglionic sympathetic and parasympathetic neurons; hydrolyzed by cholinesterase.

action potential Brief regenerative, all-or-nothing electrical potential that propagates along an axon or muscle fiber.

adaptation Decline in response of a sensory neuron to a maintained stimulus.

afferent Axons conducting impulse toward the central nervous system.

anticholinesterase Cholinesterase inhibitor (e.g., neostigmine, eserine); such agents prevent the hydrolysis of ACh and thereby allow its action to be prolonged.

area centralis In the cat, the area of retina with highest discrimination, containing cones.

areas 17, 18, 19 Neurosensory areas of the cerebral cortex subserving vision, defined by histological and physiological criteria.

autonomic nervous system Part of the nervous system supplying viscera, skin, smooth muscle, glands, and heart, consisting of two distinct divisions, parasympathetic and sympathetic.

axon The process or processes of a neuron conducting impulses, usually over long distances.

axon cap Specialized region of Mauthner cell axon surrounded by glial cells at which electrical inhibition occurs.

axon hillock Region of the cell body at which the axon originates; often the site of impulse initiation.

axoplasm Intracellular fluid within an axon.

bipolar cell Neuron with two major processes arising from the cell body; in the vertebrate retina, interposed between receptors and ganglion cells.

blood-brain barrier Term denoting restricted access of substances to neurons and glial cells within the brain.

bouton Small terminal expansion of presynaptic nerve fiber at a synapse; site of transmitter release.

α-bungarotoxin Toxin from venom of the snake, *Bungarus multicinctus;* binds to ACh receptor with high affinity.

capacitance of the membrane (C_m) Property of the cell membrane enabling electrical charge to be stored and separated and introducing distortion in the time course of passively conducted signals; measured in farads (F).

caudal Posterior (in four-legged animals).

centrifugal control Regulation of performance of peripheral sense organs by axons coming from the central nervous system.

cerebrospinal fluid (CSF) Clear liquid filling the ventricles of the brain and spaces between meninges, arachnoid, and pia. See **subarachnoid space, ventricles.**

cholinergic Neurons releasing ACh as the transmitter.

choroid plexus Folded processes rich in blood vessels, that project into ventricles of the brain and secrete cerebrospinal fluid.

conductance (g) Reciprocal of electrical resistance and thus a measure of the ability of a circuit to conduct electricity; in excitable cells a useful measure of permeability for an ion or ions.

contralateral Relating to the opposite side of the body.

convergence Coming together and making synapses; a group of presynaptic neurons on one postsynaptic neuron.

coronal Vertical section through the skull at right angles to the front-back (sagittal) axis.

cortical column Aggregate of cortical neurons sharing common properties (e.g., sensory modality, receptive field position, eye dominance, orientation, movement sensitivity).

coulomb Unit of electrical charge.

coupling potential Synaptic potential seen as a result of current spread through an electrical synapse.

dendrite Process of a neuron specialized to act as a receptor; postsynaptic region of a neuron.

depolarization Reduction of membrane potential from the resting value toward zero.

divergence Branching of a neuron to form synapses with several other neurons.

efferent Axons conducting impulses outward from the central nervous system.

electroencephalogram (EEG) Record taken of the electrical activity of the brain by external electrodes on the scalp.

electrogenic sodium pump Active transport of sodium out of a cell resulting in an outward current and an increase in membrane potential.

electroretinogram (ERG) Potential change in response to light measured by external electrodes on the eye.

electrotonic potentials Localized, graded potentials produced by subthreshold currents, determined by passive electrical properties of cells.

endothelial cells Layer of cells lining blood vessels.

end plate Postsynaptic area of vertebrate skeletal muscle fiber.

ependyma Layer of cells lining cerebral ventricles and central canal of spinal cord.

epinephrine (adrenaline) Hormone secreted by the adrenal medulla; certain of its actions resemble those of sympathetic nerves.

epp End plate potential; synaptic potential in a skeletal muscle fiber produced by ACh liberated from presynaptic terminals.

epsp Excitatory postsynaptic potential in a neuron.

equilibrium potential Membrane potential predicted from Nernst equation on the basis of concentration differences for an ion to which the membrane is permeable.

eserine Anticholinesterase (see above); also known as physostigmine.

excitation Process tending to produce action potentials.

extrafusal Muscle fibers making up the mass of a skeletal muscle (i.e., not within the sensory muscle spindles).

facilitation Greater effectiveness of synaptic transmission by successive presynaptic impulses usually due to increased transmitter release.

farad (F) Unit of capacitance; the capacity of a condensor having a charge of 1 coulomb with a potential of 1 volt between the two plates, usually used as microfarad (μF).

faraday (F) Quantity of electricity liberating 1 g equivalent of an element in electrolysis (96,500 coulombs).

field axis orientation For simple, complex, and hypercomplex cortical neurons, the angle of the long axis of the receptive field (e.g., horizontal, vertical, oblique).

fovea Central depression in the retina composed of slender cones; area of greatest visual resolution.

fusimotor Motoneurons supplying muscle fibers in muscle spindle.

g Symbol for conductance.

γ-aminobutyric acid (GABA) Inhibitory transmitter at crustacean neuromuscular synapses and a candidate in the central nervous system of vertebrates.

γ-efferent fiber Small myelinated motor axon supplying intrafusal muscle fiber. See **fusimotor.**

ganglion Discrete collection of nerve cells; ganglion cells.

gap junction Region of contact between cells at which intercellular space between adjacent membranes is reduced to about 2 nm; site of electrical coupling.

generator potential Graded, localized potential change in a sensory receptor initiated by the appropriate stimulus; the electrical sign of the transduction process.

glia See **neuroglia.**

gray matter Part of the central nervous system composed predominantly of the cell bodies of neurons and fine terminals, as opposed to major axon tracts (white matter).

horseradish peroxidase Enzyme used as a histochemical marker for tracing processes of neurons or spaces between cells.

hyperpolarization Increase in membrane potential from the resting value, tending to reduce excitability.

I Symbol for current (amperes).

impulse See **action potential.**

inactivation Gradual reduction in sodium conductance produced by depolarization.

inhibition Effect of one neuron upon another tending to prevent it from initiating impulses.

Postsynaptic inhibition is mediated through a permeability change in the postsynaptic cell, holding the membrane potential away from threshold.

Presynaptic inhibition is mediated by an inhibitory fiber upon an excitatory terminal, reducing the release of transmitter.

Electrical inhibition is mediated by currents in presynaptic fibers that hyperpolarize the postsynaptic cell and do not involve the secretion of a chemical transmitter.

initial segment Region of an axon close to the cell body; often site of impulse initiation.

integration Process whereby a neuron sums the various excitatory and inhibitory influences converging upon it and synthesizes a new output signal.

intercellular clefts Narrow fluid-filled spaces between membranes of adjacent cells; usually about 20 nm wide.

interneuron Neuron which is neither purely sensory nor motor but connects neurons.

internode Myelinated portion of a nerve axon lying between two nodes of Ranvier (see below).

intrafusal fiber Muscle fiber within a muscle spindle; its contraction initiates or modulates sensory discharge.

iontophoresis Transfer of ions by passing current through a micropipette; used for applying charged molecules with a high degree of temporal and spatial resolution.

ipsilateral On the same side of the body.

ipsp Inhibitory postsynaptic potential.

lateral geniculate nucleus Small knee-shaped nucleus; part of posteroinferior aspect of thalamus acting as a relay in the visual pathway.

length constant ($\lambda = \sqrt{r_m/r_i}$) Distance (usually in millimeters) over which a localized graded potential decreases to $1/e$ of its original size in an axon or a muscle fiber.

Mauthner cell Large nerve cell in the mesencephalon of fish and amphibians, up to 1 mm in length.

mepp Miniature end plate potential; small depolarization at neuromuscular synapse caused by spontaneous release of a single quantum of transmitter from the presynaptic terminal.

mho Reciprocal of ohm, a unit of conductance.

modality Class of sensation (e.g., touch, vision, olfactory).

motoneuron (motor neuron) A neuron that innervates muscle fibers.

motor unit A single motoneuron and the muscle fibers it innervates.

muscle spindle Fusiform end organ in skeletal muscle in which afferent sensory fibers and a few motoneurons terminate.

myelin Fused membranes of Schwann cells or glial cells forming a high-resistance sheath surrounding an axon.

neostigmine An anticholinesterase; also known as prostigmine.

neuroglia Non-neuron satellite cells associated with neurons. In the mammalian central nervous system the main groupings are astrocytes and oligodendrocytes; in peripheral nerves the satellite cells are called Schwann cells.

neuropil Network of axons, dendrites, and synapses.

node of Ranvier Localized area devoid of myelin occurring at intervals along a myelinated axon.

norepinephrine (noradrenaline) Transmitter liberated by most sympathetic nerve terminals.

ocular dominance Greater effectiveness of one eye over the other for driving simple, complex, or hypercomplex cells in the visual cortex.

Ohm's law Relates current (I) to voltage (V) and resistance (R); $I = V/R$.

optic chiasm The point of crossing or decussation of the optic nerves. In cats and primates fibers arising from the medial part of the retina cross to supply the lateral geniculate nucleus on the other side of the animal.

ouabain G-strophanthidin, a glycoside that specifically blocks the sodium-potassium coupled pump.

overshoot Reversal of the membrane potential during the peak of the action potential.

permeability Property of the membrane allowing substances to pass into or out of the cell.

quantal release Secretion of multimolecular packets (quanta) of transmitter by the presynaptic nerve terminal.

receptive field Denotes the area of the periphery whose stimulation influences the firing of a neuron. For cells in the visual pathway, the receptive field refers to an area on the retina whose illumination influences the activity of a neuron.

receptor 1. Sensory nerve terminal. 2. A molecule in the cell membrane that combines with a specific chemical substance.

reciprocal innervation Interconnections of neurons arranged so that pathways exciting one group of muscles inhibit the antagonistic motoneurons.

reflex Involuntary movement or other responses elicited by a stimulus applied to the periphery, transmitted to the central nervous system, and reflected back out to the periphery.

refractory period The time following an impulse during which a stimulus cannot elicit a second impulse.

resting potential The steady electrical potential across the membrane in the quiescent state.

reversal potential The value of the membrane potential at which a chemical transmitter produces no change in potential.

Ringer's fluid A balanced saline solution containing sodium chloride, potassium chloride, and calcium chloride; named after Sidney Ringer.

sagittal Section in the anteroposterior direction.

saltatory conduction Conduction along a myelinated axon whereby the impulse leaps from node to node.

Schwann cell Satellite cell in the peripheral nervous system, responsible for making the myelin sheath.

serotonin Also known as 5-hydroxytryptamine or 5-HT; transmitter in molluscan nervous system and a candidate in the vertebrate central nervous system.

soma Cell body.

striate cortex Also known as area 17 or visual I; primary visual region of occipital lobe marked by striation of Gennari, visible with the naked eye.

subarachnoid space Space filled by cerebrospinal fluid between two layers of connective tissue, the meninges, surrounding the brain.

supersensitivity Increase in sensitivity to chemical transmitters in neurons, gland, or muscle cells following denervation.

synapse Site at which neurons make functional contact; a term coined by Sherrington.

synaptic cleft The space between the membranes of the pre- and postsynaptic cells at a chemical synapse across which transmitter must diffuse.

synaptic vesicles Small membrane-bound sacs contained in presynaptic nerve terminals. Those with dense cores contain catecholamines and serotonin; clear vesicles are presumed to be the storage sites for other transmitters (released from terminals).

tetraethylammonium (TEA) Quaternary ammonium compound that selectively blocks the potassium conductance channel in neurons and muscle fibers.

tetrodotoxin (TTX) Toxin from puffer fish that selectively blocks the regenerative sodium conductance channel in neurons and muscle fibers.

threshold 1. Critical value of membrane potential potential or depolarization at which an impulse is initiated. 2. Minimal stimulus required for a sensation.

tight junction Site at which fusion occurs between the outer leaflets of membranes of adjacent cells, resulting in a five-layered junction. It is called a **macula occludens** if the area is a spot and a **zonula occludens** if the junction is a circumferential ring. Such complete junctions prevent the movement of substances through the extracellular space between the cells.

time constant τ A measure of the rate of buildup or decay of a localized graded potential, depending on the resistance and the capacity of the membrane.

transducer Device for converting one form of energy into another (e.g., a microphone, photoelectric cell, loudspeaker).

transmitter Chemical substance liberated by a presynaptic nerve terminal causing an effect on the membrane of the postsynaptic cell, usually an increase in permeability to one or more ions.

undershoot Transient hyperpolarization following an action potential; caused by increased potassium conductance.

ventricles Cavities within the central nervous system containing cerebrospinal fluid and lined by ependymal cells.

voltage clamp Technique for displacing membrane potential abruptly to a desired value and keeping the potential constant while measuring currents across the cell membrane; devised by Cole and Marmont.

white matter Part of the central nervous system appearing white; consisting of myelinated fiber tracts.

BIBLIOGRAPHY

Note: The bracketed numbers alongside of the citations are the numbers of the text pages on which footnote references or figure attributions occur. If there are no numbers, the entry appears only within an end-of-chapter Suggested Reading list.

Adrian, E. D. 1946. *The Physical Background of Perception.* Clarendon Press, Oxford. [5, 21]

Alexandrowicz, J. S. 1951. Muscle receptor organs in the abdomen of *Homarus vulgaris* and *Palinurus vulgaris. Q. J. Microsc. Sci. 92*:163–199. [309]

Alnaes, E., and Rahamimoff, R. 1975. On the role of mitochondria in transmitter release from motor nerve terminals. *J. Physiol. 248*:285–306. [184]

Ames, A., Higashi, K., and Nesbett, F. B. 1965. Relation of potassium concentration in choroid-plexus fluid to that in plasma. *J. Physiol. 181*: 506–515. [301]

Anderson, C. R., and Stevens, C. F. 1973. Voltage clamp analysis of acetylcholine produced end-plate current fluctuations at frog neuromuscular junction. *J. Physiol. 235*:655–691. [208]

Angeletti, R. H., Mercanti, D., and Bradshaw, R. A. 1973. Amino acid sequences of mouse 2.5 S nerve growth factor. I. Isolation and characterization of the soluble tryptic and chymotryptic peptides. *Biochemistry 12*:90–100. [391]

Armstrong, C. M., and Hille, B. 1972. The inner quaternary ammonium ion receptor in potassium channels of the node of Ranvier. *J. Gen. Physiol. 59*:388–400. [116]

Armstrong, C. M., Bezanilla, F., and Rojas, E. 1973. Destruction of sodium conductance inactivation in squid axons perfused with pronase. *J. Gen. Physiol. 62*:375–391. [120]

Armstrong, C. M., and Bezanilla, F. 1974. Charge movement associated with the opening and closing of the activation gates of Na channels. *J. Gen. Physiol. 63*:533–552. [127, 129]

Asanuma, H. 1975. Recent developments in the study of the columnar arrangement of neurons within the motor cortex. *Physiol. Rev. 55*:143–156. [72]

Attardi, G., and Sperry, R. W. 1963. Preferential selection of central pathways by regenerating optic fibers. *Exp. Neurol. 7*:46–64. [393]

Atwood, H. L. 1972. Crustacean Muscle. In G. H. Bourne (ed.). *The Structure and Function of Muscle,* Vol. I. Acadamic Press, New York, pp. 421–489.

[173, 338] Atwood, H. L., and Morin, W. A. 1970. Neuromuscular and axoaxonal synapses of the crayfish opener muscle. *J. Ultrastruct. Res.* *32*:351–369.

[339, 366] Atwood, H. L., and Bittner, G. D. 1971. Matching of excitatory and inhibitory inputs to crustacean muscle fibers. *J. Neurophysiol.* *34*:157–170.

[379, 380] Axelsson, J., and Thesleff, S. 1959. A study of supersensitivity in denervated mammalian skeletal muscle. *J. Physiol.* *147*:178–193.

[415] Baker, F. H., Grigg, P., and van Noorden, G. K. 1974. Effects of visual deprivation and strabismus on the response of neurons in the visual cortex of the monkey, including studies on the striate and prestriate cortex in the normal animal. *Brain Res.* *66*:185–208.

[247] Baker, P. F. 1972. Transport and metabolism of calcium ions in nerve. *Prog. Biophys. Mol. Biol.* *24*:177–223.

[92, 93] Baker, P. F., Hodgkin, A. L., and Shaw, T. I. 1962a. Replacement of the axoplasm of giant nerve fibres with artificial solutions. *J. Physiol.* *164*: 330–354.

[92, 93] Baker, P. F., Hodgkin, A. L., and Shaw, T. I. 1962b. The effects of changes in internal ionic concentrations on the electrical properties of perfused giant axons. *J. Physiol.* *164*:355–374.

[240] Baker, P. F., Blaustein, M. P., Keynes, R. D., Manil, J., Shaw, T. I., and Steinhardt, R. A. 1969. The ouabain-sensitive fluxes of sodium and potassium in squid giant axon. *J. Physiol.* *200*:459–496.

[249] Baker, P. F., Blaustein, M. P., Hodgkin, A. L., and Steinhardt, R. A. 1969. The influence of calcium on sodium efflux in squid axons. *J. Physiol.* *200*:431–458.

[91, 123, 247, 248] Baker, P. F., Hodgkin, A. L., and Ridgway, E. B. 1971. Depolarization and calcium entry in squid giant axons. *J. Physiol.* *218*:709–755.

[241] Baker, P. F., Foster, R. F., Gilbert, D. S., and Shaw, T. I. 1971. Sodium transport by perfused giant axons of *Loligo*. *J. Physiol.* *219*:487–506.

[312] Barker, D., Stacey, M. J., and Adal, M. N. 1970. Fusimotor innervation in the cat. *Philos. Trans. R. Soc. Lond.* (Ser. B) *258*:315–346.

[322] Barker, D., Emonet-Dénand, F., Laporte, Y., Proske, U., and Stacey, M. J. 1973. Morphological identification and intrafusal distribution of the endings of static fusimotor axons in the cat. *J. Physiol.* *230*:405–427.

[25] Barlow, H. B., Hill, R. M., and Levick, W. R. 1964. Retinal ganglion cells responding selectively to direction and speed of image motion in the rabbit. *J. Physiol.* *173*:377–407.

[43] Barlow, H. B., and Levick, W. R. 1965. The mechanism of directionally selective units in rabbit's retina. *J. Physiol.* *178*:477–504.

[384] Barnard, E. A., Wieckowski, J., and Chiu, T. H. 1971. Cholinergic receptor molecules and cholinesterase molecules at mouse skeletal muscle junction. *Nature* *234*:207–209.

[205] Barnard, E. A., Dolly, J. O., Porter, C. W., and Albuquerque, E. X. 1975. The acetylcholine receptor and the ionic conductance modulation system of skeletal muscle. *Exp. Neurol.* *48*:1–28.

Barrett, J. N., and Crill, W. E. 1974a. Specific membrane properties of cat motoneurones. *J. Physiol.* *239*:301–324.

[350] Barrett, J. N., and Crill, W. E. 1974b. Influence of dendritic location and

membrane properties on the effectiveness of synapses on cat motoneurones. *J. Physiol.* *239*:325–345.

Barron, D. H., and Matthews, B. H. C. 1935. Intermittent conduction in the spinal cord. *J. Physiol.* *85*:73–103. [369]

Baylor, D. A., and Nicholls, J. G. 1969. Changes in extracellular potassium concentration produced by neuronal activity in the central nervous system of the leech. *J. Physiol.* *203*:555–569. [280, 281]

Baylor, D. A., and Nicholls, J. G. 1971. Patterns of regeneration between individual nerve cells in the central nervous system of the leech. *Nature* *232*:268–270. [394]

Baylor, D. A., Fuortes, M. G. F., and O'Bryan, P. M. 1971. Receptive fields of cones in the retina of the turtle. *J. Physiol.* *214*:265–294. [27, 28, 29]

Baylor, D. A., and Hodgkin, A. L. 1973. Detection and resolution of visual stimuli by turtle photoreceptors. *J. Physiol.* *234*:163–198.

Beam, K. G., and Greengard, P. 1976. Protein phosphorylation and synaptic function. *Cold Spring Harbor Symp. Quant. Biol.* *40*:157–167. [175]

Bennett, M. R., Florin, T., and Woog, R. 1974. The formation of synapses in regenerating mammalian striated muscle. *J. Physiol.* *238*:79–92. [386]

Bennett, M. V. L. 1973. Function of electronic junctions in embryonic and adult tissues. *Fed. Proc.* *32*:65–75. [157]

Bennett, M. V. L. (ed.). 1974. *Synaptic Transmission and Neuronal Interaction.* Raven Press, New York. [157]

Bennett, M. V. L. 1974. Flexibility and Rigidity in Electronically Coupled Systems. In M. V. L. Bennett (ed.). *Synaptic Transmission and Neuronal Interaction.* Raven Press, New York, pp. 153–178.

Benzer, S. 1971. From the gene to behavior *J.A.M.A.* *218*:1015–1022. [374]

Berg, D. K., Kelly, R. B., Sargent, P. B., Williamson, P., and Hall, Z. W. 1972. Binding of α-bungarotoxin to acetylcholine receptors in mammalian muscle. *Proc. Natl. Acad. Sci. U.S.A.* *69*:147–151. [384]

Berg, D. K., and Hall, Z. W. 1975. Increased extrajunctional acetylcholine sensitivity produced by chronic postsynaptic neuromuscular blockade. *J. Physiol.* *244*:659–676. [379, 384]

Berlucchi, G., Gazzaniga, M. S., and Rizzolatti, G. 1967. Microelectrode analysis of visual information by the corpus callosum. *Arch. Ital. Biol.* *105*:583–596. [51]

Berlucchi, G., and Rizzolatti, G. 1968. Binocularly driven neurons in visual cortex of split-chiasm cats. *Science* *159*:308–310. [51]

Bernstein, J. 1902. Untersuchungen zur Thermodynamik der bioelektrischen Ströme. *Pflügers Arch. Physiol.* *92*:521–562. [89]

Bessou, P., and Laporte, Y. 1962. Responses from Primary and Secondary Endings of the Same Neuromuscular Spindle of the Tenuissimus Muscle of the Cat. In D. Barker (ed.). *Symposium on Muscle Receptors.* Hong Kong University Press, Hong Kong. [314]

Bignami, A., and Dahl, D. 1974. Astrocyte-specific protein and neuroglial differentiation: An immunofluorescence study with antibodies to the glial fibrillary acidic protein. *J. Comp. Neurol.* *153*:27–38. [259]

[197] Birks, R. I. 1974. The relationship of transmitter release and storage to fine structure in a sympathetic ganglion. *J. Neurocytol.* *3*:133–160.

[183] Birks, R., Katz, B., and Miledi, R. 1960. Physiological and structural changes at the amphibian myoneural junction in the course of nerve degeneration. *J. Physiol.* *150*:145–168.

[190] Birks, R., and MacIntosh, F. C. 1961. Acetylcholine metabolism of a sympathetic ganglion. *Can. J. Biochem. Physiol.* *39*:787–827.

[42] Bishop, P. O., Coombs, J. S., and Henry, G. H. 1971. Responses to visual contours: Spatio-temporal aspects of excitation in the receptive fields of simple striate neurones. *J. Physiol.* *219*:625–657.

[52] Bishop, P. O., Coombs, J. S., and Henry, G. H. 1973. Receptive fields of simple cells in the cat striate cortex. *J. Physiol.* *231*:31–60.

[406] Blakemore, C. 1974. Development of functional connexions in the mammalian visual system. *Br. Med. Bull.* *30*:152–157.

[416] Blakemore, C., and Cooper, G. F. 1970. Development of the brain depends on the visual environment. *Nature* *228*:477–478.

[409] Blakemore, C., and Van Sluyters, R. C. 1974. Reversal of the physiological effects of monocular deprivation in kittens: Further evidence for a sensitive period. *J. Physiol.* *237*:195–216.

[249] Blaustein, M. F., and Hodgkin, A. L. 1969. The effect of cyanide on the efflux of calcium from squid axons. *J. Physiol.* *200*:497–528.

[194] Bloom, F. E. 1972. Electron microscopy of catecholamine-containing structures. *Handbook Exp. Pharmacol.* *33*:46–78.

[301] Bodenheimer, T. S., and Brightman, M. W. 1968. A blood-brain barrier to peroxidase in capillaries surrounded by perivascular spaces. *Am. J. Anat.* *122*:249–268.

[340] Bodian, D. 1937. The structure of the vertebrate synapse: A study of the axon endings on Mauthner's cell and neighboring centers in the goldfish. *J. Comp. Neurol.* *68*:117–159.

[337, 343] Bodian, D. 1942. Cytological aspects of synaptic function. *Physiol. Rev.* *22*: 146–169.

[266] Bodian, D. 1966. Development of fine structure of spinal cord in monkey fetuses. I. The motoneuron neuropil at the time of onset of reflex activity. *Bull. Johns Hopkins Hosp.* *119*:129–149.

[229] Boistel, J., and Fatt, P. 1958. Membrane permeability change during inhibitory transmitter action in crustacean muscle. *J. Physiol.* *144*:176–191.

[265] Bowery, N. G., and Brown, D. A. 1972. γ-Aminobutyric acid uptake by sympathetic ganglia. *Nature (New Biol.)* *238*:89–91.

[20] Boycott, B. B., and Dowling, J. E. 1969. Organization of primate retina: Light microscopy. *Philos. Trans. R. Soc. Lond.* (*Ser.* B) *255*:109–184.

[312] Boyd, I. A. 1962. The structure and innervation of the nuclear bag muscle fibre system and the nuclear chain muscle fibre system in mammalian muscle spindles. *Philos. Trans. R. Soc. Lond.* (*Ser.* B) *245*:81–136.

[188] Boyd, I. A., and Martin, A. R. 1956. The end-plate potential in mammalian muscle. *J. Physiol.* *132*:74–91.

[90] Boyle, P. J., and Conway, E. J. 1941. Potassium accumulation in muscle and associated changes. *J. Physiol.* *100*:1–63.

Braitenberg, V., and Atwood, R. P. 1958. Morphological observations on the cerebellar cortex. *J. Comp. Neurol. 109*:1–33. [10]

Brightman, M. W. 1965. The distribution within the brain of ferritin injected into cerebrospinal fluid compartments. II. Parenchymal distribution. *Am. J. Anat. 117*:193–219.

Brightman, M. W., and Reese, T. S. 1969. Junctions between intimately apposed cell membranes in the vertebrate brain. *J. Cell Biol. 40*:648–677. [273, 293]

Brightman, M. W., Reese, T. S., and Feder, N. 1970. Assessment with the Electronmicroscope of the Permeability to Peroxidase of Cerebral Endothelium in Mice and Sharks. In E. H. Thaysen (ed.). *Capillary Permeability.* Alfred Benzon Symposium II. Munksgaard, Copenhagen. [294]

Brightman, M. W., Klatzo, I., Olsson, Y., and Reese, T. S. 1970. The blood-brain barrier to proteins under normal and pathological conditions. *J. Neurol. Sci. 10*:215–239.

Brockes, J. P., and Hall, Z. W. 1975. Acetylcholine receptors in normal and denervated rat diaphragm muscle. II. Comparison of junctional and extrajunctional receptors. *Biochemistry 14*:2100–2106. [384]

Brody, H. 1955. Organization of the cerebral cortex. IV. A study of aging in the human cerebral cortex. *J. Comp. Neurol. 102*:511–556. [376]

Brostoff, S. W., Sacks, H., and Di Paolo, C. 1975. The P_2 protein of bovine root myelin: Partial chemical characterization. *J. Neurochem. 24*:289–294. [259]

Brown, G. L. 1937. The actions of acetylcholine on denervated mammalian and frog's muscle. *J. Physiol. 89*:438–461. [378]

Brown, M. C., and Butler, R. G. 1973. Studies on the site of termination of static and dynamic fusimotor fibres within muscle spindles of the tenuissimus muscle of the cat. *J. Physiol. 233*:553–573. [322]

Buller, A. J. 1970. The neural control of the contractile mechanism in skeletal muscle. *Endeavour 29*:107–111. [389]

Buller, A. J., Eccles, J. C., and Eccles, R. M. 1960. Differentiation of fast and slow muscles in the cat hind limb. *J. Physiol. 150*:399–416. [389]

Bullock, T. H., and Hagiwara, S. 1957. Intracellular recording from the giant synapse of the squid. *J. Gen. Physiol. 40*:565–577. [178]

Bullock, T. H., and Horridge, G. A. 1965. *Structure and Function in the Nervous Systems of Invertebrates.* W. H. Freeman, San Francisco.

Bunge, R. P. 1968. Glial cells and the central myelin sheath. *Physiol. Rev. 48*:197–251. [259, 260]

Burden, S., Hartzell, H. C., and Yoshikami, D. 1975. Acetylcholine receptors at neuromuscular synapses: Phylogenetic differences detected by snake α-neurotoxins. *Proc. Natl. Acad. Sci. U.S.A. 72*:3245–3249. [204]

Burgess, P. R., English, K. B., Horch, K. W., and Stensaas, L. J. 1974. Patterning in the regeneration of type I cutaneous receptors. *J. Physiol. 236*:57–82. [392]

Burke, R. E., and Ten Bruggencate, G. 1971. Electronic characteristics of alpha motoneurones of varying size. *J. Physiol. 212*:1–20. [350]

Burnstock, G. 1972. Purinergic nerves. *Pharmacol. Rev. 24*:509–581. [234]

Byrne, J., Castellucci, V., and Kandel, E. R. 1974. Receptive fields and response properties of mechanoreceptor neurons innervating siphon skin and mantle shelf in *Aplysia. J. Neurophysiol. 37*:1041–1064. [361]

[241] Caldwell, P. C., Hodgkin, A. L., Keynes, R. D., and Shaw, T. I. 1960. The effects of injecting "energy-rich" phosphate compounds on the active transport of ions in the giant axons of *Loligo. J. Physiol. 152*:561–590.

[381] Cangiano, A. 1973. Acetylcholine supersensitivity: The role of neurotrophic factors. *Brain Res. 58*:255–259.

[378] Cannon, W. B., and Rosenblueth, A. 1949. *The Supersensitivity of Denervated Structures: A Law of Denervation.* Macmillan, New York.

[158] Cantino, D., and Mugnaini, E. 1975. The structural basis for electronic coupling in the avian cilary ganglion: A study with thin sectioning and freeze-fracturing. *J. Neurocytol. 4*:505–536.

[355] Carew, T. J., Pinsker, H. M., and Kandel, E. R. 1972. Long-term habituation of a defensive withdrawal reflex in *Aplysia. Science 175*:451–454.

Carew, T. J., and Kandel, E. R. 1974. Synaptic Analysis of the Interrelationships Between Behavioral Modifications in *Aplysia.* In M. V. L. Bennett (ed.). *Synaptic Transmission and Neuronal Interaction.* Raven Press, New York, pp. 339–383.

[285] Castellucci, V. F., and Goldring, S. 1970. Contribution to steady potential shifts of slow depolarization in cells presumed to be glia. *Electroencephalogr. Clin. Neurophysiol. 28*:109–118.

[197] Ceccarelli, B., Hurlbut, W. P., and Mauro, A. 1973. Turnover of transmitter and synaptic vesicles at the frog neuromuscular junction. *J. Cell Biol. 57*:499–524.

[190] Ceccarelli, B., and Hurlbut, W. P. 1975. The effects of prolonged repetitive stimulation in hemicholinium on the frog neuromuscular junction. *J. Physiol. 247*:163–188.

[125] Chandler, W. K., and Meves, H. 1965. Voltage clamp experiments on internally perfused giant axons. *J. Physiol. 180*:788–820.

[51] Choudhury, B. P., Whitteridge, D., and Wilson, M. E. 1965. The function of the callosal connections of the visual cortex. *Q. J. Exp. Physiol. 50*:214–219.

[411] Chow, K. L., and Stewart, D. L. 1972. Reversal of structural and functional effects of long-term visual deprivation in cats. *Exp. Neurol. 34*:409–433.

[395] Chow, K. L., Mathers, L. H., and Spear, P. D. 1973. Spreading of uncrossed retinal projection in superior colliculus of neonatally enucleated rabbits. *J. Comp. Neurol. 151*:307–322.

[189] Christensen, B. N., and Martin, A. R. 1970. Estimates of probability of transmitter release at the mammalian neuromuscular junction. *J. Physiol. 210*:933–945.

[25] Cleland, B. G., Dubin, M. W., and Levick, W. R. 1971. Sustained and transient neurones in the cat's retina and lateral geniculate nucleus. *J. Physiol. 217*:473–496.

[302] Clementi, F., and Palade, G. E. 1969. Intestinal capillaries. I. Permeability to peroxidase and ferritin. *J. Cell Biol. 41*:33–58.

[389] Close, R. I. 1972. Dynamic properties of mammalian skeletal muscles. *Physiol. Rev. 52*:129–197.

[269, 358] Coggeshall, R. E., and Fawcett, D. W. 1964. The fine structure of the central nervous system of the leech, *Hirudo medicinalis. J. Neurophysiol. 27*:229–289.

Cohen, J. B., and Changeux, J.-P. 1975. The cholinergic receptor protein in its membrane environment. *Annu. Rev. Pharmacol. 15*:83–103. [212]

Cohen, M. W., 1970. The contribution by glial cells to surface recordings from the optic nerve of an amphibian. *J. Physiol. 210*:565–580. [283, 284]

Cohen, M. W. 1972. The development of neuromuscular connexions in the presence of D-tubocurarine. *Brain Res. 41*:457–463. [388]

Cohen, M. W., Gerschenfeld, H. M., and Kuffler, S. W. 1968. Ionic environment of neurones and glial cells in the brain of an amphibian. *J. Physiol. 197*:363–380. [299, 300]

Cohen, S. 1959. Purification and metabolic effects of a nerve growth-promoting protein from snake venom. *J. Biol. Chem. 234*:1129–1137. [391]

Cohen, S. 1960. Purification of a nerve-growth promoting protein from the mouse salivary gland and its neuro-cytotoxic antiserum. *Proc. Natl. Acad. Sci. U.S.A. 46*:302–311. [391]

Cole, K. S. 1968. *Membranes, Ions and Impulses*. University of California Press, Berkeley. [109]

Collier, B., and MacIntosh, F. C. 1969. The source of choline for acetylcholine synthesis in a sympathetic ganglion. *Can. J. Physiol. Pharmacol. 47*:127–135. [194, 235]

Colquhoun, D., Henderson, R., and Ritchie, J. M. 1972. The binding of labelled tetrodotoxin to non-myelinated nerve fibres. *J. Physiol. 227*: 95–126. [126]

Cooke, J. D., and Quastel, D. M. J. 1973. The specific effect of potassium on transmitter release by motor nerve terminals and its inhibition by calcium. *J. Physiol 228*:435–458. [287]

Coombs, J. S., Eccles, J. C., and Fatt, P. 1955. The specific ion conductances and the ionic movements across the motoneuronal membrane that produce the inhibitory post-synaptic potential. *J. Physiol. 130*:326–373. [169, 171]

Cooper, J. R., Bloom, F. E., and Roth, R. E. 1974. *The Biochemical Basis of Neuropharmacology*, 2nd ed. Oxford University Press, London.

Cottrell, G. A., and Macon, J. B. 1974. Synaptic connexions of two symmetrically placed giant serotonin-containing neurones. *J. Physiol. 236*: 435–464. [223]

Couteaux, R. 1974. Remarks on the organization of axon terminals in relation to secretory processes. *Adv. Cytopharmacol. 2*:369–379.

Couteaux, R., and Pécot-Dechavassine, M. 1970. Vesicules synaptiques et poches au niveau des zones actives de la jonction neuromusculaire. *C. R. Acad. Sci. (Paris) 271*:2346–2349. [196, 199]

Cowan, W. M. 1970. Centrifugal fibres to the avian retina. *Br. Med. Bull. 26*:112–118. [329]

Cowan, W. M. 1973. Neuronal Death as a Regulation Mechanism in the Control of Cell Number in the Nervous System. In M. Rockstein (ed.). *Development and Aging in the Nervous System*. Academic Press, New York, pp. 19–41. [376]

Cowan, W. M., and Powell, T. P. S. 1963. Centrifugal fibres in the avian visual system. *Proc. R. Soc. Lond.* B *158*:232–252. [31]

Craik, K. 1943. *The Nature of Explanation*. Cambridge University Press, London. [57]

[327, 328] Critchlow, V., and von Euler, C. 1963. Intercostal muscle spindle activity and its γ-motor control. *J. Physiol* *168*:820–847.

[322] Crowe, A., and Matthews, P. B. C. 1964. The effects of stimulation of static and dynamic fusimotor fibres on the response to stretching of the primary endings of muscle spindles. *J. Physiol.* *174*:109–131.

[301] Cserr, H. F. 1971. Physiology of the choroid plexus. *Physiol. Rev.* *51*:273–311.

[299] Cserr, H., and Rall, D. P. 1967. Regulation of cerebrospinal fluid (K^+) in the spiny dogfish, *Squalus acanthias*. *Comp. Biochem. Physiol.* *21*:431–434.

[350] Curtis, D. R., and Eccles, J. C. 1959. Repetitive synaptic activation. *J. Physiol.* *149*:43–44P.

[230] Curtis, D. R., and Johnston, G. A. R. 1974. Amino acid transmitters in the mammalian central nervous system. *Ergeb. Physiol.* *69*:97–188.

[98] Curtis, H. J., and Cole, K. S. 1940. Membrane action potentials from the squid giant axon. *J. Cell. Comp. Physiol* *15*:147–157.

[243] Dahl, J. L., and Hokin, L. E. 1974. The sodium-potassium adenosine triphosphatase. *Annu. Rev. Biochem.* *43*:327–356.

[235] Dahlström, A. 1971. Axoplasmic transport (with particular respect to adrenergic neurons). *Philos. Trans. R. Soc. Lond. (Ser. B)* *261*:325–358.

Dahlström, A. 1973. Aminergic transmission: Introduction and short review. *Brain Res.* *62*:441–460.

[147] Dale, H. H. 1953. *Adventures in Physiology*. Pergamon Press, London.

[148] Dale, H. H., Feldberg, W., and Vogt, M. 1936. Release of acetylcholine at voluntary motor nerve endings. *J. Physiol.* *86*:353–380.

[38] Daniel, P. M., and Whitteridge, D. 1961. The representation of the visual field on the cerebral cortex in monkeys. *J. Physiol.* *159*:203–221.

Davson, H. 1972. The Blood-Brain Barrier. In G. H. Bourne (ed.). *The Structure and Function of Nervous Tissue,* Vol. IV. Academic Press, New York, pp. 321–345.

Daw, N. W. 1973. Neurophysiology of color vision. *Physiol. Rev.* *53*:571–611.

[230] De Groat, W. C. 1972. GABA-depolarization of a sensory ganglion: Antagonism by picrotoxin and bicuculline. *Brain Res.* *38*:429–432.

[185, 186] del Castillo, J., and Katz, B. 1954a. Quantal components of the end-plate potential. *J. Physiol.* *124*:560–573.

[189] del Castillo, J., and Katz, B. 1954b. Statistical factors involved in neuromuscular facilitation and depression. *J. Physiol.* *124*:574–585.

[184] del Castillo, J., and Katz, B. 1954c. Changes in the end-plate activity produced by presynaptic polarization. *J. Physiol.* *124*:586–604.

[164] del Castillo, J., and Katz, B. 1956. Biophysical aspects of neuro-muscular transmission. *Prog. Byophys.* *6*:121–170.

[273] Dennis, M. J., and Gerschenfeld, H. M. 1969. Some physiological properties of identified mammalian neuroglial cells. *J. Physiol.* *203*:211–222.

[151, 152, 216] Dennis, M. J., Harris, A. J., and Kuffler, S. W. 1971. Synaptic transmission and its duplication by focally applied acetylcholine in parasympathetic neurons in the heart of the frog. *Proc. R. Soc. Lond.* B *177*:509–539.

Dennis, M. J., and Miledi, R. 1974a. Electrically induced release of acetylcholine from denervated Schwann cells. *J. Physiol.* *237*:431–452. [265]

Dennis, M. J., and Miledi, R. 1974b. Characteristics of transmitter release at regenerating frog neuromuscular junctions. *J. Physiol.* *239*:571–594. [189]

DeRobertis, E. 1967. Ultrastructure and cytochemistry of the synaptic region. *Science* *156*:907–914. [194]

de Santillana, G. 1965. Alessandro Volta. *Sci. Am.* *212*:82–91. [89]

Diamond, J. 1968. The activation and distribution of GABA and L-glutamate receptors on goldfish Mauthner neurones: An analysis of dendritic remote inhibition. *J. Physiol.* *194*:669–723. [341, 344]

Diamond, J. 1971. The Mauthner Cell. In W. S. Hoar and D. J. Randall (eds.). *Fish Physiology,* Vol. 5. Academic Press, New York, pp. 265–346.

Diamond, J., and Miledi, R. 1962. A study of foetal and new-born muscle fibres. *J. Physiol.* *162*:393–408. [387]

Diamond, J., Roper, S., and Yasargil, G. M. 1973. The membrane effects, and sensitivity to strychnine, of neural inhibition of the Mauthner cell, and its inhibition by glycine and GABA. *J. Physiol.* *232*:87–111. [233]

Dodge, F. A., and Rahamimoff, R. 1967. Co-operative action of calcium ions in transmitter release at the neuromuscular junction. *J. Physiol.* *193*: 419–432.

Douglas, W. W. 1968. Stimulus-secretion coupling: The concept and clues from chromaffin and other cells. *Br. J. Pharmacol.* *34*:451–474. [180]

Dowling, J. E., and Boycott, B. B. 1966. Organization of the primate retina: Electron microscopy. *Proc. R. Soc. Lond.* B *166*:80–111. [26, 27]

Dowling, J. E., and Werblin, F. S. 1971. Synaptic organization of the vertebrate retina. *Vision Res.* *3*:1–15. [27]

Drujan, B. D., and Svaetichin, G. 1972. Characterization of different classes of isolated retinal cells. *Vision Res.* *12*:1777–1784. [32]

Du Bois-Reymond, E. 1848. Untersuchungen über thierische Electricität (Erster Band). Reimer, Berlin. [146]

Dudel, J. 1974. Nonlinear voltage dependence of excitatory synaptic current in crayfish muscle. *Pflügers Arch.* *352*:227–241. [166]

Dudel, J., and Kuffler, S. W. 1960. Excitation at the crayfish neuromuscular junction with decreased membrane conductance. *Nature* *187*:246–247. [174]

Dudel, J., and Kuffler, S. W. 1961. Presynaptic inhibition at the crayfish neuromuscular junction. *J. Physiol.* *155*:543–562. [171, 172, 189, 190, 229, 340]

Eccles, J. C. 1964. *The Physiology of Synapses.* Springer Verlag, Berlin. [172, 347]

Eccles, J. C., and Sherrington, C. S. 1930. Numbers and contraction-values of individual motor-units examined in some muscles of the limb. *Proc. R. Soc. Lond.* B *106*:326–357. [320]

Eccles, J. C., and O'Connor, W. J. 1939. Responses which nerve impulses evoke in mammalian striated muscles. *J. Physiol.* *97*:44–102. [148]

Eccles, J. C., Katz, B., and Kuffler, S. W. 1942. Effect of eserine on neuromuscular transmission. *J. Neurophysiol.* *5*:211–230. [150, 151]

Eccles, J. C., Fatt, P., and Koketsu, K. 1954. Cholinergic and inhibitory synapses in a pathway from motor-axon collaterals to motoneurones. *J. Physiol.* *126*:524–562. [222]

[171] Eccles, J. C., Eccles, R. M., and Magni, F. 1961. Central inhibitory action attributable to presynaptic depolarization produced by muscle afferent volleys. *J. Physiol. 159*:147–166.

[10] Eccles, J. C., Ito, M., and Szentágothai, J. 1967. *The Cerebellum as a Neuronal Machine.* Springer Verlag, Berlin.

[310] Edwards, C., and Ottoson, D. 1958. The site of impulse initiation in a nerve cell of a crustacean stretch receptor. *J. Physiol. 143*:138–148.

[312] Edwards, C., Terzuolo, C. A., and Washizu, Y. 1963. The effect of changes of the ionic environment upon an isolated crustacean sensory neuron. *J. Neurophysiol. 26*:948–957.

[220] Eisenstadt, M. L., and Schwartz, J. H. 1975. Metabolism of acetylcholine in the nervous system of *Aplysia californica. J. Gen. Physiol. 65*:293–313.

[147] Elliot, T. R. 1904. On the action of adrenalin. *J. Physiol. 31*:20.

[25] Enroth-Cugell, C., and Robson, J. G. 1966. The contrast sensitivity of retinal ganglion cells of the cat. *J. Physiol. 187*:517–552.

[322] Eyzaguirre, C. 1957. Functional organization of neuromuscular spindle in toad. *J. Neurophysiol. 20*:523–542.

[309, 310, 311] Eyzaguirre, C., and Kuffler, S. W. 1955. Processes of excitation in the dendrites and in the soma of single isolated sensory nerve cells of the lobster and crayfish. *J. Gen. Physiol. 39*:87–119.

[223] Falck, B., Hillarp, N.-Å., Thieme, G., and Thorp, A. 1962. Fluorescence of catecholamines and related compounds condensed with formaldehyde. *J. Histochem. Cytochem. 10*:348–354.

[384] Fambrough, D. M. 1970. Acetylcholine sensitivity of muscle fiber membranes: Mechanism of regulation by motoneurons. *Science 168*:372–373.

[205] Fambrough, D. M. 1974. Acetylcholine receptors: Revised estimates of extrajunctional receptor density in denervated rat diaphragm. *J. Gen. Physiol. 64*:468–572.

[190] Fambrough, D. M., Drachman, D. B., and Satyamurti, S. 1973. Neuromuscular junction in myasthenia gravis: Decreased acetylcholine receptors. *Science 182*:293–295.

[153] Fatt, P., and Katz, B. 1951. An analysis of the end-plate potential recorded with an intra-cellular electrode. *J. Physiol. 115*:320–370.

[183, 184] Fatt, P., and Katz, B. 1952. Spontaneous subthreshold activity at motor nerve endings. *J. Physiol 117*:109–128.

[171, 339] Fatt, P., and Katz, B. 1953. The effect of inhibitory nerve impulses on a crustacean muscle fibre. *J. Physiol. 121*:374–389.

[148] Feldberg, W. 1945. Present views on the mode of action of acetylcholine in the central nervous system. *Physiol. Rev. 25*:596–642.

[205] Fertuck, H. C., and Salpeter, M. M. 1974. Localization of acetylcholine receptor by ^{125}I-labeled α-bungarotoxin binding at mouse motor endplates. *Proc. Natl. Acad. Sci. U.S.A. 71*:1376–1378.

[329] Fex, J. 1962. Auditory activity in centrifugal and centripetal cochlear fibres in cat. *Acta Physiol. Scand. 55* (Suppl. 189): 1–68.

Fex, J. 1974. Neural Excitatory Processes of the Inner Ear. In W. D. Keidel and W. D. Neff (eds.). *Handbook of Sensory Physiology,* Vol. V, *Auditory System.* Springer Verlag, New York, pp. 585–646.

Fex, S., Sonessin, B., Thesleff, S., and Zelená, J. 1966. Nerve implants in botulinum poisoned mammalian muscle. *J. Physiol. 184*:872–882. [386]

Fields, H. L., Evoy, W. H., and Kennedy, D. 1967. Reflex role played by efferent control of an invertebrate stretch receptor. *J. Neurophysiol 30*: 859–874. [320]

Fields, W. S. (ed.). 1971. Myasthenia gravis. *Ann. N.Y. Acad. Sci. 183*:1–386. [190]

Fischbach, G. D. 1972. Synapse formation between dissociated nerve and muscle cells in low density cell cultures. *Dev. Biol. 28*:407–429. [375]

Fischbach, G. D., and Dichter, M. A. 1974. Electrophysiologic and morphologic properties of neurons in dissociated chick spinal cord cell cultures. *Dev. Biol. 37*:100–116. [266, 375]

Fleischhauer, K. 1972. Ependyma and Subependymal Layer. In G. H. Bourne (ed.). *The Structure and Function of Nervous Tissue,* Vol. IV. Academic Press, New York, pp. 1–46.

Flock, Å., and Russell, I. J. 1973. The post-synaptic action of efferent fibres in the lateral line organ of the burbot *Lota lota. J. Physiol. 235*:591–605. [329]

Flock, Å., and Lam, D. M. K. 1974. Neurotransmitter synthesis in inner ear and lateral line sense organs. *Nature 249*:142–144. [329]

Florey, E. 1961. Comparative physiology: Transmitter substances. *Annu. Rev. Physiol. 23*:501–528. [223]

Frank, E. 1973. Matching of facilitation at the neuromuscular junction of the lobster: A possible case for influence of muscle on nerve. *J. Physiol. 233*:635–658. [366]

Frank, E., Jansen, J. K. S., Lømo, T., and Westgaard, R. H. 1975. The interaction between foreign and original nerves innervating the soleus muscle of rats. *J. Physiol. 247*:725–743. [389]

Frank, E., Jansen, J. K. S., and Rinvik, E. 1975. A multisomatic axon in the central nervous system of the leech. *J. Comp. Neurol. 159*:1–14. [394]

Frank, K., and Fuortes, M. G. F. 1957. Presynaptic and postsynaptic inhibition of monosynaptic reflexes. *Fed. Proc. 16*:39–40. [171]

Frankenhaeuser, B., and Hodgkin, A. L. 1956. The after-effects of impulses in the giant nerve fibres of *Loligo. J. Physiol. 131*:341–376. [277]

Frankenhaeuser, B., and Hodgkin, A. L. 1957. The action of calcium on the electrical properties of squid axons. *J. Physiol. 137*:218–244. [123]

Fukami, Y., and Hunt, C. C. 1970. Structure of snake muscle spindles. *J. Neurophysiol. 33*:9–27. [323]

Fuller, J. L. 1967. Experiential deprivation and later behavior. *Science 158*: 1645–1652. [417]

Fuortes, M. G. F., and Poggio, G. F. 1963. Transient responses to sudden illumination in cells of the eye of *Limulus. J. Gen. Physiol. 46*:435–452. [29]

Furshpan, E. J. 1964. "Electrical transmission" at an excitatory synapse in a vertebrate brain. *Science 144*:878–880. [344]

Furshpan, E. J., and Potter, D. D. 1959. Transmission at the giant motor synapses of the crayfish. *J. Physiol. 145*:289-325. [154, 156]

Furshpan, E. J., and Furukawa, T. 1962. Intracellular and extracellular responses of the several regions of the Mauthner cell of the goldfish. *J. Neurophysiol. 25*:732–771. [340]

[161] Furshpan, E. J., and Potter, D. D. 1968. Low resistance junctions between cells in embryos and tissue culture. *Curr. Top. Dev. Biol. 3*:95–127.

[342, 346] Furukawa, T. 1966. Synaptic interaction at the Mauthner cell of goldfish. *Prog. Brain Res. 21A*:44–70.

[173, 345, 346] Furukawa, T., and Furshpan, E. J. 1963. Two inhibitory mechanisms in the Mauthner neurons of goldfish. *J. Neurophysiol. 26*:140–176.

[342] Furukawa, T., Fukami, Y., and Asada, Y. 1965. A third type of inhibition in the Mauthner cell of goldfish. *J. Neurophysiol. 26*:759–774.

[375] Garber, B. B., and Moscona, A. A. 1972. Reconstruction of brain tissue from cell suspensions. I. Aggregation patterns of cells dissected from different regions of the developing brain. *Dev. Biol. 27*:217–234.

[57] Garey, L. J., and Powell, T. P. S. 1971. An experimental study of the termination of the lateral geniculo-cortical pathway in the cat and monkey. *Proc. R. Soc. Lond.* B *179*:41–63.

[242] Garrahan, P. J., and Glynn, I. M. 1967. The incorporation of inorganic phosphate into adenosine triphosphate by reversal of the sodium pump. *J. Physiol. 192*:237–256.

[393] Gaze, R. M. 1970. *The Formation of Nerve Connections.* Academic Press, New York.

[50] Gazzaniga, M. S. 1967. The split brain in man. *Sci. Am. 217*:24–29.

Gazzaniga, M. S. 1970. *The Bisected Brain.* Appleton, New York.

[235] Geffen, L. B., and Livett, B. G. 1971. Synaptic vesicles in sympathetic neurons. *Physiol. Rev. 51*:98–157.

Gerschenfeld, H. M. 1973. Chemical transmission in invertebrate central nervous systems and neuromuscular junctions. *Physiol. Rev. 53*:1–119.

[174] Gerschenfeld, H. M., and Paupardin-Tritsch, D. 1974a. Ionic mechanisms and receptor properties underlying the responses of molluscan neurones to 5-hydroxytryptamine. *J. Physiol. 243*:427–456.

[223] Gerschenfeld, H. M., and Paupardin-Tritsch, D. 1974b. On the transmitter function of 5-hydroxytryptamine at excitatory and inhibitory monosynaptic junctions. *J. Physiol. 243*:457–481.

[159] Gilula, D. 1974. Junctions Between Cells. In R. P. Cox (ed.). *Cell Communication.* Wiley, New York, pp. 1–29.

[242] Glynn, I. M. 1968. Membrane adenosine triphosphatase and cation transport. *Br. Med. Bull. 24*:165–169.

[243] Glynn, I. M., and Karlish, S. J. D. 1975. The sodium pump. *Annu. Rev. Physiol. 37*:13–55

[266] Golgi, C. 1903. *Opera Omnia,* Vols. I, II. Hoepli, Milan.

[159] Goodenough, D. A. 1976. The structure and permeability of isolated hepatocyte gap junctions. *Cold Spring Harbor Symp. Quant. Biol. 40*:37–44.

[324] Goodwin, G. M., McCloskey, D. I., and Matthews, P. B. C. 1972. The contribution of muscle afferents to kinaesthesia shown by vibration induced illusions of movement and by the effects of paralysing joint afferents. *Brain 95*:705–748.

[148] Göpfert, H., and Schaefer, H. 1938. Uber den direkt und indirekt erregten Aktionsstrom und die Funktion der motorischen Endplatte. *Pflügers Arch. 239*:597–619.

Gorman, A. F. L., and Marmor, M. F. 1970. Contributions of the sodium pump and ionic gradients to the membrane potential of a molluscan neurone. *J. Physiol.* *210*:897–917. [244]

Grafstein, B. 1971. Transneuronal transfer of radioactivity in the central nervous system. *Science* *172*:177–179.

Grafstein, B., and Laurens, R. 1973. Transport of radioactivity from eye to visual cortex in the mouse. *Exp. Neurol.* *39*:44–57. [235]

Granit, R. 1947. *Sensory Mechanisms of the Retina.* Oxford University Press, London. [21]

Granit, R. (ed.). 1966. *Muscle Afferents and Motor Control.* Wiley, New York.

Gray, J. A. B. 1959. Initiation of Impulses at Receptors. In J. Field (ed.). *Handbook of Physiology,* Section I, Vol. I, Chap. IV, American Physiological Society, Washington, pp. 123–145. [314]

Grinnell, A. D. 1970. Electrical interaction between antidromically stimulated frog motoneurones and dorsal root afferents: Enhancement by gallamine and TEA. *J. Physiol.* *210*:17–43. [157]

Grossman, R. G., and Hampton, T. 1968. Depolarization of cortical glial cells during electrocortical activity. *Brain Res.* *11*:316–324. [276]

Guillery, R. W. 1970. The laminar distribution of retinal fibers in the dorsal lateral geniculate nucleus of the cat: A new interpretation. *J. Comp. Neurol.* *138*:339–368. [33]

Guillery, R. W. 1974. Visual pathways in albinos. *Sci. Am.* *230*:44–54. [398]

Guillery, R. W., and Stelzner, D. J. 1970. The differential effects of unilateral lid closure upon the monocular and binocular segments of the dorsal lateral geniculate nucleus in the cat. *J. Comp. Neurol.* *139*:413–422. [411]

Guillery, R. W., and Kaas, J. H. 1971. A study of normal and congenitally abnormal retinogeniculate projections in cats. *J. Comp. Neurol.* *143*:73–100. [395]

Guillery, R. W., and Kaas, J. H. 1974. The effects of monocular lid suture upon the development of the visual cortex in squirrels *(Sciureus carolinensis). J. Comp. Neurol.* *154*:443–452.

Guth, L. 1968. "Trophic" influences of nerve. *Physiol. Rev.* *48*:645–687. [379]

Gutmann, E. 1962. Denervation and disuse atrophy in cross-striated muscle *Rev. Can. Biol.* *21*:353–365. [379]

Hagiwara, S. 1974. Ca^{++-} Dependent Action Potentials. In G. Eisenman (ed.). *Membranes: A Series of Advances,* Vol. 3, *Artificial and Biological Membranes.* Dekker, New York, pp. 359–381. [124]

Hagiwara, S., and Tasaki, I. 1958. A study on the mechanism of impulse transmission across the giant synapse of the squid. *J. Physiol.* *143*:114–137. [178]

Hagiwara, S., Kusano, K., and Saito, S. 1960. Membrane changes in crayfish stretch receptor neuron during synaptic inhibition and under action of gamma-aminobutyric acid. *J. Neurophysiol.* *23*:505–515. [320]

Hall, Z. W., Bownds, M.D., and Kravitz, E. A. 1970. The metabolism of gamma aminobutyric acid in the lobster nervous system. *J. Cell. Biol.* *46*:290–299. [225, 231]

Hall, Z. W., Hildebrand, J. G., and Kravitz, E. A. 1974. *Chemistry of Synaptic Transmission.* Chiron Press, Newton, Mass.

[390] Hamburger, V. 1939. Motor and sensory hyperplasia following limb-bud transplantations in chick embryos. *Physiol. Zool. 12*:268–284.

[383] Harris, A. J. 1974. Inductive functions of the nervous system. *Annu. Rev. Physiol. 36*:251–305.

[202] Harris, A. J., Kuffler, S. W., and Dennis, M. J. 1971. Differential chemosensitivity of synaptic and extrasynaptic areas on the neuronal surface membrane in parasympathetic neurons of the frog, tested by microapplication of acetylcholine. *Proc. R. Soc. Lond.* B *177*:541–553.

[169] Harris, E. J., and Hutter, O. F. 1956. The action of acetylcholine on the movements of potassium ions in the sinus venosus of the heart. *J. Physiol. 133*:58P–59P.

[379] Harris, J. B., and Thesleff, S. 1971. Studies on tetrodotoxin resistant action potentials in denervated skeletal muscle. *Acta. Physiol. Scand. 83*:382–388.

[21] Hartline, H. K. 1940a. The receptive fields of optic nerve fibers. *Am. J. Physiol. 130*:690–699.

[21] Hartline, H. K. 1940b. The nerve messages in the fibers of the visual pathway. *J. Opt. Soc. Am. 30*:239–247.

[384] Hartzell, H. C., and Fambrough, D. M. 1972. Acetylcholine receptors: Distribution and extrajunctional density in rat diaphragm after denervation correlated with acetylcholine sensitivity. *J. Gen. Physiol. 60*:248–262.

[209, 211, 212, 214] Hartzell, H. C., Kuffler, S. W., and Yoshikami, D. 1975. Postsynaptic potentiation: Interaction between quanta of acetylcholine at the skeletal neuromuscular synapse. *J. Physiol. 251*:427–463.

[403] Held, R., and Bauer, J. A. 1974. Development of sensorially guided reaching in infant monkeys. *Brain Res. 71*:265–271.

[7] Helmholtz, H. 1889. *Popular Scientific Lectures.* Longmans, London.

[126] Henderson, R., Ritchie, J. M., and Strichartz, G. R. 1973. The binding of labelled saxitoxin to the sodium channels in nerve membranes. *J. Physiol. 235*:783–804.

Henneman, E. 1974. Organization of the Spinal Cord. In V. B. Mountcastle (ed.). *Medical Physiology,* Vol. I. Mosby, St. Louis, pp. 636–650.

[350] Henneman, E., Somjen, G., and Carpenter, D. O. 1965. Functional significance of cell size in spinal motoneurons. *J. Neurophysiol. 28*:560–580.

[391] Herrup, K., and Shooter, E. M. 1973. Properties of the β nerve growth factor receptor of avian dorsal root ganglia. *Proc. Natl. Acad. Sci. U.S.A. 70*:3884–3888.

[197] Heuser, J. E., and Reese, T. S. 1973. Evidence for recycling of synaptic vesicle membrane during transmitter release at the frog neuromuscular junction. *J. Cell Biol. 57*:315–344.

Heuser, J. E., and Reese, T. S. 1974. Morphology of Synaptic Vesicle Discharge and Reformation at the Frog Neuromuscular Junction. In M. V. L. Bennett (ed.). *Synaptic Transmission and Neuronal Interaction.* Raven Press, New York, pp. 59–77.

[196] Heuser, J. E., Reese, T. S., and Landis, D. M. D. 1974. Functional changes

in frog neuromuscular junctions studied with freeze-fracture. *J. Neurocytol.* *3*:109–131.

Hildebrand, J. G., Barker, D. L., Herbert, E., and Kravitz, E. A. 1971. Screening for neurotransmitters: A rapid radiochemical procedure. *J. Neurobiol.* *2*:231–246. [227]

Hille, B. 1970. Ionic channels in nerve membranes. *Prog. Biophys. Mol. Biol.* *21*:1–32. [116, 117]

Hille, B. 1971. The permeability of the sodium channel to organic cations in myelinated nerve. *J. Gen. Physiol.* *58*:599–619. [125]

Hille, B. 1973. Potassium channels in myelinated nerves: Selective permeability to small cations. *J. Gen. Physiol.* *61*:669–686. [125]

Hille, B. 1976. Ionic Basis of Resting and Action Potentials. In E. Kandel (ed.). *Handbook of the Nervous System,* Vol. I, Chap. 3. American Physiological Society, Bethesda, Md.

Hirsch, H. V. B., and Spinelli, D. N. 1971. Modification of the distribution of receptive field orientation in cats by selective visual exposure during development. *Exp. Brain Res.* *12*:509–527. [416]

Hitchcock, D. I. 1945. Diffusion in Liquids. In R. Höber (ed.). *Physical Chemistry of Cells and Tissues.* Blakiston, Philadelphia, pp. 7–21. [296]

Hodgkin, A. L. 1937. Evidence for electrical transmission in nerve. I, II. *J. Physiol.* *90*:183–210, 211–232. [140, 141]

Hodgkin, A. L. 1939. The relation between conduction velocity and the electrical resistance outside a nerve fibre. *J. Physiol.* *94*:560–570. [141]

Hodgkin, A. L. 1951. The ionic basis of electrical activity in nerve and muscle. *Biol. Rev.* *26*:339–409. [89]

Hodgkin, A. L. 1964. *The Conduction of the Nervous Impulse.* Liverpool University Press, Liverpool. [78, 91, 103]

Hodgkin, A. L. 1973. Presidential address. *Proc. R. Soc. Lond.* B *183*:1–19. [91]

Hodgkin, A. L., and Huxley, A. F. 1939. Action potentials recorded from inside a nerve fibre. *Nature* *144*:710–711. [98]

Hodgkin, A. L., and Rushton, W. A. H. 1946. The electrical constants of a crustacean nerve fibre. *Proc. R. Soc. Lond.* B *133*:444–479. [133, 138]

Hodgkin, A. L., and Katz, B. 1949. The effect of sodium ions on the electrical activity of the giant axon of the squid. *J. Physiol.* *108*:37–77. [98, 99]

Hodgkin, A. L., Huxley, A. F., and Katz, B. 1952. Measurement of current-voltage relations in the membrane of the giant axon of *Loligo*. *J. Physiol.* *116*:424–448. [109, 115]

Hodgkin, A. L., and Huxley, A. F. 1952a. Currents carried by sodium and potassium ions through the membrane of the giant axon of *Loligo*. *J. Physiol.* *116*:449–472. [113, 114, 115]

Hodgkin, A. L., and Huxley, A. F. 1952b. The components of membrane conductance in the giant axon of *Loligo*. *J. Physiol.* *116*:473–496.

Hodgkin, A. L., and Huxley, A. F. 1952c. The dual effect of membrane potential on sodium conductance in the giant axon of *Loligo*. *J. Physiol.* *116*:497–506. [119]

Hodgkin, A. L., and Huxley, A. F. 1952d. A quantitative description of membrane current and its application to conduction and excitation in nerve. *J. Physiol.* *117*:500–544. [123]

Hodgkin, A. L., and Huxley, A. F. 1953. Movement of radioactive potassium and membrane current in a giant axon. *J. Physiol. 121*:403–414.

[92, 239, 240] Hodgkin, A. L., and Keynes, R. D. 1955. Active transport of cations in giant axons from *Sepia* and *Loligo*. *J. Physiol. 128*:28–60.

[92] Hodgkin, A. L., and Keynes, R. D. 1956. Experiments on the injection of substances into squid giant axons by means of a microsyringe. *J. Physiol. 131*:592–617.

Hodgkin, A. L., and Horowicz, P. 1959. The influence of potassium and chloride ions on the membrane potential of single muscle fibres. *J. Physiol. 148*:127–160.

[43] Hoffman, K.-P., and Stone, J. 1971. Conduction velocity of afferents to cat visual cortex: A correlation with cortical receptive field properties. *Brain Res. 32*:460–466.

[234] Hornykiewicz, O. 1973. Dopamine in the basal ganglia: Its role and therapeutic implications (including the clinical use of L-DOPA). *Br. Med. Bull. 29*:172–178.

[280] Hotson, J. R., Sypert, G. W., and Ward, A. A. 1973. Extracellular potassium concentration changes during propagated seizures in neocortex. *Exp. Neurol. 38*:20–26.

[327] Houk, J., and Henneman, E. 1967. Responses of Golgi tendon organs to active contractions of the soleus muscle of the cat. *J. Neurophysiol. 30*:466–481.

[394] Hoy, R., Bittner, G. D., and Kennedy, D. 1967. Regeneration in crustacean motoneurons: Evidence for axonal fusion. *Science 156*:251–252.

Hubbard, J. I. 1973. Microphysiology of vertebrate neuromuscular transmission. *Physiol. Rev. 53*:674–723.

[417] Hubel, D. H. 1967. Effects of distortion of sensory input on the visual system of kittens. *Physiologist 10*:17–45.

[40, 41, 49] Hubel, D. H., and Wiesel, T. N. 1959. Receptive fields of single neurones in the cat's striate cortex. *J. Physiol. 148*:574–591.

[34, 35] Hubel, D. H., and Wiesel, T. N. 1961. Integrative action in the cat's lateral geniculate body. *J. Physiol. 155*:385–398.

[41, 44, 53, 64, 65] Hubel, D. H., and Wiesel, T. N. 1962. Receptive fields, binocular interaction and functional architecture in the cat's visual cortex. *J. Physiol. 160*:106–154.

[64] Hubel, D. H., and Wiesel, T. N. 1963a. Shape and arrangement of columns in cat's striate cortex. *J. Physiol. 165*:559–568.

[404, 405] Hubel, D. H., and Wiesel, T. N. 1963b. Receptive fields of cells in striate cortex of very young, visually inexperienced kittens. *J. Neurophysiol. 26*:994–1002.

[38, 45, 46, 53, 72] Hubel, D. H., and Wiesel, T. N. 1965a. Receptive fields and functional architecture in two non-striate visual areas (18 and 19) of the cat. *J. Neurophysiol. 28*:229–289.

[414, 415] Hubel, D. H., and Wiesel, T. N. 1965b. Binocular interaction in striate cortex of kittens reared with artificial squint. *J. Neurophysiol. 28*:1041–1059.

[51] Hubel, D. H., and Wiesel, T. N. 1967. Cortical and callosal connections concerned with the vertical meridian of visual fields in the cat. *J. Neurophysiol. 30*:1561–1573.

Hubel, D. H., and Wiesel, T. N. 1968. Receptive fields and functional architecture of monkey striate cortex. *J. Physiol. 195*:215–243. [41, 64, 66]

Hubel, D. H., and Wiesel, T. N. 1970a. The period of susceptibility to the physiological effects of unilateral eye closure in kittens. *J. Physiol. 206*: 419–436. [408, 409]

Hubel, D. H., and Wiesel, T. N. 1970b. Stereoscopic vision in macaque monkey. *Nature 225*:41–42. [48]

Hubel, D. H., and Wiesel, T. N. 1971. Aberrant visual projections in the Siamese cat. *J. Physiol. 218*:33–62. [395, 397]

Hubel, D. H., and Wiesel, T. N. 1972. Laminar and columnar distribution of geniculo-cortical fibers in the macaque monkey. *J. Comp. Neurol. 146*: 421–450. [39, 67, 68, 71]

Hubel, D. H., and Wiesel, T. N. 1974. Sequence regularity and geometry of orientation columns in the monkey striate cortex. *J. Comp. Neurol. 158*: 267–294. [64, 66, 70]

Hubel, D., Wiesel, T., and LeVay, S. 1976. Functional architecture of area 17 in normal and monocularly deprived macaque monkeys. *Cold Spring Harbor. Symp. Quant. Biol. 40*:581–589. [412]

Hughes, A., and Edgar, M. 1972. The innervation of the hind limb of *Eleutherodactylus martiniensis:* Further comparison of cell and fiber numbers during development. *J. Embryol. Exp. Morphol. 27*:389–412. [389]

Hunt, C. C. 1974. The Physiology of Muscle Receptors. In C. C. Hunt (ed.). *Handbook of Sensory Physiology,* Vol. III, *Muscle Receptors.* Springer Verlag, New York, pp. 191–234.

Hunt, C. C., and Kuffler, S. W. 1951. Further study of efferent small nerve fibres to mammalian muscle spindles: Multiple spindle innervation and activity during contraction. *J. Physiol. 113*:283–297. [324, 325]

Hunt, C. C., and Kuffler, S. W. 1954. Motor innervation of skeletal muscle: Multiple innervation of individual muscle fibres and motor unit function. *J. Physiol. 126*:293–303. [386]

Hunt, C. C., and Nelson, P. 1965. Structural and functional changes in the frog sympathetic ganglion following cutting of the presynaptic nerve fibre. *J. Physiol. 177*:1–20. [265]

Hunt, C. C., and Ottoson, D. 1975. Impulse activity and receptor potential of primary and secondary endings of isolated mammalian muscle spindles. *J. Physiol. 252*:259–281. [323]

Huxley, A. F., and Stämpfli, R. 1949. Evidence for saltatory conduction in peripheral myelinated nerve fibres. *J. Physiol. 108*:315–339. [143]

Iversen, L. L. 1967. *The Uptake and Storage of Noradrenaline in Sympathetic Nerves.* Cambridge University Press, London.

Iversen, L. L. (ed.) 1973. Catecholamines. *Br. Med. Bull. 29*:91–178.

Iversen, L. L., and Kelly, J. S. 1975. Uptake and metabolism of γ-aminobutyric acid by neurons and glial cells. *Biochem. Pharmacol. 24*:933–938. [265]

Jack, J. J. B., and Redman, S. J. 1971. An electrical description of the motoneurone and its application to the analysis of synaptic potentials. *J. Physiol. 215*:321–352. [350]

Jacobson, M. 1970. *Developmental Neurobiology.* Holt, New York. [376, 393]

[327] Jankowska, E., and Lindström, S. 1972. Morphology of interneurones mediating Ia reciprocal inhibition of motoneurones in the spinal cord of the cat. J. *Physiol.* *226*:805–823.

[314, 321] Jansen, J. K. S., and Matthews, P. B. C. 1962. The central control of the dynamic response of muscle spindle receptors. *J. Physiol.* *161*:357–378.

[317] Jansen, J. K. S., Njå, A., Ormstad, K., and Walloe, L. 1971. On the innervation of the slowly adapting stretch receptor of the crayfish abdomen: An electrophysiological approach. *Acta Physiol. Scand.* *81*:273–285.

[394] Jansen, J. K. S., and Nicholls, J. G. 1972. Regeneration and changes in synaptic connections between individual nerve cells in the central nervous system of the leech. *Proc. Natl. Acad. Sci. U.S.A.* *69*:636–639.

[287, 367] Jansen, J. K. S., and Nicholls, J. G. 1973. Conductance changes, an electrogenic pump and the hyperpolarization of leech neurones following impulses. *J. Physiol.* *229*:635–655.

[386, 387] Jansen, J. K. S., Lømo, T., Nicolaysen, K., and Westgaard, R. H. 1973. Hyperinnervation of skeletal muscle fibers: Dependence on muscle activity. *Science* *181*:559–561.

[369] Jansen, J. K. S., Muller, K. J., and Nicholls, J. G. 1974. Persistent modification of synaptic interactions between sensory and motor nerve cells following discrete lesions in the central nervous system of the leech. *J. Physiol.* *242*:289–305.

[230] Jasper, H., and Koyama, I. 1969. Rate of release of amino acids from the cerebral cortex in the cat as affected by brainstem and thalamic stimulation. *Can. J. Physiol. Pharmacol.* *47*:889–905.

[164, 379] Jenkinson, D. H., and Nicholls, J. G. 1961. Contractures and permeability changes produced by acetylcholine in depolarized denervated muscle. *J. Physiol.* *159*:111–127.

[355] Kandel, E. R., and Spenser, W. A. 1968. Cellular neurophysiological approaches in the study of learning. *Physiol. Rev.* *48*:65–134.

[27, 28, 31] Kaneko, A. 1970. Physiological and morphological identification of horizontal, bipolar and amacrine cells in goldfish retina. *J. Physiol.* *207*:623–633.

[28, 30] Kaneko, A. 1971. Electrical connexions between horizontal cells in the dogfish retina. *J. Physiol.* *213*:95–105.

[116] Kao, C. T. 1966. Tetrodotoxin, saxitoxin and their significance in the study of excitation phenomena. *Pharmacol. Rev.* *18*:997–1049.

[273] Karahashi, Y., and Goldring, S. 1966. Intracellular potentials from "idle" cells in cerebral cortex of cat. *Electroencephalog. Clin. Neurophysiol.* *20*: 600–607.

[212] Karlin, A. 1967. On the application of a plausible model of allosteric proteins to the receptor for acetylcholine. *J. Theor. Biol.* *16*:306–320.

[302] Karnovsky, M. J. 1967. The ultrastructural basis of capillary permeability studied with peroxidase as a tracer. *J. Cell Biol.* *35*:213–236.

[322] Katz, B. 1949. The efferent regulation of the muscle spindle in the frog. *J. Exp. Biol.* *26*:201–217.

[312] Katz, B. 1950. Depolarization of sensory terminals and the initiation of impulses in the muscle spindle. *J. Physiol.* *111*:261–282.

[103, 154] Katz, B. 1966. *Nerve, Muscle and Synapse.* McGraw-Hill, New York.

Katz, B. 1969. *The Release of Neural Transmitter Substances.* Liverpool University Press, Liverpool.

Katz, B., and Thesleff, S. 1957. A study of the "desensitization" produced by acetylcholine at the motor end-plate. *J. Physiol. 138*:63–80. [211]

Katz, B., and Miledi, R. 1964. The development of acetylcholine sensitivity in nerve-free segments of skeletal muscle. *J. Physiol. 170*:389–396. [383]

Katz, B., and Miledi, R. 1965. The measurement of synaptic delay, and the time course of acetylcholine release at the neuromuscular junction. *Proc. R. Soc. Lond.* B *161*:483–495. [178, 180, 211]

Katz, B., and Miledi, R. 1967a. Tetrodotoxin and neuromuscular transmission. *Proc. R. Soc. Lond.* B *167*:8–22.

Katz, B., and Miledi, R. 1967b. The release of acetylcholine from nerve endings by graded electric pulses. *Proc. R. Soc. Lond.* B *167*:23–38. [185]

Katz, B., and Miledi, R. 1967c. The timing of calcium action during neuromuscular transmission. *J. Physiol. 189*:535–544.

Katz, B., and Miledi, R. 1967d. A study of synaptic transmission in the absence of nerve impulses. *J. Physiol. 192*:407–436. [178, 287]

Katz, B., and Miledi, R. 1969. Spontaneous and evoked activity of motor nerve endings in calcium Ringer. *J. Physiol. 203*:689–706. [167, 182]

Katz, B., and Miledi, R. 1971. The effect of prolonged depolarization on synaptic transfer in the stellate ganglion of the squid. *J. Physiol. 216*:503–512. [181]

Katz, B., and Miledi, R. 1972. The statistical nature of the acetylcholine potential and its molecular components. *J. Physiol. 224*:665–699. [206, 207]

Katz, B., and Miledi, R. 1973a. The characteristics of "end-plate noise" produced by different depolarizing drugs. *J. Physiol. 230*:707–717.

Katz, B., and Miledi, R. 1973b. The binding of acetylcholine to receptors and its removal from the synaptic cleft. *J. Physiol. 231*:549–574. [208]

Katz, B., and Miledi, R. 1973c. The effect of atropine on acetylcholine action at the neuromuscular junction. *Proc. R. Soc. Lond.* B *184*:221–226. [208]

Kehoe, J. 1972. Ionic mechanisms of a two-component cholinergic inhibition in *Aplysia* neurones. *J. Physiol. 225*:85–114. [171]

Kelly, J. P., and Van Essen, D. C. 1974. Cell structure and function in the visual cortex of the cat. *J. Physiol. 238*:515–547. [39, 64, 273, 278]

Kennedy, D. 1971. Crayfish interneurons. *Physiologist 14*:5–30.

Kennedy, D. 1974. Connections among Neurons of Different Types in Crustacean Nervous Systems. In F. O. Schmitt and F. G. Worden (eds.). *The Neurosciences, Third Study Program.* MIT Press, Cambridge, Mass., pp. 379–388. [355]

Keynes, R. D., and Lewis, P. R. 1951. The sodium and potassium content of cephalopod nerve fibres. *J. Physiol. 114*:151–182. [101]

Keynes, R. D., and Rojas, E. 1974. Kinetics and steady-state properties of the charged system controlling sodium conductance in the squid giant axon. *J. Physiol. 239*:393–434. [127]

Kirkwood, P. A., and Sears, T. A. 1974. Monosynaptic excitation of motoneurones from secondary endings of muscle spindles. *Nature 252*:243–244. [325]

[220] Koike, H., Kandel, E. R., and Schwartz, J. H. 1974. Synaptic release of radioactivity after intrasomatic injection of choline-^{3}H into an identified cholingergic interneuron in abdominal ganglion of *Aplysia californica*. *J. Neurophysiol.* *37*:815–827.

[347] Korn, H., and Faber, D. S. 1975. An electrically mediated inhibition in goldfish medulla. *J. Neurophysiol.* *38*:452–471.

[224, 225] Kravitz, E. A., Kuffler, S. W., Potter, D. D., and van Gelder, N. M. 1963. Gamma-aminobutyric acid and other blocking compounds in crustacea. II. Peripheral nervous system. *J. Neurophysiol.* *26*:729–738.

[224] Kravitz, E. A., Kuffler, S. W., and Potter, D. D. 1963. Gamma-aminobutyric acid and other blocking compounds in crustacea. III. Their relative concentrations in separated motor and inhibitory axons. *J. Neurophysiol.* *26*:739–751.

[200] Kriebel, M. E., and Gross, C. E. 1974. Multimodal distribution of frog miniature end-plate potentials in adult, denervated and tadpole leg muscle. *J. Gen. Physiol.* *64*:85–103.

[370] Kristan, W. B., Jr., Stent, G. S., and Ort, C. A. 1974. Neuronal control of swimming in the medicinal leech. III. Impulse patterns of the motor neurons. *J. Comp. Physiol.* *94*:155–176.

Kristan, W. B., Jr., and Stent, G. S. 1976. Peripheral feedback in the leech swimming rhythm. *Cold Spring Harbor Symp. Quant. Biol.* *40*:663–674.

[215, 230] Krnjević, K. 1974. Chemical nature of synaptic transmission in vertebrates. *Physiol. Rev.* *54*:418–540.

[151] Kuffler, S. W. 1942. Electrical potential changes at an isolated nerve-muscle junction. *J. Neurophysiol.* *5*:18–26.

[378] Kuffler, S. W. 1943. Specific excitability of the endplate region in normal and denervated muscle. *J. Neurophysiol.* *6*:99–110.

[150] Kuffler, S. W. 1948. Physiology of neuromuscular junctions: Electrical aspects. *Fed. Proc.* *7*:437–446.

[23, 24] Kuffler, S. W. 1953. Discharge patterns and functional organization of the mammalian retina. *J. Neurophysiol.* *16*:37–68.

[316] Kuffler, S. W. 1954. Mechanisms of activation and motor control of stretch receptors in lobster and crayfish. *J. Neurophysiol.* *17*:558–574.

Kuffler, S. W. 1960. Excitation and Inhibition in Single Nerve Cells. *Harvey Lect.* 1958–1959. Academic Press, New York, pp. 176–218.

Kuffler, S. W. 1967. Neuroglial cells: Physiological properties and a potassium mediated effect of neuronal activity on the glial membrane potential. *Proc. R. Soc. Lond.* B *168*:1–21.

Kuffler, S. W. 1973. The single-cell approach in the visual system and the study of receptive fields. *Invest. Ophthalmol.* *12*:794–813.

[320, 321] Kuffler, S. W., Hunt, C. C., and Quilliam, J. P. 1951. Function of medullated small-nerve fibers in mammalian ventral roots: Efferent muscle spindle innervation. *J. Neurophysiol.* *14*:29–54.

[388] Kuffler, S. W., and Vaughan Williams, E. M. 1953. Small-nerve junctional potentials: The distribution of small motor nerves to frog skeletal muscle, and the membrane characteristics of the fibres they innervate. *J. Physiol.* *121*:289–317.

Kuffler, S. W., and Eyzaguirre, C. 1955. Synaptic inhibition in an isolated nerve cell. *J. Gen. Physiol.* *39*:155–184. [171, 317, 319]

Kuffler, S. W., and Edwards, C. 1958. Mechanism of gamma aminobutyric acid (GABA) action and its relation to synaptic inhibition. *J. Neurophysiol.* *21*:589–610. [224]

Kuffler, S. W., and Potter, D. D. 1964. Glia in the leech central nervous system: Physiological properties and neuron-glia relationship. *J. Neurophysiol.* *27*:290–320. [269, 271, 281]

Kuffler, S. W., and Nicholls, J. G. 1966. The physiology of neuroglial cells. *Ergeb. Physiol.* *57*:1–90. [267, 275]

Kuffler, S. W., Nicholls, J. G., and Orkand, R. K. 1966. Physiological properties of glial cells in the central nervous system of amphibia. *J. Neurophysiol.* *29*:768–787. [269, 270, 272, 294, 295]

Kuffler, S. W., Dennis, M. J., and Harris, A. J. 1971. The development of chemosensitivity in extrasynaptic areas of the neuronal surface after denervation of parasympathetic ganglion cells in the heart of the frog. *Proc. R. Soc. Lond.* B *177*:555–563. [385]

Kuffler, S. W., and Yoshikami, D. 1975a. The distribution of acetylcholine sensitivity at the post-synaptic membrane of vertebrate skeletal twitch muscles: Iontophoretic mapping in the micron range. *J. Physiol.* *244*: 703–730. [203, 204, 215, 216]

Kuffler, S. W., and Yoshikami, D. 1975b. The number of transmitter molecules in a quantum: An estimate from iontophoretic application of acetylcholine at the neuromuscular synapse. *J. Physiol.* *251*:465–482. [203]

Kuno, M. 1971. Quantum aspects of central and ganglionic synaptic transmission in vertebrates. *Physiol. Rev.* *51*:647–678. [349]

Kuno, M. 1974. Factors in Efficacy of Central Synapses. In M.V.L. Bennett (ed.). *Synaptic Transmission and Neuronal Interaction.* Raven Press, New York, pp. 79–86.

Kuno, M., and Rudomin, P. 1966. The release of acetylcholine from the spinal cord of the cat by antidromic stimulation of motor nerves. *J. Physiol.* *187*:177–193. [222]

Kuno, M., and Miyahara, J. T. 1968. Factors responsible for multiple discharge of neurons in Clarke's column. *J. Neurophysiol.* *31*:624–638. [367]

Lam, D. M. K. 1972. Biosynthesis of acetylcholine in turtle photoreceptors. *Proc. Natl. Acad. Sci. U.S.A.* *69*:1987–1991. [32]

Landmesser, L. 1971. Contractile and electrical responses of vagus-innervated frog sartorius muscles. *J. Physiol.* *213*:707–725. [388]

Landmesser, L. 1972. Pharmacological properties, cholinesterase activity and anatomy of nerve-muscle junctions in vagus-innervated frog sartorius. *J. Physiol.* *220*:243–256. [388]

Landmesser, L., and Pilar, G. 1974a. Synapse formation during embryogenesis on ganglion cells lacking a periphery. *J. Physiol.* *241*:715–736. [376]

Landmesser, L., and Pilar, G. 1974b. Synaptic transmission and cell death during normal ganglionic development. *J. Physiol.* *241*:737–749. [376, 390]

Landowne, D., and Ritchie, J. M. 1970. The binding of tritiated ouabain to mammalian non-myelinated nerve fibres. *J. Physiol.* *207*:529–537. [247]

Landowne, D., Potter, L. T., and Terrar, D. A. 1975. Structure-function relationship in excitable membranes. *Annu. Rev. Physiol. 37*:485–508.

[338] Lang, F., Atwood, H. L., and Morin, W. A. 1972. Innervation and vascular supply of the crayfish opener muscle. *Z. Zellforsch. 127*:189–200.

[200] Langley, J. N. 1907. On the contraction of muscle, chiefly in relation to the presence of "receptive" substances. *J. Physiol. 36*:347–384.

[388] Langley, J. N., and Anderson, H. K. 1904. The union of different kinds of nerve fibres. *J. Physiol. 31*:365–391.

Lasansky, A. 1971. Synaptic organization of cone cells in the turtle retina. *Philos. Trans. R. Soc. Lond.* (Ser. B) *262*:365–381.

[268] Lasek, R., Gainer, H., and Przybylski, R. J. 1974. Transfer of newly synthesized proteins from Schwann cells to the squid giant axon. *Proc. Natl. Acad. Sci. U.S.A. 71*:1188–1192.

[46] Lashley, K. S. 1941. Patterns of cerebral integration indicated by the scotomas of migraine. *Arch. Neurol. Psychiat. 46*:331–339.

[63, 235] La Vail, J. H., and La Vail, M. M. 1974. The retrograde intraaxonal transport of horseradish peroxidase in the chick visual system: A light and electron microscopic study. *J. Comp. Neurol. 157*:303–358.

[200] Lee, C. Y. 1972. Chemistry and pharmacology of polypeptide toxins in snake venoms. *Annu. Rev. Pharmacol. 12*:265–286.

[234] Leeman, S. E., and Mroz, E. A. 1974. Substance P. *Life Sci. 15*:2033–2044.

[320] Leksell, L. 1945. The action potential and excitatory effects of the small ventral root fibres to skeletal muscle. *Acta Physiol. Scand. 10* (Suppl. 31): 1–84.

Leusen, I. 1972. Regulation of cerebrospinal fluid composition with reference to breathing. *Physiol. Rev. 52*:1–56.

[68, 69, 411] LeVay, S., Hubel, D. H., and Wiesel, T. N. 1975. The pattern of ocular dominance columns in macaque visual cortex revealed by a reduced silver stain. *J. Comp. Neurol. 159*:559–575.

[390, 392] Levi-Montalcini, R. 1964. Growth-control of nerve cells by a protein factor and its antiserum. *Science 143*:105–110.

[392] Levi-Montalcini, R. 1975. NGF: An Uncharted Route. In F. G. Worden, J. P. Swaze, and G. Adelman (eds.). *The Neurosciences: Paths of Discovery.* MIT Press, Cambridge, Mass., pp. 245–265.

[391, 392] Levi-Montalcini, R., and Cohen, S. 1960. Effects of the extract of the mouse submaxillary salivary glands on the sympathetic system of mammals. *Ann. N.Y. Acad. Sci. 85*:324–341.

[390] Levi-Montalcini, R., and Angeletti, P. U. 1968. Nerve growth factor. *Physiol. Rev. 48*:534–569.

[174] Libet, B. 1970. Generation of slow inhibitory and excitatory postsynaptic potentials. *Fed. Proc. 29*:1945–1956.

[287] Liley, A. W. 1956. The effects of presynaptic polarization on the spontaneous activity at the mammalian neuromuscular junction. *J. Physiol. 134*:427–443.

[151] Ling, G., and Gerard, R. W. 1949. The normal membrane potential of frog sartorius fibers. *J. Cell. Comp. Physiol. 34*:383–396.

[262] Livingston, R. B., Pfenninger, K., Moor, H., and Akert, K. 1973. Specialized paranodal and interparanodal glial-axonal junctions in the peripheral and central nervous system: A freeze-etching study. *Brain Res. 58*:1–24.

Ljungdahl, A., and Hökfelt, T. 1973. Autoradiographic uptake patterns of [^{3}H]-GABA and [^{3}H]-glycine in central nervous tissues with special reference to the cat spinal cord. *Brain Res.* *62*:587–595. [265]

Llinás, R. R. 1975. The cortex of the cerebellum. *Sci. Am.* *232*:56–71. [10]

Llinás, R., Baker, R., and Sotelo, C. 1974. Electrotonic coupling between neurons in cat inferior olive. *J. Neurophysiol.* *37*:560–571. [157]

Lloyd, D. P. C., and Chang, H. T. 1948. Afferent fibers in muscle nerves. *J. Neurophysiol.* *11*:199–207. [314]

Loewenstein, W. R. 1966. Permeability of membrane junctions. *Ann. N.Y. Acad. Sci.* *137*:441–472. [160]

Loewenstein, W. R. 1974. Intercellular Communication Through Membrane Junctions and Cancer Etiology. In J. R. Schultz and R. E. Block (eds.). *Membrane Transformations in Neoplasia.* Academic Press, New York, pp. 103–120. [161]

Loewenstein, W. R., and Mendelson, M. 1965. Components of receptor adaptation in a Pacinian corpuscle. *J. Physiol.* *177*:377–397. [315]

Loewi, O. 1921. Über humorale Übertragbarkeit der Herznervenwirkung. *Pflügers Arch.* *189*:239–242. [147]

Lømo, T., and Rosenthal, J. 1972. Control of ACh-sensitivity by muscle activity in the rat. *J. Physiol.* *221*:493–513. [379, 381, 382]

Lømo, T., and Westgaard, R. H. 1976. Control of ACh sensitivity in rat muscle fibers. *Cold Spring Harbor Symp. Quant. Biol.* *40*:263–274. [382]

Longo, A. M., and Penhoet, E. E. 1974. Nerve growth factor in rat glioma cells. *Proc. Natl. Acad. Sci. U.S.A.* *71*:2347–2349. [266]

LoPresti, V., Macagno, E. R., and Levinthal, C. 1973. Structure and development of neuronal connections in isogenic organisms: Cellular interactions in the development of the optic lamina of *Daphnia. Proc. Natl. Acad. Sci. U.S.A.* *70*:433–437. [374]

Lorenz, K. 1970. *Studies in Animal and Human Behavior,* Vols. I, II. Harvard University Press, Cambridge, Mass. [417]

Lund, R. D., and Lund, J. S. 1973. Reorganization of the retinotectal pathway in rats after neonatal retinal lesions. *Exp. Neurol.* *40*:377–390. [395]

Lundberg, A. 1970. The Excitatory Control of the Ia Inhibitory Pathway. In P. Andersen and J. K. S. Jansen (eds.). *Excitatory Synaptic Mechanisms.* Universitetsforlaget, Oslo, pp. 333–340. [327]

Magleby, K. L., and Stevens, C. R. 1972a. The effect of voltage on the time course of end-plate currents. *J. Physiol.* *223*:151–171.

Magleby, K. L., and Stevens, C. R. 1972b. A quantitative description of end-plate currents. *J. Physiol.* *223*:173–197.

Magleby, K. L., and Terrar, D. A. 1975. Factors affecting the time course of decay of end-plate currents: A possible co-operative action of acetylcholine on receptors at the frog neuro-muscular junction. *J. Physiol.* *244*:467–495. [212]

Mark, R. F. 1974. Selective innervation of muscle. *Br. Med. Bull.* *30*:122–125. [389]

Marmont, G. 1949. Studies on the axon membrane. *J. Cell. Comp. Physiol.* *34*:351–382. [109]

Martin, A. R., and Pilar, G. 1963. Dual mode of synaptic transmission in the avian ciliary ganglion. *J. Physiol.* *168*:443–463.

Martin, A. R., and Wickelgren, W. O. 1971. Sensory cells in the spinal cord of the sea lamprey. *J. Physiol. 212*:65–83. [361]

Matsuda, T., Wu, J.-Y., and Roberts, E. 1973a. Immunochemical studies on glutamic acid decarboxylase (EC 4.1.1.15) from mouse brain. *J. Neurochem. 21*:159–166.

[221] Matsuda, T., Wu, J.-Y., and Roberts, E. 1973b. Electrophoresis of glutamic acid decarboxylase (EC 4.1.1.15) from mouse brain in sodium dodecyl sulphate polyacrylamide gels. *J. Neurochem. 21*:167–172.

[312, 313] Matthews, B. H. C. 1931a. The response of a single end organ. *J. Physiol. 71*:64–110.

[312] Matthews, B. H. C. 1931b. The response of a muscle spindle during active contraction of a muscle. *J. Physiol. 72*:153–174.

[312] Matthews, B. H. C. 1933. Nerve endings in mammalian muscle. *J. Physiol. 78*:1–53.

[314] Matthews, P. B. C. 1964. Muscle spindles and their motor control. *Physiol. Rev. 44*:219–288.

[325] Matthews, P. B. C. 1969. Evidence that the secondary as well as the primary endings of the muscle spindles may be responsible for the tonic stretch reflex of the decerebrate cat. *J. Physiol. 204*:365–393.

[327] Matthews, P. B. C. 1972. *Mammalian Muscle Receptors and Their Central Actions*. Edward Arnold, London.

[25] Maturana, H. R., Lettvin, J. Y., McCulloch, W. S., and Pitts, W. H. 1960. Anatomy and physiology of vision in the frog *(Rana pipiens)*. *J. Gen. Physiol. 43*:129–175.

[349] McLaughlin, B. J. 1972. The fine structure of neurons and synapses in the motor nuclei of the cat spinal cord. *J. Comp. Neurol. 144*:429–460.

[149, 192, 194, 198, 203] McMahan, U. J., and Kuffler, S. W. 1971. Visual identification of synaptic boutons on living ganglion cells and of varicosities in postganglionic axons in the heart of the frog. *Proc. R. Soc. Lond.* B *177*:485–508.

[201] McMahan, U. J., Spitzer, N. C., and Peper, K. 1972. Visual identification of nerve terminals in living isolated skeletal muscle. *Proc. R. Soc. Lond.* B *181*:421–430.

[158] McNutt, N., and Weinstein, R. S. 1973. Membrane ultrastructure at mammalian intercellular junctions. *Prog. Biophys. Mol. Biol. 26*:45–101.

[123] Meech, R. W. 1974. The sensitivity of *Helix aspersa* neurones to injected calcium ions. *J. Physiol. 237*:259–277.

[349] Mendell, L. M., and Henneman, E. 1971. Terminals of single Ia fibers: Location, density, and distribution within a pool of 300 homonymous motoneurons. *J. Neurophysiol. 34*:171–187.

[327] Merton, P. A. 1953. Speculations on the Servo-Control of Movement. In G. E. W. Wolstenholme (ed.). *The Spinal Cord*. Churchill, London, pp. 247–255.

[25] Michael, C. R. 1973. Color vision. *N. Eng. J. Med. 288*:724–728.

[379] Miledi, R. 1960a. The acetylcholine sensitivity of frog muscle fibres after complete or partial denervation. *J. Physiol. 151*:1–23.

[151] Miledi, R. 1960b. Junctional and extra-junctional acetylcholine receptors in skeletal muscle fibres. *J. Physiol. 151*:24–30.

[386] Miledi, R. 1962. Induced innervation of end-plate free muscle segments. *Nature 193*:281–282.

Miledi, R. 1969. Transmitter action in the giant synapse of the squid. *Nature* *223*:1284–1286. [167]

Miledi, R. 1973. Transmitter release induced by injection of calcium ions into nerve terminals. *Proc. R. Soc. Lond.* B *183*:421–425. [181]

Miledi, R., Stefani, E., and Steinbach, A. B. 1971. Induction of the action potential mechanism in slow muscle fibres of the frog. *J. Physiol.* *217*: 737–754. [389]

Miles, F. A. 1972. Centrifugal control of the avian retina. III. Effects of electrical stimulation of the isthmo-optic tract on the receptive field properties of retinal ganglion cells. *Brain Res.* *48*:115–129. [31, 329]

Miller, R. F., and Dowling, J. E. 1970. Intracellular responses of the Muller (glial) cells of mudpuppy retina: Their relation to b-wave of the electroretinogram. *J. Neurophysiol.* *33*:323–341. [285]

Minchin, M. C. W., and Iversen, L. L. 1974. Release of [^{3}H]-gamma-aminobutyric acid from glial cells in rat dorsal root ganglia. *J. Neurochem.* *23*: 533–540. [265]

Miyazaki, S., Takahashi, K., and Tsuda, K. 1974. Electrical excitability in the egg cell membrane of the tunicate. *J. Physiol.* *238*:37–54. [124]

Miyazaki, S., Nicholls, J. G., and Wallace, B. G. 1976. Modification and regeneration of synaptic connections in cultured leech ganglia. *Cold Spring Harbor Symp. Quant. Biol.* *40*:483–493. [370]

Molinoff, P., and Axelrod, J. 1971. Biochemistry of catecholamines. *Annu. Rev. Biochem.* *40*:465–500. [223]

Moore, J. W. 1971. Voltage Clamp Methods. In W. J. Adelman (ed.). *Biophysics and Physiology of Excitable Membranes.* Van Nostrand, New York, pp. 143–167.

Moore, J. W., Blaustein, M. P., Anderson, N. C., and Narahashi, T. 1967. Basis of tetrodotoxin's selectivity in blockage of squid axons. *J. Gen. Physiol.* *50*:1401–1411. [116]

Moore, J. W., Narashashi, T., and Shaw, T. I. 1967. An upper limit to the number of sodium channels in nerve membrane? *J. Physiol.* *188*:99–105. [126]

Mountcastle, V. B. 1957. Modality and topographic properties of single neurons of cat's somatic sensory cortex. *J. Neurophysiol.* *20*:408–434. [62]

Mountcastle, V. B. 1975. The view from within: Pathways to the study of perception. *Johns Hopkins Med. J.* *136*:109–131. [58]

Muller, K. J., and Nicholls, J. G. 1974. Different properties of synapses between a single sensory neurone and two different motor cells in the leech CNS. *J. Physiol.* *238*:357–369. [365, 366]

Muller, K. J., and McMahan, U. J. 1976. The shapes of sensory and motor neurons and the distribution of their synapses in ganglia of the leech: A study using intracellular injection of horseradish peroxidase. *Proc. R. Soc. Lond.* B (in press). [362, 363]

Mullins, L. J., and Brinley, F. J. 1967. Some factors influencing sodium extrusion by internally dialyzed squid axons. *J. Gen. Physiol.* *50*:2333–2355. [241, 242]

Nageotte, J. 1910. Phénomènes de sécrétion dans le protoplasma des cellules névrogliques de la substance grise. *C. R. Soc. Biol. (Paris)* *68*:1068–1069. [265]

[27] Naka, K. I., and Witkovsky, P. 1972. Dogfish ganglion cell discharge resulting from extrinsic polarization of the horizontal cells. *J. Physiol. 223*:449–460.

[315] Nakajima, S., and Takahashi, K. 1966. Post-tetanic hyperpolarization and electrogenic Na pump in stretch receptor neurone of crayfish. *J. Physiol. 187*:105–127.

[247, 312] Nakajima, S., and Onodera, K. 1969a. Membrane properties of the stretch receptor neurones of crayfish with particular reference to mechanisms of sensory adaptation. *J. Physiol. 200*:161–185.

[315] Nakajima, S., and Onodera, K. 1969b. Adaptation of the generator potential in the crayfish stretch receptors under constant length and constant tension. *J. Physiol. 200*:187–204.

[343] Nakajima, Y. 1974. Fine structure of the synaptic endings on the Mauthner cell of the goldfish. *J. Comp. Neurol. 156*:375–402.

[173] Nakajima, Y., Tisdale, A. D., and Henkart, M. P. 1973. Presynaptic inhibition at inhibitory nerve terminals: A new synapse in the crayfish stretch receptor. *Proc. Natl. Acad. Sci. U.S.A. 70*:2462–2466.

Narahashi, T. 1974. Chemicals as tools in the study of excitable membranes. *Physiol. Rev. 54*:812–889.

[151] Nastuk, W. L. 1953. Membrane potential changes at a single muscle endplate produced by transitory application of acetylcholine with an electrically controlled microjet. *Fed. Proc. 12*:102.

[63] Nauta, W. J. H., and Gygax, P. A. 1954. Silver impregnation of degenerating axons in the central nervous system: A modified technic. *Stain Technol. 29*:91–93.

[375] Nelson, P. G. 1975. Nerve and muscle cells in culture. *Physiol. Rev. 55*:1–61.

[378] Nicholls, J. G. 1956. The electrical properties of denervated skeletal muscle. *J. Physiol. 131*:1–12.

[287, 294, 297] Nicholls, J. G., and Kuffler, S. W. 1964. Extracellular space as a pathway for exchange between blood and neurons in the central nervous system of the leech: Ionic composition of glial cells and neurons. *J. Neurophysiol. 27*:645–671.

[272] Nicholls, J. G., and Kuffler, S. W. 1965. Na and K content of glial cells and neurons determined by flame photometry in the central nervous system of the leech. *J. Neurophysiol. 28*:519–525.

[269, 358, 359, 360] Nicholls, J. G., and Baylor, D. A. 1968. Specific modalities and receptive fields of sensory neurons in the CNS of the leech. *J. Neurophysiol. 31*:740–756.

Nicholls, J. G., and Baylor, D. A. 1969. The specificity and functional role of individual cells in a simple nervous system. *Endeavour 103*:3–7.

[364] Nicholls, J. G., and Purves, D. 1970. Monosynaptic chemical and electrical connexions between sensory and motor cells in the central nervous system of the leech. *J. Physiol. 209*:647–667.

[364, 365] Nicholls, J. G., and Purves, D. 1972. A comparison of chemical and electrical synaptic transmission between single sensory cells and a motoneurone in the central nervous system of the leech. *J. Physiol. 225*:637–656.

[230] Obata, K. 1969. Gamma-aminobutyric acid in Purkinje cells and motoneurones. *Experientia 25*:1283.

Obata, K. 1974. Transmitter sensitivities of some nerve and muscle cells in culture. *Brain Res. 73*:71–88. [230]

Obata, K., Takeda, K., and Shinozaki, H. 1970. Further study on pharmacological properties of the cerebellar-induced inhibition of Deiters neurones. *Exp. Brain Res. 11*:327–342. [231]

Orkand, P., and Kravitz, E. A. 1971. Localization of the sites of γ-aminobutyric acid (GABA) uptake in lobster nerve-muscle preparations. *J. Cell Biol. 49*:75–89. [228, 265]

Orkand, P. M., Bracho, H., and Orkand, R. K. 1973. Glial metabolism: Alteration by potassium levels comparable to those during neural activity. *Brain Res. 55*:467–471. [286]

Orkand, R. K., Nicholls, J. G., and Kuffler, S. W. 1966. Effect of nerve impulses on the membrane potential of glial cells in the central nervous system of amphibia. *J. Neurophysiol. 29*:788–806. [275, 276, 279, 280]

Otsuka, M. 1972. γ-Aminobutyric Acid in the Nervous System. In G. H. Bourne (ed.). *The Structure and Function of Nervous Tissue,* Vol. IV. Academic Press, New York, pp. 249–289. [230]

Otsuka, M., Iversen, L. L., Hall, Z. W., and Kravitz, E. A. 1966. Release of gamma-aminobutyric acid from inhibitory nerves of lobster. *Proc. Natl. Acad. Sci. U.S.A. 56*:1110–1115. [169, 228]

Otsuka, M., Kravitz, E. A., and Potter, D. D. 1967. The physiological and chemical architecture of a lobster ganglion with particular reference to gamma-aminobutyrate and glutamate. *J. Neurophysiol. 30*:725–752. [226, 227]

Otsuka, M., Obata, K., Miyata, Y., and Tanaka, Y. 1971. Measurement of γ-aminobutyric acid in isolated nerve cells of cat central nervous system. *J. Neurochem. 18*:287–295. [231]

Otsuka, M., and Konishi, S. 1976. Substance P and excitatory transmitter of primary sensory neurons. *Cold Spring Harbor Symp. Quant. Biol. 40*: 135–144. [233]

Otsuka, R., and Hassler, R. 1962. Über Aufbau und Gliederung der corticalen Sehsphäre bei der Katze. *Arch. Psychiatr. Nervenkr. 203*:212–234. [37]

Ottoson, D. 1971. The Electro-Olfactogram: A Review of Studies on the Receptor Potential of the Olfactory Organ. In L. M. Biedler (ed.). *Handbook of Sensory Physiology,* Vol. IV, *Olfaction.* Springer Verlag, New York, pp. 95–131. [329]

Overton, E. 1902. Beiträge zur allgemeinen Muskel- und Nervenphysiologie. II. Über die Unentbehrlichkeit von Natrium- (oder Lithium-) Ionen für den Kontraktionsakt des Muskels. *Pflügers Arch. 92*:346–386. [98, 238]

Palay, S. L., and Palade, G. E. 1955. Fine structure of neurons. *J. Biophys. Biochem. Cytol. 1*:69–88 [194]

Palay, S. L., and Chan-Palay, V. 1974. *Cerebellar Cortex.* Springer Verlag, Berlin. [10, 11]

Palka, J., and Edwards, J. S. 1974. The cerci and abdominal giant fibers of the house cricket. *Acheta domesticus. II.* Regeneration and effects of chronic deprivation. *Proc. R. Soc. Lond.* B *185*:105–121. [395]

Pappas, G. D., and Waxman, S. G. 1972. Synaptic Fine Structure: Morphological Correlates of Chemical and Electrotonic Transmission. In G. D. Pappas and D. Purpura (eds.). *Structure and Function of Synapses.* Raven Press, New York, pp. 1–43. [158]

[293] Pappenheimer, J. R. 1953. Passage of molecules through capillary walls. *Physiol. Rev.* *33*:387–423.

[298] Pappenheimer, J. R. 1967. The ionic composition of cerebral extracellular fluid and its relation to control of breathing. *Harvey Lect.* *61*:71–94.

[174] Parnas, I., and Strumwasser, F. 1974. Mechanisms of long-lasting inhibition of a bursting pacemaker neuron. *J. Neurophysiol.* *37*:609–620.

[268] Patterson, P. H., and Chun, L. L. Y. 1974. The influence of non-neuronal cells on catecholamine and acetylcholine synthesis and accumulation in cultures of dissociated sympathetic neurons. *Proc. Natl. Acad. Sci. U.S.A.* *71*:3607–3610.

[161] Payton, B. W., Bennett, M. V. L., and Pappas, G. D. 1969. Permeability and structure of junctional membranes at an electrotonic synapse. *Science* *166*:1641–1643.

[258] Penfield, W. 1932. *Cytology and Cellular Pathology of the Nervous System*, Vol. II. Hafner, New York.

[196] Peper, K., Dreyer, F., Sandri, C., Akert, K., and Moor, H. 1974. Structure and ultrastructure of the frog motor end-plate: A freeze-etching study. *Cell Tissue Res.* *149*:437–455.

Perez-Polo, J. R., Bamburg, J. R., De Jong, W. W. W., Straus, D., Baker, M., and Shooter, E. M. 1972. Nerve Growth Factors of the Mouse Submaxillary Gland. In E. Zaimis and J. Knight (eds.). *Nerve Growth Factor and Its Antiserum.* Athlone Press, London, pp. 19–34.

[258, 260, 261, 349] Peters, A., Palay, S. L., and Webster, H. de F. 1976. *The Fine Structure of the Nervous System.* Saunders, Philadelphia.

[375] Peterson, E. R., and Crain, S. M. 1972. Regeneration and innervation in cultures of adult mammalian skeletal muscle coupled with fetal rodent spinal cord. *Exp. Neurol.* *36*:136–159.

[349] Poritsky, R. 1969. Two and three dimensional ultrastructure of boutons and glial cells on the motoneuronal surface in the cat spinal cord. *J. Comp. Neurol.* *135*:423–452.

[197] Porter, C. W., and Bernard, E. A. 1975. The density of cholinergic receptors at the endplate postsynaptic membrane: Ultrastructural studies in two mammalian species. *J. Membrane Biol.* *20*:31–49.

[62] Powell, T. P. S., and Mountcastle, V. B. 1959. Some aspects of the functional organization of the cortex of the postcentral gyrus of the monkey: A correlation of findings obtained in a single unit analysis with cytoarchitecture. *Bull. Johns Hopkins Hosp.* *105*:133–162.

[389] Prestige, M. C. 1967. The control of cell number in the lumbar spinal ganglia during the development of *Xenopus laevis* tadpoles. *J. Embryol. Exp. Morphol.* *17*:453–471.

[330] Price, J. L., and Powell, T. P. S. 1970. An experimental study of the origin and the course of the centrifugal fibres to the olfactory bulb in the rat. *J. Anat.* *107*:215–237.

[265] Purves, D. 1975. Functional and structural changes in mammalian sympathetic neurones following interruption of their axons. *J. Physiol.* *252*:429–463.

[362] Purves, D., and McMahan, U. J. 1972. The distribution of synapses on a physiologically identified motor neuron in the central nervous system of the leech. *J. Cell Biol.* *55*:205–220.

Purves, D., and Sakmann, B. 1974a. The effect of contractile activity on fibrillation and extrajunctional acetylcholine sensitivity in rat muscle maintained in organ culture. *J. Physiol.* *237*:157–182. [381]

Purves, D., and Sakmann, B. 1974b. Membrane properties underlying spontaneous activity of denervated muscle fibres. *J. Physiol.* *239*:125–153. [378]

Raisman, G., and Field, P. M. 1973. A quantitative investigation of the development of collateral innervation after partial deafferentation of the septal nuclei. *Brain Res.* *50*:241–264. [395]

Rakič, P. 1971. Neuron-glia relationship during granule cell migration in developing cerebellar cortex: A Golgi and electronmicroscopic study in *Macacus rhesus*. *J. Comp. Neurol.* *141*:283–312. [266]

Rakič, P., and Sidman, R. L. 1973a. Sequence of developmental abnormalities leading to granule cell deficit in cerebellar cortex of weaver mutant mice. *J. Comp. Neurol.* *152*:103–132.

Rakič, P., and Sidman, R. L. 1973b. Organization of cerebellar cortex secondary to deficit of granule cells in weaver mutant mice. *J. Comp. Neurol.* *152*:133–162. [374]

Ramón y Cajal, S. 1955. *Histologie du système nerveux,* Vol. II. C.S.I.C., Madrid. [9, 10, 11, 38, 39]

Rang, H. P., and Ritchie, J. M. 1968. On the electrogenic sodium pump in mammalian non-myelinated nerve fibres and its activation by various external cations. *J. Physiol.* *196*:183–221.

Ransom, B. R., and Goldring, S. 1973a. Ionic determinants of membrane potential of cells presumed to be glia in cerebral cortex of cat. *J. Neurophysiol.* *36*:855–868.

Ransom, B. R., and Goldring, S. 1973b. Slow depolarization in cells presumed to be glia in cerebral cortex of cat. *J. Neurophysiol.* *36*:869–878. [276, 277]

Rasminsky, M., and Sears, T. A. 1972. Internodal conduction in undissected demyelinated nerve fibres. *J. Physiol.* *227*:323–350. [263]

Rawlins, F. 1973. A time-sequence autoradiographic study of the in vivo incorporation of [1,2-^{3}H] cholesterol into peripheral nerve myelin. *J. Cell Biol.* *58*:42–53. [263]

Redfern, P. A. 1970. Neuromuscular transmission in new-born rats. *J. Physiol.* *209*:701–709. [389]

Reese, T. S., and Karnovsky, M. J. 1967. Fine structural localization of a blood-brain barrier to exogenous peroxidase. *J. Cell Biol.* *34*:207–217. [301]

Rice, S. O. 1944. Mathematical analysis of random noise. *Bell Syst. Tech. J.* *23*:282–332. [206]

Riesen, A. H., and Aarons, L. 1959. Visual movement and intensity discrimination in cats after early deprivation of pattern vision. *J. Comp. Physiol. Psychol.* *52*:142–149. [403]

Ritchie, J. M. 1973. Energetic aspects of nerve conduction: The relationships between heat production, electrical activity and metabolism. *Prog. Biophys. Mol. Biol.* *26*:149–187.

Roberts, A., and Bush, B.M.H. 1971. Coxal muscle receptors in the crab: The receptor current and some properties of the receptor nerve fibres. *J. Exp. Biol.* *54*:515–524. [311]

[341] Robertson, J. D. 1963. The occurrence of a subunit pattern in the unit membranes of club endings in Mauthner cell synapses in goldfish brains. *J. Cell Biol.* *19*:201–221.

[341, 343] Robertson, J. D., Bodenheimer, T. S., and Stage, D. E. 1963. The ultrastructure of Mauthner cell synapses and nodes in goldfish brains. *J. Cell Biol.* *19*:159–199.

Rodieck, R. W. 1973. *The Vertebrate Retina: Principles of Structure and Function.* W. H. Freeman, San Francisco.

[421] Rogers, E. M. 1960. *Physics for the Enquiring Mind.* Princeton University Press, Princeton, N.J.

Rushton, W. A. H. 1951. A theory of the effects of fibre size in medullated nerve. *J. Physiol.* *115*:101–122.

[239] Russell, J. M., and Brown, A. M. 1972. Active transport of chloride by the giant neuron of the *Aplysia* abdominal ganglion. *J. Gen. Physiol.* *60*:499–518.

[286] Salem, R. D., Hammerschlag, R., Bracho, H., and Orkand, R. K. 1975. Influence of potassium ions on accumulation and metabolism of [^{14}C]-glucose by glial cells. *Brain Res.* *86*:499–503.

Salzberg, B. M., Davila, H. V., and Cohen, L. B. 1973. Optical recording of impulses in individual neurones of an invertebrate central nervous system. *Nature* *246*:508–509.

[34] Sanderson, K. J., Bishop, P. O., and Darian-Smith, I. 1971. The properties of the binocular receptive fields of lateral geniculate neurons. *Exp. Brain Res.* *13*:178–207.

[292] Scharrer, E. 1944. The blood vessels of the nervous tissue. *Q. Rev. Biol.* *19*:308–318.

[173] Schmidt, R. F. 1971. Presynaptic inhibition in the vertebrate central nervous system. *Ergeb. Physiol* *63*:20–101.

[264, 265] Schon, F., and Kelly, J. S. 1974. Autoradiographic localization of [^{3}H] GABA and [^{3}H] glutamate over satellite glial cells. *Brain Res.* *66*:275–288.

Schon, F., and Kelly, J. S. 1975. Selective uptake of [^{3}H]-β-alanine by glia: Association with the glial uptake system for GABA. *Brain Res.* *86*:243–257.

[389] Scott, S. A. 1975. Persistence of foreign innervation on reinnervated goldfish extraocular muscles. *Science* *189*:644–646.

[327] Sears, T. A. 1964. Efferent discharges in alpha and fusimotor fibres of intercostal nerves of the cat. *J. Physiol.* *174*:295–315.

[328] Serverin, F. V., Orlovskii, G. N., and Shik, M. L. 1967. Work of the muscle receptors during controlled locomotion. *Biophysics* *12*:575–586.

[311] Shaw, S. R. 1972. Decremental conduction of the visual signal in barnacle lateral eye. *J. Physiol.* *220*:145–175.

Shepherd, G. M. 1974. *The Synaptic Organization of the Brain.* Oxford University Press, New York.

[405] Sherk, H., and Stryker, M. P. 1976. Quantitative study of cortical orientation selectivity in visually inexperienced kittens. *J. Neurophysiol.* *39*:63–70.

[411] Sherman, S. M., Hoffmann, K.-P., and Stone, J. 1972. Loss of a specific cell type from dorsal lateral geniculate nucleus in visually deprived cats. *J. Neurophysiol.* *35*:532–541.

Sherman, S. M., and Stone, J. 1973. Physiological normality of the retina in visually deprived cats. *Brain Res. 60*:224–230. [407]

Sherrington, C. S. 1933. *The Brain and Its Mechanism.* Cambridge University Press, London. [86]

Sherrington, C. S. 1947. *Integrative Action of the Nervous System.* Yale University Press, New Haven. [47, 86]

Sherrington, C. S. 1951. *Man on his Nature.* Cambridge University Press, London. [59]

Sidman, R. L., Dickie, M. M., and Appel, S. H. 1964. Mutant mice (quaking and jimpy) with deficient myelination in the central nervous system. *Science 144*:309–311. [263]

Sidman, R. L., and Rakic, P. 1973. Neuronal migration with special reference to developing human brain: A review. *Brain Res. 62*:1–35. [266, 374]

Skou, J. C. 1957. The influence of some cations on an adenosine triphosphatase from peripheral nerves. *Biochim. Biophys. Acta 23*:394–401. [238]

Skou, J. C. 1964. Enzymatic aspects of active linked transport of Na^+ and K^+ through the cell membrane. *Prog. Biophys. Mol. Biol. 14*:133–166. [241]

Sokolove, P. G., and Cooke, I. M. 1971. Inhibition of impulse activity in a sensory neuron by an electrogenic pump. *J. Gen. Physiol. 57*:125–163. [315]

Somjen, G. G. 1975. Electrophysiology of neuroglia. *Annu. Rev. Physiol. 37*:163–190.

Sotelo, C., and Taxi, J. 1970. Ultrastructural aspects of electrotonic junctions in the spinal cord of the frog. *Brain Res. 17*:137–141. [157]

Sotelo, C., Llinás, R., and Baker, R. 1974. Structural study of inferior olivary nucleus of the cat: Morphological correlates of electrotonic coupling. *J. Neurophysiol. 37*:541–559. [158]

Specht, S., and Grafstein, B. 1973. Accumulation of radioactive protein in mouse cerebral cortex after injection of ^{3}H-fucose into the eye. *Exp. Neurol. 41*:705–722. [63]

Sperry, R. W. 1944. Optic nerve regeneration with return of vision in anurans. *J. Neurophysiol. 7*:57–69. [393]

Sperry, R. W. 1945. Restoration of vision after crossing of optic nerves and after contralateral transplantation of eye. *J. Neurophysiol. 8*:15–28. [393]

Sperry, R. W. 1970. Perception in the Absence of the Neocortical Commissures. In D. A. Hamburg, K. H. Pribram, and A. J. Stunkard (eds.). *Perception and its Disorders.* Williams & Wilkins, Baltimore, pp. 123–138. [50, 51]

Stein, R. B. 1974. Peripheral control of movement. *Physiol. Rev. 54*:215–243.

Steinbach, J. H., Harris, A. J., Patrick, J., Schubert, D., and Heineman, S. 1973. Nerve-muscle interaction in vitro: Role of acetylcholine. *J. Gen. Physiol. 62*:255–270. [387]

Stevens, C. F. 1976. Molecular basis for postjunctional conductance increases induced by acetylcholine. *Cold Spring Harbor Symp. Quant. Biol. 40*:169–174.

Steward, O., Cotman, C. W., and Lynch, G. S. 1973. Reestablishment of electrophysiologically functional entorhinal cortical input to the dentate gyrus deafferented by ipsilateral entorhinal lesions: Innervation by the contralateral entorhinal cortex. *Exp. Brain Res. 18*:396–414. [395]

[25] Stone, J., and Hoffmann, K.-P. 1972. Very slow-conducting ganglion cells in the cat's retina: A major new functional type? *Brain Res. 43*:610–616.

[393] Stone, L. S., and Zaur, I. S. 1940. Reimplantation and transplantation of adult eyes in the salamander (*Triturus viridescens*) with return of vision. *J. Exp. Zool. 85*:243–269.

[355] Strumwasser, F. 1973. Neural and humoral factors in the temporal organization of behavior. *Physiologist 16*:9–42.

[416] Stryker, M. P., and Sherk, H. 1975. Modification of cortical orientation selectivity in the cat by restricted visual experience: A reexamination. *Science 190*:904–906.

[361, 362] Stuart, A. E. 1970. Physiological and morphological properties of motoneurones in the central nervous system of the leech. *J. Physiol. 209*:627–646.

[231] Susz, J. P., Haber, B., and Roberts, E. 1966. Purification and some properties of mouse brain L-glutamic decarboxylase. *Biochemistry 5*:2870–2877.

[26] Svaetichin, G. 1953. The cone action potential. *Acta Physiol. Scand. 29*:565–600.

[201] *The Synapse.* 1976. Cold Spring Harbor Symp. Quant. Biol. *40*.

[288] Sypert, G. W., and Ward, A. A., Jr. 1971. Unidentified neuroglia potentials during propagated seizures in neocortex. *Exp. Neurol. 33*:239–255.

[33] Szentágothai, J. 1973. Neuronal and Synaptic Architecture of the Lateral Geniculate Nucleus. In H. H. Kornhuker (ed.). *Handbook of Sensory Physiology*, Vol. VI, *Central Visual Information.* Springer Verlag, Berlin, pp. 141–176.

[234] Takahashi, T., and Otsuka, M. 1975. Regional distribution of substance P in the spinal cord and nerve roots of the cat and the effect of dorsal root section. *Brain Res. 87*:1–11.

[164] Takeuchi, A., and Takeuchi, N. 1960. On the permeability of the end-plate membrane during the action of transmitter. *J. Physiol. 154*:52–67.

[229, 338] Takeuchi, A., and Takeuchi, N. 1965. Localized action of gamma-aminobutyric acid on the crayfish muscle. *J. Physiol. 177*:225–238.

[230] Takeuchi, A., and Takeuchi, N. 1966. On the permeability of the presynaptic terminal of the crayfish neuromuscular junction during synaptic inhibition and the action of γ-aminobutyric acid. *J. Physiol. 183*:433–449.

[169, 229] Takeuchi, A., and Takeuchi, N. 1967. Anion permeability of the inhibitory post-synaptic membrane of the crayfish neuromuscular junction. *J. Physiol. 191*:575–590.

[229] Takeuchi, A., and Takeuchi, N. 1969. A study of the action of picrotoxin on the inhibitory neuromuscular junction of the crayfish. *J. Physiol. 205*:377–391.

[167] Takeuchi, A., and Onodera, K. 1973. Reversal potentials of the excitatory transmitter and L-glutamate at the crayfish neuromuscular junction. *Nature (New Biol.) 242*:124–126.

[167] Takeuchi, N. 1963. Effects of calcium on the conductance change of the end-plate membrane during the action of transmitter. *J. Physiol. 167*:141–155.

[38] Talbot, S. A., and Marshall, W. H. 1941. Physiological studies on neural mechanisms of visual localization and discrimination. *Am. J. Ophthalmol. 24*:1255–1264.

Tasaki, I. 1959. Conduction of the Nerve Impulse. In J. Field (ed.). *Handbook of Physiology*, Section 1, Vol. I, Chap. III. American Physiological Society, Bethesda, Md., pp. 75–121. [142, 143]

Tauc, L. 1967. Transmission in invertebrate and vertebrate ganglia. *Physiol. Rev. 47*:521–593. [355]

Terzuolo, C. A., and Washizu, Y. 1962. Relation between stimulus strength, generator potential and impulse frequency in stretch receptor of *Crustacea. J. Neurophysiol. 25*:56–66. [312]

Thesleff, S. 1960. Supersensitivity of skeletal muscle produced by botulinum toxin. *J. Physiol. 151*:598–607. [386]

Thomas, R. C. 1969. Membrane current and intracellular sodium changes in a snail neurone during extrusion of injected sodium. *J. Physiol. 201*:495–514. [244, 245]

Thomas, R. C. 1972. Electrogenic sodium pump in nerve and muscle cells. *Physiol. Rev. 52*:563–594. [247]

Thomas, R. C. 1972. Intracellular sodium activity and the sodium pump in snail neurones. *J. Physiol. 220*:55–71. [244]

Tomita, T. 1965. Electrophysiological study of the mechanisms subserving color coding in the fish retina. *Cold Spring Harbor Symp. Quant. Biol. 30*:559–566. [26]

Towe, A. L. 1973. Somatosensory Cortex: Descending Influences on Ascending Systems. In A. Iggo (ed.). *Handbook of Sensory Physiology*, Vol. II, *Somatosensory System*. Springer Verlag, New York, pp. 700–718. [330]

Truex, R. C., and Carpenter, M. B. 1969. *Human Neuroanatomy*, 6th ed. Williams & Wilkins, Baltimore.

Vallbo, A. B. 1971. Muscle spindle response at the onset of isometric voluntary contractions in man: Time difference between fusimotor and skeletomotor effects. *J. Physiol. 218*:405–431. [328]

Valverde, F. 1970. The Golgi Method: A Tool for Comparative Structural Analysis. In W. J. H. Nauta and S. O. E. Ebbeson (eds.). *Contemporary Research Methods in Neuroanatomy*. Springer Verlag, New York, pp. 12–31. [411]

Van der Loos, H., and Woolsey, T. A. 1973. Somatosensory cortex: Structural alterations following early injury to sense organs. *Science 179*:395–398. [413]

Van Essen, D. C. 1973. The contribution of membrane hyperpolarization to adaptation and conduction block in sensory neurones of the leech. *J. Physiol. 230*:509–534. [368, 369]

Van Essen, D., and Jansen, J. K. S. 1974. Reinnervation of rat diaphragm during perfusion with α-bungarotoxin. *Acta Physiol. Scand. 91*:571–573. [388]

Van Essen, D., and Jansen, J. K. S. 1976. Repair of specific neuronal pathways in the leech. *Cold Spring Harbor Symp. Quant. Biol. 40*:495–502. [394]

Vera, C. L., Vial, J. D., and Luco, J. V. 1957. Reinnervation of nictitating membrane of cat by cholinergic fibers. *J. Neurophysiol. 20*:365–373. [388]

Verveen, A. A., and DeFelice, L. J. 1974. Membrane noise. *Prog. Biophys. Mol. Biol. 28*:189–265.

Vigh, B., Vigh-Teichmann, I., Koritsanszky, S., and Aros, B. 1970. Ultrastruktur der Liquorkontaktneurone des Ruckenmarkes von Reptilien. *Z. Zellforsch. 109*:180–194. [298]

Villegas, J. 1972. Axon-Schwann cell interaction in the squid nerve fibre. *J. Physiol. 225*:275–296. [273]

[256] Virchow, R. 1859. *Cellularpathologie.* Trans. F. Chance. Hirschwald, Berlin.

[410] Von Senden, M. 1960. *Space and Sight: The Perception of Space and Shape in the Congenitally Blind before and after Operation.* Free Press, Glencoe, Ill.

[169] Wachtel, H., and Kandel, E. 1971. Conversion of synaptic excitation to inhibition at a dual chemical synapse. *J. Neurophysiol. 34*:56–68.

[73] Walker, C., and Woolsey, T. A. 1974. Structure of layer IV in the somatosensory neocortex of the rat: Description and comparison with the mouse. *J. Comp. Neurol. 158*:437–454.

[330] Wall, P. D. 1973. Dorsal Horn Electrophysiology. In A. Iggo (ed.). *Handbook of Sensory Physiology,* Vol. II, *Somatosensory System.* Springer Verlag, New York, pp. 253–270.

[234] Wallace, B. G., Talamo, B. R., Evans, P. D., and Kravitz, E. A. 1974. Octopamine: Selective association with specific neurons in the lobster nervous system. *Brain Res. 74*:349–355.

[374] Ward, S., Thomson, N., White, J. G., and Brenner, S. 1975. Electron microscopical reconstruction of the anterior sensory anatomy of the nematode *Caenorhabditis elegans. J. Comp. Neurol. 160*:313–338.

Waxman, S. G. 1972. Regional differentiation of the axon: A review with special reference to the concept of the multiplex neuron. *Brain Res. 47*:269–288.

[174] Weight, F. F., and Padjen, A. 1973. Acetylcholine and slow synaptic inhibition in frog sympathetic ganglion cells. *Brain Res. 55*:225–228.

Werblin, F. S., and Dowling, J. E. 1969. Organization of the retina of the mudpuppy *Necturus maculosus.* II. Intracellular recording. *J. Neurophysiol. 32*:339–355.

[233] Werman, R., Davidoff, R. A., and Aprison, M. H. 1968. Inhibitory action of glycine on spinal neurons in the cat. *J. Neurophysiol. 31*:81–95.

[189] Wernig, A. 1972. Changes in statistical parameters during facilitation at the crayfish neuromuscular junction. *J. Physiol. 226*:751–759.

[43] Whitsel, B. L., Roppolo, J. R., and Werner, G. 1972. Cortical information processing of stimulus motion on primate skin. *J. Neurophysiol. 35*:691–717.

[197] Whittaker, V. P. 1970. The Vesicle Hypothesis. In P. Andersen and J. K. S. Jansen (eds.). *Excitatory Synaptic Mechanisms.* Universitetsforlaget, Oslo, pp. 67–76.

[224, 338] Wiersma, C. A. G., and Ripley, S. H. 1952. Innervation patterns of crustacean limbs. *Physiol. Comp. Oecol. 2*:391–405.

[23] Wiesel, T. N. 1960. Receptive fields of ganglion cells in the cat's retina. *J. Physiol. 153*:583–594.

[411] Wiesel, T. N., and Hubel, D. H. 1963a. Effects of visual deprivation on morphology and physiology of cells in the cat's lateral geniculate body. *J. Neurophysiol. 26*:978–993.

[407] Wiesel, T. N., and Hubel, D. H. 1963b. Single-cell responses in striate cortex of kittens deprived of vision in one eye. *J. Neurophysiol. 26*:1003–1017.

Wiesel, T. N., and Hubel, D. H. 1965a. Comparison of the effects of unilateral and bilateral eye closure on cortical unit responses in kittens. *J. Neurophysiol. 28*:1029–1040. [407]

Wiesel, T. N., and Hubel, D. H. 1965b. Extent of recovery from the effects of visual deprivation in kittens. *J. Neurophysiol. 28*:1060–1072.

Wiesel, T. N., and Hubel, D. H. 1974. Ordered arrangement of orientation columns in monkeys lacking visual experience. *J. Comp. Neurol. 158*:307–318. [404, 405, 406, 414]

Wiesel, T. N., Hubel, D. H., and Lam, D. M. K. 1974. Autoradiographic demonstration of ocular-dominance columns in the monkey striate cortex by means of transneuronal transport. *Brain Res. 79*:273–279. [67]

Williams, P. L., and Warwick, R. 1975. *Functional Neuroanatomy of Man.* Saunders, Philadelphia.

Wilson, D. M. 1966. Central nervous mechanisms for the generation of rhythmic behavior in arthropods. *Symp. Soc. Exp. Biol. 20*:199–228. [370]

Woolsey, T. A., and Van der Loos, H. 1970. The structural organization of layer IV in the somatosensory region (S1) of mouse cerebral cortex. *Brain Res. 17*:205–242. [413]

Wright, E. M. 1972. Mechanisms of ion transport across the choroid plexus. *J. Physiol. 226*:545–571. [301]

Yoon, M. 1972. Transposition of the visual projection from the nasal hemiretina onto the foreign rostral zone of the optic tectum in goldfish. *Exp. Neurol. 37*:451–462. [394]

Young, J. A. C., Brown, D. A., Kelly, J. S., and Schon, F. 1973. Autoradiographic localization of sites of [^{3}H] γ-aminobutyric acid accumulation in peripheral autonomic ganglia. *Brain Res. 63*:479–486. [265]

Young, J. Z. 1936. The giant nerve fibres and epistellar body of cephalopods. *Q. J. Microsc. Sci. 78*:367–386. [90]

Zeki, S. M. 1973. Color coding in rhesus monkey prestriate cortex. *Brain Res. 53*:422–427. [57, 65]

Zeki, S. M. 1974a. Functional organization of a visual area in the posterior bank of the superior temporal sulcus of the rhesus monkey. *J. Physiol. 236*:549–573. [38, 65]

Zeki, S. M. 1974b. Cells responding to changing image size and disparity in the cortex of the rhesus monkey. *J. Physiol. 242*:827–841.

Zucker, R. S. 1973. Changes in the statistics of transmitter release during facilitation. *J. Physiol. 229*:787–810. [189]

INDEX

ABOUT THE BOOK

This book is set in Palatino, a face designed by the contemporary German typographer Hermann Zapf. Inspired by the typography of the Italian Renaissance, Palatino letters derive their elegance from the natural motion of the edged pen. Palatino is a highly readable face especially suited for extended use in books. Titles are in Helvetica bold.

Drawings for the book were prepared by Laszlo Meszoly, who worked in close collaboration with the authors. Many of the charts and graphs were rendered by Vantage Art, Inc.

Type was set by Linotype at V & M Typographical, and the book was manufactured at the Murray Printing Company.